Flotation Chemistry of Tungsten Minerals and Its Application

Wei Sun · Haisheng Han · Zhiyong Gao

Flotation Chemistry of Tungsten Minerals and Its Application

冶金工业出版社
Metallurgical Industry Press

Springer

Wei Sun
Central South University
Changsha, China

Engineering Research Center of Ministry of Education for Carbon Emission Reduction in Metal Resource Exploitation and Utilization
Central South University
Changsha, China

State Key Laboratory of Mineral Processing
Changsha, China

Haisheng Han
Central South University
Changsha, China

Engineering Research Center of Ministry of Education for Carbon Emission Reduction in Metal Resource Exploitation and Utilization
Central South University
Changsha, China

State Key Laboratory of Mineral Processing
Changsha, China

Zhiyong Gao
Central South University
Changsha, China

Engineering Research Center of Ministry of Education for Carbon Emission Reduction in Metal Resource Exploitation and Utilization
Central South University
Changsha, China

State Key Laboratory of Mineral Processing
Changsha, China

ISBN 978-981-96-5076-7 ISBN 978-981-96-5077-4 (eBook)
https://doi.org/10.1007/978-981-96-5077-4

Jointly published with Metallurgical Industry Press
The print edition is not for sale in China (Mainland). Customers from China (Mainland) please order the print book from: Metallurgical Industry Press.

This Springer imprint is published by the registered company Springer Nature Singapore Pte Ltd.
The registered company address is: 152 Beach Road, #21-01/04 Gateway East, Singapore 189721, Singapore

Contents

Chapter 1
Tungsten Resources and Its Strategic Significance

Haisheng Han, Wei Sun, and Wenjuan Sun

Abstract Tungsten is an indispensable material in modern national economic life, as well as an important material in modern aviation and military fields, and is therefore listed as a strategic mineral resource by many countries. This paper summarizes the application areas, resource distribution, production and consumption of tungsten ore worldwide in recent years. China is the largest country in the world in terms of tungsten resources reserves, production, export and consumption. As a major strategic mineral, the efficient mining and utilization of tungsten ore resources should receive attention.

Keywords Tungsten · Strategic mineral · Resource distribution

1.1 Introduction

Tungsten was first discovered in 1781 by Scheele, a Swedish chemist, and was successfully extracted [1] two years later in its elemental form. Tungsten has high melting point, high hardness, super heat resistance, excellent high temperature mechanical properties, almost the lowest expansion coefficient, good electrical and thermal conductivity, low heat dissipation coefficient, so it is widely used in electric power, electronics, petrochemical, military, metal and wood processing, construction, mining, abrasion resistance and other fields. It is often used in cutting tools, high-speed tool steel, filament, electrode, contact material, magnetron, tungsten carbide, catalyst and other products, known as the "teeth of industry" [2]. In the 1750s,

---

H. Han (✉) · W. Sun · W. Sun
Central South University, Changsha, China

Engineering Research Center of Ministry of Education for Carbon Emission Reduction in Metal Resource Exploitation and Utilization, Central South University, Changsha, China

State Key Laboratory of Mineral Processing, Changsha, China

H. Han
Technology Research Center of Hunan Province for Comprehensive Utilization of Associated Fluorite and Fluorine Chemical Engineering, Chenzhou, China

W. Sun et al., *Flotation Chemistry of Tungsten Minerals and Its Application*,
https://doi.org/10.1007/978-981-96-5077-4_1

chemists discovered that tungsten could significantly affect the properties of steel. In the late 19th and early twentieth centuries, tungsten steel began to be mass-produced and widely used. High speed steel was first exhibited at the 1900 World's Fair in Paris, marking a major technological advance in the field of metal cutting. Since then, the extraction and processing industry of tungsten has developed rapidly, and tungsten has become the most important alloying element [3] in the world. In 1909, tungsten was widely applied in electric vacuum technology by pressure processing based on powder metallurgy. Since then, tungsten wire has become an important material in lighting industry. The cemented carbide developed in 1928 with tungsten carbide as the main component has been widely used [4, 5] in modern technology since its properties in all aspects exceed those of the best tool steel. At present, tungsten is an important part of modern high-tech new materials. A series of electronic optical materials, special alloys, new functional materials and organometallic compounds all need to use tungsten with unique properties. Tungsten is widely used in the fields of contemporary communication technology, electronic computers, aviation, military, medicine and health, photosensitive materials, photoelectric materials, energy materials and catalytic materials.

Tungsten is widely distributed in the Earth's crust. At present, more than 20 kinds of tungsten ores have been found, but only wolframite and scheelite have industrial significance. The main types of tungsten deposits that have been mined in the world are quartz vein wolframite, skarn scheelite and porphyry scheelite. In 2022, the world's tungsten reserves of 3.8 Mt (tungsten metal) are mainly concentrated in China, Russia, Vietnam, and Spain, which together account for more than 60% of the world's tungsten reserves. China is the world's richest country in tungsten resources, as well as the world's largest producer and consumer of tungsten ore [6]. As a non-renewable resource, tungsten plays an irreplaceable role in aerospace, military, materials and other fields. With the development of society and science and technology, the world's consumption of tungsten is increasing year by year, and tungsten resource has become an important strategic resource. Based on the main properties and applications of tungsten, this chapter introduces the current distribution, exploitation, production, and consumption of tungsten resources in the world, so as to provide reference for the efficient development and utilization of tungsten resources.

1.2 Basic Properties and Applications of Tungsten

1.2.1 Basic Properties of Tungsten

Tungsten is a high melting point rare metal with a steel-like exterior, low vapor pressure and low evaporation rate. Its main physical properties are shown in Table 1.1. Tungsten has excellent physical and mechanical properties, and its density and melting point rank the best among all metals; Tungsten has the lowest vapor pressure of all metals, extremely high compression modulus and elastic modulus, extremely high resistance to thermal creep and extremely high thermal and electrical conductivity [3].

Tungsten has strong chemical stability. At room temperature, tungsten does not react with water, but above 600 °C, tungsten easily reacts with steam to form oxides. At room temperature bulk tungsten does not react with any concentration of hydrochloric, sulfuric or hydrofluoric acid; Tungsten can react with hot nitric acid or a mixture of nitric acid and hydrofluoric acid to form WO_3, and only in the presence of an oxidizing agent can aqua regia erode tungsten. In the absence of oxygen, tungsten is not corroded by aqueous solutions of alkali or ammonia, and tungsten even resists attack by molten sodium hydroxide or potassium hydroxide. At lower temperatures, tungsten can interact with NH_3 to form nitrides and CO to form carbides; At higher temperatures, N_2O, NO, and NO_2 can erode tungsten, and at 1200 C, CO_2 can interact with it [2, 3]. Tungsten also has a strong ability to emit electrons, making it the most important metal for thermal emission applications. Tungsten carbides have extremely high silicon microhardness and strength. Based on these properties, tungsten, tungsten alloys, and some tungsten compounds have many important applications in different fields of modern technology and cannot be easily replaced by other metals. Tungsten is also known as the "teeth of industry".

Table 1.1 Main physical properties of tungsten [3, 7–11]

Physical properties	Numerical values
Atomic number	74
Relative atomic mass	183.85
Atomic volume/$cm^3 \cdot mol^{-1}$	9.53
Density/$g \cdot cm^{-3}$	19.35
Crystal structure and lattice constant/nm	Alpha-W: Body-centered Cubic, Alpha = 3.16524 (25 °C)
	β-W: Cubic crystal system, Alpha = 5.046 (stable below 630 °C)
Melting point/°C	3410 ± 20
Boiling point/°C	5927
The latent heat of fusion/kg·mol	40.13 ± 6.67
Sublimation heat/$kg \cdot mol^{-1}$	847.8 (25 °C)
Evaporation heat/$kg \cdot mol^{-1}$	823.85 ± 20.9 (boiling point)
Resistance temperature coefficient/$°C^{-1}$	0.00482
Electron escape work per eV	4.55
Elasticity modulus of/MPa	35,000–38,000 (for silk)
Torque modulus/MPa	−36,000
Compressibility/$cm \cdot kg^{-1}$	2.917–7

1.2.2 Applications of Tungsten

Concerns about the effect of tungsten on steel properties began as early as the 1750s. High speed steel containing up to 20% tungsten was first exhibited at the 1900 World's Fair in Paris, marking a major technological advance in metal cutting that revolutionized industrial development in the early twentieth century. In 1906, the American mechanical engineer (F.W. Taylor) and metallurgical engineer (M. White) established the best composition of high speed steel for cutting $W_{18}Cr_4V$ (C 0.75%, W 18%, Cr 4.0%, V 1.0%), and the cutting speed was increased by more than ten times. Up to now, this kind of steel is still used in almost all machine shops in the world. The first tungsten bulb was patented in 1904 and quickly replaced the less efficient carbon filament lamps on the lighting market, revolutionizing artificial lighting. In 1923, Karl Schroter, a German, used powder metallurgy method for the first time to add 10–20% Co–WC powder as a binder, and developed WC–Co hard alloy, which is second only to diamond in hardness. Since then, hard alloy with high hardness, high strength and good toughness has become an important material in the field of engineering materials, and is widely used in [3, 12, 13] modern science and technology.

Tungsten is increasingly widely used in modern society. Table 1.2 lists several main applications of tungsten. Tungsten is mainly used in metal processing, machinery, electronics, mining, heavy manufacturing, petrochemical and other national economic fields, with 85% of the total consumption in this part of the application; Military, nuclear energy, aerospace and other defense and military applications account for 15% [8, 10, 14]. Among them, the main uses of tungsten in industry are the production of hard alloy and tungsten steel.

Tungsten is used the most in the cemented carbide field, accounting for more than 50% of total tungsten consumption. Cemented carbides are made by powder metallurgy, which is a process of high hardness refractory metal hard carbide (WC, TiC) and binder metal (Co or Ni, Mo) in proportion to each other through powder preparation, pressing forming and sintering at high temperature. Hard alloy has a series of advantages such as high hardness and strength, good toughness and wear resistance, high elastic modulus and compressive strength, strong chemical stability (corrosion resistance), high heat resistance and impact toughness, low coefficient of thermal expansion, thermal conductivity and conductivity close to iron and its alloy. Carbide is widely used in cutting tools, such as the production of car, milling, drilling, planing, boring and other tools, used for cutting common metal and non-metallic materials, but also cutting difficult materials such as heat-resistant steel, tool steel, high manganese steel. The cutting speed can reach 100–300 m/min, far more than high-speed steel tools, and the lifetime is several to ten times that of high-speed steel tools. Hard alloy is widely used in mining rock drilling tools and extraction tools, drilling tools for oil drilling, armament and measuring instruments for defense and military industry, and wear-resistant parts [4].

In recent years, with the development of science and technology, tungsten is increasingly widely used in the high-tech industry. For example, oxides are applied

Table 1.2 Main applications of tungsten products [8, 10, 14]

Fields	Industry	Uses
National defense field	Aerospace	High grade, high precision hard alloy: gyroscope rotor material, (space aircraft, submarine, rocket, etc.) inertial rotating components, aircraft ailerons, steering rudder, instrument and engine counterweight components, cutting blades, turning tools, milling cutters, welding blades, clip blades, CNC blades
	Military field	Tungsten-based alloy, hard alloy: materials for making various munitions, armor-piercing cores, bullet heads
		Tungsten-based alloy, hard alloy: shielding material used to protect against radiation, as in nuclear reactors
National economy	Electrical industry	Electrothermal processing materials
	Automotive industry	High quality and precision cutting tools, hole machining tools: balancers, shock absorbers, shockproof tungsten wire, tungsten contact materials, etc
	Petrochemical industry	Tungsten catalysts: petroleum refining catalyst WS_2, carbide drill bit
	Mining, smelting	Tungsten carbide drill bit, extraction drill bit, impact drill ball tooth, brazing sheet; Mine equipment lining, fan blades, bearings, etc
	Medical instruments	Radiation barrier, γ-ray knife
	Iron and steel industry	Tungsten for hard alloy rolls, dies, cutting tools and special steel production
	Machinery manufacturing	High performance, high precision carbide, etc.; CNC blades, welding blades, milling cutters, etc
	Electric power energy	Tungsten-based composites, tungsten contacts, and high-performance tungsten alloys
	Lighting industry	Tungsten light bulbs, electronic devices such as electronic tubes, picture tubes, X-ray tubes and electric heating filaments, short-arc mercury lamps and xenon lamps

to the electrode materials of supercapacitors because of their high specific capacity, high density and low cost. In addition, tungsten oxide is suitable for the manufacture of electrochromic components [15] because of its short switching time, significant color change and stable electrochemical properties. Tungsten and tungsten alloy are considered as the most suitable plasma surface materials because of their excellent properties such as high melting point, high thermal conductivity, high density, low coefficient of thermal expansion, low vapor pressure, low tritium retention, low sputtering yield and high automatic sputtering threshold. They are considered as potential materials [16, 17] in future commercial nuclear fusion devices. Due to its surface reactivity and catalytic properties, tungsten-based catalysts have been used in catalytic reduction of NO, detection of gaseous pollutants, photocatalytic degradation and other environmental fields [18]. Tungsten oxide materials have strong light absorption and semiconductor properties in a wide wavelength range in the near infrared region, so they are widely used in photo electrotherapy, solar cells, near infrared shielding, thermoelectricity, photocatalysis and biological applications [19–21]. In order to meet the requirements of modern industry and national defense for new high-performance materials, by coating tungsten and tungsten alloy coating has become an important method to enhance the mechanical, physical and chemical properties of all kinds of matrix materials, such as rocket engine and space power system throat lining materials, X-ray and gamma ray device materials, a variety of weapons coating and new pharmaceutical cover materials.

It is worth noting that special tungsten alloy and tungsten steel play an irreplaceable role in the modern military field. Tungsten is an essential element in the manufacture and production of gun barrels, rocket nozzles, armor-piercing shells, electromagnetic guns and other modern military weapons equipment in the military field of national defense. During the two World Wars in history, the demand and value of tungsten increased greatly. Therefore, tungsten is also known as the "war metal". As early as 1822, German engineers added tungsten to gun steel to increase the barrel's resistance to the corrosive effects of gunpowder combustion and extend the barrel's service [22] life. During World War I, tungsten became an important strategic resource for nations to scramble for. Now, tungsten steel is one of the most widely used weapons, such as PCrW and PCrNiW, which are used to manufacture large-caliber and thick-walled gun barrels. Tungsten helps to overcome the tempering brittleness and improve the strength of steel. Tungsten alloy shell penetrator core has stronger compression and deformation resistance and greater penetration power, which has become one [23] of the main threats to contemporary tanks. In recent years, with the successful test launch of the ultra-high speed electromagnetic railgun in the United States, the tungsten alloy coating, which is used as the surface of rail gun materials with wear-resisting and high-temperature ablation resistance, has gradually become the focus of the world's attention [24] and research. In the future, tungsten will be more widely used in high-tech industry and national defense industry, and its important position of strategic mineral resources will be further highlighted.

1.3 Tungsten Ores and Tungsten Resources

1.3.1 Tungsten Ores

Tungsten is a widely distributed element found in almost all kinds of rocks, but the content is low, through related geological processes can be enriched to form deposits as commercial ore mining. With an average concentration of 0.00013% in the earth's crust and 0.00016% in granite, tungsten ranks 18th among metals and 26th among elements in its relative abundance. Tungsten is mainly a hexavalent cationic state in nature. Due to the small radius of W^{6+} ion, high electricity price, strong polarization ability and easy to form complex anions, tungsten mainly exists in the form of complex anions. WO_4^{2-} combines with Fe^{2+}, Mn^{2+}, Ca^{2+} and other cations in the solution to form wolframite or scheelite precipitation. Under the super genetic action, due to the stability of tungsten ores, often formed placer, but under acidic conditions, tungsten ores can be decomposed, and dissolved in the form of WO_3 in surface water, under certain conditions to form some tungsten secondary minerals, sometimes in the form of mineral particles or ions by clay or iron manganese oxides adsorption and gathered in [3] shale, argillaceous fine sandstone and iron manganese ore layer.

Important minerals of tungsten are tungstate, in the process of mineralization can be combined with complex anions of only a few cations: Fe^{2+}, Mn^{2+}, Ca^{2+}, Pb^{2+}, Cu^{2+}, Zn^{2+}, etc., so tungsten ore species are limited. At present, more than 20 kinds of tungsten ores and tungsten-bearing minerals are found in the crust, namely wolframite group: tungsten iron ore, tungsten manganese iron ore, ferrotungsten manganese ore, manganese tungsten iron ore, tungsten manganese ore; scheelite group: scheelite (calcite), molybdenum scheelite, copper scheelite; tungstenite minerals: tungstenite, hydrotungstite, ferritungstite, yttrium tungstenite, copper tungstenite, anthoinite; uncommon tungstenrals: tungsten lead ore, raspite, molybdenum tungsten lead ore, tungsten zinc ore, tungsten bismuth ore, etc. [2, 3, 25]. Table 1.3 lists the physical properties of the major tungsten ores, of which only wolframite and scheelite are economically valuable for mining.

1.3.2 Tungsten Resources

1.3.2.1 World Tungsten Resources

According to the U.S. Geological Survey, there are 3.8 Mt of tungsten reserves in the world in 2022 [6]. As is shown in Fig. 1.1, the world's tungsten reserves are concentrated in China (1.8 Mt), Russia (400,000 t), Vietnam (100,000 t) and Spain (56,000 t), which together account for more than 60% of the world's tungsten reserves. Other tungsten resource countries include the Democratic People's Republic of Korea, the United States, Bolivia, Rowanda, Austria and Portugal, etc. (Fig. 1.1). Other countries with tungsten resource potential include Australia, Mongolia, Brazil, Myanmar [6, 26, 27], Kazakhstan, Turkey, Thailand, etc.

The main tungsten deposits in the world are skarn, porphyry and quartz vein, among which the quartz vein wolframite deposit is the most important tungsten deposit in the world at present. Except China, there are almost no large mining areas in other countries and regions. The ore minerals are mainly wolframite, with a small amount of scheelite. The second is skarn scheelite deposit, ore reserves are more concentrated, often form a large mining area. Porphyry tungsten deposits account for 1/4 of the total tungsten ore reserves, and nearly half of the ore minerals are

Table 1.3 Main tungsten ores and their properties [3]

Minerals	Formula	WO_3 content/%	Density/ $g\cdot cm^{-3}$	Mohs hardness	Color	Magnetism
Wolframite	(Fe, Mn) WO_4	76.5	7.3	5–5.5	Hazel, iron black, etc	Weak magnetism
Scheelite	$CaWO_4$	80.6	5.9–6.2	4.5–5	White and pale yellow	Non-magnetic
Tungsten iron ore	$FeWO_4$	76.3	7.5	5	Black	Weak magnetism
Tungsten manganese ore	$MnWO_4$	76.6	7.2	4–4.5	Brown, reddish-brown and black	Weak magnetism
Tungstite	$WO_3\cdot H_2O$	86.2	5.5	2.5–3	Bright yellow or light yellowish green	
Tungsten copper ore	$CuWO_4$	74.5	3–3.5	4.5–5	Olive green	
Chillagite	$PbWO_4$	50.9	7.8–8.34	2.7–3	Green, brownish red and light yellowish gray	
Russellite	Bi_2WO_6		6.8–7.8	3–3.5	Light white, yellow, green yellow, yellow-gray	
Tungsten-zinc ore	$ZnWO_4$	74.02	6.7		Dark brown, black	
Tungstenite	WS_2		7.18–7.4	2.5	Lead gray	

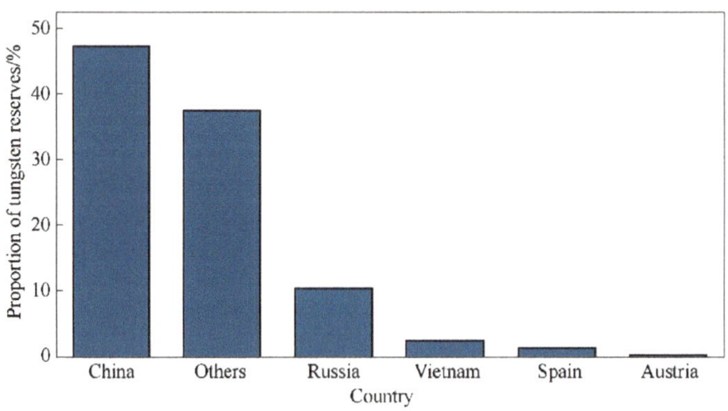

Fig. 1.1 Distribution of global tungsten resources in 2022 [6, 26]

wolframite and scheelite, with low grade (0.1% WO_3 content). In addition to the above-mentioned tungsten deposit, there are other tungsten deposits of sandstone type, pegmatite type and thermal spring type, etc. These deposit types are of little [10, 28] industrial significance. Most super-large tungsten deposits in the world are skarn type. There are mainly Mactung and Cantung in Canada, Pine Creek in the United States, Sangdong in the Republic of Korea, King Island in Australia, Uludag skarn scheelite deposits in Turkey, etc. The world's tungsten deposits are mainly distributed in the Pacific Rim metallogenic belt (Canada, the United States, Bolivia, the Democratic People's Republic of Korea, China's southeast coast), the north coast of the Mediterranean (Turkey, France, Austria, Germany, etc.), the South Ural, Central Asia and western China, Xinjiang, Gansu and other places, among which the Pacific Rim metallogenic belt tungsten deposits account for more than 50% of the world's total tungsten deposits. Many large and famous tungsten deposits are located here, such as Shizhuyuan tungsten in China, Sisson and Northern Dancer tungsten in Canada, and Nui Phao tungsten in Vietnam, etc. [9, 10, 26, 29].

The most important tungsten deposits in Europe are located in the United Kingdom, Spain and Austria. The Drakelands tungsten (formerly known as Hemerdon Tungsten) in the UK has proven ore reserves of 35.7 Mt with an average grade of 0.18% tungsten trioxide and 64,260 t tungsten trioxide reserves. The La Parrilla tungsten in Spain is currently the largest tungsten tin in production in Spain, with ore reserves of 29.7 Mt, an average grade of 0.0931% of tungsten trioxide and 27,500 t of tungsten trioxide reserves. Barruecopardo tungsten, once the largest tungsten in Spain, is a large open-pit tungsten with nearly a century of mining history, with ore reserves of 8.69 Mt, tungsten trioxide with an average grade of 0.30% and tungsten trioxide reserves of 26,100 t. Mittersill tungsten in Austria is the largest scheelite ore mountain in Europe today, with estimated ore reserves of 5 Mt, tungsten trioxide average grade of 0.5%, tungsten trioxide reserves of 25,000 t, annual mining stone of 400,000 t, and annual output of tungsten of 1800 t. France and Norway also have large reserves of tungsten resources. Other potential tungsten resource countries are in the Commonwealth of Independent States, especially Russia and Kazakhstan. The Boguty tungsten in Kazakhstan has proved ore reserves of 126,029,500 t, the average grade of tungsten trioxide is 0.226%, and the tungsten trioxide reserves of 284,900 t, ranking fifth in the world. The Bakuta tungsten project will be put into operation in 2022. It is expected that the output of tungsten concentrate will reach about 8% of the world's output, making it the world's largest monomer tungsten [26] under construction.

Bolivia used to be the world's third largest tungsten producer, but production has fallen sharply in recent years. Ckoljlla has large deposits but low ore grade. Canada is the most likely new source of tungsten for the Americas. Sisson tungsten molybdenum deposit in Canada has large reserves of about 334.4 Mt, but low ore grade, with an average grade of 0.066% tungsten trioxide and 0.021% molybdenum, and 222,000 t of tungsten trioxide reserves. Mactung tungsten in Canada has 8.5 Mt of underground ore reserves with an average grade of 1.082% tungsten trioxide and 91,997 Mt of tungsten trioxide reserves. In 2015, North American Tungsten Co. filed for bankruptcy reorganization; In August 2019, two tungsten deposits owned by North American Tungsten, Mactung and Cantung, sought to be sold, which could significantly affect tungsten supply in the Americas.

In the Asia–Pacific region, Nui Phao tungsten polymetallic is the largest tungsten in Vietnam, with ore reserves of 66 Mt, tungsten trioxide average grade of 0.18% and tungsten trioxide reserves of 118,800 t. After being put into operation in 2013, the annual output of tungsten concentrate can reach 6000 t. Nui Phao tungsten polymetallic has low production cost, but due to the export tariff of tungsten concentrate in Vietnam is as high as 20%, so currently all tungsten concentrate is used for domestic production APT. the Republic of Korea's Sangdong tungsten, the largest tungsten mining project in the Republic of Korea, has proven mineral resources of 8.334 Mt, an average grade of 0.49% tungsten trioxide, and 40,800 t of tungsten trioxide reserves. The mine was one of the world's major suppliers of tungsten concentrate from the 1950s to the 1990s, but it was closed in 1992 due to low tungsten prices. In 2020, Almonty Industries Inc. began construction of the Upper East Project with a designed annual output of 3600 t of 65% tungsten concentrate. Thailand also has abundant tungsten resources, though they are usually associated with tin resources.

The Dolphin tungsten in Golden Island, Australia, has proven ore reserves of 3.14 Mt, the average grade of tungsten trioxide is 0.73%, the tungsten trioxide reserves are 22,922 t, and the designed annual capacity is 3100 t of tungsten concentrate (equivalent to about 2000 t of tungsten trioxide). In addition, there are a number of major tungsten projects in production or expected to come on stream within two years, such as the Drozhillovskoye, Severniy Katpar and Verkhniye Kayraktinskoe tungsten deposit in Kazakhstan, the RHA tungsten in Zimbabwe and Tyrnyauz tungsten in Russia.

1.3.2.2 China's Tungsten Resources

China is the country with the most abundant tungsten resources in the world, with 1.8 Mt of tungsten reserves in 2022, accounting for 47.37% of the world's tungsten reserves, which is the largest in the world (Fig. 1.2). Tungsten deposits in China are widely distributed, produced in many places, of large scale, of high grade, with complex associated components and various ore types. China is especially famous for its highly complex quartz vein type tungsten deposits, and four of the world's top ten mines are located in China [5, 25, 30].

China's tungsten deposits are generally distributed in the eastern coastal areas of Guangdong on both sides of China's Nanling Mountains, especially in Jiangxi Province, with reserves of more than 50% of the country's total reserves. China's tungsten resources are widely distributed in 21 provinces (autonomous regions and municipalities directly under the Central Government). As is shown in Fig. 1.3, Jiangxi, Hunan and Henan provinces rank the top three in the country in terms of tungsten resources reserves, of which Hunan and Jiangxi provinces account for 83.52% of the country in terms of tungsten resources reserves. Hunan is dominated by scheelite and Jiangxi is dominated by wolframite, and its wolframite resources account for 42.40% [6] of the country's total wolframite resources. In addition, Guangxi, Yunnan,

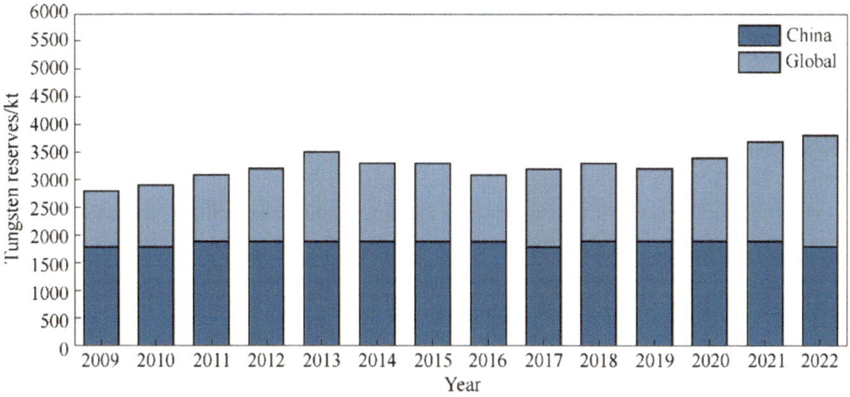

Fig. 1.2 Global and China tungsten reserves, 2005–2022 [5, 6, 26, 27]

Fujian, Gansu, Xinjiang, Heilongjiang, Guangdong and other provinces also have a certain tungsten ore deposit, tungsten reserves of more than 20 kt.

In China, scheelite accounts for 68.7%, wolframite 20.9% and mixed of ores of wolframite and scheelite 10.4%. Due to the low grade of scheelite, more lean ore and less rich ore in China, the composition of the mixed tungsten ore is complex and difficult to be separated and smelted. In recent centuries, wolframites are mainly mined because of its high-grade, easy-to-use, easy-to-choose characteristic. As wolframite consumption rate is fast, the reserves are constantly decreasing, wolframite has been basically consumed. Tungsten ore reserves are mainly scheelite. Scheelite reserves are mainly distributed in the central region (81.6%), there are 29 large scheelite deposits in China, of which there are three super-large scheelite deposits, namely Xiangnan Shizhuyuan (mixed ores of scheelite and wolframite) and Huangshaping (scheelite) tungsten-tin bismuth-molybdenum polymetallic deposit, West

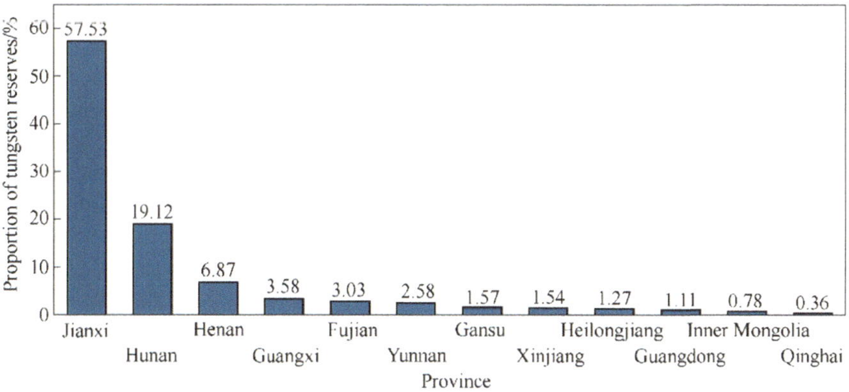

Fig. 1.3 Distribution of tungsten reserves in China in 2021

Henan Luanchhuan Sandaozhuang (scheelite) tungsten-molybdenum deposit and west Fujian Qingliu Xingluokeng (mixed ores of scheelite and wolframite) tungsten, accounting for more than 55% of the total scheelite reserves. China's scheelite mining area has a large scale and concentrated reserves, with 10 mining areas with more than 100 kt reserves. However, tungsten grade in scheelite mining areas is generally not high, and 80% mining areas have ore grade below [28] 0.5%.

Tungsten deposits in China are mostly symbiotic and associated ore deposits, with few single ore deposits, which are usually associated with copper, molybdenum, antimony, tin, bismuth, niobium, tantalum and gold and silver. China's associated tungsten reserves account for about 20% [3] of the total reserves. In tungsten deposits in Hunan, Jiangxi and Guangdong provinces, the ratio of tungsten reserves to the total reserves of tin, bismuth, molybdenum, copper and other associated metals is about 1:1. Comprehensive recovery of tungsten resources while developing the main ore deposits is very beneficial for rational utilization of mineral resources and improvement of economic benefits of enterprises. At present, Yunnan Gejiu tin ore, Daye Longjiaoshan copper mine, Jiangxi Yongping copper mine and some molybdenum mines have generally realized the comprehensive development and utilization of main metal and associated tungsten resources.

China has a large potential of tungsten resources. Since 2010, the newly discovered tungsten resources in a number of large and super-large tungsten in China have exceeded the total discovered in [14] history. At the beginning of 2012, WO_3 1,053,450 t of tungsten resources were discovered in the northern and southern sections of Dahutang, which ranks the first in the country as a single ore deposit. In recent years, millions of tons of tungsten resources were newly added in the surrounding working areas, and the discovered tungsten resources in Dahutang ore field have exceeded 2 Mt [31] so far. Dahutang mining area has refreshed the resource reserve scale of single tungsten ore body, which changed the composition proportion of skarn type, quartz vein type and granite disseminated type of tungsten resource reserves in the country, and consolidated China's dominant position in tungsten resources in the world. In the prospecting work outside Zhuxi Copper mine in Jiangxi province, a hidden super-large skarn scheelite deposit was found. WO_3 2.91 million with an average WO of 0.689% and 100,000 t of symbiotic copper, among which 600 thousand tons of rich ore resources with an average WO grade of 1.54% were discovered. In 2021, 58,400 t of tungsten ore was discovered in Fujiashan area, Daye City, reaching the scale of large-scale tungsten ore. Table 1.4 lists the newly discovered tungsten resources [32] in recent years. In the old mining areas of southern Jiangxi, Northern Guangdong and southern Hunan, through the bottom prospecting, thick and large associated tungsten bodies have been found in the deep edge of Xihuashan, Maoping, Taoxiceng, Hukeng, Yaogangxian, Wuning, Dajiashan and Fenyi tungsten mining areas in Jiangxi Province. The skarn associated tungsten ore has been discovered in Huangshaping lead–zinc mine in Hunan Province and some large non-ferrous mines in the middle and lower reaches of the Yangtze River copper iron ore belt. At the same time, new breakthroughs [33] have been made in recent years in southern Anhui (western Zhejiang), Jizhong-Yanbian, Northern Qilian, Xinjiang-Qinghai Qimantag and northern Xizang Jiagang.

Table 1.4 The most recently discovered tungsten deposits [33]

Regions	Neo tungsten deposits	Resource class/size
Fuliang County, Jiangxi Province	Zhuxi concealed super large skarn tungsten scheelite deposit	WO_3 resources of 2.86 Mt
Wuning County, Jiangxi province	Dahutang superlarge granite vein disseminated scheelite deposit	WO_3 resources of 1.06 Mt
Malipo County, Yunnan Province	Nanwenhe super-large stratified skarn scheelite deposit	WO_3 resources of 530,000 t
Lechang County, Guangdong Province	Heshangtian superlarge tungsten-tin deposit	Tungsten-tin resources of 300,000 t
Dunhuang, Gansu Province	Xiaodushan superlarge tungsten deposit	WO_3 resources of 200,000 t
Shixing County, Guangdong Province	Nanshancang-liangyuan tungsten rare polymetallic deposit	WO_3 resources are forecast to be more than 200,000 t
Chongren County, Jiangxi Province	Xiangyuan large leveine zone scheelite deposit	WO_3 has 130,000 t of resources
Yunfu County, Guangdong Province	Dajinshan large tungsten-tin deposit	Tungsten resources of 85,000 t and tin resources of 19,000 t
Cangwu, Guangxi	Shedong large tungsten polymetallic deposit	WO_3 resource reserves of 80,000 t and molybdenum reserves of 10,000 t
Mozhu Gongka County, Xizang Province	Dalong Tungsten large tungsten polymetallic deposit	WO_3 resource reserves of 50,000 t
Hami, Xinjiang	Shadong large scheelite deposit	WO_3 resource reserves of 30,000–50,000 t

It is worth noting that although China ranks first in the world in the reserves of tungsten resources, the degree of tungsten ore security is very severe. There are 107 tungsten deposits with designed production capacity in China, of which 62% have a static life of less than 10 years, and only 27 large mines have a life of more than 40 years [10, 11, 34]. China's tungsten resources account for less than 60% of the world's reserves, but the exploitation of more than 80% of the world's tungsten resources, the resource advantage is gradually weakening. China's ratio of storage and production dropped from 14.6 in 2008 to 10.8 in 2019, which is far lower than the global level of 27.1 years, and the speed of resource consumption is too fast [26]. Although the identified reserves of tungsten resources have increased, the basic reserves have decreased significantly, and the resource endowment of newly discovered exploration is relatively poor, and the economic benefits of development and utilization are expected to decrease. With the rapid consumption of domestic tungsten resources, China's advantage in tungsten resources is likely to be reversed. While strengthening detailed investigation and exploration of tungsten resources, it is essential to firmly implement the policy of controlling the total consumption of tungsten resources.

1.3.2.3 Secondary Tungsten Resources

As a non-renewable resource, the secondary recycling of tungsten resources is an important source of tungsten resources in many countries. The secondary utilization of tungsten resources mainly comes from hard alloy, followed by tungsten material, alloy steel, tungsten catalyst, tungsten contact material, high specific gravity alloy, etc. [35]. China is the world's largest tungsten consumer, but the recycling of tungsten waste is less than 10%, well below the global average of 30%. The main secondary use of tungsten in China is the recovery of cemented carbide, which currently amounts to 1500 t per year, with great potential for growth [36]. The tungsten content (WO) of the secondary recycling tungsten products is more than 10 times that of the raw ore$_3$. The world's secondary utilization of tungsten resources accounts for about 1/3 of the total demand, so it is highly valued by the United States, Japan and other large tungsten resource consuming countries [37]. Skoda, a German company, specializes in purchasing tungsten waste from tungsten plants around the world at low prices for a long time and building a special production line in the Gesla plant, whose processing capacity accounts for 30% of tungsten raw materials. Russia has set up a special institute for recycling non-ferrous metals to engage in metal recycling, and its industrial treatment technology has very promising application prospects. In Japan, a tungsten recovery committee has been set up to study the recycling ways, treatment technologies and reprocessing and utilization of tungsten waste. 100 t of hard alloy can recover 31 t of waste. In the United States, the amount of hard alloy recycling accounts for about 30% of the whole proportion, metallurgical industry companies from waste alloy recycling WC500 t/a, pure tungsten alloy additives for 100 t/a [38]; Huachang company zinc smelting recovery 500 t of hard alloy/year. The UK's Serkasian company is a specialized tungsten waste processing company [36, 39]. China's tungsten recycling technology started late, mainly focusing on the recovery of secondary resources such as hard alloys and catalysts.

1.4 Production and Consumption of Tungsten Products

Since 2017, the output of tungsten deposits in the world has remained above 80,000 t, of which China's tungsten output has always accounted for more than 70% of the global tungsten output. In 2022, the output of tungsten deposits in China is 71,000 t, accounting for 84.52% of the world's tungsten output, and China remains the world's largest tungsten producer. Other major tungsten deposits in China include Vietnam (4800 t), Russia (2300 t), Bolivia (1400 t), Austria (900 t), Portugal (500 t), etc. (Fig. 1.4) [5, 6, 26, 40].

China is the world's largest producer of tungsten ore, with an annual output of more than 60,000 t in recent years, taking an absolute advantage in the world. Enterprises producing tungsten ore in China are located in 11 provinces (autonomous regions), and the output of tungsten concentrate is mainly concentrated in Jiangxi, Hunan and Henan provinces, with the output of tungsten concentrate accounting for 58.2%

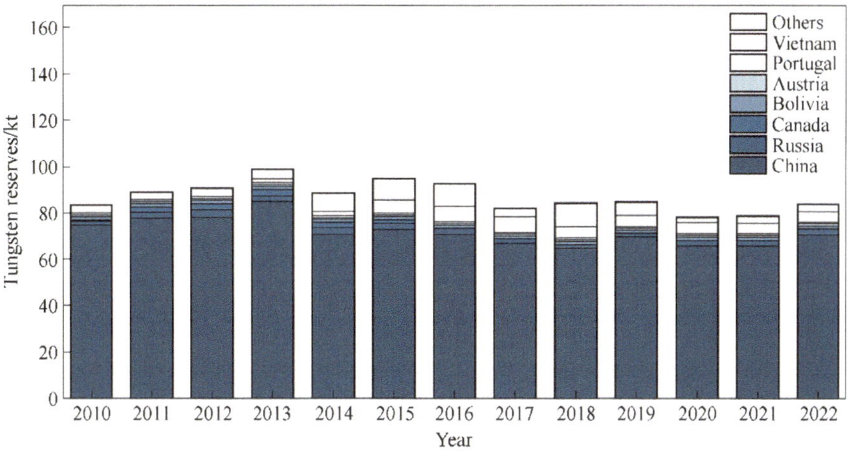

Fig. 1.4 Production of tungsten deposits in various countries from 2010 to 2022 [5, 6, 26, 40]

[5, 26] of the national output. Most of China's tungsten exports are upstream raw materials such as tungsten concentrate and APT. Although the export volume of high value-added oxides, tungsten powder, tungsten carbide and hard alloy has increased in recent years, the growth rate is small, and there is still a long distance [11] between China and the United States, Japan, Europe and other tungsten processing countries. At present, China implements the policy of protective exploitation of tungsten, the total quantity control of tungsten production, and the total quantity control and quota license system for tungsten export.

In 2015, Vietnam produced 5600 t of tungsten, making it the world's second largest tungsten producer, overtaking Canada and Russia. Most of Vietnam's tungsten comes from the Nui Phao tungsten polymetallic mine, which Masan Resources estimates accounts for 30% of the world's total tungsten supply. It is also the largest single fluorite mine outside China and Mexico, accounting for 6–7% [41] of the world's total fluorite supply. Russia is the world's leading tungsten producer, with the Primorsky and Lermontov tungsten deposits being the two major Russian tungsten deposits [42]. The main tungsten producing companies are JSCA&IR Mining Company (whose ore comes mainly from the Primorsky tungsten), KGUP Primteploenergo Mining Company, Russian Ore Mining Company (RGRK), Artel Quartz Limited and Wolfram Company. Canada was once a significant tungsten producer in the world, with the Cantung Mine, Mactung Mine and Glotone being the three scheelite deposits with the largest reserves in Canada, and Mount Pleasant being an important wolframite mine in Canada. North American Tungsten Corp. Ltd., Northcliff Exploration LTD., and Adex Mining Company are the three largest tungsten mining companies [34, 35] in Canada. In 2015, North American Tungsten filed for bankruptcy reorganization, and the Cantung Mine was discontinued, while the Mactung Mine was put on hold until now. In 2018, the proved tungsten and molybdenum reserves of the North Katpal deposit in Karaganda, Kazakhstan are about 90,000 t and 13,000 t, and

Kazakhstan's tungsten production is expected to rise to the second place in the world. At the end of 2019, Tungsten West acquired the Hemerdon tungsten ore deposit in the UK with a raw ore volume of 40,100 t and an average grade of 0.13%. In 2020, the Sangdong Tungsten Iron-denumProject$_3$ in the Republic of Korea, owned by Canada's Almonty Industries Inc, entered the semi-industrial test stage. The tungsten reserves of the mine amounted to about 51,000 t. The grade of WO_3 is 0.41% [26].

The world's major tungsten consumption countries are China, the United States, Japan and Europe, and other important tungsten consumption countries and regions are Russia, the Republic of Korea, South Africa, India, Brazil, Israel and Pakistan, etc. China has become the global center of tungsten consumption. Since 2002, China's tungsten consumption has continued to grow, becoming the world's largest tungsten consumer, accounting for about 80% [26] of the global tungsten consumption. In 2019, the world consumed 98,000 t of tungsten, of which China accounted for 57%, the European Union 16%, the United States 11% and Japan 8%, accounting for 92% of the world's tungsten consumption. As shown in Fig. 1.5, China, the United States, Japan and the European Union are the major countries and regions of original tungsten consumption. In 2019, China, the European Union, the United States and Japan together accounted for 91% of the global consumption of primary tungsten. Carbide manufacturing is the largest tungsten consumption sector. According to the International Tungsten Association (ITIA), 59% of the world's tungsten is used in the production of cemented carbide, 19% is used in special steels and alloys, 16% is used in tungsten materials, and 6% is used in chemicals and other sectors. The consumption structure of tungsten in China is roughly as follows: 51% for cemented carbide, 30% for special steel and alloys, 13% for tungsten, and 6% for chemical and other sectors.

Per capita resource consumption, as an important indicator to measure the degree of industrialization, can reflect the relationship between economic development and resource consumption level. As is shown in Fig. 1.5, The United States and the European Union have completed the industrialization process, and the per capita tungsten consumption level shows a stable trend. At present, China is in the process of rapid industrialization and urbanization, and the per capita tungsten consumption is growing gradually, gradually surpassing the level of the United States and the European Union after 2009. It can be predicted that the per capita tungsten demand of China will continue to show a stable growth trend in the future, and China will remain the largest [43] tungsten resource demand country in the next 30 years.

1.5 Strategic Significance of Tungsten Resources

Hard alloy is the main consumption area of tungsten, accounting for as high as 72% in the downstream demand structure of tungsten in developed countries and only 45% in China. There is still a large room for growth. At present, the consumption of hard alloy in China is more than 6400 t/a, of which raw tungsten consumption

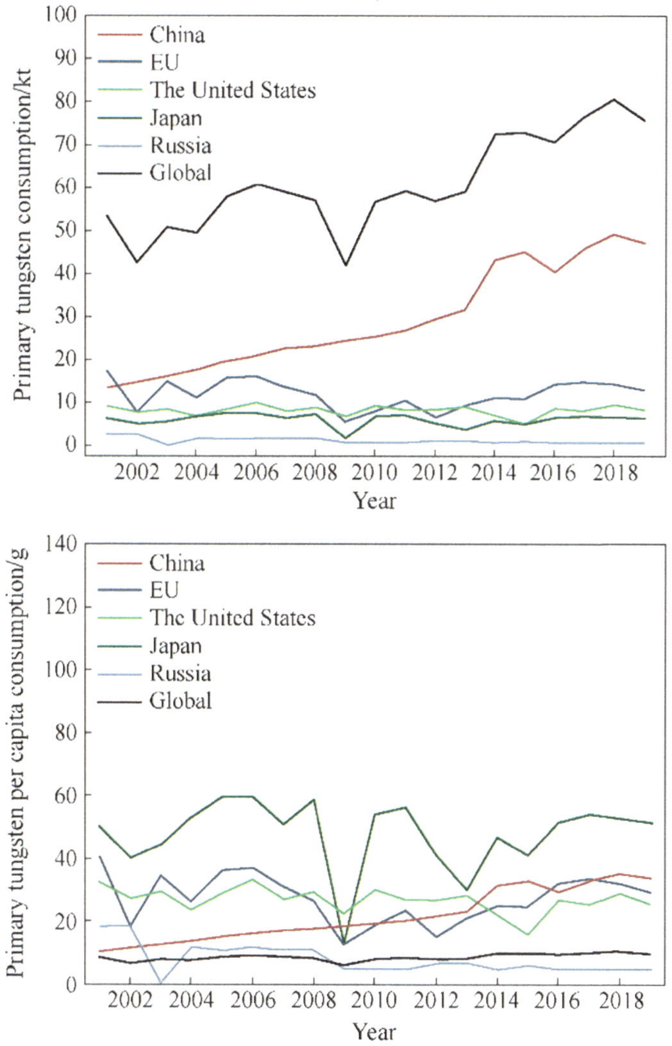

Fig. 1.5 Tungsten consumption levels in countries around the world

accounts for 69% and recycled metal consumption accounts for 31%. As the production of wear-resistant parts, mineral alloys, and cutting blades continues to grow, the consumption of tungsten resources will continue to increase. The ratio of special steel to crude steel in China is much lower than that in developed countries, and industrial upgrading is driving the growth of tungsten consumption in the special steel sector. Tungsten is the main alloying element of high-speed tool steel, alloy structural steel and spring steel. In 2021, China's special steel output will only account for 13.4% of the total crude steel output, far from Japan's 20%. The demand for special steel will expand rapidly with China's industrial upgrading. In terms of tungsten, according to the relevant statistics of China Lighting Electrical Appliances Association, the production capacity is more than 1000 t/a, among which the production capacity of tungsten wire is more than 10 billion meters, with more than 200 enterprises. The domestic consumption of tungsten is more than 530 t/a, of which the output is more than 735 t/a, the export is about 39 t/a, and the import is about 185 t/a [7, 8, 10]. China has become the world's largest tungsten supplier, supplying about 80% of the international market demand for tungsten products, which could supply the world for another 26 years based on China's 1.8 Mt of reserves. In the short term, the global supply of tungsten resources is sufficient, while in the medium [26] and long term, the global supply of tungsten ore resources is tight.

At present, there is no alternative material in the main application field of tungsten, so its importance is in a strategic position. China, the United States, Russia, the Republic of Korea and other countries have established corresponding strategic reserves [44]. In recent years, with the increasingly prominent irreplaceability of tungsten ore in the high-tech industry, the strategic protection policies for tungsten ore in various countries have become more comprehensive. In May 2014, the European Commission issued the List of Key Raw materials, which expanded the scope of the list from 14 to 20, and tungsten was included in it. At the end of 2016, China issued the National Mineral Resources Plan (2016–2020), which listed tungsten as a national strategic mineral catalog, as a key object of macro-control, supervision and management of mineral resources, and provided differentiated management [31] in resource allocation, fiscal revenue, mining land and other aspects. In 2018, the United States published a list of 35 key mineral products, including tungsten, in order to reduce its dependence on imported mineral resources, prevent strategic economic and military vulnerability and adverse foreign government actions, natural disasters and other events that may interfere with the supply of key minerals, and took a series of measures to address its dependence on overseas supplies. To encourage domestic companies to build an integrated domestic supply chain in the production, refining, materials science, metallurgy, intellectual property, research and development, and commercial and defense applications of key minerals such as tungsten ores. In 2020, the Queensland Government of Australia announced an investment of 10 million Australian dollars to encourage mining enterprises to explore tungsten and other "New Economy Minerals" and technological innovation in mineral development, so as to ensure the security of strategic resource reserves. In 2021, King Island, a local enterprise, will be directly funded to redevelop the "Dolphin Project" tungsten in Tasia, which has been out of production for about 30 years, so as to reduce its dependence on China's tungsten. In June 2021, the Brazilian government announced a list of minerals of special importance to the national interest, hoping to expand the production of these minerals and metals through the recently formulated policies. Tungsten is included in the list as a mineral needed for high technology products and processes. In the same year, the Korean government decided to double its reserves of key rare metals such as tungsten, molybdenum and rare earth to secure 100 days of consumption, and to develop and cultivate 100 key rare metal companies by 2025. The Japanese government added industries related to important minerals such as tungsten, molybdenum and rare earth minerals to the list of key foreign investment targets under the revision of the Foreign Exchange Law, aiming to minimize the vulnerability of the country's supply chain of key raw materials, while improving the international competitive position of various industries such as automobiles, batteries and aircraft carriers.

1.6 Summary

Tungsten is an essential material in modern national economic life, as well as a vital material in modern aviation and military fields. China is the country with the largest reserves, production, export, and consumption of tungsten resources, which are mainly distributed in Jiangxi and Hunan provinces. Recently, a number of new-medium and large-sized tungsten deposits have been surveyed in many places. In the short term, domestic tungsten resources are sufficient, but it faces the hidden danger of long-term resource shortage. In addition to China, Russia, Vietnam, and Spain are the world's main tungsten resources reserve countries, while Russia, Vietnam and Bolivia are the world's main tungsten producers. With the acceleration of China's industrialization and urbanization, China's demand for tungsten resources will further increase in the future, which will put great pressure on China's tungsten resources supply. Due to its irreplaceability in many key fields, tungsten has been listed as a strategic resource by many countries such as the Republic of Korea, Japan, the United States and Australia. China will further investigate and explore tungsten resources, and at the same time implement a total consumption control policy for tungsten resources to ensure the safety of domestic tungsten resources. Based on the current situation of tungsten resources in the world, it can be predicted that China's tungsten mining work will focus on the following three aspects in the future:

(1) Efficient utilization of low-grade associated lean fine tungsten resources. The grade of tungsten ore is generally reduced in our country, and the proportion of mixed associated ore is high, so it is necessary to develop more efficient and environmentally friendly technology and smelting technology and equipment.
(2) Recycling and utilization of secondary tungsten resources. The secondary tungsten resource recovery starts late in our country, and the low utilization rate is required to speed up the tungsten recycling technology.
(3) Develop and produce high value-added tungsten products to improve the competitiveness of tungsten products in the international market.

Acknowledgements This work was supported by National key research center and development Program of the 14th Five-Year Plan (No. 2022YFC2905105 and No. 2022YFC2905104), National Natural Science Foundation of China (NSFC) (No. 52122406and No. 52374291), Hunan High-tech Industry Technology Innovation Leading plan (No. 2022GK4056), Hunan Innovative Province construction special Project (No. 2020RC3001).

References

1. Li H (1990) Rare metal metallurgy. Rare Metal Metallurgy
2. Ma D (2009) Research progress of tungsten geochemistry. Geol J China Universities 15(01):19–34
3. Qiu X, Dong T (2012) Modern tungsten ore dressing. Modern Tungsten Ore Dressing
4. Chu K (2011) Development status and prospect of Chinese cemented carbide industry. Rare Metals Cemented Carbide 39(01):52–56
5. Ministry of Land and Resources of the People's Republic of China (2020) China Mineral Resources Report. Beijing
6. Mineral commodity summaries 2021. Reston, VA, 2021200
7. Cao F, Yang H, Wang W et al (2018) Overview and supply and demand analysis of global tungsten resources. Conserv Utilization Miner Resourc 02:145–150
8. Seddon M (2001) Global tungsten resources and future supply. China Tungsten Industry Z1:136–138
9. Cai G-P, Wu Y-B, Chen S-P (2009) Analysis of world tungsten resources. World Nonferrous Metals (04):62–65
10. Zhang H (2017) Analysis of world tungsten resource supply and demand. China University of Geosciences (Beijing)
11. Zhao Z, Sun F, Yang J et al (2019) Current situation and prospect of tungsten resources, technology and industry in China. Chinese J Nonferrous Metals 29(09):1902–1916
12. Bose A (2011) A perspective on the earliest commercial pm metal-ceramic composite: cemented tungsten carbide. Int J Powder Metall 47(2):31–50
13. Mehrotra PK, Mizgalski KP, Santhanam AT (2007) Recent advances in tungsten-based hardmetals. Int J Powder Metall lancet (2):33–40
14. Wang M, Jia M, Xiao Y et al (2014) Status quo and sustainable development countermeasures of tungsten resources in China. Nonferrous Metals Eng 4(02):76–80
15. Han W, Shi Q, Hu R (2021) Advances in electrochemical energy devices constructed with tungsten oxide-based nanomaterials. Nanomaterials 11(3):692
16. Liu D, Liang Z, Luo L et al (2018) An overview of oxidation-resistant tungsten alloys for nuclear fusion. J Alloy Compd 765:299–312
17. Ding XY, Luo LM, Huang LM et al (2015) Research progress on radiation damage of tungsten and tungsten alloys used in nuclear fusion reactor. Rare Metals 39(12):1139–1147
18. Can F, Courtois X, Duprez D (2021) Tungsten-Based catalysts for environmental applications. Catalysts 11(6)
19. Wu C, Naseem S, Chou M et al (2019) Recent advances in tungsten-oxide-based materials and their applications. Front Mater 6:49
20. Bentley J, Desai S, Bastakoti BP (2021) Porous tungsten oxide: recent advances in design, synthesis, and applications. Chem European J 27(36):9241–9252
21. Mardare CC, Hassel AW (2019) Review on the versatility of tungsten oxide coatings. Physica Status Solidi Appl Mater Sci 216(12)
22. She J, Qiu Y (2001) Application of tungsten—from electronic materials to military ammunition. China Tungsten Industry 02:39–41
23. Wang F, Zhao M (1997) High density tungsten alloy and its application in military industry. Powder Metall Mater Sci Eng 02:114–120
24. Zhang X, Lin C, Cui S et al (2013) Research status of tungsten and its alloy coatings. Acta Armamentarii 34(3):365–372
25. Cui Z, Guo G, Zhao J et al (2019) Research status and progress of tungsten deposits in China. J Hebei Univ Geosci 42(01):27–36
26. Yu Z, Su G (2021) Development report of China tungsten industry in 2020. China Tungsten Industry 36(282(02)):1–9
27. Anonymous (2006) Mineral resource of the month: Tungsten. Geotimes 51(2):45
28. Yang X, Liu Q (2008) Status and progress of resource distribution and mineral processing of scheelite in China. Mining Express 04:6–9

29. Han Z, Golev A, Edraki M (2021) A review of tungsten resources and potential extraction from mine waste. Minerals 11(7)
30. Liu Z, Xia Q, Wang X et al (2014) Mineral resources distribution and metallogenic belt division of tungsten deposits in China. Miner Deposits Geol 33(S1):947–948
31. Zang J (2012) Approach to the world's largest tungsten deposit—Jiangxi Dahutang. Earth 07:44–45
32. Ouyang Y, Rao J, Wei J (2020) Preliminary study on multi-stage mineralization of Zhuxi super large Tungsten-Copper Deposit
33. Wei X (2016) New discoveries and implications of tungsten exploration in China in recent years. China Tungsten Industry 31(03):1–7
34. Shen J (1999) The depletion of wolframite resources and its countermeasures. Cemented Carbide (02):126–129
35. Ahn HH, Lee MS (2018) Hydrometallurgical processes for the recovery of tungsten from ores and secondary resources. J Korean Instit Resourc Recycl 27(6):3–10
36. Research on regeneration and properties of secondary resource tungsten (2012)
37. Xiao LS (2013) Progress and prospect of tungsten extraction and metallurgy technology in China. Nonferrous Metals Sci Eng 4(05):6–10
38. Mishra D, Sinha S, Sahu KK et al (2017) Recycling of secondary tungsten resources. Trans Indian Inst Met 70(2):479–485
39. Wentang X (2006) Secondary resources of tungsten and its development prospect. Renew Resourc Res 01:11–17
40. China Tungsten Industry Development Plan (2016–2020). China Tungsten Industry 32(01):9–15 (2017)
41. Tabor Company of Canada developing tungsten deposit in Vietnam. Mining Research and Development (03):15 (2004)
42. Zhao Q (2004) New progress of tungsten and molybdenum metallurgy in Russia. Rare Metals Hard Metals 04:59
43. The Gray Prediction of World Tungsten Demand and Some Suggestions (2009)
44. Gao B, Li Z (2016) Early warning for tungsten resources industrial security in China. In: 3rd International conference on management, education technology and sports science (METSS), vol 25, pp 111–114

Chapter 2
Geochemistry and Mineralization of Tungsten Ore

Haisheng Han, Wei Sun, and Junhao Fu

Abstract As an important strategic resource, tungsten ore is a superior resource in China. Flotation technology has been the main technology for the separation of complex tungsten resources. The feasibility and efficiency of flotation separation depend on the mineralogy characteristics of ore, which is closely related to the geological mineralization. From the perspective of geochemistry, the principle of tungsten mineralization and its relationship with flotation are summarized in this paper. In the mineralization of tungsten, granite magmatic differentiation, high F content, and surrounding rock properties are very crucial to the enrichment and mineralization of tungsten, and determine the migration and mineralization of tungsten and deposit types. Different tectonic–magmatic mechanisms contribute to the different deposit types and there are great differences in the types and contents of gangue minerals in these different deposits. Accordingly, a series of tungsten ore flotation technologies were developed, such as fatty-acid flotation of the quartz vein type, fatty acid roughing-heating cleaning of the greisen type, and chelating collector roughing—heating cleaning of the skarn type. Especially for complex high-calcium scheelite and wolframite mixed ore, the flotation technology of complex collector greatly improves the efficient exploitation and utilization of tungsten resources. The geological ore-forming principles of tungsten ore not only determine the essential characteristics of the ore, but also provide a new train of thought for the development of tungsten utilization.

Keywords Tungsten deposit · Magmatic differentiation · Mineral flotation · Hydrometallurgy

H. Han (✉) · W. Sun · J. Fu
Central South University, Changsha, China

Engineering Research Center of Ministry of Education for Carbon Emission Reduction in Metal Resource Exploitation and Utilization, Central South University, Changsha, China

State Key Laboratory of Mineral Processing, Changsha, China

H. Han
Technology Research Center of Hunan Province for Comprehensive Utilization of Associated Fluorite and Fluorine Chemical Engineering, Chenzhou, China

W. Sun et al., *Flotation Chemistry of Tungsten Minerals and Its Application*,
https://doi.org/10.1007/978-981-96-5077-4_2

2.1 Introduction

For a long time, the research on metallogenic geochemistry of tungsten ore has been the focus of geologists, and has made great contributions to the theory and practice of deep detection of tungsten ore. In fact, the geochemical mineralization process determines the mineralogical characteristics of ore, and then affects the process flow of beneficiation, even the efficiency of tungsten metallurgical extraction. Flotation technology has become the main technology for separation of complex tungsten resources. The feasibility and efficiency of flotation depend on the mineralogy characteristics of ore, which is closely related to the geological and metallogenic conditions of tungsten ore. Geochemical mineralization process determines the ore's "genetic characteristics" (tungsten mineral phase, crystal structure characteristics, paragenetic and associated mineral components, distribution relations, etc.), and then determines the feasibility of its separation. This genetic characteristic not only provides support for the determination of the separation process, but also provides certain guidance for the development of new flotation agents and new technologies.

This chapter summarizes the ore-forming process and beneficiation technology of different types of tungsten ore deposits, and traces the principle of geochemical mineralization of the corresponding type of ore deposits, to explore the relationship between the rules of geochemical mineralization and tungsten flotation and even metallurgical technology, in order to enrich and develop the principle of flotation separation, to open up new ideas for the development of new technology of tungsten flotation.

2.2 Occurrence State and Geochemical Characteristics of Tungsten Element

Tungsten has an atomic number of 74 and an electron configuration of $5d^4 6s^2$ in its outer layer. It mainly exists in the form of +6 valence in nature.

Tungsten exists in nature in the following four main forms [1]:

(1) Independent minerals of tungsten. More than 20 kinds of tungsten minerals and tungsten-bearing minerals have been found, including wolframite group, scheelite group and tungstite minerals. However, the main minerals with economic value are wolframite and scheelite. Wolframite accounts for about 30% of the world's total tungsten resources, while scheelite accounts for about 70%.

(2) It enters into lithogenic minerals and vice minerals through its isotropy with Mo, Nb, Ta and other elements (e.g., extensive isotropy substitution between Mo and W caused by the close radius of the complex anion of $[MoO_4]^{2-}$ and $[WO_4]^{2-}$).

(3) Tungstic acid or various complexes exist in various natural fluids, such as natural water, magma, and intergranular fluids.

(4) In the state of ionic adsorption exists in the surface of fine debris, and colloids (Table 2.1).

The geochemistry of tungsten determines the distribution characteristics and enrichment of the element in the various spheres of the Earth. In the early Earth's evolutionary reduction environment, more than 90% of tungsten entered the Earth's core as a medium siderophile [2]. Due to the introduction of tungsten from the Earth's core to the crust or even the surface through tectonic movements and magmatic activities in the early stage of the Earth's evolution, the crust has a higher tungsten content than the mantle, while the mantle has the lowest tungsten content (Fig. 2.1). The significant reduction of tungsten content in the mantle during the upward migration from the core may reflect the possibility of tungsten moving up the silica-aluminum sphere due to differentiation in the early stage of the Earth's evolution.

2.3 Geological and Metallogenic Mechanism of Tungsten

The tungsten-rich geochemical areas and tungsten deposits have experienced a long and complex process of geochemical action, which is closely related to the development history of the crust and the corresponding history of granite diagenetic evolution. The formation of tungsten ore (as shown in Fig. 2.2) has experienced multiple stages of mineralization, such as deep source material spillage, weathering denudation and deposition, multiple evolutions of stratigraphic remelting diagenesis and granite, magmatic hydrothermal precipitation, alkali metasomatism and acid leaching extraction of mother rock, hydrothermal migration, accumulation precipitation, etc. Each stage is of great significance to the final formation of tungsten ore [4].

2.3.1 Tungsten Enrichment in the Original Formation

Because tungsten appears as a moderate siderophile element under strong reduction conditions, >90% of tungsten elements entered the Earth's core during Earth accretion and core-mantle differentiation early in Earth evolution, resulting in the highest tungsten abundance in the Earth's core [5]. At the stage when the continental crust grew and evolved from the mantle, tungsten was an extremely incompatible element, with a partition coefficient of much less than 1 between the petrogenic minerals (olivine, garnet, amphibole, pyroxene, etc.) and the melts (basaltic, basalt melts, etc.) [6]. Therefore, during this stage, tungsten was released as an incompatible element due to partial melting. It became part of the continental crust through magmatism, during which tungsten was depleted in the mantle and relatively enriched in the crust.

With the intensification of tungsten depletion in the mantle and enrichment in the crust, as well as the migration of tungsten from the core to the crust or even the

Table 2.1 Geochemical parameters of tungsten [1]

Elements	Atomic number	Atomic weight	Atomic volume/ $cm^3 \cdot g^{-1}$	Atomic density/ $g \cdot cm^{-3}$	Melting point/°C	Boiling point/°C
W	74	184	9.53	19.35	3380	5927
Peripheral electron configuration	Electronegativity	Crustal abundance/%	Atomic radius (12-coordination)/ nm	Covalent radius/ nm	Ionic radius (6 coordination)/nm	Ionization potential energy/eV
$5d^4 6s^2$	2.0 (+6)	1.5×10^{-4}	0.1371	0.13	0.068	7.98

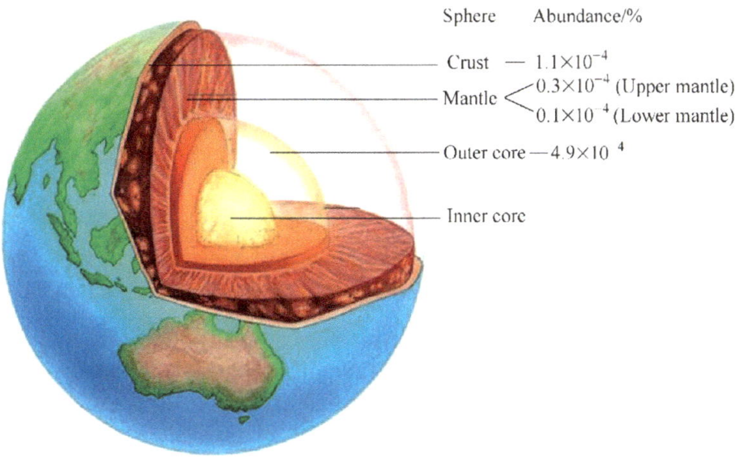

Fig. 2.1 Abundance of tungsten in the various spheres of the Earth [3]

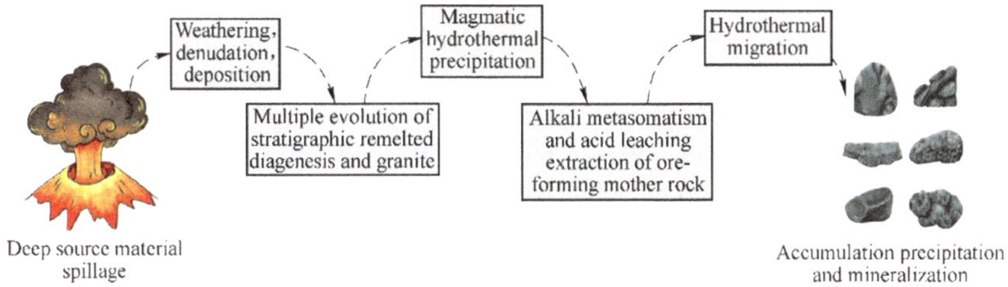

Fig. 2.2 Schematic diagram of the geological ore-forming process of tungsten ore

surface through tectonic movements and magmatic activities, the initial enrichment of tungsten, tin, bismuth and other elements in the tungsten ore concentration area was gradually formed in the crust layer, and the original tungsten structure (i.e., source layer) was gradually formed in the crust layer [3].

2.3.2 *Initial Enrichment of Tungsten Under Deposition*

Under the domination of regional tectonic activities, tungsten elements are locally enriched in some strata through denudation-gravity differentiation deposition (such as the formation of modern tungsten placers), chemisorption of tungsten by organic-rich and Fe/Mn oxide layers, and biochemical processes [7].

During this process, tungsten can be enriched to a certain extent through adsorption of Fe/Mn-rich clay during weathering, the dissolution of sulfuric acid water generated by the oxidation of paragenetic sulfide, or migration and enrichment of secondary tungsten minerals after dissolution in surface water, etc. The tungsten elements in the derived tungsten-bearing structures are derived from the weathering denudation and gravity deposition of the original tungsten-bearing structures, as well as volcanic activities and magmatic intrusion through the original tungsten-bearing structures, and are mainly terrigenous detritus in composition.

2.3.3 Crustal Remelting and Multiple Evolvements of Granite

The formation and migration of granitic magmas dominate the diffusion and enrichment of ore-forming elements in the crust. Studies show that the abundance of tungsten in all kinds of magmatic rocks gradually increases from ultrabasic rocks to basic rocks to acidic rocks (Fig. 2.3). There is a positive correlation between the content of granite and the abundance of tungsten, indicating that the formation of tungsten deposits is mainly related to granite [8]. Compared with the mantle abundance, tungsten is enriched by 50–60 times in the crust, while it is enriched by 2–3 orders of magnitude in granite [9]. In terms of origin, most of the granites associated with tungsten mineralization were formed by the remelting of the Precambrian sedimentary crust ($^{86}Sr/^{87}Sr$ is 0.719–0.741) and belong to the remelting type of granite [10]. The distribution of ore-forming tungsten elements is uneven in the vast majority. If the granitic magma formed by crustal remelting is melted by the strata with high tungsten abundance, the original tungsten content of the magma will be higher, which is conducive to the formation [11]

Within the continental crust, the extreme incompatibility of tungsten leads to further enrichment of minerals in residual magma during ascending emplacement of granitic melts, following the formation of remelting-type granite and large-scale granite multiple activity. With the selective melting and differentiation of multi-cycle and multi-stage granitization, tungsten and other oxyphilic elements (Nb, Ta, Sn, Be)

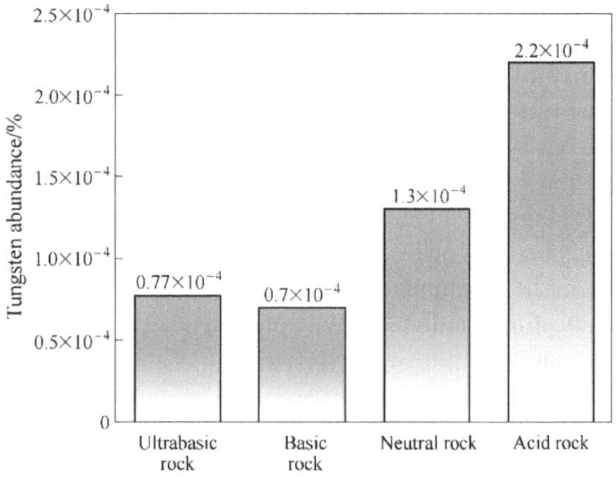

Fig. 2.3 Abundance of tungsten in various magmatic rocks [12]

gradually concentrated in the upper layer. Tungsten was continuously precipitated and entered the upper surrounding rock during granitization, forming rocks with higher tungsten content than the pre-granite, and more tungsten was precipitated from these rocks during the next stage of granitization. As a result, the tungsten content becomes higher and higher, and eventually reaches the maximum enrichment in the crust, forming tungsten-rich granite, which basically lays the material foundation for tungsten mineralization [13].

2.3.4 *Formation of Ore-Forming Hydrothermal Fluid and Alkaline Metasomatism*

In fact, the tungsten-rich granite formed by stratigraphic remelting and granite evolution is far from enough to form tungsten deposits, and there is still a long way to go before the grade and tonnage can reach the level of high-grade ore mining. At this time, tungsten is still mainly concentrated in the liquid phase, and a small amount of tungsten is scattered in the rock mass as a minor mineral, or exists in the biotite and other dark minerals in the rock mass as a quasi—like image.

Alkaline petrification of tungsten rich granite is the basic condition of tungsten mineralization. After tungsten-rich granite and gasification-hydrothermal alkalinity (potassic feldspathization and albitization) in late rock mass under magma hydrothermal action, tungsten in the form of sub-minerals and tungsten in mica and dark minerals can be released from the rock mass to form tungsten-rich ore-forming fluids [14]. In this process, gasification-hydrothermal fluid is not only the main ore-bearing medium, but also the important ore-forming factor. They play an important role in extracting, carrying, transporting and precipitating minerals.

Widely developed, planar distributed alkali-metasomatism altered granites are found on the surface or in the deep areas of all tungsten deposits in Nanling area, which indicates that the tungsten deposits all experienced different levels of alkali-metasomatism before forming [15]. At the same time, during the process of alkali metasomatism, the iron "by-products" formed when dark minerals were removed, thus producing magnetite mineralization. Granite-related skarn scheelite deposits in South China are often associated with more magnetite, and even magnetite ore bodies are formed, which also indicates that alkalic metasomatism is closely related to tungsten mineralization. In addition, during the process of alkaline metasomatism, the magmatic hydrothermal solution will leach and extract part of tungsten from the surrounding rock. Li Chongyou and Liu Yingjun et al. proved that part of tungsten from the surrounding rock was added in the granite alteration stage of tungsten ore, indicating that the surrounding rock also provided a rich source of tungsten for the formation of tungsten ore during the alteration process [16].

2.3.5 *Accumulation and Precipitation of Tungsten Mineralization*

The hydrothermal formation lays the foundation for the transfer and accumulation of ore-forming materials, but the hydrothermal accumulation and precipitation mineralization also need suitable ore liquid migration channel and ore storage space [17]. Tectonic movements provide the migration channel of mineral liquid and delineate the final accumulation place of mineral liquid, through the formation of micro-fissure, fissure, fault, etc., so as to promote the formation of ore body. Therefore, in the final stage of mineralization, under the favorable conditions provided by various geological structures for tungsten mineralization (fissure and fine cleavage structure), metallogenic hydrothermal liquid will be driven to open space by pressure difference. In these spaces, the fluids undergo enrichment, mineralization, and then release ore-forming elements in the delineated position, so that tungsten and other ore-forming elements precipitate more intensively to form ore bodies, and finally produce industrial tungsten ore deposits [18].

The migration forms of tungsten in the hydrothermal may include tungstic acid and its salts or monomer tungstate forms (such as H_2WO_4, $Na_2[WO_4]$ and $[WO_4]^{2-}$, etc.), fluoride and its complexes (such as WF_6, $[WO_2F_4]^{2-}$, and $[WO_3F_2]^{2-}$, etc.), as well as isopolytungstic acid (such as $H_6W_6O_{21}$, etc.) and heteropoly tungstic acid (such as $H_4[Si(W_3O_{10})_4]$, $H_3[P(W_3O_{10})_4]$ and $H_3[Sb(W_3O_{10})_4]$ etc. [19]. This metallogenic regularity plays an important guiding role in modern tungsten smelting technology (alkali process, acid process), and provides theoretical support for the development of new smelting and extraction technology.

To sum up, the geological mineralization of tungsten mainly experienced the enrichment of tungsten in the original strata—the local enrichment of tungsten in some strata through sedimentary evolution (weathering—denudation—migration—sedimentation—diagenesis under different external geological conditions)—the further enrichment of tungsten by magmatism—the mineralization hydrothermal alkali metasomatism, and finally the accumulation and precipitation to form tungsten ore.

2.4 Key Factors in Tungsten Mineralization Process

2.4.1 *Influence of Magmatic Source*

The magmas in different magmatic source regions have different source depth, channel material composition and different degree of differentiation, which have important influence on tungsten mineralization. In fact, the original crust of the Earth was formed by the solidification of primary magma, and the composition of the molten material trapped by the Earth's core is itself uneven. Therefore, the content of tungsten in the source layer must be uneven, with high and low levels of tungsten in different layers and spatial distribution [20].

In the research on the source of tungsten ore-forming materials, the view that the ore-forming materials come from magma has always been dominant [21]. Early scholars believe that the magmatic rocks related to tungsten mineralization have the characteristics of highly evolved type I or type S granite, and the source area is mainly crust-source material, with no or trace of mantle source material [15]. However, a large number of isotopic studies (such as Sr–Nd–Hf, etc.) have shown that the isotopic composition of granite related to tungsten mineralization in many tungsten-metallogenic areas indicate relative crustal depletion, which indicates that the source of tungsten mineralization are not only crustal sources, but also a small number of mantle-derived materials [22].

Magmatic source areas also affect tungsten deposit types, because the precipitation form of tungsten in magmatic hydrothermal solution itself depends on the relative concentrations of Fe^{2+}, Mn^{2+} and Ca^{2+}in the hydrothermal solution, and the differences of magmatic sources will directly affect the differences of the above ion concentrations in the subsequent hydrothermal solution. For example, scheelite deposits are formed when the surrounding rock is a silica-alumina rock with high calcium content. If most of the rock mass in the magmatic source area is also low-calcium granite itself, the main precipitation tungsten-manganite, thus forming wolframite deposits.

2.4.2 *Magmatic Evolution of Granites*

Academician Xu Keqin from Nanjing University noted earlier those different types of granites have different metallogenic specific properties. The granites related to tungsten mineralization mainly belong to the modified genetic series ($^{86}Sr/^{87}Sr$ is 0.719–0.741). They are formed by the differentiation and evolution of continental crust materials caused by multiple granitization and selective melting in the continental crust [23]. The results show that the Nb/Ta ratio of rock mass decreases with the process of separation and crystallization, and has a negative correlation with the content of SiO_2 [24]. As shown in Fig. 2.4, the granite associated with tungsten mineralization has a lower Nb/Ta ratio (Nb/Ta ratio in the crust is about 10) than the granite of other ore deposits, indicating that the magmatic evolution of the granite plays a significant role in the formation of tungsten ore [25].

The magmatic evolution of the granites promoted the enrichment of tungsten during the mineralization process. As tungsten is a highly incompatible lithophilic element, tungsten elements continuously precipitate and enter the surrounding rock during granitization, forming rocks with higher tungsten content than pre-granitization rocks, and more tungsten will be precipitated in the next granitization of these rocks. This repeatedly promoted the enrichment of tungsten in the magma, resulting in higher and higher tungsten content, which laid the foundation for subsequent ore-forming hydrothermal accumulation and precipitation mineralization.

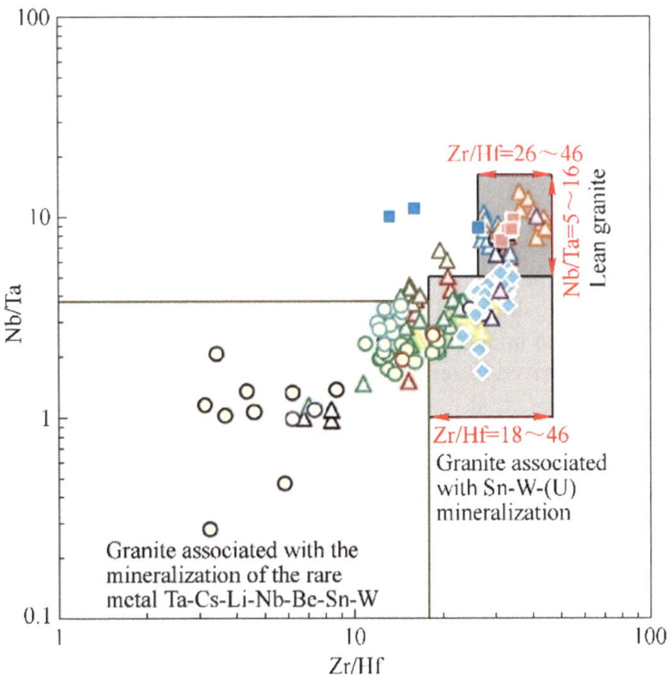

Fig. 2.4 Nb/Ta ratios of tungsten related granites with Zr/Hf [18]

In addition, the magmatic evolution of granite also affects the mineralogical composition of tungsten deposits. Generally, the granite I closely related to vein-type tungsten generally contains dispersed disseminated wolframite, arsenopyrite, pyrite, molybdenite, sphalerite and chalcopyrite sulfide by minerals, and the content of W, Mo, Bi, Cu, Pb, Zn and other elements is significantly higher than that of the granite II related to skarn mineralization, indicating that ore-forming materials have a large degree of residual in the granite I. This is mainly related to the poor magmatic differentiation of granite I, relatively low volatile content in magma, closed magmatic-metallogenic system, and the failure of effective separation of ore-forming materials from magma. In contrast, the low content of the above elements in granite II does not contain related by minerals, not because of the low content of these elements in the original magma, but the higher concentration of volatile components in the magma, high differentiation of magma, and relatively open magma-metallogenic system, which is conducive to the effective separation of a large number of fluids and ore-forming materials from the magma, and less residual ore-forming materials in the magma [26, 27].

2.4.3 *Influence of Properties of Surrounding Rock*

The properties of surrounding rock are closely related to the formation of tungsten deposits. On the one hand, surrounding rock may directly control the source of tungsten ore-forming materials. On the other hand, it will affect the tungsten mineralization, thereby controling the mineralization and enrichment, alteration metasomatism of ore bodies [28].

The physical and chemical properties, conditions and mineral composition of surrounding rock can affect whether wolframite or scheelite is formed eventually, among which the most obvious is the influence of the composition of calcareous surrounding rock [29]. The calcareous rocks can influence the composition and chemical reactions of intrusive rocks and mineral fluids through assimilation or selective metasomatism to intrusive magma. Numerous studies and facts show that whether tungsten eventually forms scheelite or wolframite deposit depends mainly on the amount of calcium in the formation of tungsten ore and the rate of tungsten precipitation. When the surrounding rock is calcium-rich skarn, tuff, carbonate rocks, or fluid calcium is rich due to close to alkali metasomatism metallogenic mother rock, generally formed scheelite deposit; and when more far away from the metallogenic mother rock by alkali metasomatism, and the surrounding rock is high silica aluminum rock (low calcium), generally formed wolframite deposit [30].

The properties of the surrounding rock not only affect the type of tungsten minerals, but also the type of associated minerals. For example, when the surrounding rock is porous, brittle, and calcium-rich skarn, scheelite is formed while an acidic hydrothermal reaction occurs with the surrounding rock, where mineralizers such as fluorine, chlorine, and boron interact with calcium to produce minerals such as fluorite. When in the closed condition of the silica-aluminum surrounding rock, the hydrothermal solution cannot escape, the formation of wolframite at the same time fluorine element and aluminum silicon combined to form topaz or mica minerals.

In addition, the surrounding rock conditions also have an important effect on the morphology of orebodies. For example, in granite and metasandstone, the relative inertia of surrounding rock is not conducive to metasomatism, the hydrothermal solution is mainly precipitated by the change of physical conditions, and the mineral material is mainly filled, which makes it easy to form vein-type tungsten ore. The properties of carbonate surrounding rock are relatively active, and the general hydrothermal fluid is easy to metasomatose along the surrounding rock structure, forming fine-vein scheelite along the structural metasomatism. In this case, the structure of the surrounding rock directly determines the morphology of the ore body to some extent. Because the formation of surrounding rock is controlled by comprehensive factors such as formation lithology, confining pressure and heat escape conditions, its morphology is different. Mineralization in carbonate rocks is affected by the shape of structural fissures, and its vein shape is more complex than that of silica-alumina rock, rarely forming narrow and straight vein bodies. Therefore, the general carbonate tungsten body shape is complex, and it tends to forming a columnar (sac-like) orebody distributed along the intersection of one main structural line and another (group) structural line. And if the rock with high calcium content (such as relatively small thickness of carbonate interlayer, calcareous sandstone, tuff) is under closed conditions, tungsten precipitates into layered tungsten ore. This type of tungsten ore may be stratified farther away from the rock mass and is often referred to as stratified or depositional remolding tungsten ore [31].

2.4.4 *Influence of Temperature*

The mineralized precipitation of tungsten minerals is actually the process by which ligands (mainly mineralizing agent elements) in the various complexes escape from solution with decreasing temperature, resulting in the decomposition of the complexes. Temperature can directly affect the solubility of tungsten minerals in the natural molten inclusions and the migration form of tungsten in the hydrothermal solution, thus affecting the mineralization process of tungsten minerals [32].

A large number of studies on inclusion temperature, gas and liquid phase composition have shown that a large number of tungsten mineralization precipitation basically occurs under the condition of lower temperature of ore-forming medium [32]. When the temperature of ore hydrothermal solution is low and the acidity is high, WO_4^{2-} can undergo an exothermic polymerization reaction to form isoacid complexes: $[WO_4]^{2-} \rightarrow H[WO_4]^- \rightarrow H[W_6O_{21}]^{5-} \rightarrow H_2[W_{12}O_{40}]^{6-}$, affecting the migration form of tungsten.

On the other hand, the temperature span of multi-stage tungsten mineralization should be different, which causes the zonation of formed ore deposit [33]. It is proved that the temperature variation trend of gas–liquid inclusions in quartz vein wolframite is consistent with the original zonation phenomenon of the deposit [34]. The temperature of ore-forming fluids in skarn-type tungsten deposits varies widely during the ore-forming process. The fluid inclusions in early skarn minerals generally inherit the high-temperature characteristics of magmatic fluids and have relatively high homogenization temperatures. The ore-forming fluids in the late stage of degeneration and alteration have relatively low temperatures (200–400 °C). Scheelite deposits can be developed in both the early skarn stage and the late degenerative alteration stage, and reach the peak when accompanied by degenerative [35, 36]. In addition, during the formation of tungsten deposits, the high temperature can reactivate and migrate the dispersed tungsten elements in the source layer and enrich again in the favorable ore-forming site, thus forming the sedimentary metamorphic modified tungsten deposits.

2.4.5 *Influence of Migration Mineralizer*

The influence of migration mineralizers fluorine, chlorine and boron (especially fluorine) is very great in the process of tungsten mineralization. During the hydrothermal metamorphism of tungsten deposit formation in Nanling area, fluorination, tourmaline and topaz are more commonly occurred, which reflects the participation of migration mineralizers fluorine, chlorine and boron in the mineralization process. Compared with the granite associated with Cu–Mo mineralization and the granite without mineralization, the granite associated with tungsten mineralization in South China usually has a higher F content, indicating that there is a certain positive correlation between the tungsten content and F content in some rock mass [28].

F has an obvious effect on the granite melt structure, and F^- and OH^- have similar ionic radii and electricity prices. Studies showed that when F enters the melt, it can break Si–O and Al–O bonds, destroy the tetrahedral structure, and make the silicate melt produce decomposition [37], thus reducing the melt viscosity, density and solid-phase line temperature, which is conducive to the diffusion of tungsten in the magma, so as to improve the degree and speed of crystallization differentiation of the magma. Therefore, the high content of F in magma is an important factor in promoting the degree of magma evolution, which promotes the enrichment and mineralization of highly incompatible tungsten elements in the later period of magma evolution.

On the other hand, tungsten mineralization mainly occurs in the gasification-hydrothermal stage. Tungsten and fluorine form WO_3F^- and $WO_2F_4^{2-}$, which exists and migrates in the middle and low temperature fluorine-rich hydrothermal fluids. At the same time, a large number of experimental studies on the fluid/melt distribution of F in ore-forming hydrothermal fluids have shown that D_F is less than 1.0, indicating that F tends to enter the melt phase in the granitic magmatic-hydrothermal liquid system. Therefore, it can be inferred that F content can further affect the melt accumulation and precipitation mineralization process.

2.5 Geochemical Characteristics and Flotation of Rocks in Typical Tungsten Deposit Areas

The basic geological characteristics of tungsten deposits in China are well known for their wide distribution, many areas of origin, large scale, high grade, complex associated components and diverse deposit types. Tungsten has strong oxygen affinity and generally forms tungstate ore deposits. Tungsten deposits can be divided into many types according to different classification criteria. For example, tungsten deposits are mainly divided into scheelite and wolframite deposits according to ore mineral types. The formation of scheelite or wolframite is generally determined by the composition of surrounding rock, physical and chemical properties of ore-forming fluids and other factors. Generally, the surrounding rock composition of wolframite deposit is mostly high silica aluminum rocks, while that of scheelite deposit is mostly carbonate rocks or other calcareous rocks. Whether the tungsten eventually forms scheelite or wolframite deposits depends mainly on the amount of calcium in the formation of tungsten and the precipitation rate of tungsten. When the surrounding rock is calcium-rich skarn, tuff, carbonate rocks, or due to its proximity to alkali metasomatism metallogenic mother rock and the fluid is rich in calcium, the scheelite deposit is mainly formed; when it is far away from the alkali metasomatic site of the ore-forming mother rock, and the surrounding rock is poor in calcium, wolframite deposit is generally formed [29].

According to deposit genesis, tungsten deposits can be divided into four types: magmatic deposit, sedimentary deposit, (volcanic) sedimentation-metamorphic transformation deposit and modern superficial deposit. Magmatic tungsten deposit is an important type of deposit for tungsten storage. According to the deposit characteristics, it can be further divided into different subclasses, among which the subclasses with economic significance mainly include quartz vein type, greisen type, porphyry type, and skarn type, as shown in Fig. 2.5. The following is a further overview of surrounding rock characteristics, material sources, metallogenic patterns and other aspects for the four subclasses.

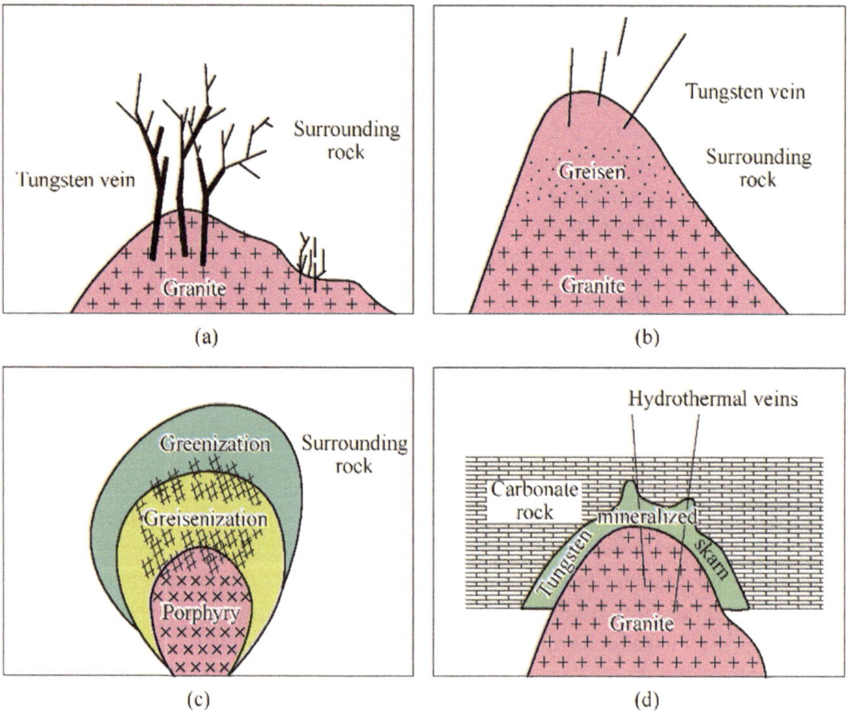

Fig. 2.5 Metallogenetic pattern of tungsten deposit **a** quartz vein type; **b** quartzite type; **c** porphyry type; **d** skarn type [36]

2.5.1 Quartz Vein Type Tungsten Deposit

Quartz vein type tungsten deposit is the most widely distributed, the earliest developed, the largest deposit, the largest reserve and the most deeply studied type of deposit [38]. The Xihuashan tungsten and Piaotang tungsten deposit are the typical deposits of this type in China [14]. These tungsten deposits are located in areas with strong tectonic activities, especially folds, faults and joints. The superposition of multiple tectonic movements and frequent magmatic activities create conditions for the formation of ore-forming hydrothermal fluids. The ore bodies are mostly formed in the cracks in the contact zone between the inner and outer rock mass, most of which are distributed in the form of veins. Some ore bodies are also formed in the surrounding rock at the top of the rock mass (Fig. 2.5a). The veins of quartz vein type tungsten deposits are mainly composed of wolframite-quartz veins, and a few contain scheelite.

The tungsten ore-forming materials mainly come from the remelted granite-regenerated magma, and some ore-forming materials come from the extraction of some surrounding rocks with high ore-forming element abundance by magmatic-hydrothermal solution in the up-invading process [26]. Therefore, surrounding rock lithology is the main factor controlling tungsten ore deposit type and ore body morphology. The surrounding rock of quartz vein type deposit is mainly Cambrian metamorphic rocks. For example, the surrounding rock of Xihuashan tungsten deposit is Cambrian metamorphic rock, composed of SLATE, schist and metamorphic sandstone [39]; The surrounding rocks of Yaoling tungsten ore deposit in Guangdong Province and Lingjiao tin-tungsten ore deposit in Lianzhou City are mainly Cambrian epimetamorphic quartz sandstone, SLATE with siliceous conglomerate and Ordovician and Devonian metamorphic rocks [40].

Due to the relatively large difference in flotation behavior between scheelite and silicate minerals, for scheelite quartz type tungsten ores, fatty acids are usually used as collectors and sodium silicate as inhibitors under room temperature conditions, namely, room temperature fatty acid method, which can realize the efficient separation of scheelite and gangue minerals. Fatty acid method uses the collector for fatty acid and its derivatives, such as oleic acid, tar oil, naphthenic acid, oxidized paraffin soap, among which oleic acid (sodium oleate) and oxidized paraffin soap are the most widely used in flotation, mainly in the form of chemical adsorption on the surface of scheelite [41].

2.5.2 Greisen Type Tungsten Deposit

Some representative greisen-type deposits are the tungsten deposit in Dangping deposit, Jiangxi Province and Daping deposit in Tongcheng, Hubei Province [42, 43]. This type of ore deposit is distributed in the upper part of the granite rock mass and the top of graywacke, sandstone, shale and other surrounding rocks. The top or edge of the rock bell is banded, lenticular and cystic. The metal minerals are disseminated and veined. The edge of the rock mass or the outer contact zone is veined (Fig. 2.5b). The ore-forming mother rock is biotite monzogranite, muscovite, lepidolite or ferrimatolite alkali feldspar granite [44]. In the metallogenic process of the greisen type tungsten deposit, the role of structure on the mineralization is often more obvious. Taking the tectonic ore-controlling mechanism of the greisen-type tungsten deposit in Dangping tungsten-type tungsten deposit, Jiangxi Province as an example, the strike slip overthrust, acting as a water-barrier, controls the southern boundary of Yanshanian granite in the mining area, and also blocks the loss of hydrothermal solution (ore-forming solution) in the post-magmatic period, so that tungsten can accumulate in the inner contact zone between granite and faultage, creating conditions for the formation of the greisen-type tungsten deposit [45, 46].

The common phenomena of potassic feldspathization and greisenization alteration in the surrounding rock prove that the mineralization of this kind of deposit is generally high-temperature hydrothermal process. In the same deposit, the greisen type, skarn type and quartz vein type often exist simultaneously. The metallic minerals of the greisen type tungsten deposit are generally wolframite, scheelite, molybdenite, cassiterite, pyrite, etc., and the non-metallic minerals are quartz, muscovite, biotite, fluorite, feldspar, etc. [47]. In fact, when the greisen vein is interspersed from the inner contact zone of granite to the outer contact zone, other types of alteration often occur: when the outer contact zone is carbonate rock, the deposit will develop skarn mineralization, resulting in the formation of calcium gangue minerals in the ore body. Therefore, in the greisen scheelite deposit. In addition to the silicate gangue mainly formed by quartz and mica formed by greisenization, there are also calcium-bearing gangue minerals formed by metasomatism with calcium-magnesite rocks. Therefore, the actual flotation of greisen scheelite must face the problem of separating scheelite from other calcium bearing gangue (fluorite, calcite, etc.).

Fatty acid flotation technology at room temperature has poor adaptability to ores. For the greisenization tungsten ore, fatty acid method—heating flotation process, namely "Petrov" method, is usually used. The principle of "Petrov" method is that the desorption degree of the surface collector of scheelite and associated calcium-bearing minerals (calcite, fluorite, etc.) varies greatly under the condition of high temperature and concentration of the slurry and the action of water glass. Under the conditions of "Petrov" method (a large amount of water glass and 85–95 °C), fluorite and calcite are inhibited by a large amount of desorption of surface collector, while scheelite still maintains good floatability, so that scheelite can be separated from calcium-bearing gangue minerals [48].

The process is widely used in tungsten flotation practice, the main process of which is: first through sodium carbonate or sodium hydroxide to adjust the pH to alkaline, and then add sodium silicate as an inhibitor, fatty acid as collector, after roughing process to get rough concentrate [48]; And then the rough concentrate is concentrated to the concentration of 50–70% of the thick pulp, adding a lot of sodium silicate, and at the same time in 85–95 °C steam heating conditions stirring slurry, then dilution flotation to produce high grade tungsten concentrate.

2.5.3 Porphyry Type Tungsten Deposit

Porphyry type tungsten deposits mainly occur in the inner and outer contact zone of the top or margin of acid-medium acid porphyry. The representative porphyry type tungsten deposits in China are Dahutang ultra-large tungsten polymetallic deposit in Jiangxi Province, Xingluokeng deposit in Fujian Province, Dongyuan tungsten molybdenum deposit in Anhui Province and Lianhuashan tungsten deposit in Guangdong Province.

Porphyry type tungsten deposit is mainly metallogenic in the middle and late period of Yanshan. Ore-forming magmas are mixed crust-mantle, weakly acidic to acidic magmas of the middle and late period of Yanshan, and are late differentiation of continental regional magma intrusion-eruption activity. It's controlled by deep crustal fractures, with zonation of enclosing alteration, mainly sodic and mica [49]. The ore bodies of porphyry type tungsten deposits are generally relatively shallow, mainly distributed in the rock mass, and some distributed in and near the contact zone between porphyry intrusion and surrounding rock. Tungsten minerals are disseminated or veined disseminated in the middle and upper part of the rock mass and the inner/outer contact zone, with planar distribution characteristics. The orebodies are stratiform, lenticular and irregular, and have no obvious boundary with the surrounding rock.

The intrusive rock mass related to mineralization of this type of deposit is porphyry, which mainly includes granodiorite porphyry, monzonite porphyry, granite porphyry, quartz porphyry and diorite porphyrite. Porphyry type tungsten deposit has low grade and large scale, and is often accompanied by molybdenite. The main mineral symbiosis association of this type tungsten deposit is wolframite (scheelite)-niobium tantalite (fine crystallite), and the metal minerals in the ore are commonly scheelite, wolframite, molybdenite, chalcopyrite, sphalerite and so on, among which the amount of metal sulphide tends to be small. The main gangue minerals in the porphyry type tungsten deposit are quartz, feldspar and a small amount of mica, and these light minerals account for more than 99%. Tungsten minerals are coarsely distributed and have a large density difference with gangue minerals, so it is suitable to concentrate wolframite and scheelite at the same time by gravity separation [50].

2.5.4 Skarn Type Tungsten Deposit

The skarn type tungsten deposit occurs in carbonate rocks or calcium-bearing clastic rocks by magmatic hydrothermal action, and is the most important type of tungsten deposit in the world, with about half of the total reserves. Skarn type tungsten deposits usually have lower grade, more uniform mineralization, and are easy to form larger mining areas. The typical skarn type tungsten deposits in China include Xintianling scheelite deposit, Zhuxi tungsten-copper polymetallic deposit and Shizhuyuan tungsten-molybdenum polymetallic deposit [51].

Their formation and distribution are closely related to the meso-epigenetic intermediate-acid magmatic rocks, and orebodies are mainly produced in the contact metamorphic zone between rock mass and carbonate rocks, or the surrounding rocks (Fig. 2.5d) [52]. The carbonate surrounding rock is relatively active in the ore-bearing surrounding rock, and the general ore-forming hydrothermal fluid is easy to carry on the metasomatic reaction to skarn, and the form deposits through enrichment in the late complex skarn stage. Ore-forming hydrothermal undergo metasomatism with surrounding rock along the tectonic fracture or contact zone, resulting in ore in the vein or fracture and granite contact zone of carbonate rocks in the form of disseminated granular development. The morphology of ore bodies is more complex, most of them are irregular sacs, lentils and lenticular, and some are stratiform, stratiform-like and vein-like.

Because in the skarn deposit, tungsten mineralization process occurs in the alteration stage after the formation of skarn, the alteration is a kind of decalcification. The solution has elevated concentrations of calcium and iron, and at the same time the concentration of H_2S is high. Therefore, wolframite cannot be stable formed in the mineralization process of skarn type tungsten ore, and scheelite is formed. Therefore, tungsten minerals in the ore are mainly scheelite, with a small amount of wolframite, often associated with molybdenite, cassiterite, chalcopyrite, sphalerite, galena, magnetite, pyrrhotite, arsenopyrite and fluorite, contributing to a polymetallic combination deposit [53].

The calcium-bearing minerals all have the same surface-active point Ca^{2+}, and their surface chemical properties are similar, which makes it very difficult to separate scheelite and calcium-bearing gangue minerals [54]. The mutual conversion among scheelite, calcite and fluorite make the flotation behavior of each mineral more complicated. In addition, during the mineralization of skarn-type tungsten ores, the hydrothermal alter surrounding rocks along the tectonic fracture or contact zone (Fig. 2.6), resulting in tungsten minerals developing in the vein or fracture near the contact zone between granite and carbonate rocks in the form of disseminated particles. The ore bodies are more complex in shape, mostly irregular sacs, lentils, lenticular, laminar, etc. and tungsten minerals are distributed in finer granularity. Tungsten minerals are difficult to dissociate from monomer in Skarn type tungsten ore, which is a relatively refractory type of ore deposit.

The nature of this type of ore is complex, and often wolframite and scheelite are co-associated, making flotation more difficult. Mineral processing workers adopt a new tungsten flotation process based on chelating collectors [55]. Chelating collector mainly includes hydroxamic acid, arsenic acid and cupferron, etc., which can form stable complexes with calcium, iron, manganese and other metal species on the mineral surfaces to adsorb on the mineral surface [56, 57]. The most typical chelating collector flotation process is "GY method" tungsten beneficiation process. In the 1990s, the "GY method" with chelating collectors as the core was successfully applied in Shizhuyuan tungsten mine, making the tungsten flotation technology reach an advanced level in the world. The core of this method involves rough selection using lead nitrate as an activator, benzohydroxamic acid (GYB) as the main collector, and modified fatty acids as auxiliary collectors to achieve the flotation

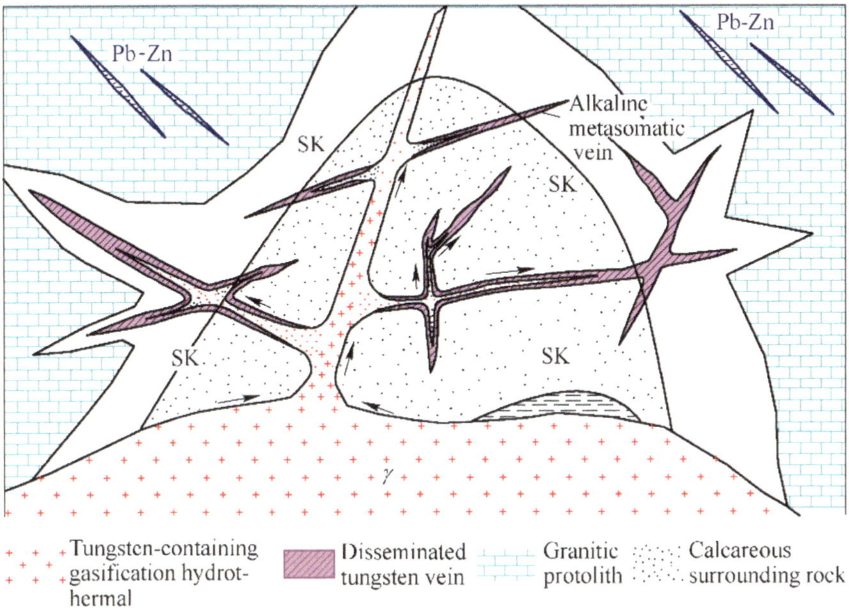

Fig. 2.6 Schematic diagram of a metallogenic model of skarn type deposit

of mixed scheelite and wolframite. Furthermore, the rough concentrate undergoes strong magnetic separation to separate scheelite from wolframite. The rough scheelite and wolframite concentrates are then further refined to enhance their grades [58]. The mineral processing team of Central South University invented a highly selective collector, a lead complex of—benzohydroxamic acid (Pb-BHA), which takes "metal base" as a functional group, the action site of the mineral surface is the anionic group WO_3^{2-}, and can be strongly adsorbed on the surface of scheelite and wolframite minerals, so as to achieve efficient synchronous flotation enrichment of scheelite and wolframite [59]. Therefore, a new technology of mixed flotation of complex scheelite and wolframite resources at room temperature was developed [60]: After desulfurized tailings were pulped with sodium carbonate, Pb-BHA complex was used as the collector, a small amount of Al–Na_2SiO_3 was used as the selective inhibitor, and high grade scheelite and wolframite mixed concentrate was obtained through multiple selective and scavenging under normal temperature conditions. This technology has been applied in large-scale mines such as Shizhuyuan, Huangshaping and Xingluokeng. The comprehensive recovery rate of tungsten in Shizhuyuan has increased from 60% to over 70%. The comprehensive recovery rate of Huangshaping was increased from 50 to 68%. In Xingluokeng, tungsten concentrate grade increased from 18% to more than 40% by using this process instead of centrifugation [61].

2.6 New Technology of Tungsten Metallurgy Based on the Principle of Metallogenic Geochemistry of Tungsten Ore

In 2003, Railsback summarized the geological information of a large number of elements in the earth's natural environment based on the theory of hard soft acid base and ion potential, and proposed the " the Earth Scientist's Periodic Table", which was widely concerned by the field of earth science and geochemistry [62]. A large number of metallurgical researchers refer to metallogenic geochemistry, and found a large number of similarities and connections between geochemistry and metallurgy from the perspective of interdisciplinary.

During the process of geochemical mineralization, the mineralization precipitation of tungsten minerals is actually the result of the decomposition of various complexes in hydrothermal fluids due to a decrease in temperature and pressure, followed by aggregation and precipitation to form ore deposits [63]. In fact, the process of tungsten hydrometallurgy can be approximated as the reverse process of tungsten mineralization precipitation: tungsten minerals are in contact with aqueous solution or other liquid phase, the tungsten element into the liquid phase, and then tungsten in the liquid phase undergo separation and enrichment [64].

Traditional alkali tungsten smelting technology requires high grade of tungsten concentrate, restricting the comprehensive recovery of tungsten, and phosphorus, molybdenum, tin and other elements are difficult to utilize. At the same time, a large number of alkali slag is produced, causing serious environmental pollution. The team of academician Zhao Zhongwei from Central South University developed the synergistic decomposition and extraction technology of tungsten concentrate using sulfuric and phosphoric acids based on the geochemical law of tungsten element and referring to the migration forms of heteropoly acids ($H_4[Si(W_3O_{10})_4]$, $H_3[P(W_3O_{10})_4]$ and $H_3[Sb(W_3O_{10})_4]$, etc.) of tungsten element in the hydrothermal migration and mineralization [65]. The technology uses cheap sulfuric acid as the main leaching agent, and uses a mixture of phosphoric acid and sulfuric acid with low corrosiveness and no volatility to decompose scheelite efficiently. By introducing the phosphorus component, the chemical conditions similar to the geochemical migration of tungsten are established, so that tungsten is transferred into solution in the form of heteropoly acid (see Eq. (2.1)). The thermodynamic impetus in the leaching process is large, and it can be carried out under atmospheric pressure [66]. When tungsten ore is decomposed efficiently, stable and single $CaSO_4 \cdot 2H_2O$ crystals, which are easy to filter and wash, are generated at the same time, and the issues of encapsulation affecting the decomposition rate of tungsten ore during the decomposition process and the difficulty of filtration and washing of decomposition residues leading to tungsten losses have been resolved. Calcium sulfate leaching slag conforms to building materials standards, and can be used as cement coagulant, solve the problem of slag stacking.

$$\begin{aligned} &12CaWO_4 + H_3PO_4 + 12H_2SO_4 + 12nH_2O \\ &= 12CaSO_4 \cdot nH_2O + H_3[PW_{12}O_{40}] + 12H_2O \end{aligned} \tag{2.1}$$

Based on the research and development of a series of complete sets of technologies, such as "sulfuric and phosphoric acid decomposition—controlled formation of leaching slag—crystallization extraction of tungsten—mother liquor cycle—comprehensive recovery of associated elements molybdenum and phosphorus", a new process of scheelite decomposition technology has been formed, as shown in Fig. 2.7. The new process can treat low-grade co-associated tungsten ore in a clean and efficient way under atmospheric pressure. The synergistic decomposition under the sulfuric acid and phosphoric acid realizes the comprehensive utilization of tungsten, molybdenum and phosphorus. The mother liquor circulation realizes the near zero discharge of wastewater in the production process, and the high-quality gypsum slag is used as building materials to realize the resource utilization of slag, and the processing cost is reduced by more than 25%. It is superior to the traditional autoclave NaOH/Na_2CO_3 autoclave technology in economic and environmental indicators. At present, this technology has been successfully applied in industrialization under the cooperation with Xiamen Tungsten Industry Group. At present, it has built the world's largest 10,000 t APT production line [67].

Based on the principle of metallogenic geochemistry of tungsten ore, the new technology of "collaborative extraction using sulfuric and phosphorus acid" in tungsten metallurgy has established the connection between metallogenic geochemistry of tungsten ore and tungsten metallurgy from the perspectives of element affinity and ion coordination law, which provides new idea and methods for the development of tungsten metallurgy technology.

2.7 Summary

The characteristics and geochemical characteristics of tungsten determine the distribution characteristics and enrichment degree of tungsten in each sphere of the earth, and have an important influence on the metallogenic mechanism of tungsten ore. Tungsten mineralization has undergone complex and multi-stage geological processes, and is closely related to the evolution of strata, magmas, granites and tectonics. Strata and magmas provide the source of ore-forming materials, the formation and migration of granitic magma dominated the diffusion and enrichment of ore-forming elements in the crust, and the evolution of granite further promoted the activation, transfer and enrichment of tungsten. The structure is conducive to the full metasomatism and hydrothermal mineralization between magma and surrounding rock, and provides the ore liquid channel and metallogenic space for the formation of tungsten deposits. The position and shape of the ore bearing structural fissure control the variation of the deposit type. Magma source area, surrounding rock conditions, crystallization temperature, magma evolution and migration mineralizers are all key factors in the formation of tungsten deposits, which contribute to more efficient research on tungsten geochemistry.

In the process of tungsten mineralization, strong granite magmatic differentiation, high F content, and surrounding rock properties are very important for tungsten

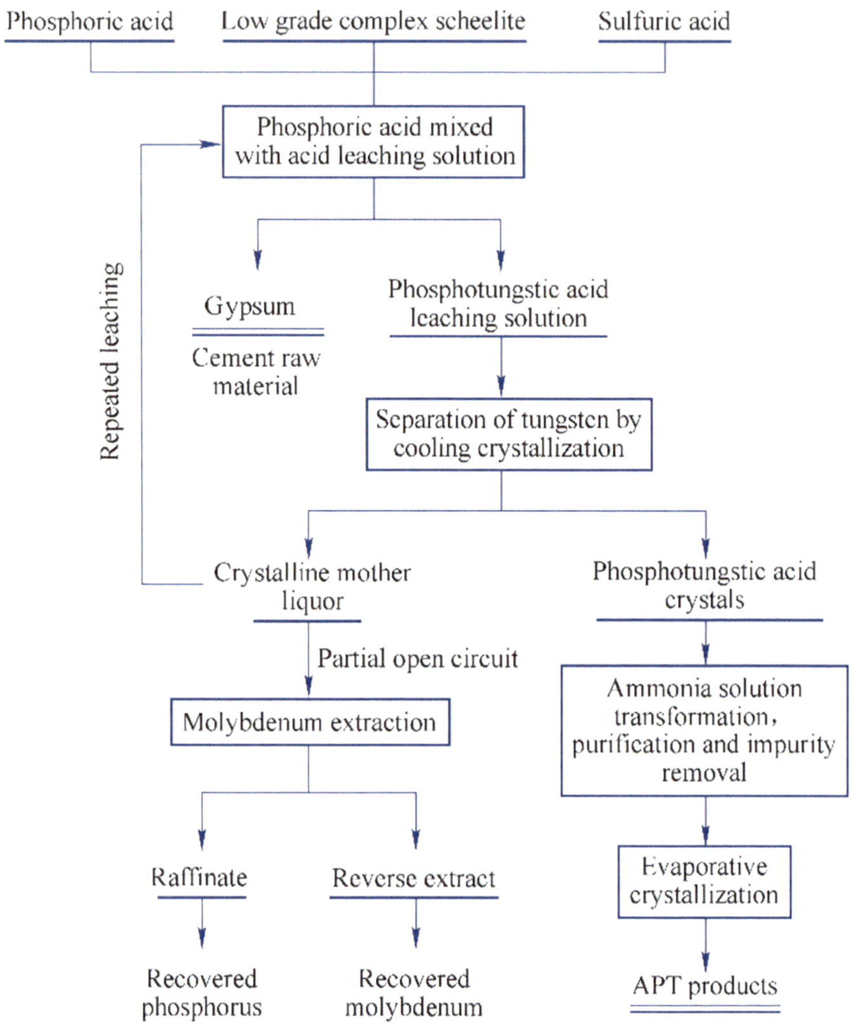

Fig. 2.7 Principle flow of collaborative decomposition technology using sulfuric and phosphorus acid for scheelite concentrate [68]

enrichment and mineralization, which determine tungsten migration mineralization and deposit types. The quartz vein type, greisen type, porphyry type and skarn type tungsten deposits are relatively significant for economic development at present. Different tectonic–magmatic mechanisms have contributed to the deposit types with different characteristics, and these different types of deposits have great differences in gangue mineral types and contents. According to the type and characteristics of ore deposits, mineral miners have developed a series of tungsten flotation technologies. For example, fatty acid flotation technology at room temperature for quartz vein type, fatty acid rough and heating cleaning technology for greisen type, chelating collector rough and heating cleaning technology for skarn type. In particular, for the complex high-calcium scheelite and wolframite mixed tungsten resources, the normal temperature flotation technology using Pb-BHA complex as a collector has greatly improved the level of efficient development and utilization of tungsten resources. Throughout the history of tungsten ore development, geochemistry has played an important role in the development of the new technology of beneficiation. Tungsten ore geological metallogenic law not only determines the essential characteristics of ore, at the same time can provide new solutions for the development of flotation agents and technology, as well as the design of metallurgical process, promoting the efficient and clean utilization of complex tungsten ore resources.

Acknowledgements This work was supported by National key research center and development Program of the 14th Five-Year Plan (No. 2022YFC2905105 and No. 2022YFC2905104), National Natural Science Foundation of China (NSFC) (No. 52122406and No. 52374291), Hunan High-tech Industry Technology Innovation Leading plan (No. 2022GK4056), Hunan Innovative Province construction special Project (No. 2020RC3001).

References

1. Liu Y (1984) Element geochemistry. Science Press
2. König S, Münker C, Hohl S et al (2011) The Earth's tungsten budget during mantle melting and crust formation. Geochim Cosmochim Acta 75(8):2119–2136
3. Wei SL, Jia B, Zeng Q (2006) Study on metallogenic mechanism of tungsten deposit in Nanling Area. Resourc Investig Environ 27(2):103–109
4. Liu X, Xie T (2017) Study on theory and technology of mineral geological exploration. World Nonferrous Metals 8:2
5. Zhu H, Zhang L, Du L et al. Geochemical properties of tungsten and origin of tungsten deposits in South China. Acta Petrologica Sinica 20, 36(01):13–22
6. Fulmer EC, Nebel O, Westrenen WV (2009) High-precision HFSE partitioning between garnet, amphibole, and alkaline melt, Kakanui, New Zealand. Geochmica et Cosmochimica Acta 73(13):403
7. Mu Z, Huang F, Lu D (1988) Petrology, mineralogy and geochemistry of tungsten-bearing granitic rocks in Zuitan, Subei, Gansu Province. J Petromineral (2):109–117
8. He Y, Yue K, Dong Z et al (2003) W content in mantle rocks of eastern China and its significance. Geochimica 32(6):5
9. Ma D (2009) Progress in the study of tungsten geochemistry. Geol J China Universities 15(01):19–34
10. Ni Z, Wei G (1996) Source rocks of crustal remelting granites, constraints of structure and physicochemical conditions. Northwest Geol 000(001):6–10
11. Long X, Wang X (2009) Geological characteristics and genesis of Hukeng Tungsten Deposit in Wugongshan Area, Jiangxi Province. Nonferrous Metals Sci Eng 23(4):3–7
12. Popov BH, Kuceri EH, Xiong Shaoren (1962) Formation of radon base number and abnormal concentration in groundwater and its significance for general survey. Uranium Geol (03):18–22
13. Li G, Yang B, Liu Q et al (2014) Comprehensive research on tungsten mineralization in the periphery of miaoershan pluton in the western section of nanling metallogenic belt. Land Resourc Guide 11(10):55–62
14. Zhang R, Lu J, Lehmann B et al (2017) Combined zircon and cassiterite U-Pb dating of the Piaotang granite-related tungsten–tin deposit, southern Jiangxi tungsten district, China. Ore Geol Rev 82:268–284

15. Jingwen M, Hongyan L (1995) Evolution of the qianlishan granite stock and its relation to the shizhuyuan polymetallic tungsten deposit. Int Geol Rev (1):63–80
16. Editor of the Editorial Office of the Ministry of Geology and Mineral Resources. Proceedings of the Discussion on Tungsten Geology. Geological Publishing House (1984)
17. Qin Y, Wang D, Li Y et al (2010) Diagenetic and metallogenic age of Baizhangyan tungsten-molybdenum deposit in Qingyang, Anhui Province and its geological significance. Earth Science Frontiers
18. Li J, Li X (2020) Research progress on the mineralization of Skarn type tungsten deposit. Mineral Deposits 039(002):256–272
19. Jiang G (2004) Graduate University of Chinese academy of sciences. Institute of Geochemistry.
20. Wang Y, Ma X, Chei W et al (2018) Discovery and significance of late cretaceous post-collision magma tungsten mineralization in the middle part of Ban-Nu Belt. Mineral Deposits 4:886–889
21. Kang Y, Li C (1991) Geological characteristics, types and distribution of tungsten deposits in China. Mineral Deposits 01:19–26
22. Chen B, Ma X, Wang Z (2014) Origin of the fluorine-rich highly differentiated granites from the Qianlishan composite plutons (South China) and implications for polymetallic mineralization. J Asian Earth Ences 93(oct.15):301–314
23. Department of Geology, Nanjing University. Granitoids of Different Ages and their Relationship with Mineralization in South China. Granitoids of Different Ages and their Relationship with Mineralization in South China (1981)
24. Huang H, Niu Y, Zhao Z et al (2009) Differentiation of Nb/Ta, Zr/Hf ratios during magmatism: 2009 National Symposium on Petrology and Geodynamics
25. Chen J, Yang J (2015) Genesis of the highly differentiated I-type granite in Fogang, Shandong Province: constraints from elements such as Nb-Ta-Zr-Hf. Acta Petrologica Sinica 31(03):846–854
26. Wang Y (2014) Study on magmatic-hydrothermal evolution and tungsten mineralization of early Yanshanian Granitogenesis in Southeast Hunan Province. China University of Geosciences, Beijing
27. Zhu X, Wang Y, Cheng X et al (2014) Study on granitic magmatic evolution and separation and enrichment mechanism of ore-forming materials in Tungsten Mining Area. Mineral Deposits 33(S1):361–362
28. Huang LC, Jiang SY (2014) Highly fractionated S-type granites from the giant Dahutang tungsten deposit in Jiangnan Orogen, Southeast China: geochronology, petrogenesis and their relationship with W-mineralization. Lithos 202–203:207–226
29. Xu T, Gao H, Li Y et al (2012) Discussion on metallogenic characteristics of tungsten deposits in China. China Tungsten Industry 27(03):1–5
30. Qin Y, Wang DH, Li YH et al (2010) Rock-forming and ore-forming ages of the Baizhangyan tungsten-molybdenum ore deposit in Qingyang, Anhui Province and their geological significance. Earth Science Frontiers 17(2):170–177
31. Zhang Q, Jin W, Li C et al (2014) Theory of "magmatic thermal field" and its metallogenic Significance. Gansu Geology
32. Che X, Wang R, Linnen R et al (2013) Mineralogy and experimental geochemistry of wolframite during magmatic period
33. Zhu X, Wang J, Wang Y et al (2014) Geology and exploration50(05):947–960 (Wang Y, Wang Y, Wang Y et al. Stable isotope geochemistry of the tungsten deposit, Yaogang, Hunan Province)
34. Huang H, Chang H, Tan J et al (2015) Comparative study on fluid inclusions in symbiosis woltzite and quartz by infrared microthermometry: a case study of Xihuashan quartz vein tungsten deposit in Jiangxi Province. J Petrol
35. Zhao X, Guo Z, Bai Y (2015) Research progress of skarn-type scheelite deposit. Geol Surv China 2(01):9–13
36. Stepanov, Aleksandr S et al (2016) Nb-Ta fractionation in peraluminous granites: a marker of the magmatic hydrothermal transition. Geology 44(7):e394
37. Xiong X, Zhu J, Rao B (1997) The phase relationship of granite-H2-HF system and the effect of Fluorine on the structure of granitic melt. Chinese J Geol 1:1–10

38. Wang X, Ni P, Jiang S et al (2008) Research on fluid inclusion in Piotang Tungsten Deposit, Southern Jiangxi. Acta Petrologica Sinica 09:2163–2170
39. Wu Y (1987) Geology of Xihuashan tungsten deposit. Geological Publishing House
40. Wang Y, Chen M, Li M et al (2006) Mineral resources and geology 000(005):334–339
41. Zhang Q, Liu R, Cao X et al (2013) Study on the relationship between structure and properties of fatty acid scheelite collectors. Nonferrous Metals Sci Eng 000(5):85–90
42. Wang L, Komg F, Wei L et al (2014) Dolomite type mineralization of Shirenzhang tungsten deposit and its research significance. Nonferrous Metals (Mine Section) 66(04):47–51
43. Yin J, Xu X, Zhang W et al (2019) Geological characteristics and prospecting direction of Daping Tungsten deposit in Tongcheng County, Hubei Province. China Tungsten Industry 34(06):1–8
44. Liu Y, Liu Y (2006) Metallogenic structure system and its prospecting significance for tungsten Deposit. Resourc Investigation Environ 27(2):110–119
45. Wu K, Gu L, Zhu Z (2009) Discussion on structural ore control mechanism of dolomite type tungsten deposit in Jiuxi Mining area, Dangping Tungsten Mine. China Tungsten Industry 24(01):21–23
46. Fu JM, Cheng SB, Lu YY et al (2012) Geochronology of the greisen-quartz-vein type tungsten-tin deposit and its host granite in Xitian, Hunan Province. Geology and Exploration
47. Wang M, Jia M, Xiao Y et al (2014) Status quo and sustainable development countermeasures of tungsten resources in China. Nonferrous Metals Eng 4(02):76–80
48. Deng L, Zhou X, Guan T et al (2016) Research into benefication process of a law-grade and complex refractory scheelite ore. China Mining 025(006):133–138
49. Zhang Y, Liu Y, Gao S et al (1990) Rare earth geochemical characteristics of tungsten minerals-identification of genetic types of ore deposits. Geochimica 11–20
50. Liang D, Xu X, Yu L et al (2006) Study on process mineralogy of porphyry tungsten ore. China Tungsten Industry (04):24–26
51. Bi C (1987) Geological characteristics of skarn scheelite deposit in China. J Chinese Acad Geol Sci
52. Guiqing X, Jingwen M, Leon B et al (2018) Mineralogy and titanite geochronology of the Caojiaba W deposit, Xiangzhong metallogenic province, southern China: implications for a distal reduced skarn W formation. Mineralium Deposita 1–14
53. Wu S, Mao J, Ireland TR et al (2019) Comparative geochemical study of scheelite from the Shizhuyuan and Xianglushan tungsten skarn deposits, South China: implications for scheelite mineralization. Ore Geol Rev
54. Zhang Y (2012) Study on performance and mechanism of flotation separation inhibitor for scheelite and calcium-bearing gangue minerals. Central South University
55. Chen X, He W, Liao D (2020) Research on high efficiency flotation of desulphurized low-grade scheelite in a mine. Mining Res Dev 40, 238(05):140–142
56. Han H, Hu Y, Sun W et al (2017) Fatty acid flotation versus BHA flotation of tungsten minerals and their performance in flotation practice. Int J Miner Process 159:22–29
57. Gao Y, Qiu X, Xia Q et al (2001) Study on the mechanism of interaction between benzohydroxamic acid and wolframite. Guangdong J Nonferrous Metals 11(2):92–95
58. Zhang Z, Zhang X, Ye Z et al (1999) Research on new tungsten flotation technology of shizhuyuan polymetallic ore by GY method. Min Metall Eng (4):22–25
59. Sun W, Han H, Hu Y et al. Theory and research progress of molecular assembly flotation for metal ion coordination control. Chinese J Nonferrous Metals 20, 30(04):927–941
60. Han H, Liu W, Hu Y et al (2017) A novel flotation scheme: selective flotation of tungsten minerals from calcium minerals using Pb–BHA complexes in Shizhuyuan. Rare Met 4(6):533–540
61. Hu Z, Huang S, Zhou H (2021) Experimental study on improving the recovery rate of a tungsten polymetallic ore. Min Metall Eng 41(05):75–78
62. Sun F, Zhao Z (2020) An interdisciplinary perspective from the earth scientist's periodic table: similarity and connection between geochemistry and metallurgy. Engineering 6(6):707–715
63. Zhao Z (2013) Theory and application of tungsten smelting. Theory Appl Tungsten Smelting

64. Zhongwei Z, Xingyu C, Xuheng L et al (2017) Challenges and development of tungsten extraction and metallurgy under the new situation. Conserv Utilization Min Resourc 1:98–102
65. Zhao Z, Chen X, Ren H. A method for extracting tungsten by decomposing woltzite or scheelite and wolframite mixed ore under pressure with phosphorous acid
66. He L, Zhao Z, Yang J (2017) A new generation of green tungsten metallurgy process: scheitrotungsten-sulfur-phosphoric acid co-decomposition technology. China Tungsten Industry 32(03):49–53
67. Chen X, Zhao Z. A method for preparing tungsten products by acid decomposition of mixed scheelite and wolframite
68. Zhao Z, Li J, Chen X et al (2013) Current status and development of scheelite Tungsten smelting technology in China. Nonferrous Metals Sci Eng 4(5):11–14

Chapter 3
Crystal Chemistry of Tungsten Minerals

Zhiyong Gao and Wei Sun

Abstract Scheelite and wolframite, the national strategic mineral, are usually enriched by froth flotation basing on the different physical and chemical properties among tungsten ore and other gangue minerals. Based on the crystal structures of scheelite and wolframite, their anisotropic physical and chemical properties such as surface broken bond, cleavage and exposed surface, surface energy, wettability, surface electrokinetic and adsorption properties were studied and summarized. The relationships between different surface properties, and the separation mechanism of scheelite were explained. Surface broken bond density (D_b) was positively correlated with surface energy. A crystal surface with lower D_b and surface energy will be easily exposed when the comminution of the minerals. For wolframite, the main exposed surfaces are {111} and {100}, while {112}, {101} and {001} for scheelite. The adsorption strength of sodium oleate on crystal surfaces of a mineral is primarily influenced by the number of broken bonds present on the unit crystal surface. The anisotropy of surface electrical properties and adsorption could be explained by the differences in the active oxygen, calcium sites and their broken bonds. The exposure of highly active crystal surfaces and shape parameters of scheelite particles can be tuned by optimizing the grinding process, and hence enhancing its flotation.

Keywords Tungsten minerals · Crystal structure · Crystal surface · Grinding · Flotation

Z. Gao (✉) · W. Sun
Central South University, Changsha, China

Engineering Research Center of Ministry of Education for Carbon Emission Reduction in Metal Resource Exploitation and Utilization, Central South University, Changsha, China

State Key Laboratory of Mineral Processing, Changsha, China

W. Sun et al., *Flotation Chemistry of Tungsten Minerals and Its Application*,
https://doi.org/10.1007/978-981-96-5077-4_3

3.1 Introduction

When wolframite and scheelite particle size composition is coarser, it can be effectively recovered [1] by gravity separation. The weak magnetism of wolframite makes it can also be concentrated by magnetic separation [2, 3]. Geochemistry shows that tungsten ore in nature is usually closely associated with other calcium-bearing minerals (such as calcite, fluorite, apatite, etc.), and the distribution size is fine. Flotation has become the most important method [4–7] to separate and recover tungsten ore in this kind of ore. Flotation is a physicochemical method [8] that takes advantage of the differences in surface properties (especially wettability) between different minerals for separation. However, tungsten ore (especially scheelite) and these calcium-bearing gangue minerals are polar salt minerals, with the same or similar surface-active particles, similar surface physical and chemical properties [5–7, 9], so that the flotation separation between scheelite and calcium-bearing gangue minerals is more difficult, is a worldwide problem in the field of flotation.

The precondition to solve this separation problem is to accurately understand the microscopic differences in the surface properties of calcium-bearing minerals. Therefore, the study of surface physicochemical properties and anisotropy of scheelite (especially scheelite), such as cleavage properties, surface energy, surface electricity, wettability, surface adsorption, etc., can provide a deeper understanding of the mechanism of action between scheelite and aqueous solution and flotation reagents, which is of great significance for the efficient flotation separation of scheelite from other calcium minerals. This chapter studies and summarizes the research results of anisotropy of surface properties of scheelite at home and abroad in recent years, in order to provide certain theoretical support for efficient flotation separation of scheelite resources.

3.2 Typical Structure of Tungsten Ore

3.2.1 Crystal Structure of Tungsten Ore

The crystal structure of scheelite belongs to tetragonal system with intergroup $I4_1/a$, cell parameters are as follows: $a = b = 0.5243$ nm, $c = 1.1376$ nm, $\alpha = \beta = \gamma = 90°$, $Z = 4$ [10]. As shown in Fig. 3.1a, the complex anion in the crystal is the W—O_4 tetrahedron, which is formed by combining W^{6+} with four O, and Ca^{2+} combines with eight O^{2-} on the top angle of the W—O_4 tetrahedron to form a Ca—O_8 cube, where the bond length of Ca—O is 0.2436 nm or 0.2481 nm, and the bond length of W—O is 0.1784 nm. As shown in Fig. 3.1b, the most common exposed surfaces of scheelite are {112}, {001} and {101}, where cleavage {001} is blocked, {112} is clearly visible, and {101} is relatively fuzzy [11, 12].

There are relatively few studies on the crystal structure of wolframite, which belongs to the monoclinic system and whose crystal structure is shown in Fig. 3.2.

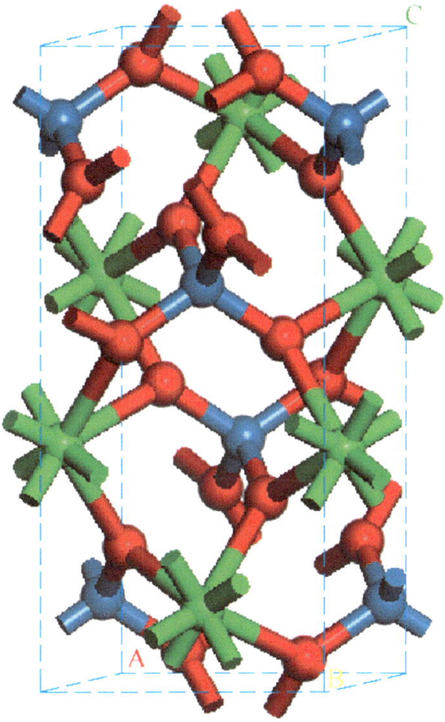

Fig. 3.1 Schematic diagram of the crystal structure of scheelite [13] (Ca in green, O in red, W in gray), reprinted from ref. [13], copyright 2018, with permission from Elsevier

O^{2-} forms a tightly packed octagonal deformation structure, so that half of the octahedral void is filled by W^{6+}, and the other part is filled by iron and manganese plasma, in which iron and manganese ions are infinitely replaceable, in a disorderly arrangement [14]. Wolframite has a set of perfectly cleavage parallel {010} planes, the {100} planes have longitudinal stripes [14, 15] parallel to the C-axis, and the {111} plane is also one of the common exposed surfaces of wolframite [16].

3.2.2 *Crystal Plane Structure of Wolframite*

The crystal surface structure of scheelite has been studied extensively at home and abroad, among which the {112}, {001} and {101} surfaces of scheelite are the three crystal surfaces that have been studied most, and the surface structure of scheelite is shown [10, 11, 17] in Fig. 3.3. On the {112} surface of scheelite, one tungstate ion is bound to $O_2{}^{2-}$, while another tungstate ion is bound to two $O_1{}^{2-}$; On the {001} surface, the tungstate ion binds two $O_1{}^{2-}$; The {101} plane has one $O_1{}^{2-}$ in one tungstate ion and two $O_1{}^{2-}$ [10] in the other tungstate ion.

There are relatively few researches on the crystal surface structure of wolframite, among which the {100}, {010} and {111} planes are the three commonly exposed

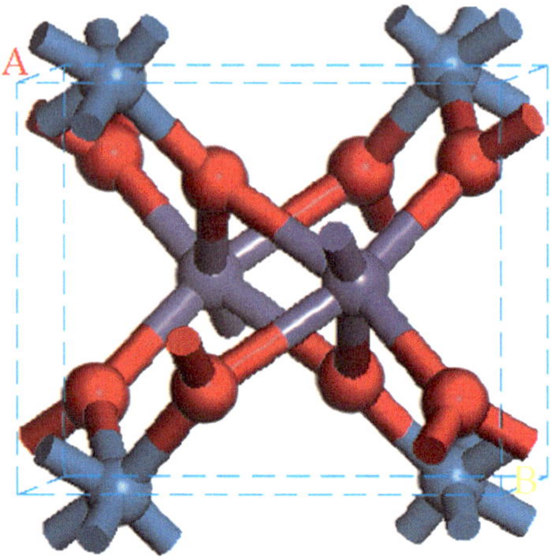

Fig. 3.2 Crystal structure of wolframite [14] (Mn in purple, O in red, W in blue), reprinted from ref. [14], copyright 2020, with permission from Elsevier

planes of wolframite, and their crystal surface structures are shown in Fig. 3.4. On the wolframite {100} surface, there are two tungstate ions, and each tungstate ion is bound to one $O_1{}^{2-}$ Similarly on the wolframite {010} plane, each tungstate ion binds two $O_1{}^{2-}$; On the {111} surface, one of the tungstate ions binds to two $O_1{}^{2-}$. Wang Dianzuo et al. also carried out a series of studies on the structure of wolframite {001}, {010} and {100} crystal faces [14, 15], and the results showed that the broken bonds number of Fe(Mn)–O bond N_b on the {001}, {010} and {100} wolframite crystal faces decreases gradually, and the generation of cleavage plane and exposed plane becomes easier and easier.

3.3 Anisotropy of Surface Properties of Tungsten Ore

3.3.1 Surface Broken Bond Properties of Tungsten Ore

Minerals under external forces usually break in a certain direction and expose crystal faces. The atoms on the crystal surface of minerals, due to the absence of the original coordination atoms in the crystal, broken bonds exist, so that the crystal surface is in a high energy state, with surface relaxation, wettability, electrical, adsorption and other reaction characteristics. It can be seen that the surface broken bond is the most initial property of mineral crystal surface, and its characteristics determine the reactivity of mineral crystal surface, and ultimately determine the surface interface chemical behavior and flotation characteristics [18–21] of minerals. For crystals with only one chemical bond type, the surface broken bond density Db can be calculated by the following formula:

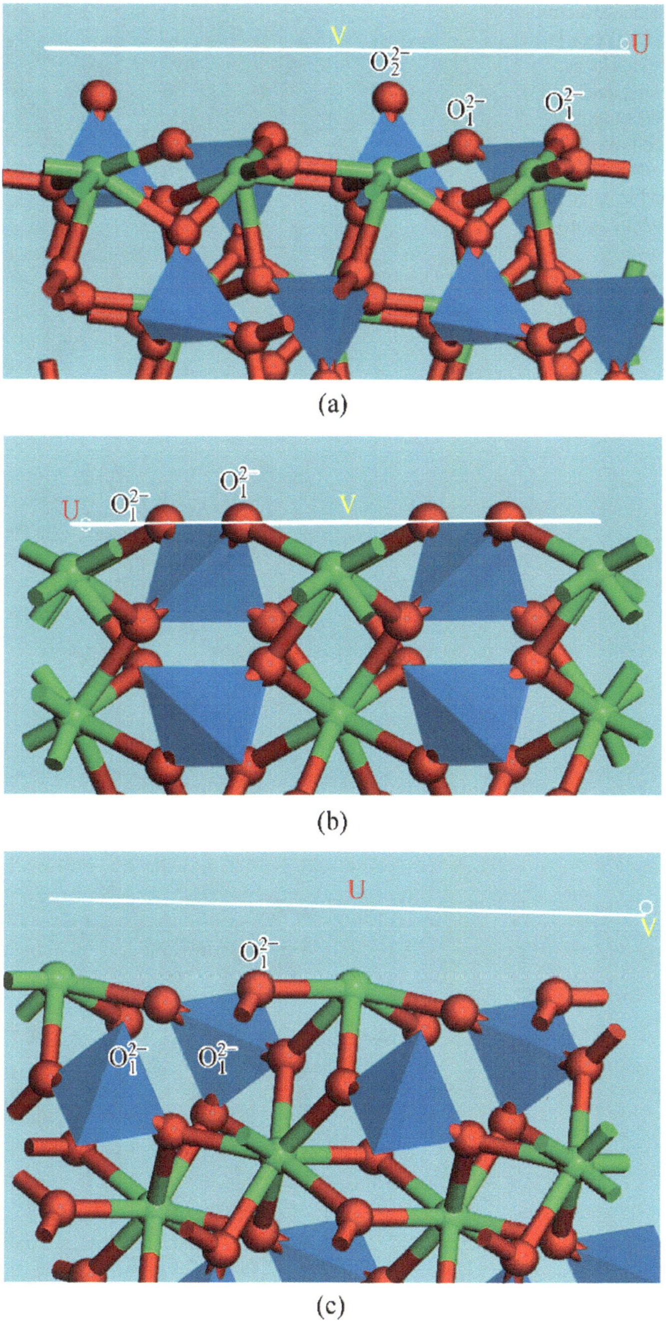

Fig. 3.3 Surface structure of scheelite {112} (**a**), {001} (**b**), {101} (**c**) surfaces, reprinted from ref. [10], copyright 2016, with permission from American Chemical Society (2 crystal cells; each cell exposes 1 Ca atom)

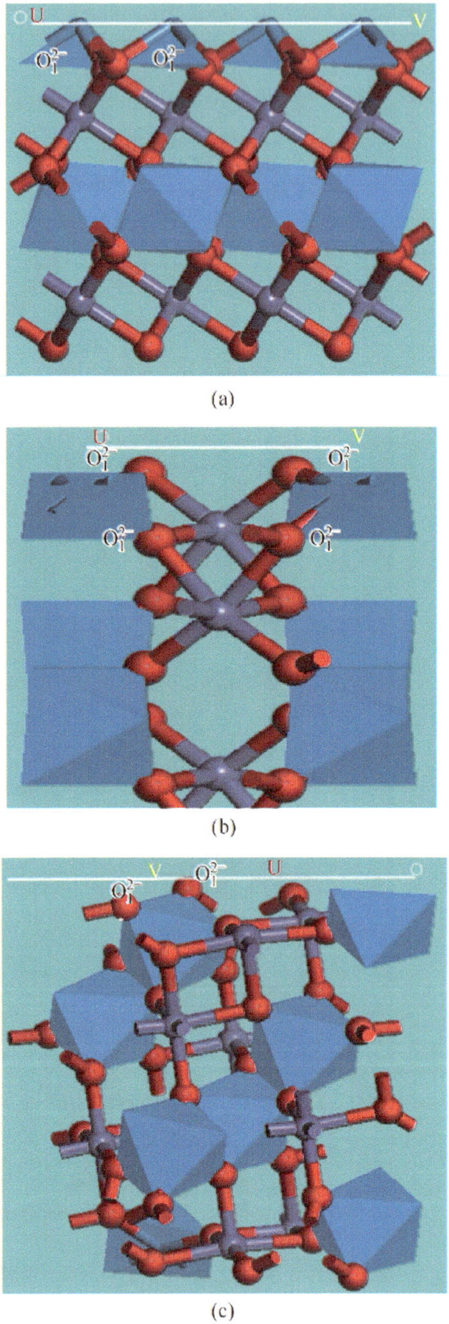

Fig. 3.4 Surface structure of wolframite {100} (**a**), {010} (**b**), {111} (**c**) surfaces. **a**{100} (2 cells; each cell exposes 1 Fe/Mn atom); **b** {010} (1 cell; each cell exposes 1 Fe/Mn atom); **c** {111} (2 cells; each cell exposes 1 Fe/Mn atom)

$$D_b = \frac{N_b}{A} \tag{3.2.34}$$

where D_b represents the number of broken bonds per unit cell area, N_b represents the number of broken bonds within a unit range of a crystal surface, and A represents the unit area of the crystal surface.

Based on the surface structure of scheelite in Fig. 3.3, Gao Zhiyong et al. [18, 20] calculated the surface broken bonds of different crystal faces, and the results are shown in Table 3.1. As can be seen from Table 3.1, the relationship of surface broken bond density (number of broken bonds per unit area) of three common scheelite crystal faces is {101}>{112}>{001}. Further, based on the formula of surface broken bond density and the surface structure of wolframite in Fig. 3.4, surface broken bonds of common exposed surfaces of wolframite are calculated, and the results are shown in Table 3.2. The relationship between the surface broken bond density of the three commonly exposed surfaces of wolframite is {010}>{100}>{111}.

The results show that the cleavage properties and common exposed surfaces of minerals can be predicted by studying the surface broken bond properties, and the reactivity of surface atoms of different minerals can be predicted, which is beneficial to guide the design of flotation agents and realize the efficient separation [22] of minerals.

Table 3.1 Calculation of surface broken bonds of scheelite crystal faces [18, 20] (Gao Zhiyong 2013 #271)

Crystal face	Unit cell area of crystal face A/nm^2	Number of broken bonds per unit cell N_b	Surface broken bonds density D_b/nm^{-2}	The space between adjacent atomic layers d/nm
{112}	0.5038	4	15.88	0.3106
{001}	0.2749	4	14.55	0.2844
{101}	0.3284	6	18.27	0.1190

Table 3.2 Calculation of surface broken bonds of wolframite crystal faces (Gao Zhiyong 2013 #271)

Crystal face	Unit cell area of crystal face A/nm^2	Number of broken bonds per unit cell N_b	Surface broken bonds density D_b/nm^{-2}	The space between adjacent atomic layers d/nm
{100}	0.2877	4	13.90	0.4730
{010}	0.2412	4	16.58	0.5703
{111}	0.4556	4	8.78	0.2933

3.3.2 Cleavage and Exposed Surface Properties of Tungsten Ore

Mineral crystals crack in a certain direction under the action of external forces and produce smooth crystal faces. This property is called cleavage, and the smooth plane formed is called cleavage [23] plane. In the grinding stage before flotation, the mineral mainly along the crystal structure of the bonding force is the weakest between the surface network broken bonds, such as along the distance between the surface network, two layers of the same number of ions adjacent to each other, Yin and Yang ions electrically neutral surface network, weak bond connection surface network and along the crack or the area of impurities gathered in the lattice crack.

The scheelite {112} surface is shown in Fig. 3.3). As can be seen from Fig. 3.3a and Fig. 3.5a, along the direction of {112} plane, adjacent ions are spaced very wide between layers, and Ca^{2+} and WO_4^{2-} groups are arranged in the same layer in a ratio of 1:1. Combined with the data in Table 3.1, the surface broken bond density of {112} plane is 15.88, less than that of {101} plane. Therefore, {112} is the most common cleavage surface and exposed surface [20] of scheelite under external force. Mogilevsky et al. [24] found by XRD that scheelite crystals are easy to grow along the {112} plane to form crystal planes, and also easy to form cleavage planes along the interlayer broken bonds of the scheelite network. This conclusion is the same with the results of Gao Zhiyong et al. [11, 20].

The scheelite {001} surface is shown in Fig. 3.3b. The Ca^{2+} ion layer is sequentially arranged with the WO_4^{2-} ion layer. The {001} plane has the smallest surface unsaturated bond density and surface energy in the scheelite crystal plane, and the crystal plane is an electrical neutral plane, and the layer spacing is larger than the {101} plane, which can be seen [20] from Table 3.1 and Fig. 3.5c. When scheelite is subjected to external forces, it is easy to cleavage along the {001} plane, making the {001} plane the most commonly exposed surface of scheelite. Studies by Chaudhuri and Phaneuf have shown that scheelite crystals tend to cleavage [20, 24–26] along the {001} surface network.

Although scheelite {101} planes have been reported abroad, it is difficult to observe scheelite {101} planes in nature. However, the {101} surface was found after grinding the pure scheelite sample, as shown in Fig. 3.6. Strong diffraction peaks appear at $d = 0.475871$ nm and $d = 0.310514$ nm, corresponding to {101} and {112} planes, respectively, indicating that {101} planes are common exposed surfaces of scheelite powder samples. Gao Zhiyong et al. [20] believed that when the external force is small (such as tapping), it is difficult to generate cleavage along this direction due to the small layer spacing, and the {101} surface is difficult to be exposed. Only when the external force is large, such as crushing and fine grinding, it will split along this direction to form a large number of {101} cleavage planes, as shown in Fig. 3.5b. This also explains, to some extent, that {101} surface is still a common exposed surface of scheelite, although {101} surface spacing is small and the surface broken bond density is the largest.

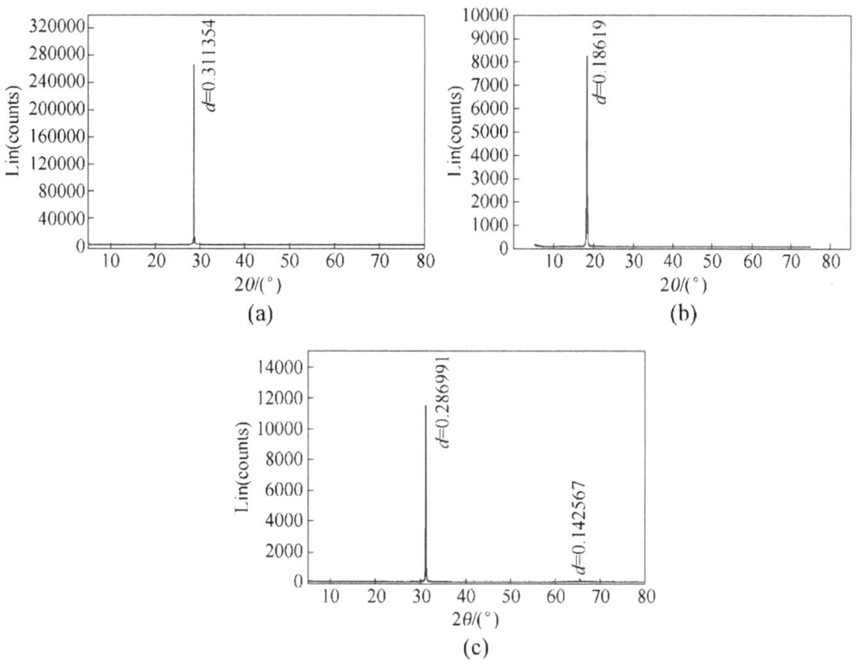

Fig. 3.5 XRDpatterns of scheelite crystals {112} (**a**), {101} (**b**) and {001} (**c**) surfaces [10] (*d* layer spacing, unit: nm)

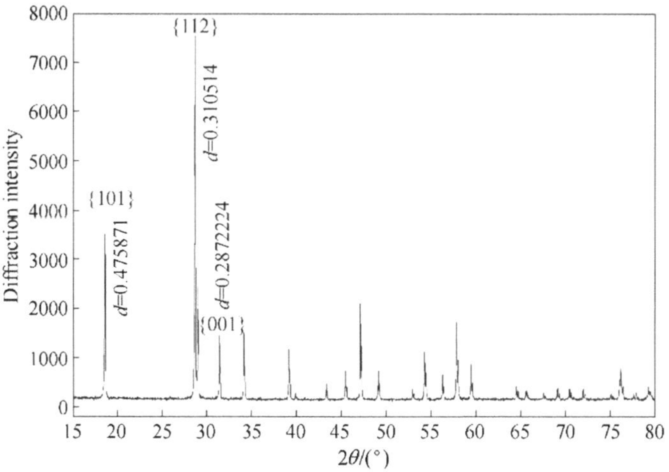

Fig. 3.6 XRD pattern of scheelite pure mineral powder sample [20] (*d* is layer spacing, unit: nm)

3.3.3 *Surface Energy Properties of Scheelite*

The surface energy of a mineral is the cleavage energy, and its magnitude depends on the strength of the atomic interaction between the two exposed surfaces. The smaller the surface energy of the crystal face, the more likely cleavage occurs along the direction of the crystal face, and the resulting cleavage plane is more stable [27].

Different research groups at home and abroad have calculated the surface energy of several common scheelite crystal faces, and the results are shown in Table 3.3. It can be seen from Table 3.2 that the surface energies of scheelite crystal faces are related as follows: {101} > {110} > {111} > {112} > {001}, {001} and {112} are the least surface energies, and {101} are the most [20, 25, 28] surface energies. This also corresponds to the cleavage and exposed surface properties in the previous section. Scheelite is easy to form {001} and {112} surfaces and exist stably. Although {101} surfaces are mentioned in the literature, they hardly cleavage along the {101} mesh surface in nature, and the {101} cleavage surface can only be formed by a strong external force (crushing and milling).

Based on the surface energy data of scheelite common crystal surface in Table 3.3 and the prediction of surface broken bond density in Fig. 3.7, it is shown that there is a strong correlation between surface broken bond density and surface energy for scheelite. As can be seen from Fig. 3.7, the surface energy of scheelite is in a positive proportional relationship with the surface broken bond density, and the surface energy increases with the increase of the surface broken bond density. Therefore, the number of broken bonds per unit surface area of crystal is an important factor that affects or even determines the surface energy of minerals.

In conclusion, the surface broken bond density and surface energy can be used as the basis to judge the stability and properties of mineral surface. However, the calculation and measurement process of surface broken bond density is too complicated and time-consuming, so compared with the calculation of surface broken bond density, the latter is simpler and faster, and the calculation formula is simpler and more concise, which can be used as a reference factor for the relative accuracy of the calculated value of crystal surface energy. For scheelite crystals, the smaller the surface broken bond density, the smaller the surface chemical bond strength, the smaller the surface energy, along the direction of the crystal surface is more likely to produce cleavage and fracture, so under natural conditions {001} and {112} surface

Table 3.3 Surface energy calculation results of scheelite crystal faces reported in the literature

Crystal plane	Gao et al. [20]	Cooper et al. [28]	Longo et al. [25]
{112}	0.773	–	–
{001}	0.686	0.64	0.64
{111}	0.848	0.88	0.88
{110}	0.927	–	0.98
{101}	–	1.19	1.19

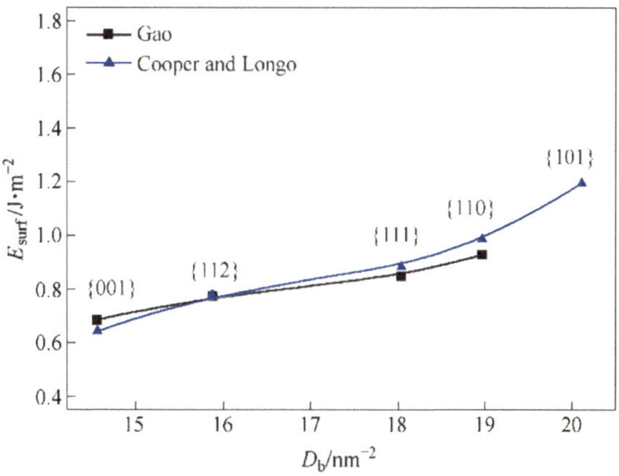

Fig. 3.7 Relationship between scheelite surface broken bond density and surface energy, reprinted from ref. [25, 28], copyright 2011 and 2004, with permission from American Chemical Society

is often view and exposed surface, {101} need to be produced under the action of a strong external force (such as broken). The exposure degree of three scheelite crystal faces is {001} > {112} > {101}.

Scheelite, as a polar salt mineral, exhibits relatively strong ionic surface bonding, which makes the mineral surface have high free energy and is easy to react with water molecules, thus possessing hydrophilicity. Therefore, scheelite has a certain degree of dissolution in water. According to the literature, the dissolved product of scheelite exists equilibrium in water at 25 °C and pH > 6. The equilibrium constant is 8.87×10^{-9} [29] or 4.9×10^{-10} [30], and the $\lg k_{sp}$ varies between −8.05 and −9.31. When pH is less than 6, the solubility of scheelite increases; when pH is more than 6, the molar concentration of WO_4^{2-}in the solution is much larger than Ca^{2+}, resulting in differential dissolution [29, 30]. The reason for this differential dissolution may be the different dissolution rates of different crystal faces and the differences [10, 31] in the ability of Ca^{2+} and WO_4^{2-} to bond with water. Some scholars have studied the dissolution rates of different crystal faces of fluorite, showing that the larger the surface energy of the crystal faces, the faster the dissolution rate, and the arrangement of cation and anion groups on the cleavage surface will also lead to the differential change [31] of dissolution rates. However, the dissolution rate of scheelite crystal faces is rarely tested, and this differential dissolution results in a negative potential of scheelite in the range of pH = 3 − 11, and the mineral electronegativity increases with the increase of temperature [32].

As for the surface energy properties of wolframite, according to Young's equation, Wang Dianzuo et al. [15] believed that the wettability of wolframite mainly depends on the surface energy of the solid–liquid interface, and is related to the number of broken bonds generated per unit area when the mineral surface is broken. According to the crystal structure of wolframite, the surface energy of each crystal surface is calculated as follows: {001} > {010} > {100}, which also corresponds to the difficulty of producing cleavage and exposed surface of wolframite. Similarly, the order of the surface energy of each wolframite crystal surface has a strong correlation with its surface broken bond density, that is, the surface broken bond density of {010} surface is greater than that of {100} surface, so the surface energy of {010} surface is higher.

3.3.4 Surface Wettability of Tungsten Ore

Flotation is carried out in water medium, water molecules are strongly polar molecules, unsaturated bonds on the fracture surface of mineral crystals may have a compensation effect with water (hydration), hydration for different properties of mineral surface strength is different, which also has differences in the surface wettability of minerals. General mineral natural wettability is divided into four types, namely strong hydrophilic, weak hydrophobic, hydrophobic and strong hydrophobic. Adding flotation agents during flotation will change the free energy of mineral–water-air three-phase interface. Adjusting the free energy can expand the difference and realize separation. At this time, the wettability and floatability of mineral surface will change [33]. Of course, different sample origin and testing means, preparation methods and so on May lead to differences in the value of contact angle.

Data show that the contact Angle of the common exposed surface of untreated scheelite in distilled water is about 60°–80°. Gao Zhiyong et al. calculated the contact angle of {112} surface and {001} surface of scheelite and found that the contact angle of distilled water on {112} and {001} surface without polishing was 62.7° and 73.1°, respectively. After polishing, the contact angle of cleavage surface was reduced by 30° – 40° to 31.6° and 42.5°. The reason for the reduction of the contact angle may be that the fresh cleavage surface formed after grinding treatment has high surface reactivity and is more hydrophilic, but the order of size does not change, the contact angle of {001} surface is always larger [20] than {112} surface.

The wettability of mineral surface is related to the unsaturated bond density [34] of active particle on mineral surface and the adhesion [35] work between mineral surface and liquid. Moon et al. [34] found the difference of contact angles between spodumene {001} and {110} surfaces and sodium oleate, and explained it by the number of unsaturated coordination bonds of Al active particles on the surface. Wang et al. [35] found a similar rule by comparing the adhesion work, contact angle and the number of surface active groups between the surface of graphite, graphene and its oxidized products and distilled water. Therefore, the stronger the surface unsaturated bond density and surface particle reactivity, the stronger the interaction with water molecules, the greater the adhesion with distilled water, and the smaller the contact angle measurement value caused by mineral hydrophilic.

Further, by studying the contact angles between scheelite {112} surface and {001} surface and sodium oleate, it is found that the contact angles between sodium oleate and {112} surface are greater than those between {001} surface in the whole concentration range, as shown in Fig. 3.8. Although there is no relevant literature on the wettability of {101}, {112} and {001} surfaces at the same time, it is found that the contact Angle of sodium oleate on {101} surface is greater than that on {001}

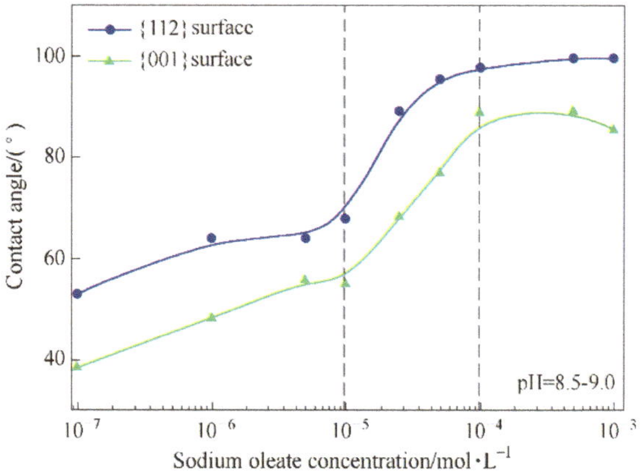

Fig. 3.8 Measurement of contact angles on {112} and {001} surfaces of scheelite after immersion with sodium oleate solution of different concentrations by suspension drop method

surface according to the data related to surfaceelectricity and adsorption [11, 17, 20, 25, 28].

Scheelite is mainly composed of {101}, {112} and {001} cleavage planes, and the three cleavage planes mainly determine the physical and chemical properties of the surface [20]. The wettability of the three common crystal surfaces determines the hydrophobicity of scheelite pure minerals, and also indirectly determines the flotation recoveryrate of scheelite pure minerals [17, 20]. Therefore, exposing more {101} and {112} surfaces and reducing the exposure ratio of {001} surfaces has become one of the important measures to improve the flotation recovery.

Similarly, anisotropy exists in hydrophobicity of different cleavage planes of wolframite. By calculation, Wang Dianzuo et al. [15] found that the (Fe, Mn)–O bond number broken on the wolframite unit crystal plane caused a difference in the contact angle with sodium oleate, and a large contact angle appeared in the pH range of 7.0–9.0, and the flotation effect was the best. The research results are shown in Fig. 3.9, which is also consistent with the research results of Qi et al. [19, 36] That is, in the whole concentration range, the contact angle between sodium oleate and {010} surface is greater than {100} surface. Therefore, for wolframite flotation operation, exposing more {010} surface is one of the important measures to improve the floatability of wolframite.

3.3.5 Surface Electrical Properties of Tungsten Ore

Due to the differential dissolution of scheelite in solution, the concentration of WO_4^{2-} in solution is always greater than Ca^{2+}, so scheelite is negatively charged over almost the entire pH range, as shown in Fig. 3.10 [10, 32, 37]. When Ca^{2+} is present in high concentrations in solution (under hard water conditions), scheelite tends to the isoelectric point [38]. Hu is the only author [39] who measured the isoelectric point of scheelite when pH = 2. Recent studies also show that the surface charge

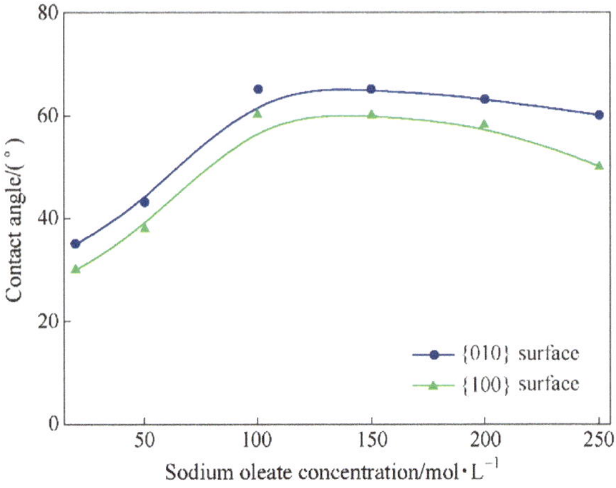

Fig. 3.9 Measurement of contact angle between {100} surface and {010} surface of wolframite by suspension drop method after immersion of different concentrations of sodium oleate solution

of scheelite is not evenly [13, 40, 41] distributed on the mineral surface, and with the gradual increase of pH, the charge difference between different crystal surfaces becomes more obvious, as shown in Fig. 3.11 [20]. At the natural pH of scheelite, the three commonly exposed surfaces are all negatively charged, of which {101} crystal surface has the highest electronegativity, about −37 mV, {001} and {112} crystal surface are −24 mV and −21 mV [10], respectively.

Through hypothesis and test, Gao et al. found that the electrical properties of different crystal faces of scheelite may be related to surface reactive oxygen species, as shown in Table 3.4 [10]. As the pH value of the solution increases, the concentration of OH^- in the solution will increase correspondingly, and the magnitude of negative surface potential will also increase. OH^- adsorbs to surfaces with a higher density of reactive oxygen atoms, resulting in more negatively charged surfaces. Thus, consistent with the results in Fig. 3.8, surface {101} has the largest negative

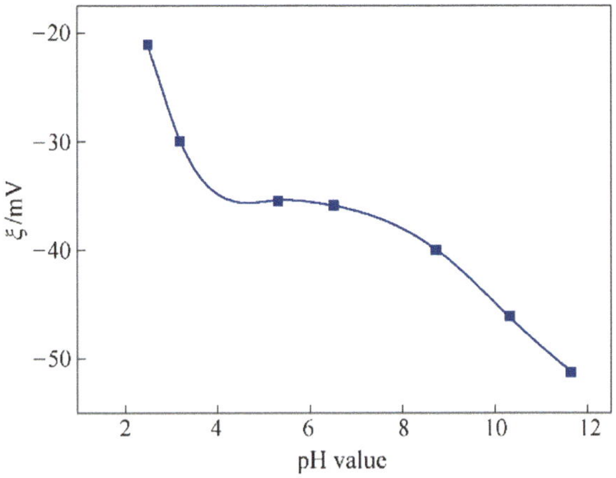

Fig. 3.10 Relation between surface potential of scheelite and pH in distilled water, reprinted from ref. [10], copyright 2016, with permission from American Chemical Society

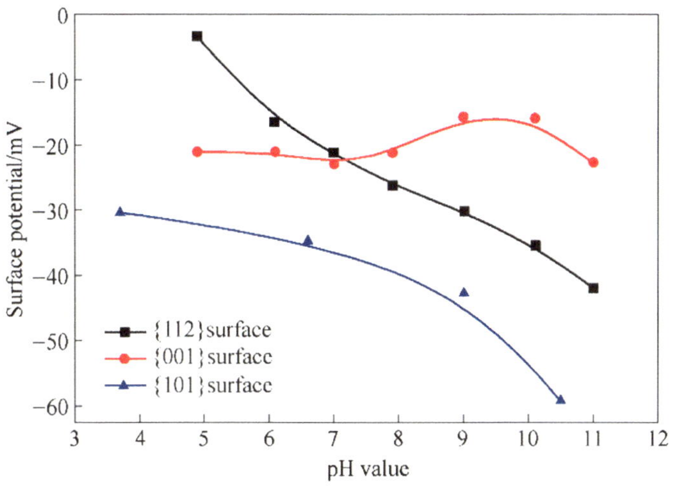

Fig. 3.11 Effect of pH on surface charges of scheelite {001}, {101}, and {112} surfaces, reprinted from ref. [10], copyright 2016, with permission from American Chemical Society

surface potential, followed by {112}, then {001} surfaces. Surface {001} has the lowest density of surface reactive oxygen atoms, so that OH^- adsorption on the surface can almost reach saturation levels at low OH^- concentrations (low pH). The surface charge of {001} changes only slightly as pH increases.

In summary, the anisotropy of scheelite surface charge may be related to the surface density of active oxygen atoms, i.e., the density of oxygen with one or two broken bonds in the topmost surface tungstic acid ion. As the pH of the solution increases, a surface with a larger density of active oxygen atoms may attract a large number of hydroxide ions, thus increasing the surface negative charge. The surface with the highest density of reactive oxygen atoms, {101}, shows the highest surface potential of the three surfaces. Conversely, the surface {001} exhibits the lowest density of active oxygen atoms and the lowest negative surface potential. In addition,

Table 3.4 Amount of WO_4^{2-} or reactive oxygen species on the surface of scheelite {112}, {101} and {001}

Crystal plane	{112}	{101}	{001}
Unit cell area/nm^2	0.504	0.328	0.275
Number of tungstates per unit cell (units)	2	2	1
Number of active oxygen atoms per unit cell (in pieces)	3	3	2
Active oxygen atomic density/nm^{-2}	6.0	9.1	3.6

Note At surface {112}, one tungstate ion contains 1 $O_2{}^{2-}$ and the other tungstate ion contains 1 $O_1{}^{2-}$; On the {101} plane, one tungstate ion has 1 $O_1{}^{2-}$ and another tungstate ion has 2 $O_1{}^{2-}$; On the {001} plane, the tungstate ion has 2 $O_1{}^{2-}$'s

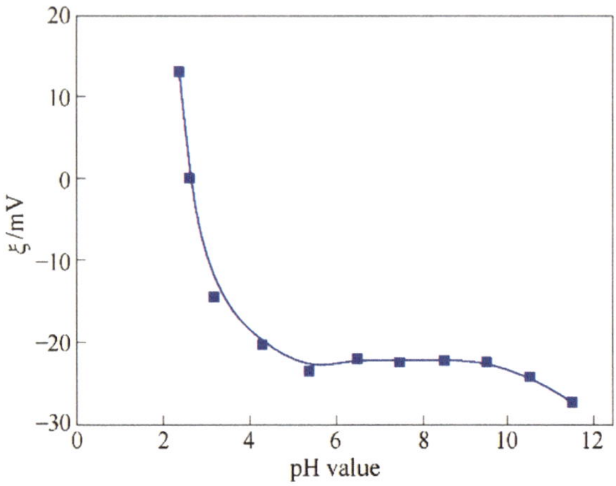

Fig. 3.12 Relationship between the surface potential of wolframite and pH in distilled water

crystal lattice distortion, point defect and lattice substitution may also cause the change of surface electrical properties.

Wolframite surface dissolution characteristics are different from scheelite, when pH < IEP, wolframite dissolved surface with positive charge; When IEP < pH < 6.5, there is a maximum electronegativity extreme point of wolframite; When pH > 6.5, the electronegativity of wolframite increases gradually [36]. It is reported that the flotation effect of wolframite is the best when the flotation pH is between 6 and 10. The surface of wolframite shows low electronegativity, and the located ions are Mn^{2+} and Fe^{2+}, which is not only conducive to the electrostatic adsorption of agents, but also conducive to the chemical bond cooperation between agents and ore, as shown in Fig. 3.12 [15, 19].

Wang Dianzuo et al. [15] proved through theory and experiment that $MnWO_4$ has stronger electronegativity than $FeWO_4$, that is, the zero electric point of $MnWO_4$ is higher than that of $FeWO_4$. The main factors affecting the zero electric point of $MnWO_4$ and $FeWO_4$ are the length of Me–O bond and the stabilization energy of crystal field. The Mn–O bond in $MnWO_4$ is larger than the length of Fe–O bond in $FeWO_4$. Although both Mn and Fe are highly spin octahedrons, when they are complexed with water as ligands, the number of electrons in *D*-layer of Mn^{2+} is 5, and the stabilization energy of crystal field is zero. While Fe has crystal field stabilization energy. These two factors result in higher zero electric point of $MnWO_4$, as shown in Table 3.5. Through calculation, Wang Dianzuo et al. believed that the theoretical zero electric point of $MnWO_4$ and $FeWO_4$ is 2.75 and 1.95, respectively.

Table 3.5 PZC theoretical calculation results of wolframite [15]

Minerals	Me–O average bond length/m	Crystal field stabilization energy/kJ·mol^{-1}	YSD equation calculated value	Calculate the value graphically	Test value
$MnWO_4$	2.31×10^{-10}	0	2.75	2.8	2.8
$FeWO_4$	2.05×10^{-10}	48.98	1.95	2.0	2.1

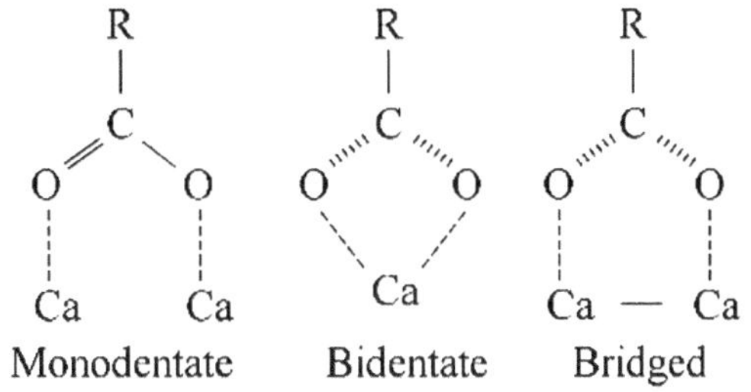

Fig. 3.13 Schematic diagram of three configurations of oleic acid ions interacting with Ca particles on the surface of calcium-containing minerals

3.3.6 *Surface Adsorption of Tungsten Ore*

Fatty acid is the most common collector in scheelite flotation, and its adsorption behavior on scheelite surface has been the focus of mineral processing and related fields.

De Leeuw et al. [42] studied the adsorption behavior of formic acid with similar carboxylic groups to oleic acid on Ca particles on the surface of calcium-containing minerals by means of atomic simulation, and found that the adsorption mode and adsorption energy were anisotropic. The results showed that the carboxyl groups of fatty acid collectors could form three coordination modes with Ca particles on the surface of minerals. They could be called monodentate, bidentate and bridged coordination [43], as shown in Fig. 3.13.

It has been reported in the literature that the coverage area of a single oleic acid ion bonding with Ca particles on the mineral surface is 0.33 nm or 0.32 nm^2 [37, 44]. Through calculation, Gao Zhiyong et al. [20] found that there was one Ca particle on the {001} surface of scheelite {001}, and the unit cell area was 0.2749 nm^2, which was smaller than the covering area of a single oleic acid ion, making sodium oleate occupy at least two units of the cell. Moreover, due to the spatial obstruction of adjacent WO_4 groups and the electrostatic repulsion of O ions on the surface, oleic acid ions can only have single or double coordination with one of the Ca particles. Therefore, the coverage of sodium oleate on the cell is 50% and it is a single coordination configuration.

The unit cell area of scheelite {112} surface is 0.5038 nm^2, and there are two Ca particles within the unit cell range. The distance between the two Ca particles is 0.386 nm, and there is no electrostatic repulsion of O ion. Therefore, the two O atoms of oleic acid ion interact with the two adjacent Ca particles on the surface respectively in the form of bidentate, and the coverage is 100% [20]. Therefore, the theoretical prediction of the interaction intensity between the three crystal surfaces and oleic acid molecules is as follows: {101} > {112} > {001}, which is consistent with the research results of copper et al. [28].

The number of Ca particles per unit cell of the {112}, {101} and {001} planes and the number of broken bonds N_b of each Ca particle are shown in Table 3.6. Gao et al. believe that the larger the number of broken bonds of Ca particles, the stronger the reactivity, and the Ca particle of {101} surface has stronger reactivity than the Ca particle of {112} surface. Therefore, oleic acid molecules can interact more with {101} plane Ca particles when oleic acid molecules also form bidentate with {101} and {112} plane Ca particles.

Figure 3.14 shows the interaction energy between oleic acid and scheelite {001} and {112} plane. The interaction energy of {112} plane is −361.8 kcal/mol(−1543.5 kJ/mol), which is about −70 kcal/mol(−304.6 kJ/mol) less than that of {001} plane, indicating stronger chemical interaction with the former. This is because oleic acid has a more stable bidentate on {112} plane than the monodentate on {001} plane, with stronger chemical interaction and higher adsorption energy, which is also consistent with previous theoretical predictions.

Table 3.6 Calculation results of Ca particles and broken bonds of scheelite crystal faces

Crystal face	Ca particle number of unit cell range N_{Ca}	Number of broken bonds of per Ca particle N_b	Interaction configuration with a single oleic acid molecule (predicted by theory)
{001}	1	2	Monodentate
{101}	1	3	Bidentate
{112}	2	2	Bidentate

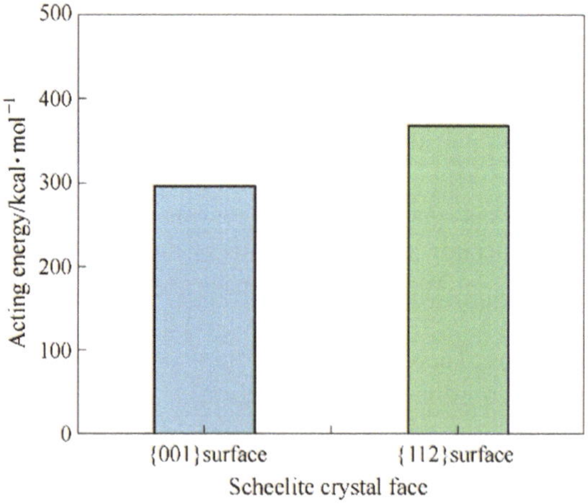

Fig. 3.14 Comparison of interaction energies between oleic acid ions and scheelite {001}

3.4 Separation Principle of Scheelite Based on Crystal Chemical Differences

Scheelite and wolframite are brittle minerals with Mohs hardness of about 4–4.5, which are easy to overgrind and cause a series of problems for flotation. Therefore, grinding and flotation are the key points to achieve effective separation of tungsten ore.

Grinding is an extremely complex process, grinding media, environment and feed will affect the final product. The traditional ball mill mainly grinds ore through point contact, which has poor selectivity, serious over-grinding phenomenon and wide distribution of product particle size, which is not conducive to flotation. Cylindrical media such as rod grinding belongs to line contact, the output energy is relatively low, not easy to over grinding phenomenon, uniform product particle size; other shape grinding media such as double plane short column, hexagonal prism have been introduced. According to the difference of surface properties of tungsten ore and the existing literature, the exposure degree of common crystal surface and shape index of tungsten ore will be different under different grinding conditions. Combined with the surface broken bond, cleavage surface and surface energy properties, scheelite {101} surface is extremely difficult to expose, according to the surface wettability and absorption of tungsten ore, scheelite {101} and {112} crystal surface and sodium oleate adsorption is more stable, the configuration is more firm, wolframite {001} and {010} surface and sodium oleate contact angle is larger, stronger hydrophobicity. Therefore, a suitable grinding system to produce a specific crystal surface may be the means to achieve effective separation.

Zhu Guangli and Xu Longhua et al. [45, 46] studied the grinding of spospoene and found that the effect of wet grinding is better than that of dry grinding, and increasing the particle size will expose more {110} surface, and {110} surface has stronger adsorption with sodium oleate. Gao Zhiyong et al. [17, 47] established the relationship between the shape index of the grinding product and its grinding method and the exposure ratio of the crystal surface. After rod grinding, more {101} surfaces of scheelite are exposed and the particle elongation is larger. More {001} surfaces are exposed by ball milling, and the roundness of the product is greater [17]. The XRD diffraction results of the product are shown in Fig. 3.15, and the model diagram is shown in Fig. 3.16. Similarly, compared with scheelite ball milling products, the particle elongation of the product after segmentate media grinding is larger, and more {101} and {112} surfaces are exposed. Similar conclusions were also proved in the fluorite grinding test, rod grinding exposed more {310} and {110} surface, the particle elongation is greater; ball milling exposed more {111} surfaces and larger grain roundness [47]. Elongation too large will cause the hydration layer to break, exposed particle surface and pharmaceutical combination; higher roundness prevents the "armor action" of bubbles, agents, and minerals. By comparing the crystal surface exposure ratio of rod mill and ball mill products and the above summary, it can be predicted that more {101} surfaces of scheelite produced by rod mill will lead to higher flotation recovery; ball milling produces more {001} surfaces with weak

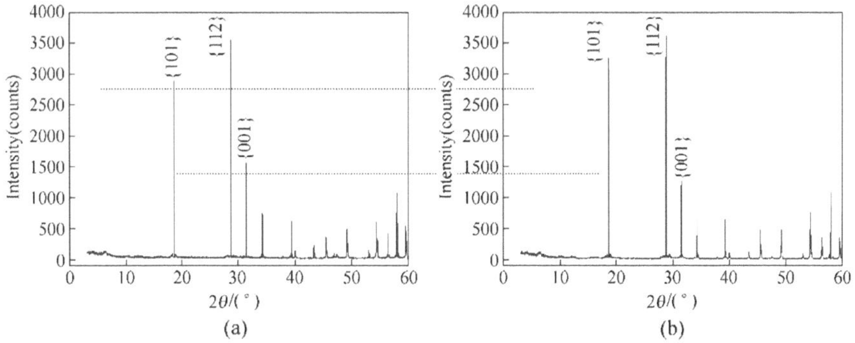

Fig. 3.15 XRD patterns of scheelite under ball milling (**a**) and rod grinding (**b**) conditions, reprinted from ref. [17], copyright 2014, with permission from Elsevier

hydrophobicity and low recovery rate, which is exactly the same as the flotation results of Gao Zhiyong et al. In the whole pH range, the flotation recovery rate of rod milling products is about 20% higher than that of ball milling products, as shown in Fig. 3.17 [17]. Similarly, compared with scheelite ball mill products, section mill products have a higher flotation recovery, and the flotation recovery of scheelite section mill products is about 10% higher than that of ball mill products under the best conditions when Pb-BHA collector is selected. These research results achieve selective grinding to a certain extent, provide convenience for subsequent flotation, and provide research ideas for subsequent scientific research work.

Since the twenty-first century, many flotation reagents have been studied by concentrators [48], including fatty [49, 50] acids, chelates [51], sulfonic [52] acid, phosphonic acid arsenic acid [53], cationic collector and amphoteric collector [54]. The main inhibitors are sodium silicate and its modified products. At present, the vast majority of tungsten concentrators in the world still adopt the classic "Petrov" heating flotation method, that is, fatty acid—sodium silicate system [13]. "Petrov" heating flotation method needs to be carried out under high temperature conditions, high energy consumption, poor environment, high consumption of sodium silicate, resulting in products and tailings concentration, filtration difficulties. Replacing the "Petrov" heating flotation technology is one of the important directions of the development of scheelite flotation technology in recent decades. The breakthrough of this technology depends on the design and development of highly selective collectors and inhibitors. Crystal chemistry provides a solid theoretical basis for this process.

3.5 Summary

The surface atom arrangement and surface broken bond properties of scheelite and wolframite crystal face have anisotropy, scheelite surface broken bond density is proportional to the surface energy.

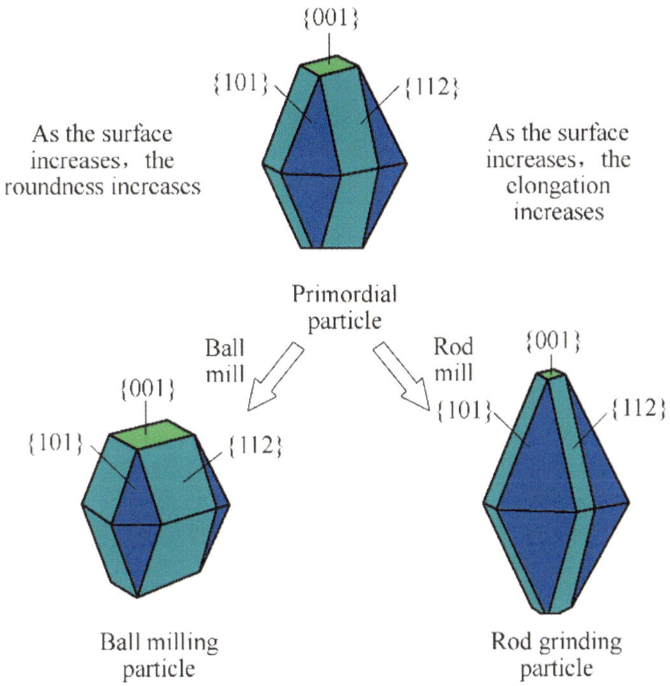

Fig. 3.16 Schematic diagram of scheelite rod milling and ball milling

Fig. 3.17 Recovery rate of scheelite in different grinding media, reprinted from ref. [17], copyright 2014, with permission from Elsevier

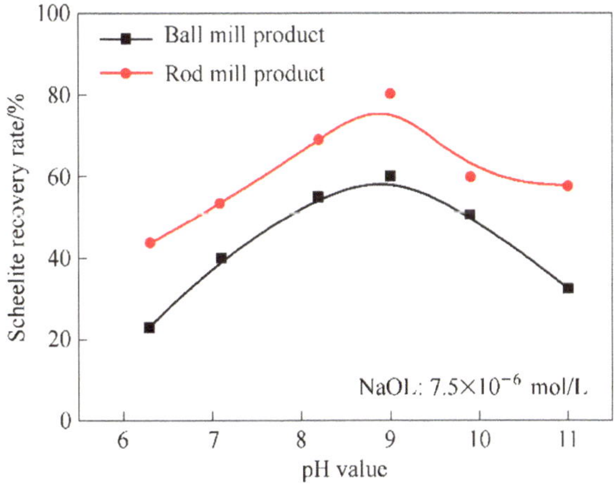

The surface broken bond density can be used to predict the cleavage properties and common exposed surfaces of minerals. The higher the surface broken bond density of scheelite, the higher the surface energy, the more likely cleavage and fracture will occur. The scheelite cleavage is easiest along the {112} surface, followed by

{001} surface, and {101} surface requires a larger external force to cleavage. The number of Fe(Mn)–O bonds broken along the wolframite unit plane can also predict the cleavage properties of wolframite and the common exposed surfaces.

The surface broken bond density and surface energy affect the surface wettability and adsorption properties of scheelite and wolframite. The higher the surface broken bond density, the higher the surface energy, the stronger the surface solubility and the stronger the surface hydrophilicity.

The anisotropy of scheelite's surface charge may be related to the surface density of active oxygen atoms, i.e. the density of oxygen with one or two broken bonds in the topmost surface tungstic acid ion. The main factors affecting the zero electric point of wolframite are the length of Me–O bond and the stabilization energy of crystal field. In addition, the external environment and testing means (AFM, electrophoresis) will also affect the test results.

The number of Ca particles per unit cell of scheelite, the number of N_b broken bonds of each Ca particle, the distance between adjacent Ca particles and the distance between Ca and collector O, and other differences affect the adsorption behavior of flotation reagent on the surface. Solution pH and reagent concentration also have great influence on the adsorption process.

Scheelite flotation needs to expose more {101} and {112} surface, increase the elongation of crystal surface, to avoid the enlargement of the roundness; More exposed {010} surfaces are needed in wolframite flotation, and different Mn and Fe contents in wolframite will also lead to different flotation recovery rates. The present research shows that rod grinding or section grinding may be the method to produce more specific cleavage planes.

References

1. Yang X (2018) Beneficiation studies of tungsten ores—a review. Miner Eng 125:111–119
2. Huang J, Chen L, Ding L (2014) Influence of cylindrical rod media arrangement and combination on high gradient magnetic separation index. Min Metall 23:9–13
3. Li H, Zhang C, Zhang Y et al (2008) Application research of dry permanent magnet high intensity magnetic separator in wolframite separation. China Mining Industry 64–66
4. Habashi F (1997) Precious metals, refractory metals, scattered metals, radioactive metals, rare earth metals. Wiley-VCH
5. Wang JZ (2015) Study on strengthening and mechanism of flotation separation of complex refractory scheelite. Northeastern University
6. Wu Y (2014) Flotation separation and mechanism of scheelite from calcite and fluorite. Jiangxi University of Science and Technology
7. Hu W (2014) Study on mechanism and application of combined collector Flotation for fine wolframite. Jiangxi University of Science and Technology
8. Huang H (1987) Flotation of micro-fine minerals. Metal Mine 10:46–49
9. Napiermunn TJ (1988) Wills' mineral processing technology: an introduction to the practical aspects of ore treatment and mineral recovery. Pergamon Press
10. Gao Z, Hu Y, Sun W et al (2016) Surface-charge anisotropy of scheelite crystals. Langmuir 32(25):6282–6288
11. Gao Z, Sun W, Hu Y et al (2013) Surface energies and appearances of commonly exposed surfaces of scheelite crystal. Trans Nonferrous Met Soc China 23(7):2147–2152

12. Baek S, Jeon H (2018) Application of jig separation for pre-concentration of low-grade scheelite ore. Mater Trans M2017150
13. Kupka N, Rudolph M (2018) Froth flotation of scheelite—a review. Int J Min Sci Technol 28(3):373–384
14. Huang H, Qiu T, Ren S et al (2020) Research on flotation mechanism of wolframite activated by Pb(II) in neutral solution. Appl Surf Sci 530(12):156–179
15. Li Y, Peng M, Wang D (1990) Relationship between crystal structure characteristics and floatability of wolframite. Nonferrous Metals
16. Zhang S (2022) Study on flotation performance and action mechanism of a new diligand hydroxamic acid for tungstelite. Jiangxi University of Science and Technology
17. Li C, Gao Z (2017) Effect of grinding media on the surface property and flotation behavior of scheelite particles. Powder Technol 322:386–392
18. Gao Z, Sun W, Liu X et al (2010) Difference of broken bond between scheelite and calcite crystal surfaces and its effect on cleavage and surface properties of minerals. Acta Mineralogica Sinica 30:69–74
19. Qi M, He G, Luo X et al (2014) Research status of the relationship between surface characteristics and floatability of tungsten ore. Nonferrous Met Sci Eng 5:73–76
20. Gao Z (2013) Study on the relationship between crystal anisotropy and flotation behavior of three calcium-containing minerals. Central South University, Changsha
21. Gao Z, Sun W, Hu Y (2014) Mineral cleavage nature and surface energy: anisotropic surface broken bonds consideration. Trans Nonferrous Met Soc China 24(9):2930–2937
22. Gao Z, Sun W, Hu Y et al (2012) Anisotropy of surface broken bond properties and wettability of calcite and fluorite crystals. Chinese J Nonferrous Met 22:1203–1208
23. Tang J (1985) Interpretation of terms for identification of mineral Chinese medicinal materials (continued). Correspondence of Traditional Chinese Medicine
24. Mogilevsky P (2005) Identification of slip systems in $CaWO_4$ scheelite. Philos Mag A 85(30):3511–3539
25. Longo VMG, Lourdes, Stroppa, Daniel G (2011) A joint experimental and theoretical study on the nanomorphology of $CaWO_4$ crystals. J Phys Chem C 115(41):20113–20119
26. Hay RS (2008) Monazite and scheelite deformation mechanisms
27. Gao X-S, Gao Z-Y, Sun W (2016) Research progress on anisotropy of fluorite surface properties. Chinese J Nonferrous Met 26:415–422
28. Cooper TG, De Leeuw NH (2004) A computer modeling study of the competitive adsorption of water and organic surfactants at surfaces of the mineral scheelite. Langmuir 20(10):3984–3994
29. Marinakis KI, Kelsall GH (1987) The surface chemical properties of scheelite ($CaWO_4$) I. The scheelite/water interface and $CaWO_4$ solubility. Colloids and Surfaces 25(2–4):369–385
30. Atademir MR, Kitchener JA, Shergold HL (1979) The surface chemistry and flotation of scheelite. I. Solubility and surface characteristics of precipitated calcium tungstate. J Colloid Interface Sci 71(3):466–476
31. Godinho JRA, Piazolo S, Evins LZ (2012) Effect of surface orientation on dissolution rates and topography of CaF^2. Geochimica et Cosmochimica Acta 86(none):392–403
32. Ozcan ANB (1993) Electrokinetic, infrared and flotation studies of scheelite and calcite with oxine, alkyl oxine, oleoyl sarcosine and quebracho. Int J Miner Process 39(3–4):275–290
33. Long X (2016) Study on density functional effects of water molecules on surface properties, oxidation and molecular adsorption of sulfide minerals. Guangxi University
34. Moon KS, Fuerstenau DW (2003) Surface crystal chemistry in selective flotation of spodumene ($LiAl[SiO_3]_2$) from other aluminosilicates. Int J Miner Process 72(1):11–24
35. Wang S, Zhang Y, Abidi N et al (2009) Wettability and surface free energy of graphene films. Langmuir 25(18):11078–11081
36. Qi M (2014) Study on the Relationship between Surface Characteristics and Floatability of Wolframite. Jiangxi University of Science and Technology
37. Hanumantha Rao K, Forssberg KSE (1991) Mechanism of fatty acid adsorption in salt-type mineral flotation. Miner Eng 4(7–11):879–890

38. Hicyilmaz C, Ozbayoglu G (1992) The effects of amine and electrolytes on the zeta-potential of scheelite from Uludag, Turkey. Min Eng 5(8):945–951
39. Hu Y, Xu Z (2003) Interactions of amphoteric amino phosphoric acids with calcium-containing minerals and selective flotation. Int J Miner Process 72(1–4):87–94
40. Nesset JE, Zhang W, Finch JA (2012) A benchmarking tool for assessing flotation cell performance. In: Proceedings of the 44th annual meeting of the canadian mineral processors (CIM)
41. Rudolph M, Hartmann R (2017) Specific surface free energy component distributions and flotabilities of mineral microparticles in flotation—an inverse gas chromatography study. Colloids Surf, A 513:380–388
42. De Leeuw NH, Parker SC, Rao KH (1998) Modeling the competitive adsorption of water and methanoic acid on calcite and fluorite surfaces. Langmuir 14(20):5900–5906
43. Rai B (2012) Molecular modeling for the design of novel performance chemicals and materials || Basic concepts in molecular modeling. Int J Artif Organs, 1–26. https://doi.org/10.1201/b11590
44. Gomari KAR, Denoyel R, Hamouda AA (2006) Wettability of calcite and mica modified by different long-chain fatty acids (C18 acids). J Colloid Interface Sci 297(2):470–479
45. Xu L, Hu Y, Wu H et al (2016) Surface crystal chemistry of spodumene with different size fractions and implications for flotation. Sep Purif Technol 169:33–42
46. Xu L, Tian J, Wu H et al (2017) New insights into the oleate flotation response of feldspar particles of different sizes: anisotropic adsorption model. J Colloid Interface Sci 505:500–508
47. Li C, Gao Z (2018) Tune surface physicochemical property of fluorite particles by regulating the exposure degree of crystal surfaces. Miner Eng 128:123–132
48. Li Shiliang (2010) Study on separation of scheelite and gangue minerals by flotation with cationic collector. Central South University
49. Jiang Q, Dai Z (2012) Effect of mixed fatty acids on separation of schetungsten from fluorite and calcite. Min Metall Eng 42–44
50. Zhang Q, Liu R, Cao X et al (2013) Study on the Relationship between structure and properties of fatty acid scheelite collectors. Nonferrous Metals Sci Eng 000:85–90
51. Huang J (2013) Effect of hydroxamic acid Collector on the flotation of scheelite and wolframite. Central South University
52. Wang J, Yin W, Zhang Y et al (2018) Study on strengthening and mechanism of scheelite flotation of sodium oleate and sodium dodecylbenzene sulfonate. Nonferrous Metals: Min Process Section 001:106–110
53. Sun W, Han H, Tang H. A novel amino phosphonic acid compound and its preparation and application method
54. Hu Y, Wang D (1989) Research on flotation of fluorite, barite and scheelite by a new amphoteric collector. Nonferrous Metals 001:10–13

Chapter 4
Solution Chemistry of Tungsten Flotation

Haisheng Han, Wei Sun, and Ruolin Wang

Abstract Solution chemistry directly reflects the flotation behavior. Wolframite and scheelite are the typical tungsten ores, and their surface dissolution and transformation are inevitably occurred, which influence the adsorption of regents and the effect of flotation. In this chapter, the solution chemistry of wolframite and scheelite was firstly studied and summarized to release the characteristic of tungsten minerals. Then, the flotation solution chemistry of reagents, such as regulators and collectors, was also researched and analyzed to provide a reference for the selection of tungsten flotation. Finally, the coordination reaction of metal ions with flotation reagents was systematically concluded to improve the theoretical system of flotation.

Keywords Flotation · Solution chemistry · Wolframite · Scheelite · Reagent

4.1 Introduction

Flotation is an effective means of separating tungsten ore from other associated minerals (calcite, fluorite, barite, etc.), involving gas–liquid-solid three-phase systems. Once solid mineral particles come into contact with aqueous solution, a series of reactions will occur, which will affect the flotation process. In the flotation pulp system of tungsten ore, tungsten ore and its associated minerals will inevitably dissolve, the ions dissolved and the ions in the solution can react with the mineral surface, change the surface properties of tungsten ore and associated minerals, and

H. Han (✉) · W. Sun · R. Wang
Central South University, Changsha, China

Engineering Research Center of Ministry of Education for Carbon Emission Reduction in Metal Resource Exploitation and Utilization, Central South University, Changsha, China

State Key Laboratory of Mineral Processing, Changsha, China

H. Han
Technology Research Center of Hunan Province for Comprehensive Utilization of Associated Fluorite and Fluorine Chemical Engineering, Chenzhou, China

W. Sun et al., *Flotation Chemistry of Tungsten Minerals and Its Application*,
https://doi.org/10.1007/978-981-96-5077-4_4

then affect the flotation behavior; on the other hand, the solution chemistry of tungsten ore affects the adsorption and analytical behavior of flotation reagents. In this chapter, from the perspective of solution chemistry, the dissolution of tungsten and its surface transformation mechanism with associated minerals are analyzed, and the adsorption mechanism of flotation reagents on the surface of minerals is clarified, which provides a theoretical basis for the development of tungsten flotation technology.

4.2 Solution Chemistry of Wolframite and Its and Flotation Significance

Wolframite is the main tungstate mineral of manganese and iron, mainly including $MnWO_4$ and $FeWO_4$. In aqueous solution, manganese (Mn^{2+}) and iron (Fe^{2+}) on the surface of wolframite can undergo dissolution and hydration reactions to form various complex ions and intermediate products. In this section, the solution chemistry of these two main wolframite minerals, $MnWO_4$ and $FeWO_4$, are studied.

4.2.1 Solution Chemistry and Surface Electrical Properties of $MnWO_4$

4.2.1.1 Dissolution Behavior of $MnWO_4$

In the saturated solution of $MnWO_4$, the reaction occurs as shown in Table 4.1. By analyzing the data in the table, it can be seen that H_2WO_4 precipitation and $Mn(OH)_2$ precipitation can occur in $MnWO_4$ under certain conditions, which is determined by the pH value of the solution.

The pH value of $Mn(OH)_2$ precipitation generated from $MnWO_4$ saturated solution can be calculated by Eq. (4.13).

$$MnWO_4(s) + 2OH^- \rightleftharpoons Mn(OH)_2(s) + WO_4^{2-} \quad K_{sp} = 10^{-9.3} \tag{4.13}$$

It can be obtained that, when pH > 9.9, $MnWO_4$ dissolves and transforms into $Mn(OH)_2$ and other components are in equilibrium. The concentrations of each component are shown in Table 4.2.

The pH of H_2WO_4 precipitate produced by a $MnWO_4$ ore saturated solution can be calculated by Eq. (4.14).

$$MnWO_4(s) + 2H^+ \rightleftharpoons H_2WO_4(s) + Mn^{2+} \quad K_{sp} = 10^{5.2} \tag{4.14}$$

Table 4.1 Dissolution reaction and reaction constants of $MnWO_4$ [1–3]

Reaction formula	lg*K*	Formula serial number
$MnWO_4(s) \rightleftharpoons Mn^{2+} + WO_4^{2-}$	−8.85	(4.1)
$Mn^{2+} + OH^- \rightleftharpoons MnOH^+$	3.5	(4.2)
$Mn^{2+} + 2OH^- \rightleftharpoons Mn(OH)_2(aq)$	5.8	(4.3)
$Mn^{2+} + 3OH^- \rightleftharpoons Mn(OH)_3^-$	7.2	(4.4)
$Mn^{2+} + 4OH^- \rightleftharpoons Mn(OH)_4^{2-}$	7.3	(4.5)
$2Mn^{2+} + OH^- \rightleftharpoons Mn_2OH^{3+}$	4.13	(4.6)
$2Mn^{2+} + 3OH^- \rightleftharpoons Mn_2(OH)_3^+$	16.53	(4.7)
$Mn(OH)_2(s) \rightleftharpoons Mn^{2+} + 2OH^-$	−12.6	(4.8)
$H^+ + WO_4^{2-} \rightleftharpoons HWO_4^-$	3.5	(4.9)
$H^+ + HWO_4^- \rightleftharpoons H_2WO_4(aq)$	4.6	(4.10)
$WO_3 + H_2O \rightleftharpoons WO_4^{2-} + 2H^+$	−14.05	(4.11)
$WO_3 + H_2O \rightleftharpoons H_2WO_4(aq)$	5.95	(4.12)

Table 4.2 Solution components and concentrations of $MnWO_4$ at pH > 9.9 [1–3]

Composition	lg*c*
Mn^{2+}	15.4 − 2pH
WO_4^{2-}	−24.25 + 2pH
$MnOH^+$	4.94 − pH
$Mn(OH)_2(aq)$	−6.8
$Mn(OH)_3^-$	−19.4 + pH
$Mn(OH)_4^{2-}$	−32.5 + 2pH
$Mn_2(OH)_3^+$	5.33 − pH
HWO_4^-	−20.75 + pH

It can be obtained that H_2WO_4 precipitation will be generated when pH < 4.8. The concentrations of other components are shown in Table 4.3.

When the pH is between 4.8 and 9.9, neither H_2WO_4 nor $Mn(OH)_2$ precipitates. At this time, the solution components and their concentrations are shown in Table 4.4.

Table 4.3 Solution components and concentrations of $MnWO_4$ when pH < 4.8 [1–3]

Component	lg*c*
WO_4^{2-}	−14.05 + 2pH
HWO_4^-	−10.55 + pH
$H_2WO_4(s)$	−5.95
Mn^{2+}	5.2 − 2pH
$Mn_2(OH)_3^+$	−5.26 − pH
Mn_2OH^{3+}	0.53 + pH

Table 4.4 Solution components and concentrations of $MnWO_4$ at $4.8 < pH < 9.9$ [1–3]

Component	lg$c$
Mn^{2+}	$0.5(\lg K_1 + \lg \alpha(WO_4) - \lg \alpha(Mn))$
WO_4^{2-}	$0.5(\lg K_1 + \lg \alpha(Mn) - \lg \alpha(WO_4))$
$MnOH^+$	$-10.46 + pH + \lg c(Mn^{2+})$
$Mn(OH)_2(aq)$	$-22.2 + 2pH + \lg c(Mn^{2+})$
$Mn_2(OH)_3^+$	$-25.57 + 3pH + 2\lg c(Mn^{2+})$
$Mn(OH)_3^-$	$-34.8 + 3pH + \lg c(Mn^{2+})$
HWO_4^-	$3.5 - pH + \lg c(WO_4^{2-})$
$H_2WO_4(aq)$	$8.1 - 2pH + \lg c(WO_4^{2-})$

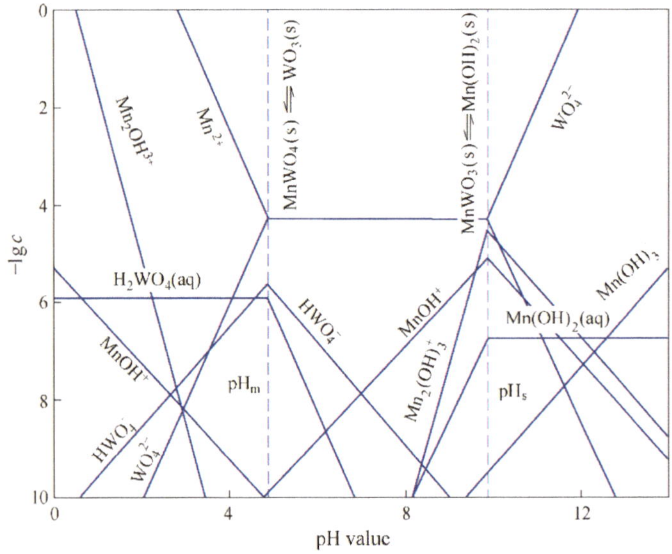

Fig. 4.1 Logarithmic diagram of concentration of dissolved components in saturated solution of $MnWO_4$ [1]

The concentration logarithm of dissolved components of $MnWO_4$ can be drawn according to the formulas for calculating components and their concentrations shown in Tables 4.2, 4.3 and 4.4 (Fig. 4.1).

4.2.1.2 Dissolution and Surface Electrical Properties of $MnWO_4$

The dissolution of $MnWO_4$ will cause the change of solution and surface components, so as to change the surface potential of $MnWO_4$.

When $pH < 2.8$, a large amount of Mn^{2+} on the surface of the $MnWO_4$ is dissolved, and the surface localization ions are $MnOH^+$ and HWO_4^-, and the concentration of

MnOH is higher than that of HWO_4^-, making the $MnWO_4$ surface charged positively. When pH = 2.8, $c(MnOH^+) = c(HWO_4^-)$, that is, the theoretical isoelectric point is 2.8.

2.8 < pH < 4.8, $c(WO_4^{2-}) > c(MnOH^+)$ and $c(HWO_4^-) > c(MnOH^+)$, the surface of $MnWO_4$ is negatively charged, until pH = 4.8, $c(HWO_4^-)$ has a maximum, corresponding to the maximum of negative ζ potential. With the increase of pH, $c(HWO_4^-)$ decreases and $c(MnOH^+)$ increases, and the negative ζ potential decreases.

When pH is between 6.0 and 9.9, the dissolved Mn^{2+} on the surface is greatly reduced, and the main ion concentration relationship is $c(Mn^{2+}) = c(WO_4^{2-})$, which is much larger than $c(HWO_4^-)$and $c(MnOH^+)$. At this time, the ζ potential of the $MnWO_4$ ore surface depends on Mn^{2+} and WO_4^{2-}, but Mn^{2+} is more easily dissolved from the surface than WO_4^{2-}, so the $MnWO_4$ ore surface still has a small negative value in this pH range, which is also called "near zero electric point region". When the pH is greater than 9.9, $c(WO_4^{2-})$ increases sharply, and the negative potential on the surface of the $MnWO_4$ ore also increases sharply. The positioning ion is WO_4^{2-}.

4.2.2 *Solution Chemistry and Surface Electricity of $FeWO_4$*

4.2.2.1 Dissolution Behavior of $FeWO_4$

Similar to $MnWO_4$, the reaction occurs in the saturated solution of $FeWO_4$ as shown in Table 4.5. Under certain conditions, H_2WO_4 precipitation and $Fe(OH)_2$ precipitation can also occur in $FeWO_4$. Again, this precipitation reaction is controlled by the pH of the solution.

Also in the $FeWO_4$ saturated solution, the conditions for $Fe(OH)_2$ precipitation and H_2WO_4 precipitation can be calculated by Eqs. (4.21) and (4.22) as pH > 9.3 and pH < 4.3, respectively, and the equilibrium concentrations of each component

Table 4.5 $FeWO_4$ dissolution reaction and reaction constants [1–3]

Reaction formula	lg*K*	Serial number
$FeWO_4(s) \rightleftharpoons Fe^{2+} + WO_4^{2-}$	−11.04	(4.15)
$Fe^{2+} + OH^- \rightleftharpoons FeOH^+$	4.5	(4.16)
$Fe^{2+} + 2OH^- \rightleftharpoons Fe(OH)_2(aq)$	7.2	(4.17)
$Fe^{2+} + 3OH^- \rightleftharpoons Fe(OH)_3^-$	11.0	(4.18)
$Fe^{2+} + 4OH^- \rightleftharpoons Fe(OH)_4^{2-}$	10.0	(4.19)
$Fe(OH)_2(s) \rightleftharpoons Fe^{2+} + 2OH^-$	−14.95	(4.20)
$H^+ + WO_4^{2-} \rightleftharpoons HWO_4^-$	3.5	(4.9)
$H^+ + HWO_4^- \rightleftharpoons H_2WO_4(aq)$	4.6	(4.10)
$WO_3 + H_2O \rightleftharpoons WO_4^{2-} + 2H^+$	−14.05	(4.11)
$WO_3 + H_2O \rightleftharpoons H_2WO_4(aq)$	−5.95	(4.12)

Table 4.6 Solution components and concentration of $FeWO_4$ when pH > 9.3 [1–3]

Composition	lg*c*
Fe^{2+}	$13.05 - 2pH$
$WO_4{}^{2-}$	$-24.09 + 2pH$
$FeOH^+$	$3.55 - pH$
$Fe(OH)_2(aq)$	-7.75
$Fe(OH)_3{}^-$	$-17.95 + pH$
$Fe(OH)_4{}^{2-}$	$-32.95 + 2pH$
$HWO_4{}^-$	$-20.59 + pH$

Table 4.7 Solution components and concentration of $FeWO_4$ at pH < 4.3 [1–3]

Composition	lg*c*
$WO_4{}^{2-}$	$-14.05 + 2pH$
$HWO_4{}^-$	$-10.55 + pH$
$H_2WO_4(s)$	-5.95
Fe^{2+}	$3.01 - 2pH$
$FeOH^+$	$-6.49 - pH$

are shown in Tables 4.6 and 4.7.

$$FeWO_4(s) + 2OH^- \rightleftharpoons Fe(OH)_2(s) + WO_4^{2-} \quad K_{sp} = 10^{3.75} \tag{4.21}$$

$$FeWO_4(s) + 2H^+ \rightleftharpoons H_2WO_4(s) + Fe^{2+} \quad K_{sp} = 10^{3.01} \tag{4.22}$$

At 4.3 < pH < 9.3, neither H_2WO_4 nor $Fe(OH)_2$ precipitates in the $FeWO_4$ saturated solution. At this time, the solution components and their concentrations are shown in Table 4.8.

Table 4.8 Solution components and concentrations of $FeWO_4$ at 4.3 < pH < 9.3 [1–3]

Composition	lg*c*
Fe^{2+}	$0.5(\lg K_2 + \lg\alpha(WO_4) - \lg\alpha(Fe))$
$WO_4{}^{2-}$	$0.5(\lg K_2 + \lg\alpha(Fe) - \lg\alpha(WO_4))$
$FeOH^+$	$-9.5 + pH + \lg c(Fe^{2+})$
$Fe(OH)_2(aq)$	$-20.8 + 2pH + \lg c(Fe^{2+})$
$Fe(OH)_3{}^-$	$-31 + 3pH + \lg c(Fe^{2+})$
$Fe(OH)_4{}^{2-}$	$-46 + 4pH + \lg c(Fe^{2+})$
$HWO_4{}^-$	$3.5 - pH + \lg c(WO_4{}^{2-})$
$H_2WO_4(aq)$	$8.1 - 2pH + \lg c(WO_4{}^{2-})$

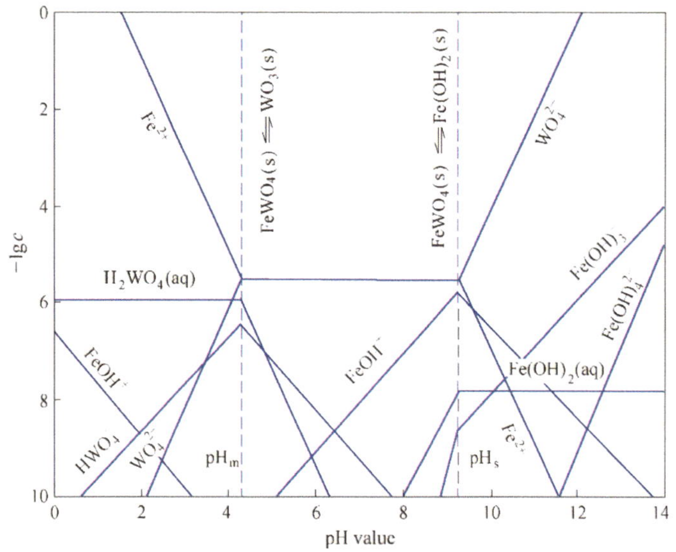

Fig. 4.2 Logarithm diagram of the concentration of dissolved components of $FeWO_4$ [1]

According to the solution components and concentrations at different pH values shown in Tables 4.6, 4.7 and 4.8, the concentration logarithm of dissolved components of $FeWO_4$ as shown in Fig. 4.2 can be drawn.

4.2.2.2 Dissolution and Surface Electrical Properties of $FeWO_4$

Similar to $MnWO_4$, the dissolution of $MnWO_4$ will directly affect the surface positioning ion and surface potential.

When pH < 2.0, Fe^{2+} on the surface of $FeWO_4$ is dissolved in large quantities, the surface charge is positive, the surface positioning ions are $FeOH^+$ and HWO_4^-, and the theoretical isoelectric point is pH = 2.0 of $c(FeOH^+) = c(HWO_4^-)$.

When pH > 2.0, the concentration of $HWO_4^-$is greater than that of $FeOH^+$, and the surface of the $FeWO_4$ is negatively charged, and the maximum values of $c(HWO_4^-)$and negative ζ potential appear at pH = 4.3. When pH < 6.0, the negative ζ potential decreases with $c(HWO_4^-)$.

The near-zero electric point region appears in the pH = 6.0–9.3 range. When pH is greater than 9.9, the negative potential on the surface of $FeWO_4$ increases with the increase of the concentration of located ion WO_4^{2-}.

4.2.3 *Solution Chemistry and Flotation Behavior of Wolframite*

According to the solution chemical calculation of wolframite ($MnWO_4$ and $FeWO_4$) and the influence of its dissolution on the surface potential, it can be seen that in aqueous solution or pulp, wolscheelite will inevitably undergo hydrolysis reaction, release Mn^{2+} and Fe^{2+}, and cause a series of dissolution reactions controlled by pH, changing the electrical properties of mineral surface and the positioning ions of mineral surface. According to the dissolved component diagram of wolframite, the favorable pH range of wolframite flotation can be accurately analyzed. As shown in Fig. 4.3a, the theoretical IEP of $MnWO_4$ is 2.8, and that of $FeWO_4$ is 2.0. The IPE of actual wolstenite fluctuates between 2.0 and 2.8 due to the influence of iron and manganese content. When pH < IEP, a large number of Mn^{2+} and Fe^{2+} are dissolved from the surface of wolframite, and the surface positioning ions are $MnOH^+$, $FeOH^+$ and $HWO_4{}^-$, and the concentrations of $MnOH^+$ and $FeOH^+$ are higher than the concentration of $HWO_4{}^-$. The surface of wolframite is positively charged, but the anionic collector is not dissociated under this condition, so it is difficult to adsorb on the surface of wolframite by electrostatic force. When IEP < pH < 6, the surface of wolframite negative potential appeared extreme value, this is because the concentration of $HWO_4{}^-$ increased with the increase of pH and then decreased. The high negative charge on wolframite surface and the positioning ions $MnOH^+$, $FeOH^+$, $HWO_4{}^-$ are not conducive to the electrostatic adsorption and chemisorption of anionic collector; when 6 < pH < 9.5, the surface of wolframite is low electronegativity, and the positioning ions, Mn^{2+} and Fe^{2+}, are conducive to the anionic collector through electrostatic adsorption. At the same time, the collector also through the chemical bond with the positioning ion used in wolframite surface chemical adsorption. When pH > 9.5, as (Fe, Mn)WO_4 and OH^- Generating $Mn(OH)_2$ and $Fe(OH)_2$ precipitates, $WO_4{}^{2-}$ as a positioning ion, the surface negative potential increases dramatically, causing the anion collector difficult to adsorb. As shown in Fig. 4.3b, whether by electrostatic adsorption of anionic collectors or by bonding with iron and manganese ions on the surface of wolframite and chemically adsorbed collectors, the best flotation effect was found in the pH interval (6–9.5) of the "Near Zero Power Point Area" of the wolframite.

On this basis, the mechanism of flotation reagents on wolframite surface can be explained by solution chemical analysis combined with collector. As shown in Fig. 4.4a, the OL^-, $(OL)_2^{2-}$ and $H(OL)_2^-$ increased significantly in the pH range of 6.0–9.5, where oleic acid flotation effect was good, indicating that oleic acid ion and ion-molecular association in the solution played a leading role in wolframite flotation process. As shown in Fig. 4.4b, when pH = 6.0–9.5, the negative standard free energies of $Fe(OL)_2$ and $Mn(OL)_2$ generated by the lattice cation on the surface of wolframite and the oleic acid collector anion are the highest, which indicates that oleic acid is adsorbed on the surface of wolframite through chemical reaction with surface Mn and Fe particles to form oleate. As shown in Fig. 4.4c, d, the pH range with the highest concentration of $Fe(OL)_2$ and $Mn(OL)_2$ components is consistent with

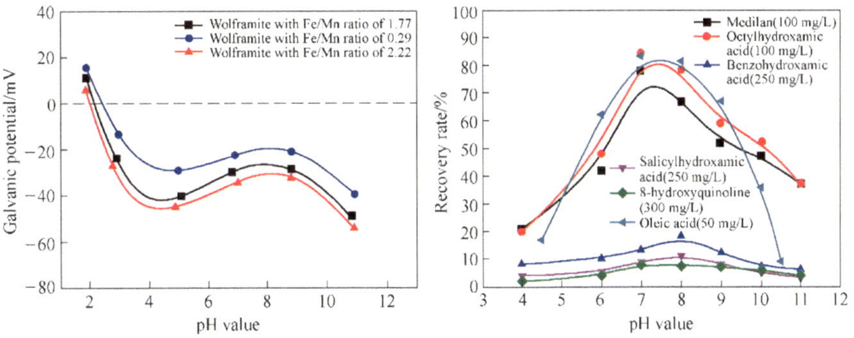

Fig. 4.3 Relationship between zeta–potential, flotation recovery and pH value of wolframite [4, 5]

the pH range with the best flotation effect of wolframite, indicating that $Fe(OL)_2$ and $Mn(OL)_2$ precipitation generated by surface adsorption of oleic acid and wolframite may be the main form of oleic acid adsorption on wolframite surface [6].

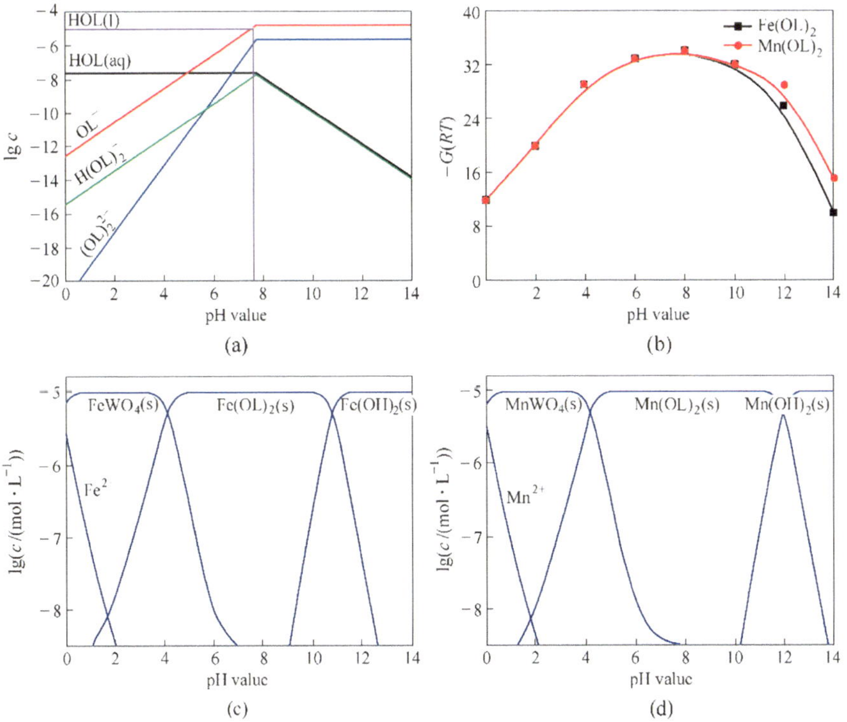

Fig. 4.4 Oleic-wolframite flotation system **a** oleic acid lg$c$-pH diagram (10^{-5} mol/L), **b** wolframite surface oleate-pH diagram and **c** ΔG^{Θ}, **d** metal cation lg$c$-pH diagram [6]

Feng et al. [7, 8] analyzed the adsorption mechanism of octyl hydroxamic acid on wolframite surface by means of solution chemistry calculation. As shown in Fig. 4.5, the flotation effect of octyl hydroxamic acid on woltungsten is closely related to the precipitation of metal octyl hydroxamic acid $Fe(OHA)_2$ and $Mn(OHA)_2$ generated on the mineral surface. Under acidic conditions, the concentration of metal octyl hydroxamic acid precipitates are lower, and correspondingly, the recovery rate of wolframite and the adsorption amount of collector are lower. This is mainly because the solubility of woltungsten is low under acidic conditions, and OHA exists mainly in the form of molecules, the presence of OHA ions is less, unable to form enough metal octyl hydroxamic acid precipitate on the surface of woltungsten, resulting in the decrease of the adsorption of agents and mineral floatability on the surface of woltungsten. In the range of pH value of 8–10, the concentration of metal octyl hydroxamic acid precipitation is higher than metal tungstate and metal hydroxyl compounds, at this time metal octyl hydroxamic acid is the main component of wolscheelite surface. The octyl hydroxamic acid anion is adsorbated to wolscheelite surface by exchanging the OH^- ions of $MeOH^-$ and the ions of $MeWO_4$, thus improving the wolscheelite flotsability WO_4^{2-} [9]. When the pH exceeds 10, the concentration of metal octyl hydroxamic acid precipitates decreases and the concentration of metal hydroxyl compounds increases. When pH > 11, the concentration of metal-hydroxyl compounds is higher than that of metal-octyl hydroxamic acid. At this time, the surface of woltungsten is mainly precipitated by metal-hydroxyl compounds, so the hydrophilicity increases and the floatability decreases.

Wang Dianzuo et al. [1] studied the effect of complex ions generated by flotation reagents and metal ions on wolframite flotation. As shown in Fig. 4.6, in the flotation solution citric acid reacted with Mn^{2+} dissolved from wolframite and Ca^{2+} dissolved from calcite to form MnL^- and CaL^- complexation ions. In the range where the concentration of these complex ions increases significantly, the flotation recovery of wolframite and calcite drops sharply. Under the same dosage of citric acid inhibitor, CaL^- is easier to form than MnL^-. Therefore, in flotation, calcite can be preferentially inhibited by controlling citric acid dosage in a certain concentration range, without affecting the flotation effect of wolframite.

4.3 Solution Chemistry and Flotation Significance of Scheelite

4.3.1 Solution Chemistry and Surface Electricity of Scheelite

Scheelite is a typical salt mineral, with the common characteristic of salt minerals solubility [10]. In an aqueous solution, scheelite dissolves and releases calcium ions (Ca^{2+}) and tungstate ions ($WO_4{}^{2-}$). The dissolution reaction of scheelite in aqueous solution and the corresponding reaction constants are shown in Table 4.9. The anion in the chemical component of scheelite is tungstate $WO_4{}^{2-}$, which belongs to the weak

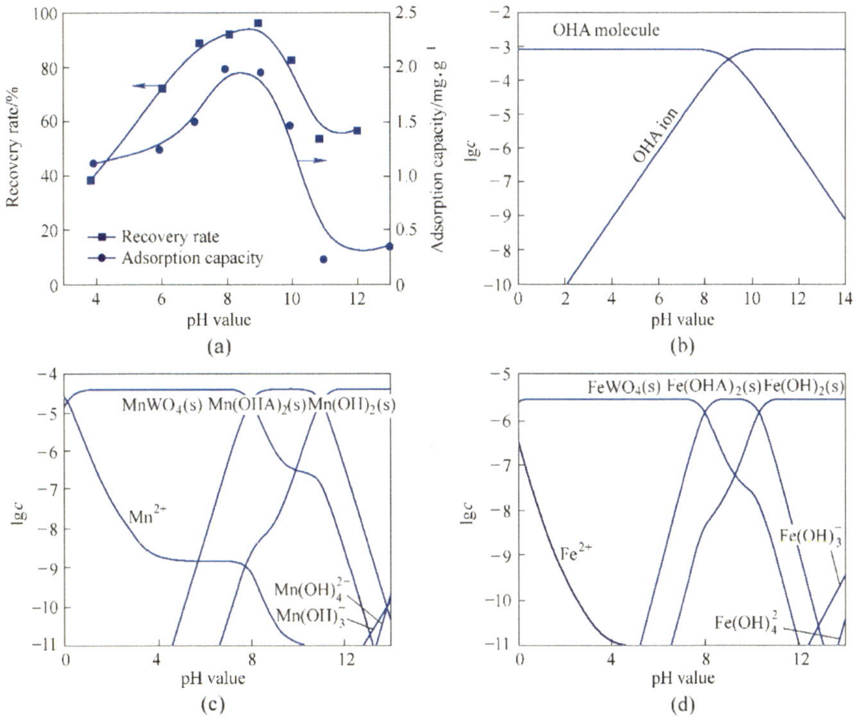

Fig. 4.5 Octylhydroxamic acid-wolframite flotation system **a** flotation recovery and reagent adsorption, **b** collector lg*c*-pH and **c**, **d** metal cation lg*c*-pH [7, 8]

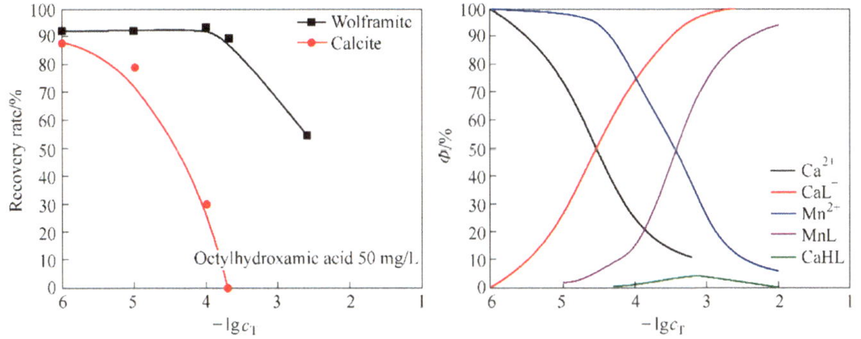

Fig. 4.6 Flotation recovery of wolframite and calcite and distribution of complex ions in flotation solution when citric acid is the inhibitor [1]

acid ion and will undergo hydrolysis reaction with water molecules after dissolution and release.

According to a series of reactions of scheelite in aqueous solution and corresponding equilibrium constants, the diagram of dissolved components of scheelite in saturated aqueous solution as shown in Fig. 4.7 can be obtained. It can be seen that the dissolution of scheelite in aqueous solution is controlled by the pH of the solution, which is a function of pH [13]. When pH < 1.3, the solution contains $Ca(OH)^+$and HWO_4^-, then $c(Ca(OH)^+) > c(HWO_4^-)$, in addition to the solution will obviously generate a trace of HWO precipitation; When pH = 1.3, the concentration of $Ca(OH)^+$ in the solution is the same as that of HWO_4^-, and the isoelectric point IEP = 1.3. When the pH value of the solution is higher than 2 but lower than 4.7, there is a large amount of free H^+ in the saturated aqueous solution of schetungsten, and the hydrolysis reaction Eqs. (4.27) and (4.28) are dominant, so that H_2WO_4 of tungstate exists in the whole solution system in the form of precipitation. In addition, the concentrations of HWO_4^- and WO_4^{2-} increase with the increase of pH. And the concentration of Ca^{2+} decreases with the increase of pH. When the pH value of the solution is higher than 13.7, there is a large amount of free OH^- in the saturated aqueous solution of schetungsten, and the excess OH^- in the solution will combine with the preferential dissociation of Ca^{2+}, forming calcium hydroxide $Ca(OH)_2$ precipitation, and the Eq. (4.25) is the main side reaction. When the pH of the solution is between 4.7 and 13.7, the dominant reactions are Eqs. (4.24) and (4.27). At this time, there is neither H_2WO_4 precipitation nor $Ca(OH)_2$ precipitation in the scheelite saturated solution, and the existing ions are mainly Ca^{2+}, WO_4^{2-}, HWO_4^- and $CaOH^+$ [11, 14, 15].

It is worth emphasizing that the dissolution behavior of scheelite in aqueous solution also involves the preferential dissociation of WO_4^{2-}, the tungstate anion to which calcium ion binds. Studies have shown that when scheelite comes into contact with water, the priority of Ca^{2+} dissociation is higher than that of WO_4^{2-} dissociation, that is, calcium ion Ca^{2+} occurs before tungstate ion WO_4^{2-}, which results in the absence or insufficient Ca^{2+} and excess WO_4^{2-} on the surface of scheelite, resulting in a negative surface charge of scheelite [16].

Table 4.9 Dissolution reaction and reaction constants of scheelite in aqueous solution [1, 11, 12]

Reaction formula	lg*K*	Serial number
$CaWO_4(s) \rightleftharpoons Ca^{2+} + WO_4^{2-}$	−9.3	(4.23)
$Ca^{2+} + OH^- \rightleftharpoons CaOH^+$	1.40	(4.24)
$Ca^{2+} + 2OH^- \rightleftharpoons Ca(OH)_2(aq)$	2.77	(4.25)
$Ca(OH)_2(aq) \rightleftharpoons Ca^{2+} + 2OH^-$	−5.22	(4.26)
$H^+ + WO_4^{2-} \rightleftharpoons HWO_4^-$	8.5	(4.27)
$H^+ + HWO_4^- \rightleftharpoons H_2WO_4(aq)$	4.6	(4.28)
$WO_3 + H_2O \rightleftharpoons WO_4^{2-} + 2H^+$	−14.05	(4.29)
$CaWO_4(s) + 2OH^- \rightleftharpoons Ca(OH)_2(s) + WO_4^{2-}$	−4.08	(4.30)

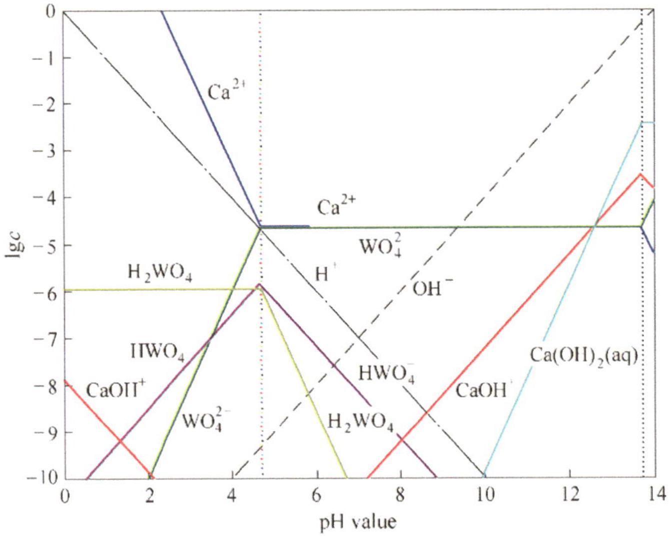

Fig. 4.7 Diagram of dissolved components in scheelite saturated solution [1, 10]

4.3.2 Solution Chemistry and Flotation Behavior of Scheelite

According to the calculation of scheelite's solution chemistry, in aqueous solution or pulp, scheelite will inevitably undergo hydrolysis reaction, release Ca^{2+} and cause a series of dissolution reactions controlled by pH, changing the electrical properties of mineral surface and the positioning ion of mineral surface. According to the dissolved component diagram of scheelite, the theoretical isoelectric point of scheelite IEP = 1.3. When pH < 1.3, the surface positioning ions of scheelite are $Ca(OH)^+$ and HWO_4^-, and then $c(Ca(OH)^+) > c(HWO_4^-)$, and the scheelite surface is charged positive point. When 1.3 < pH < 4.7, $c(HWO_4^-)$ and $c(WO_4^-)$ increase gradually, and when pH = 4.7, the maximum value appears, and the negative potential value of scheelite surface reaches the maximum value. When 4.7 < pH < 13.7, $c(Ca^{2+}) = c(WO_4^-) > c(Ca(OH)^+)$and $c(HWO_4^-)$, the surface potential value of scheelite remains unchanged. At this time, the scheelite surface positioning ions are Ca^{2+} and WO_4^{2-}, among which Ca^{2+} is easier to dissolve from the surface than WO_4^{2-}, so that there is excess WO_4^{2-} on the scheelite surface, forming a negatively charged positioning ion layer. When the anionic collector is used to flotation scheelite, it is not conducive to the electrostatic adsorption of the collector. At this time, the flotation of scheelite by the anionic collector is mainly through the chemical bonding with the Ca^{2+} dissolved on the surface [17].

The surface potential of scheelite is more negative than that of calcite, fluorite and other calcium-containing minerals, so the separation of scheelite from other calcium-containing minerals can be realized by using cationic collector. The pH range of floatation of scheelite by cationic collector such as alkyl amine can be analyzed by solution chemistry calculation, and the flotation practice can be guided. As shown in Fig. 4.8, when pH < 9.5, dodecylamine has a better flotation effect on scheelite. Combined with the analysis of the chemical components of dodecylamine solution, it can be seen that the main component are RNH_3^+, $(RNH_3^+)_2^{2+}$ and other amine root ions. For fluorite and calcite, ion-molecular associations play a major role. According to the calculation of solution chemistry, Wang and Hu et al. [18] studied the solution chemical behavior of alkyl amines and the collection performance of alkali earth metal salt minerals such as scheelite. Alkyl amines adsorbed on the surface of negatively charged minerals through electrostatic action, and then chemically reacted with mineral lattice anions to generate ammonium salt precipitation. As shown in Fig. 4.9 and Table 4.10, the upper pH value of flotation of salt minerals by alkyl amines corresponds to the critical value of precipitation of amine molecules.

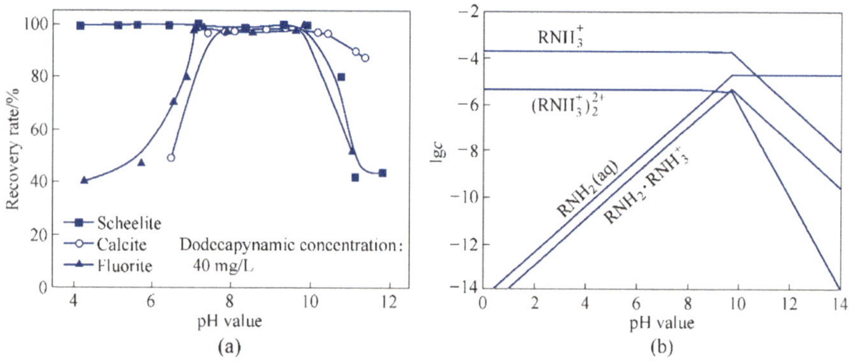

Fig. 4.8 Recovery rate of calcium-bearing minerals in flotation of dodecylamine solution (**a**) and logarithmic concentration of its dissolved component (**b**) [19]

Fig. 4.9 Relationship between pH and recovery rate of scheelite flotation by different alkyl amines [19]

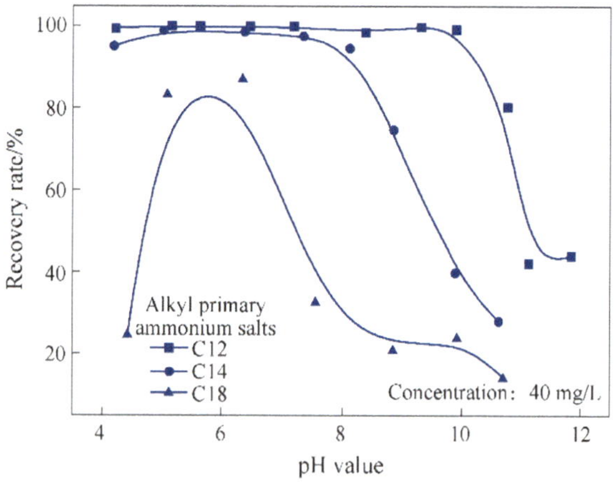

Table 4.10 pH values of precipitating alkyl amines ($c_T = 5 \times 10^{-5}$ mol/L) [18]

Alkyl amines	Decamines	Dodecamine	Tetradecylamine	Cetylamine	Octadecamide
Solubility/ $mol \cdot L^{-1}$	$10^{-3.3}$	$10^{-4.7}$	$10^{-8.0}$	$10^{-7.1}$	$10^{-8.3}$
Precipitation pH	–	10.5	8.9	7.9	6.6
Dissociation constant p*K*a	10.64	10.63	10.60	10.60	10.60

Based on the analysis of solution chemistry of scheelite and flotation reagent, the components and mechanism of flotation reagent in the process of scheelite flotation can be explained. Qiu et al. [20] analyzed the flotation mechanism of scheelite by benzohydroxamic acid in combination with the chemical calculation in flotation solution. As shown in Fig. 4.10, in the range of pH = 7–10, benzohydroxamic acid has the strongest capturing capacity of scheelite, because the concentration of anion produced by the dissociation of benzohydroxamic acid is similar to that of neutral molecules, and co-adsorption occurs on the surface of scheelite. At the same time, the metal hydroxamic acid is re-adsorbed on the mineral surface in the interface region [21], both of which promote the adsorption of benzohydroxamic acid, the thickness of the adsorption layer of the reagent increases, and the hydrophobicity of scheelite is enhanced.

In scheelite flotation, sodium silicate is often used as an inhibitor to inhibit gangue minerals such as fluorite and calcite. Sun et al. [22] calculated the component concentration diagram of mineral-sodium silicate-water system, and explained the mechanism of using sodium silicate inhibitor to selectively flotation scheelite from fluorite from another angle. As shown in Fig. 4.11, when the concentration of sodium silicate exceeds $10^{-3.34}$ mol/L in the scheelite system, calcium silicate precipitation begins to form, while in the fluorite system, calcium silicate precipitation only needs 10 mol/L

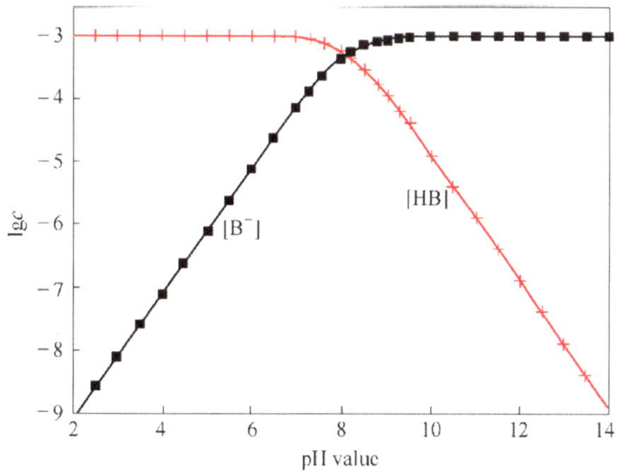

Fig. 4.10 Logarithmic diagram of the concentration of each component of benzohydroxamic acid solution (15 °C, $c_{BHA} = 1 \times 10^{-3}$ mol/L) [20]

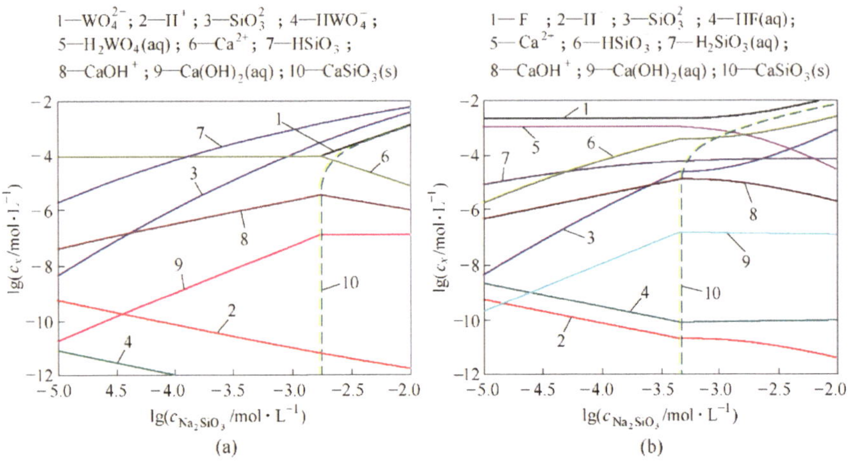

Fig. 4.11 Concentration distribution of components in scheelite-sodium silicate-water system and fluorite-sodium silicate-water system [22]

sodium silicate concentration. Sodium silicate is more likely to form calcium silicate precipitation on the surface of fluorite, reducing its flotation effect.

4.3.3 Surface Conversion and Flotation Behavior of Calcium-Containing Minerals

In monomineral systems, the dissolution of salt minerals will dissociate different ions and change the composition of the solution [23–25]. In the system where two or more minerals coexist, the ions dissociated by the dissolution of one mineral will adsorb on the surface of another or several minerals and undergo chemical reactions, resulting in certain changes in the composition and properties of the mineral surface [23, 26]. If the reaction time is sufficient and the reaction conditions are sufficient, the affected mineral surface will gradually approach to the dissolved mineral. And even transform into dissolved minerals, that is, to complete the modification and transformation process of the mineral surface.

Scheelite is often closely associated with calcium-containing minerals such as calcite and fluorite, which all have the same surface-active point Ca^{2+} and similar surface chemical properties. At the same time, scheelite, calcite and fluorite all belong to soluble salt minerals with high solubility, and there is mutual conversion between calcium-containing minerals, which makes the flotation behavior of each mineral very complicated and the flotation separation between calcium-containing minerals more difficult. On the one hand, the ions produced by the dissolution of minerals will be hydrolyzed, affecting the adjustment of pulp pH value; on the other hand, $WO_4{}^{2-}$, $F^-$and $CO_3{}^{2-}$ produced by the dissolution of scheelite, fluorite and calcite can chemically react with the calcareous points on the mineral surface, forming another calcium-containing salt mineral on the surface, resulting in the mutual transformation of the three mineral surfaces.

4.3.3.1 Surface Conversion of Scheelite and Fluorite

The hydrolytic reaction of scheelite with fluorite in an aqueous solution is shown below.

$$CaWO_4 \rightleftharpoons Ca^{2+} + WO_4^{2-} \quad K_{sp}(CaWO_4) = 10^{-9.3} \tag{4.31}$$

$$CaF_2 \rightleftharpoons Ca^{2+} + 2F^- \quad K_{sp}(CaF_2) = 10^{-10.41} \tag{4.32}$$

By combining Eqs. (4.31) and (4.32), the reaction equilibrium formula of fluorite-scheelite system can be obtained:

$$CaF_2 + WO_4^{2-} \rightleftharpoons CaWO_4 + 2F^- \tag{4.33}$$

Based on the equation and solubility product constants K_{sp} of CaF_2 and $CaWO_4$, the equilibrium constant K of the Eq. (4.33) can be found:

$$K = \frac{c^2(F^-)}{c\left(WO_4^{2-}\right)} = \frac{c(Ca)c^2(F^-)}{c(Ca)c\left(WO_4^{2-}\right)} = \frac{K_{sp}(CaF_2)}{K_{sp}(CaWO_4)} = 10^{-1.11} \tag{4.34}$$

Combined with thermodynamics, the Gibbs free energy of the fluorite-calcite surface transformation reaction can be determined, as shown in Eq. (4.35).

$$\Delta G = -RT\ln K + RT\ln\frac{c^2(F^-)}{c\left(WO_4^{2-}\right)} \tag{4.35}$$

If $\Delta G < 0$, the Eq. (4.33) automatically proceeds to the right, and schewolite dissolved $WO_4{}^{2-}$ will react with fluorite surface CaF_2 to form $CaWO_4$ under the condition of

$$-RT\ln K + RT\ln\frac{c^2(F^-)}{c\left(WO_4^{2-}\right)} < 0 \tag{4.36}$$

i.e.

$$\lg c\left(WO_4^{2-}\right) > 1.11 + 2\lg c^2\left(F^-\right) \tag{4.37}$$

In the pulp where fluorite and scheelite coexist, if the concentration of $WO_4{}^{2-}$ produced by the dissolution of scheelite meets the condition of Eq. (4.37), then the

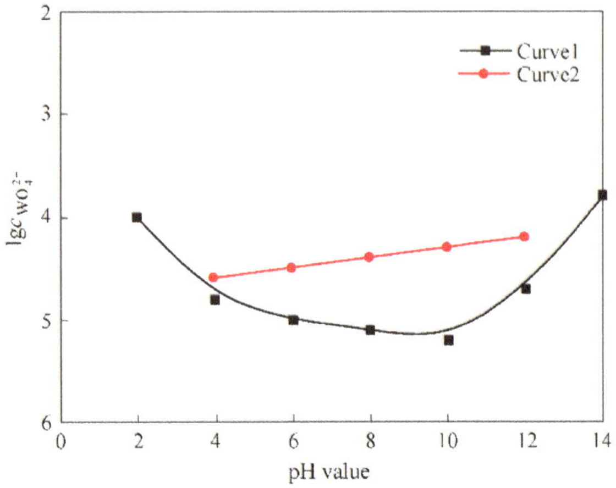

Fig. 4.12 Surface transformation equilibrium diagram of fluorite-scheelite system (1: fluorite; 2: scheelite) [12]

Eq. (4.33) proceeds to the right and $CaWO_4$ precipitates will be generated on the surface of fluorite. According to the conditions determined by the Eq. (4.37), the critical surface conversion curve of the fluorite-scheelite system can be obtained, as shown in Fig. 4.12. When the curve of the relationship between $WO_4{}^{2-}$concentration and pH value in actual pulp is above curve 1, then the condition of Eq. (4.37) is satisfied, and the surface conversion of fluorite to scheelite occurs. Curve 2 in Fig. 4.12 shows the relationship between $WO_4{}^{2-}$ concentration and pH in the clarified liquid from tungsten ore measured and calculated. It can be seen that in the range of pH > 4, curve 2 is always above curve 1, indicating that the $WO_4{}^{2-}$concentration of scheelite dissolved in fluorite-scheelite pulp meets the condition of Eq. (4.37) when pH > 4, and $WO_4{}^{2-}$reacts on the surface of fluorite to form $CaWO_4$. Therefore, pH = 4 is the critical pH value for the conversion of fluorite to scheelite. For the reversible reaction to the left of Eq. (4.33), the critical curve can also be obtained, but for the fluorite-scheelite system, the reaction cannot occur at any pH measured by experiments, so the conditions and curve for the reversible reaction to the left of Eq. (4.33) are not listed here [11]. This shows that the mutual conversion behavior of minerals is closely related to the concentration of ions in the solution.

4.3.3.2 Surface Conversion of Scheelite and Calcite

The hydrolysis reaction formula for calcite is shown below.

$$CaCO_3 \rightleftharpoons Ca^{2+} + CO_3^{2-} \quad K_{sp}(CaCO_3) = 10^{-8.35} \tag{4.38}$$

By combining Eqs. (4.33) and (4.38), the solution chemical reaction equilibrium formula of the calcite—scheelite system can be obtained, as shown in Eq. (4.39).

$$CaCO_3 + WO_4^{2-} \rightleftharpoons CaWO_4 + CO_3^{2-} \tag{4.39}$$

Based on the equation and solubility product constants K_{sp} of $CaCO_3$ and $CaWO_4$, the equilibrium constant K of the Eq. (4.39) can be found:

$$K = \frac{c(CO_3^{2-})}{c(WO_4^{2-})} = \frac{c(Ca^{2+})c(CO_3^{2-})}{c(Ca^{2+})c(WO_4^{2-})} = \frac{K_{sp}(CaCO_3)}{K_{sp}(CaWO_4)} = 10^{0.95} \tag{4.40}$$

Combined with thermodynamics, the Gibbs free energy of scheelite-calcite surface conversion reaction can be determined, as shown in Eq. (4.41).

$$\Delta G = -RT \ln K + RT \ln \frac{c(CO_3^{2-})}{c(WO_4^{2-})} \tag{4.41}$$

If $\Delta G < 0$, the Eq. (4.39) automatically proceeds to the right, and scheelite dissolved $WO_4{}^{2-}$ will react with calcite surface $CaCO_3$ to form $CaWO_4$ under conditions of

$$-RT \ln K + RT \ln \frac{c(CO_3^{2-})}{c(WO_4^{2-})} < 0 \tag{4.42}$$

i.e.,

$$\lg c(WO_4^{2-}) > \lg c(CO_3^{2-}) - 0.95 \tag{4.43}$$

In the pulp where calcite and scheelite coexist, if the concentration of $WO_4{}^{2-}$ produced by scheelite dissolution meets the condition of Eq. (4.43), then the Eq. (4.39) goes to the right, $CaWO_4$ precipitates will be generated on the surface of fluorite. Conversely, if the $WO_4{}^{2-}$concentration produced by the dissolution of scheelite does not satisfy the condition of Eq. (4.43), the Eq. (4.39) proceeds to the left.

In an open system, Eq. (4.44) can be obtained.

$$\lg c(CO_3^{2-}) = -21.64 + 2\text{pH} \tag{4.44}$$

Thus Eq. (4.45)

$$\lg c(WO_4^{2-}) > -22.59 \overset{\leftarrow}{+} 2\text{pH} \tag{4.45}$$

According to Eq. (4.45), the critical surface conversion curve of the calcite—scheelite system can be obtained, as shown in Fig. 4.13. It can be seen from the figure that the intersection of critical curve 1 with $WO_4{}^{2-}$ concentration in scheelite clarifying liquid and critical curve 2 with $CO_3{}^{2-}$concentration in calcite clarifying liquid is pH = 8.8. When pH < 8.8, the $WO_4{}^{2-}$concentration is always higher than the critical curve 1, indicating that the $WO_4{}^{2-}$concentration satisfies the Eq. (4.45), and $CaWO_4$ precipitates will occur on the surface of calcite in the scheelite clarifying

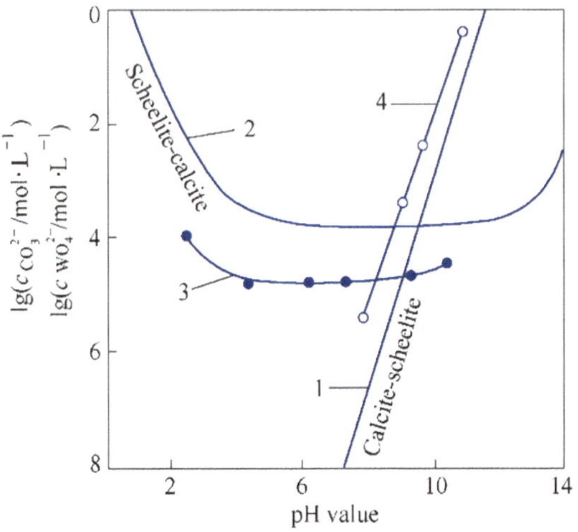

Fig. 4.13 Surface conversion equilibrium diagram of calcite-scheelite system (critical curves of 1, 2-surface conversion, $WO6_4^{2-}$ concentration in 3-scheelite clarifying liquid, CO_3^{2-} concentration in 4-calcite clarifying liquid) [14]

solution. When pH > 8.8, the concentration of $CO_3^{2-}$is always higher than the critical curve 2, indicating that the concentration of CO_3^{2-}meets the reverse reaction of Eq. (4.39), at this time, $CaCO_3$ precipitation will be generated on the surface of scheelite in the calcite clarifying solution.

4.3.3.3 Surface Conversion of Fluorite to Calcite

Scheelite is usually accompanied by fluorite and calcite. Surface conversion can occur not only between scheelite and fluorite and calcite, but also between fluorite and calcite. The main hydrolysis reactions of fluorite and calcite in aqueous solution are shown in Eqs. (4.32) and (4.38). Combining the hydrolysis reactions of these two associated minerals, the surface conversion equation of fluorite and calcite can be obtained as shown below.

$$CaF_2 + CO_3^{2-} \rightleftharpoons CaCO_3 + 2F^- \tag{4.46}$$

According to the equation and solubility product constant K_{sp} of CaF_2 and $CaCO_3$, the equilibrium constant K of Eq. (4.46) can be obtained:

$$K = \frac{c^2(F^-)}{c(CO_3^{2-})} = \frac{c(Ca)c^2(F^-)}{c(Ca)c(CO_3^{2-})} = \frac{K_{sp}(CaF_2)}{K_{sp}(CaCO_3)} = 10^{-2.06} \tag{4.47}$$

According to the above solution chemistry calculation, when the concentration of carbonate increases, the surface conversion reaction of fluorite and calcite moves in the direction of calcite formation, and the surface of fluorite gradually transforms into calcite structure. However, when the concentration of fluoride ion increases, the reaction goes in the direction favorable to fluorite, and the surface of calcite gradually transforms into the structure of fluorite. The forward reaction and the reverse reaction increase and decrease, and finally achieve dynamic equilibrium.

4.3.3.4 Significance of Surface Conversion of Calcium-Containing Minerals for Flotation Separation

1. **The change of mineral flotation behavior**

The surface transformation of scheelite and associated minerals directly affects the flotation behavior of minerals. Hu and Zhang et al. [23, 24, 26, 27] found that the surface conversion of salt minerals caused by the surface dissolution balance and the new products generated by it make the surface characteristics of salt minerals tend to be the same and complicated, which is the main reason why flotation separation is difficult in the coexistence system of a variety of salt minerals.

Wang et al. [28] studied the flotation behavior after surface conversion of scheelite, calcite and fluorite, and found that the conversion reaction of fluorite on the surface of scheelite basically does not occur, while the conversion reaction of fluorite on the surface of calcite is easy and obvious. The conversion of fluorite on the surface of calcite will generate calcium fluoride film on the surface of calcite, so that the flotation behavior of calcite is close to that of fluorite. In this study, natural fluorite minerals were used to convert calcite surface into CaF_2 by taking advantage of the fact that Pb-BHA coordination collector has good selectivity and basically has no ability to collect fluorite [29, 30], thus achieving the separation of scheelite and calcite [31]. The results are shown in Fig. 4.14.

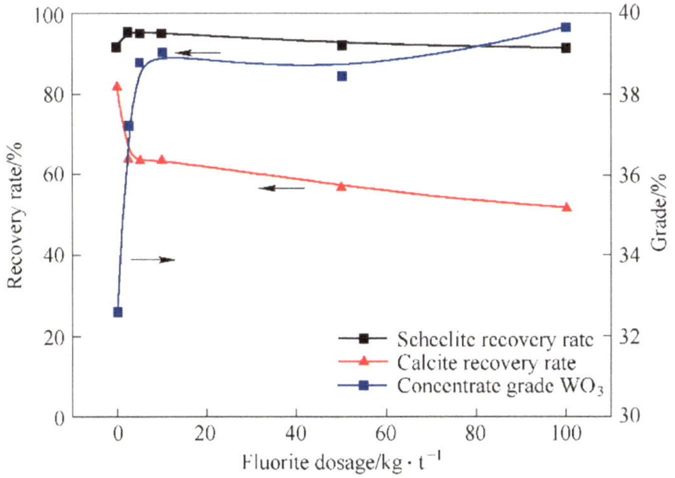

Fig. 4.14 Effect of fluorite on flotation separation of scheelite-calcite mixed ore, reprinted from ref [12, 28], copyright 2024, with permission from Minerals Engineering

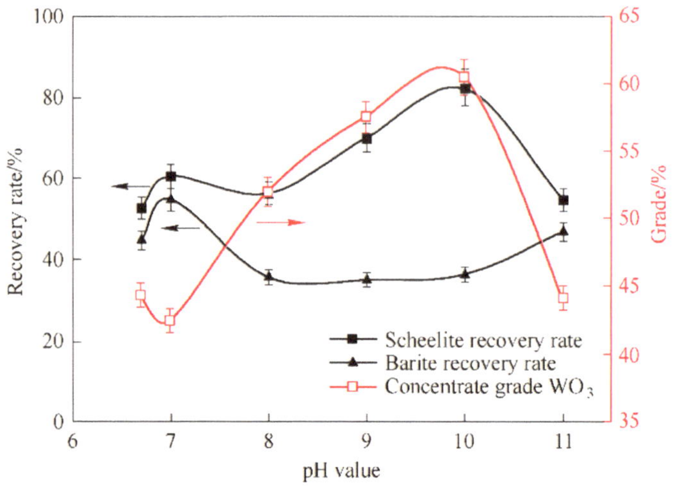

Fig. 4.15 Effect of fluorite on flotation separation of scheelite-barite mixed ore, reprinted from ref. [32], copyright 2024, with permission from Minerals Engineering

Similarly, Wang et al. [32] studied the flotation behavior after three surface conversions of scheelite, barite and fluorite, and found that the conversion of fluorite on the surface of barite is more likely to occur than that of fluorite on scheelite surface to form fluoride BaF_2 film. At the same time, the fluoride BaF_2 formed after surface conversion will not be captured under the Pb-BHA coordination collector system. Based on this, the floatation separation of scheelite and barite was realized by using Pb-BHA as collector and fluorite as inhibitor. The results are shown in Fig. 4.15.

2. **Changes in the electrical properties of the mineral surface**

In a system where two or more salt minerals coexist, the surface electrical properties of one or more minerals are affected by the dissolved components of the other. When a surface transformation occurs, that is, when other minerals are formed on the surface of the transformed mineral, the surface electrical properties of the mineral will tend to be similar to the surface electrical properties of the generated mineral, and the surface properties will change significantly.

The zero electric point of scheelite is 1.8, and the zero electric point of fluorite is 7.0, but the zero electric point of fluorite in scheelite clarified liquid decreases to 4.5. The change of Zeta potential with pH value is gradually close to the electrokinetic behavior of scheelite in deionized water, which is mainly attributed to the tungstate ion WO_4^{2-} dissociated from the surface of scheelite. On the surface of the fluorite adsorption and chemical reaction, the formation of calcium tungstate $CaWO_4$ precipitate, so that the chemical composition of the surface of the fluorite and surface electrical properties have scheelite like properties. But scheelite in the saturated solution of fluorite, at any pH measured by experiments, it is difficult for the reverse reaction of Eq. (4.33) to occur, that is, F^- is difficult to chemically react

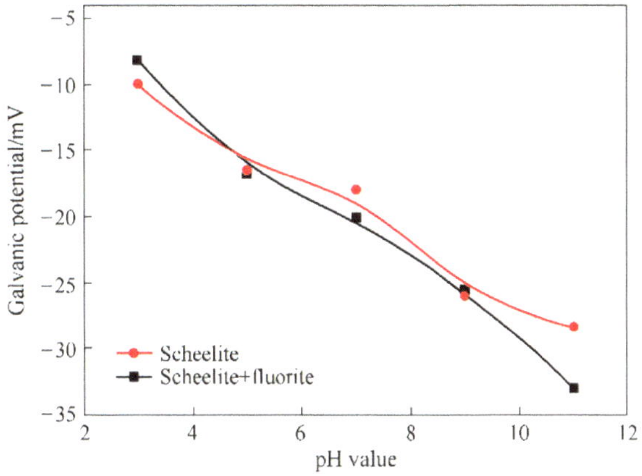

Fig. 4.16 Effect of surface conversion on scheelite electrokinetic potential, reprinted from ref. [28], copyright 2024, with permission from Minerals Engineering

on the surface of scheelite to produce CaF_2. Figure 4.16 shows the effect of fluorite on scheelite's electrokinetic potential. It can be seen that the addition of fluorite to scheelite has little effect on its surface potential, and the fluorite and dissolved F^- in the solution have little adsorption on the mineral surface.

On the contrary, the surface potential of calcite can be significantly changed by the conversion of fluorite on the surface of calcite. As shown in Fig. 4.17, the surface potential of calcite without treatment gradually changes from positive to negative with the increase of pH, and the zero electric point is in the pH range of 8–9. When fluorite was added, the curve of calcite shifted to the left obviously, and the curve was between the zeta potential of fluorite and the zeta potential of calcite. The zeta potential decreased significantly, that is, the surface potential of calcite tended to approach the galvanic potential of fluorite. Similarly, the conversion of fluorite onto the barite surface can also significantly change its original surface potential, as shown in Fig. 4.18. The zeta potential of barite is negative in the whole pH range of alkaline flotation, and after the conversion of fluorite on its surface, the zeta potential shifts positively as a whole. The isoelectric point is between 8.0 and 9.0, which is closer to the surface potential of BaF_2. This indicates that the surface potential of barite is obviously changed by the conversion of fluorite to barite.

3. **Energy changes of mineral surface elements**

(1) Surface transformation of scheelite and fluorite

The scheelite, fluorite and surface transformed samples have different characteristic peaks. The characteristic peaks of scheelite samples were Ca 299 eV, W 179 eV, W 1745 eV and O 518 eV under Auger electron spectroscopy (AES), while the characteristic peaks of fluorite samples were Ca 292 eV and F 654 eV under the same detection. However, the characteristic peaks of Ca and F in the scheelite clarifying solution shifted to 299 eV and 660 eV, and the characteristic peaks of new elements W 178 eV, W 1732 eV and O 519eV appeared, which proved that the chemical

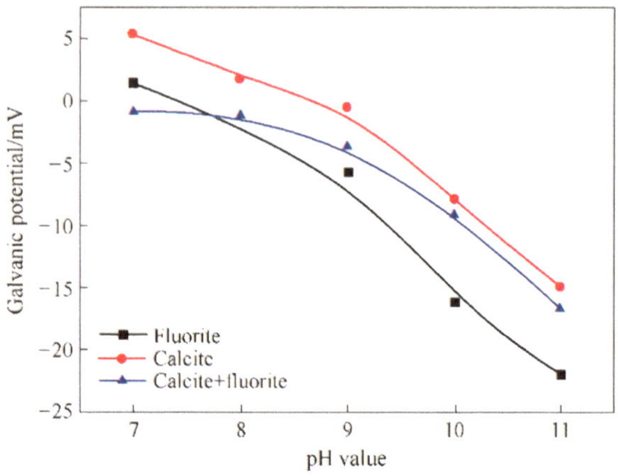

Fig. 4.17 Effect of surface conversion on galvanic potential of calcite, reprinted from ref. [28], copyright 2024, with permission from Minerals Engineering

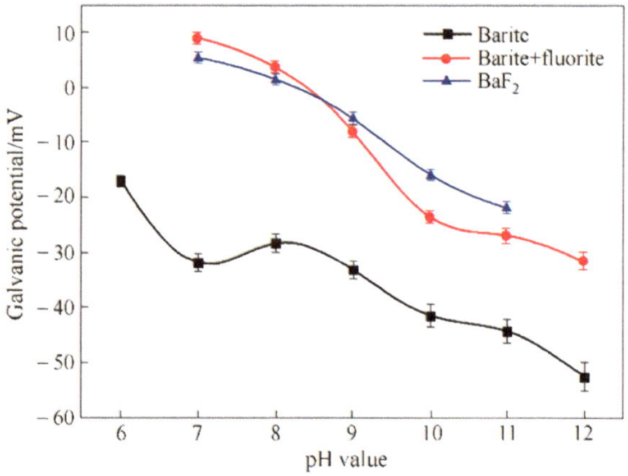

Fig. 4.18 Effect of surface conversion on the galvanic potential of barite, reprinted from ref. [32], copyright 2024, with permission from Minerals Engineering

reaction between fluorite and WO_4^{2-} in the scheelite clarifying solution resulted in the formation of $CaWO_4$ on the surface.

X-ray photoelectron spectroscopy (XPS) can accurately detect the atomic weight changes of different elements on the mineral surface and the degree of shift of binding energy corresponding to their peaks. Figure 4.19a shows the Ca 2*p* band of scheelite. The two peaks in the figure correspond to binding energies of 346.79 eV and 350.34 eV, which in turn correspond to Ca $2p_{3/2}$ and Ca $2p_{1/2}$ [29], respectively. After the addition of fluorite, the peaks of Ca $2p_{3/2}$ and Ca $2p_{1/2}$ only shift very slightly, reaching 346.84 eV and 350.39 eV respectively, indicating that fluorite has little effect on Ca 2*p*. As can be seen from Fig. 4.19b, the binding energy peaks of O1*s* appear at 530 eV and 532 eV, corresponding to W–O and Ca–O groups, respectively. When fluorite is added, the binding energies of the two groups are shifted from 530.24 eV and 531.89 eV to 530.29 eV and 531.79 eV, respectively, with offsets of 0.05 eV and 0.1 eV. The small shift of O1*s* peak on scheelite surface indicates that

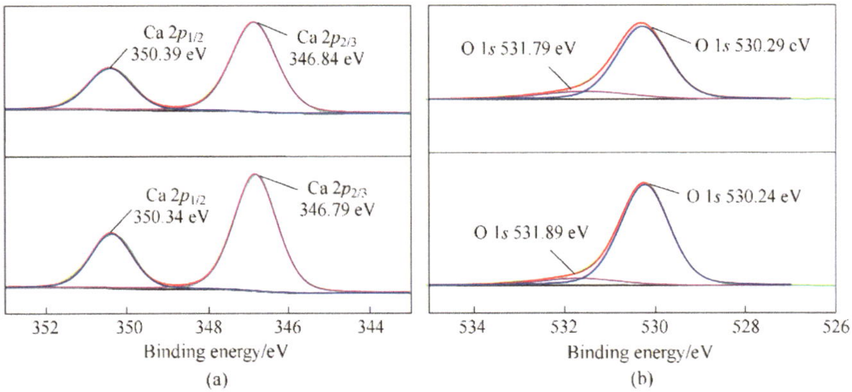

Fig. 4.19 Spectra of Ca 2*p* and O1*s* of scheelite without (bottom) and with (top) fluorite (**a**: Ca2*p*; **b**: O1*s*), reprinted from ref. [28], copyright 2024, with permission from Minerals Engineering

fluorite has little influence on scheelite surface, and the binding energies of W–O and Ca–O groups remain stable after fluorite treatment. This further confirms that the conversion of fluorite on scheelite surface is very weak and difficult to occur [28].

(2) Surface conversion of scheelite to calcite

Similar to the surface conversion of scheelite and fluorite, the solution chemical calculation of the surface conversion of scheelite to calcite shows that the surface conversion of scheelite to calcite can also be divided into two types: scheelite to calcite and calcite to scheelite. Both calcite and scheelite can be converted to each other, that is, both the forward reaction and the reverse reaction of the Eq. (4.39) can occur under conventional flotation conditions, and the pH is 8.8 as the boundary.

The characteristic peaks of elements C, Ca and O appear on the surface of calcite, corresponding to energy values of 276 eV, 299 eV and 519 eV, respectively. As shown in Fig. 4.20, the characteristic peaks of elements C, Ca and O on the surface of calcite were shifted to 277 eV, 298 eV and 515 eV in the scheelite clarification solution. The presence of characteristic peaks 204 and 225 eV of W at the same time proves that $CaWO_4$ precipitates are generated on the surface of calcite in the scheelite clarification solution [14].

The characteristic peaks of scheelite surface elements are Ca(299 eV), W(174 eV), O(518 eV). As shown in Fig. 4.21, the characteristic peaks of these elements are shifted to Ca(294 eV), W(179 eV), O(551 eV) in the calcite clarification solution. At the same time, the characteristic peak of element C appears at 272 eV, indicating that the surface of scheelite reacts with CO_3^{2-} in the calcite clarification solution to generate $CaCO_3$ precipitation.

(3) Surface conversion of fluorite and calcite

Fluorite has great influence on the calcium active sites of calcite. The spectra of Ca 2*p* on the calcite surface are shown in Fig. 4.22a, and the binding energies of Ca $2p_{2/3}$

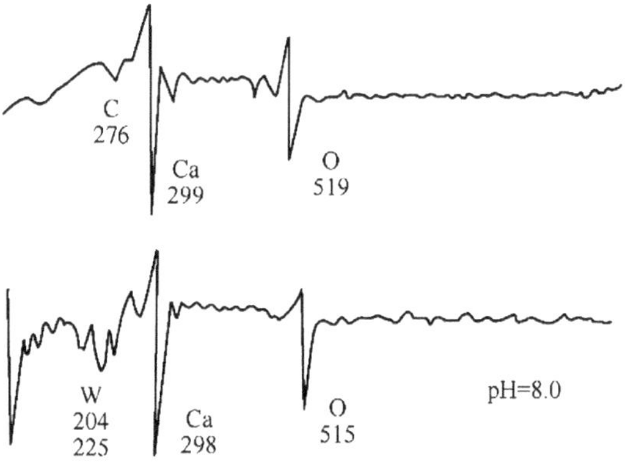

Fig. 4.20 Calcite (top) and its characteristic peaks in the scheelite clarified solution (bottom) [14]

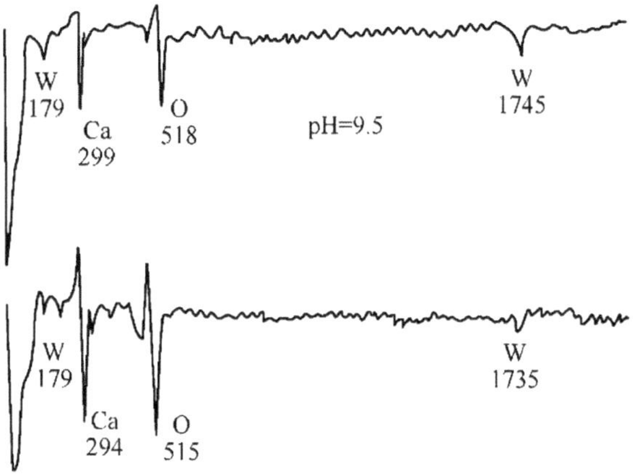

Fig. 4.21 Scheelite (top) and its characteristic peaks in the calcite clarified solution (bottom) [14]

and Ca $2p_{1/2}$ without fluorite treatment are 346.84 eV and 350.39 eV, respectively. After the addition of fluorite treatment, the binding energy of Ca $2p_{2/3}$ was shifted by 0.11 eV, and the binding energy of Ca $2p_{1/2}$ was shifted by 0.12 eV. The O 1 *s* spectra of the calcite surface are shown in Fig. 4.22b, and the binding energies of the C–O and Ca–O groups appear at 531.19 eV and 532.39 eV, respectively. With the addition of calcium fluoride, the O 1 *s* peak at 531.19 eV remains unchanged, while the O 1 *s* peak at 532.39 eV moves to 532.19 eV with a shift of 0.2 eV, indicating that the Ca–O group is greatly influenced by fluorite and electron transfer occurs between calcite and calcium fluoride. In conclusion, the surface of calcite can provide calcium ion active site, and the active site has a stronger adsorption effect on fluoride ions produced by fluorite dissolution.

Figure 4.23 shows the results of further peg-splitting fitting of calcite's Ca 2*p* spectrum, which can further determine the influence of surface conversion of fluorite on the binding energy spectrum of Ca 2*p*. After calcite is treated with fluorite surface conversion, the peaks of Ca $2p_{2/3}$ and Ca $2p_{1/2}$ can be fitted again into two more suitable peaks. These two peaks contain the peaks of Ca–F groups, which appear at

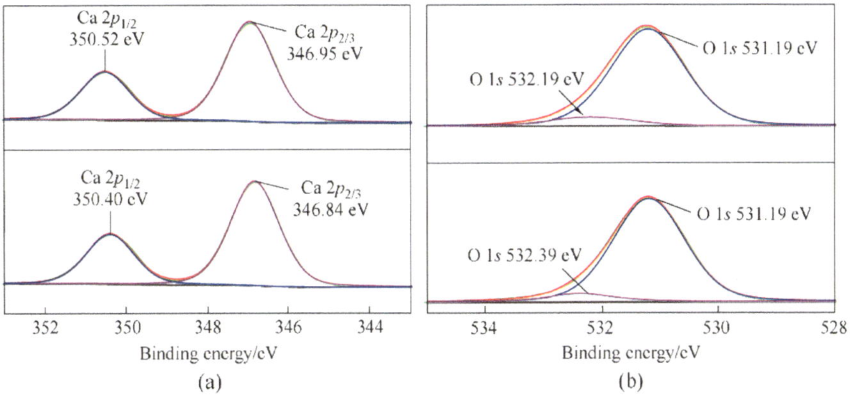

Fig. 4.22 Spectra of Ca 2*p* and O 1 *s* of calcite with no (bottom) and with (top) fluorite addition (**a**: Ca 1 *s*; **b**: O 1 *s*), reprinted from ref. [28], copyright 2024, with permission from Minerals Engineering

the binding energy of 347.65 eV and 350.89 eV. The difference in binding energy of Ca $2p_{2/3}$ peaks between calcite and fluorite is close to 0.9 eV, indicating the formation of fluorite (CaF_2) on the calcite surface.

4. **The change of mineral surface morphology**

Atomic force microscopy (AFM) can intuitively observe the changes of mineral surface in two and three dimensions, and can reveal the microstructure of solid surface or interface. Figure 4.24 shows AFM images of the surface changes of scheelite before and after treatment with fluorite clarification solution. It can be seen that before the treatment, the scheelite surface is relatively flat, with surface roughness

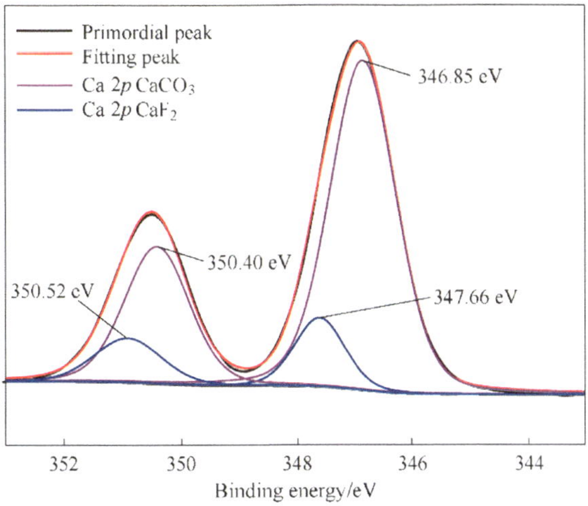

Fig. 4.23 Sub-peaks of Ca 2*p* under the influence of surface transformation on calcite surface, reprinted from ref. [28], copyright 2024, with permission from Minerals Engineering

$R_a = 3.78$ nm and $R_q = 4.76$ nm in a given area of 1.21 μm^2. Although the 3D images showed significant convex or concave changes in the scheelite surface morphology after the addition of fluorite clarification solution treatment, the roughness increased slightly to $R_a = 4.30$ nm and $R_q = 5.22$ nm in the area of 1.05 μm^2, with very little increment. This indicates that it is difficult for F^- to chemically react to generate CaF_2 on scheelite surface, that is, under conventional flotation conditions and no external conditions (pressure, temperature), the Eq. (4.33) can only take place forward, and the reverse reaction will not take place.

Different from scheelite, the surface morphology of calcite is significantly different before and after treatment with fluorite clarifying solution. As shown in Fig. 4.25, the surface of calcite is also relatively flat before treatment, with roughness $R_a = 5.12$ nm and $R_q = 7.63$ nm in the region of 3.79 μm^2. However, after the treatment with fluorite clarifying solution, it can be intuitively seen from the 3D image that the mineral surface morphology fluctuates more obviously, with very obvious gully depression and mountain convex. The roughness R_a and R_q of the mineral surface increased to 30.6 nm and 37.6 nm, respectively, with corresponding increases of 25.48 and 29.97, which supported the conclusion of the above-mentioned conversion between fluorite and calcite.

Fig. 4.24 AFM images of scheelite surface before (**a**, **b**) and after (**c**, **d**) surface conversion treatment, reprinted from ref. [28], copyright 2024, with permission from Minerals Engineering

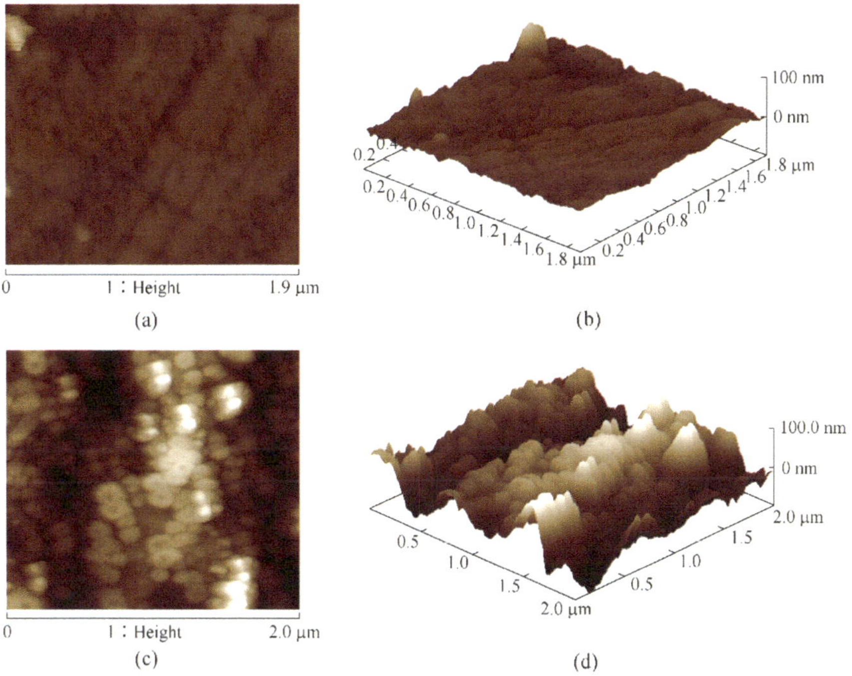

Fig. 4.25 AFM images of calcite surface before (**a**, **b**) and after (**c**, **d**) surface conversion treatment, reprinted from ref. [28], copyright 2024, with permission from Minerals Engineering

4.4 Flotation Reagents and Their Adsorption Solution Chemistry at the Solid–Liquid Interface

4.4.1 Inorganic Regulator and Its Adsorption at the Solid–Liquid Interface

Sodium carbonate is widely used in tungsten flotation process as an alkaline regulator. It ionizes and hydrolyzes in the pulp to produce OH^-, HCO_3^- and CO_3^{2-} etc. These ions buffer a small amount of H^+ and OH^-, so the pH value of sodium carbonate is relatively stable and can keep the pH value of the pulp in the range of 8–10. In addition, sodium carbonate has the function of dispersing slime, for micro-fine hematite [33], micro-fine aluminum silicate minerals [34], serpentine slime [35] and so on have a good dispersion effect.

Under different pH values, sodium carbonate has a great effect on the dispersion behavior of minerals, which is related to the chemical behavior of the solution. Sodium carbonate has the following equilibrium reactions in aqueous solution [10]:

$$Na_2CO_3 \rightleftharpoons 2Na^+ + CO_3^{2-} \tag{4.48}$$

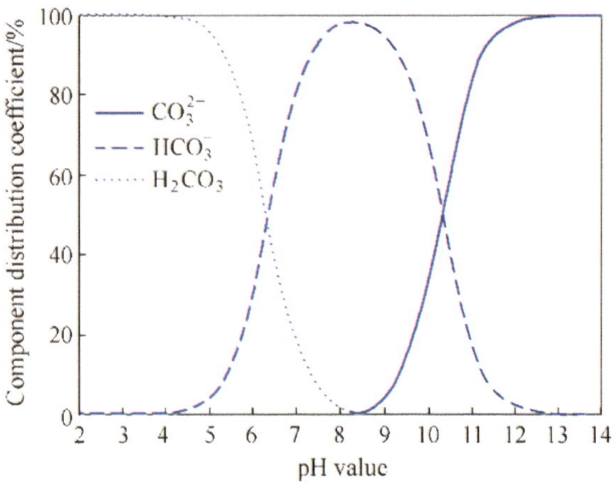

Fig. 4.26 Distribution coefficients of each component of sodium carbonate in relation to pH value [36]

$$CO_3^{2-} + H_2O \rightleftharpoons OH^- + HCO_3^- \quad K_1 = 1 \times 10^{10.33} \tag{4.49}$$

$$HCO_3^- + H_2O \rightleftharpoons OH^- + H_2CO_3 \quad K_2 = 1 \times 10^{6.35} \tag{4.50}$$

The relationship between Na_2CO_3 hydrolyzed components and pH value was drawn by Eqs. (4.48) to (4.50), as shown in Fig. 4.26. As can be seen from Fig. 4.26, when pH value is less than 6, the dominant group of sodium carbonate in solution is divided into H_2CO_3; when pH value is 6–10, the dominant group is divided into HCO_3^-. When pH is greater than 10, the dominant group is divided into CO_3^{2-}, the concentration of decreases with HCO_3^- with the increase of pH, and the concentration of CO_3^{2-} increases with the increase of pH.

Zhu et al. [37] believed that when the pH value of sodium carbonate was adjusted, the ions in the CO_3^{2-} pulp and calcium ions generated calcium carbonate covering the surface of scheelite, scheelite ions were characteristic adsorption to CO_3^{2-}, and there was competitive adsorption between CO_3^{2-} and SiO_3^-, so the floatability of scheelite could be improved. Chen et al. [38] believed that when sodium carbonate was used as the regulator, the adsorption strength of oleate ion on scheelite surface was enhanced, and that of fluorite surface was weakened, while that of calcite was almost unchanged. This is because within a certain concentration range of CO_3^{2-}, it chemically reacted with Ca^{2+} produced by the dissolution of calcium-containing minerals in the pulp to precipitate respectively, which reduced the precipitation of Ca^{2+} on sodium oleate and eliminated the adverse effect of Ca^{2+} on flotation, thus enhancing the effect of sodium oleate and scheelite surface. It is beneficial to improve the flotation recovery rate of mineral. Up to now, there is no accepted conclusion on the specific action mechanism of sodium carbonate in tungsten flotation.

Sodium silicate is a strong base and weak acid salt, in aqueous solution will occur a series of hydrolysis reactions, generating various components of the hydrolytic products, the hydrolysis reaction process can be expressed by the following equation

[39].

$$SiO_2(s,\ \text{amorphous}) + 2H_2O \rightleftharpoons SiO_4 \quad K = 10^{-2.7} \tag{4.50}$$

$$SiO_2(s,\ \text{"amorphous"}) + 2H_2O \rightleftharpoons SiO_4 \quad K = 10^{-2.7} \tag{4.51}$$

$$Si(OH)_4 \rightleftharpoons SiO(OH)_3^- + H^+ \quad K = 10^{-9.46} \tag{4.52}$$

$$SiO(OH)_3^- \rightleftharpoons SiO_2(OH)_2^2 + H^+ \quad K = 10^{-12.56} \tag{4.53}$$

The components obtained by sodium silicate hydrolysis are very complex. The $Si(OH)_4$ produced by hydrolysis undergo a series of ionization in the aqueous solution to produce a large number of $SiO(OH)_3^-$ $SiO_2(OH)_2^{2-}$ $Si_4O_6(OH)_6^{2-}$ or other components of various forms, as well as a large number of H_2SiO_3 and SiO_3^{2-} and other components of various forms. Under different pH conditions, the direction of hydrolysis and ionization reaction will also change, and the main components formed in the solution will also change. According to Eqs. (4.51)–(4.53), the contents of main components in sodium silicate solution under different pH conditions can be drawn, as shown in Fig. 4.27.

The solution chemistry study of sodium silicate shows that: when pH < 9.4, the dominant component in sodium silicate solution is 9.4 < pH < 12.6, the dominant component is $Si(OH)_4$ When pH > 12.6, the dominant component is $SiO_2(OH)_2^{2-}$. Sodium silicate began to inhibit the pH value of fluorite and calcite, corresponding to the pH value with $SiO(OH)_3^-$ formed at the beginning, so it can be known that $SiO(OH)_3^-$ is an effective component to inhibit fluorite and calcite [1].

It is believed that the inhibition mechanism of sodium silicate on calcium-containing minerals is due to the chemical reaction between the hydrolyzed components and the calcium points on the mineral surface to generate calcium silicate

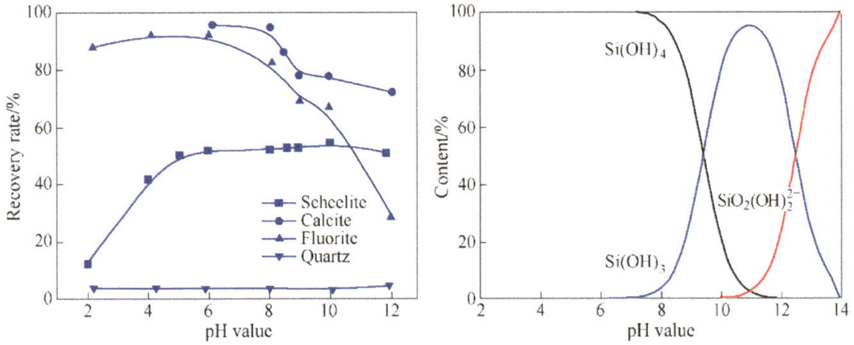

Fig. 4.27 Inhibition performance of sodium silicate on minerals under different pH conditions and its solution component content ($c_{\text{sodium silicate}}$ = 100 mg/L, c = 100 mg/L) [40, 41]

precipitation, which inhibits the hydrophilicity on the mineral surface. The reaction between sodium silicate and calcium-containing mineral surface to produce calcium silicate involves the following reactions [17].

$$Ca^{2+} + SiO_3^{2-} = CaSiO_4 \quad K = 10^{-11.08} \tag{4.54}$$

$$Ca^{2+} + OH^- = CaOH^+ \quad K = 10^{1.4} \tag{4.24}$$

$$Ca^{2+} + 2OH^- = Ca(OH)_2 \quad K = 10^{2.77} \tag{4.25}$$

$$H^+ + SiO_3^{2-} = HSiO_3^- \quad K = 10^{12.56} \tag{4.55}$$

$$H^+ + HSiO_3^- = H_2SiO_4 \quad K = 10^{12.56} \tag{4.56}$$

According to the formula, the relationship between the calcium silicate conditional concentration volume and the pH of the solution is obtained, as shown in Curve 1 in Fig. 4.28. Curves 2, 3 and 4 show the product of total calcium ion concentration and total concentration $-\lg$ in scheelite, fluorite and calcite saturated solutions at different pH values, respectively. $c(Ca^{2+})_T$ and SiO_3^{2-} total concentration $c(SiO_3^{2-})_T$ product $-\lg(c(Ca^{2+})_T \cdot c(SiO_3^{2-})_T)$ When it is above curve 1, it indicates that calcium silicate precipitates are generated in the solution. As can be seen from the figure, the intersection points of curves 2, 3 and 4 and curve 1 are pH = 8.6, 8.8 and 9.5, respectively, indicating that with the gradual increase of pH, $SiO(OH)_3^-$ calcium silicate precipitates occur on the mineral surface through chemisorption in the order of calcite, fluorite and scheelite. Therefore, by controlling the pH between 8.6 and 9.50, sodium silicate can generate calcium silicate precipitate on the surface of calcite and fluorite, selectively inhibit calcite and fluorite, and then flotation scheelite [17]. Related research on modified sodium silicate is discussed in detail in the following chapters.

Dong et al. [42] used sodium fluosilicate (Na_2SiF_6) as an inhibitor to study the selective inhibition mechanism of schetungite and calcite Na_2SiF_6. The dissociation reaction of sodium fluosilicate occurs as follows in solution:

$$SiF_6^{2-} + 2H^+ + 4H_2O \rightleftharpoons Si(OH)_4 + 6HF \quad K = 10^{-26.27} \tag{4.57}$$

$$HF \rightleftharpoons H^+ + F^- \quad K = 10^{-3.13} \tag{4.58}$$

$$Si(OH)_4 \rightleftharpoons SiO(OH)_3^- + H^+ \quad K = 10^{-9.46} \tag{4.52}$$

$$SiO(OH)_3^- \rightleftharpoons SiO_2(OH)_2^{2-} + H^+ \quad K = 10^{-12.56} \tag{4.53}$$

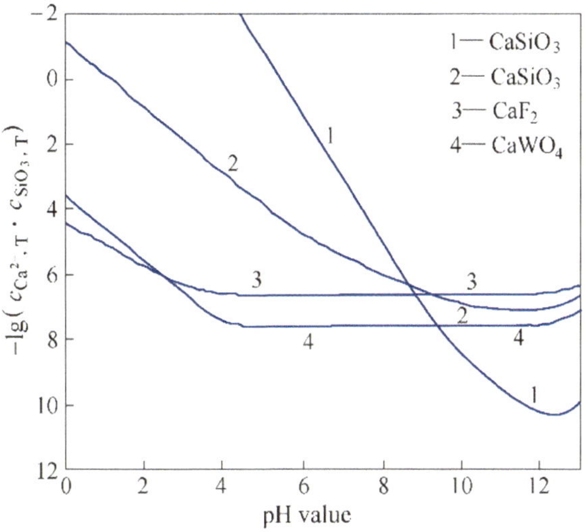

Fig. 4.28 Solution chemical calculation of the conditions for calcium silicate formation on the surface of different calcium-containing minerals [17]

Figure 4.29 shows the solution composition diagram of sodium fluosilicate. As shown in the figure, under alkaline conditions, sodium fluosilicate has a strong selective inhibition on calcite, which can achieve the selective separation of scheelite and calcite. At this point. The main solution groups in sodium fluosilicate solution are divided into negatively charged F^-, $SiO_2(OH)_2^{2-}$ and $SiO(OH)_3^-$, which can be selectively adsorbed on aircraft surfaces by electrostatic action. In addition, $SiO_2(OH)_2^{2-}$ and $SiO(OH)_3^-$ also act on the surface of calcite through hydrogen bonding, forming a hydrophilic film, preventing the further adsorption of the collector, and thus inhibiting the floating of calcite.

Phosphate inhibitors have strong inhibition ability to silicate and some calcium-containing minerals, so they are also widely used in scheelite flotation practice. In

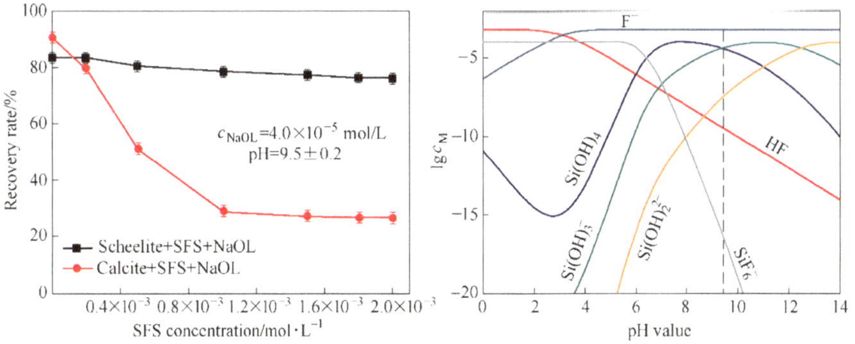

Fig. 4.29 Flotation recovery of scheelite and calcite under the action of sodium fluosilicate (SFS) and solution component diagram ($c = 1 \times 10^{-3}$ mol/L), reprinted from ref. [42], copyright 2019, with permission from Minerals Engineering

the case of sodium hexametaphosphate, the following hydrolysis reactions will occur in solution.

$$(NaPO_3)_6 + 6H_2O \rightarrow 6Na^+ + 12H^+ + 6PO_4^{3-} \quad (4.59)$$

$$H^+ + PO_4^{3-} \rightleftharpoons HPO_4^{2-} \quad K_1^H = 10^{12.35} \quad (4.60)$$

$$H^+ + HPO_4^{2-} = H_2PO_4^- \quad K_2^H = 10^{7.20} \quad (4.61)$$

$$H^+ + H_2PO_4^- = H_3PO_4 \quad K_3^H = 10^{2.15} \quad (4.62)$$

As shown in Fig. 4.30, when pH = 4–6, the main groups in the solution are HPO_4^{2-}, and when pH = 9–11, the main groups in the solution are HPO_4^{2-}. These phosphate anions are easy to react with Ca^{2+} ions on the surface of calcite and adsorb $Ca_3(PO_4)_2$, which makes the surface of calcite strongly hydrophilic and thus inhibited.

Feng et al. [44] found that after adding sodium hexametaphosphate in the process of calcite flotation, the concentration of calcium ions in the flotation solution increased significantly. They believed that sodium hexametaphosphate reacted with calcium ions on the surface of calcite to form chelates with high stability and water solubility, which made the calcium ions on the surface of calcite fall off into the flotation solution, so that the adsorption site of collector on the surface of calcite was significantly reduced. As a result, it is difficult for the collector to adsorb and float up.

Sodium pyrophosphate ($Na_4P_2O_7$) also has strong selective inhibition on calcium-bearing gangue minerals and silosilicate, especially according to the "mirror symmetry rule", sodium pyrophosphate has selective inhibition on apatite, so it is often used in the flotation separation of scheelite and apatite [45]. As shown in

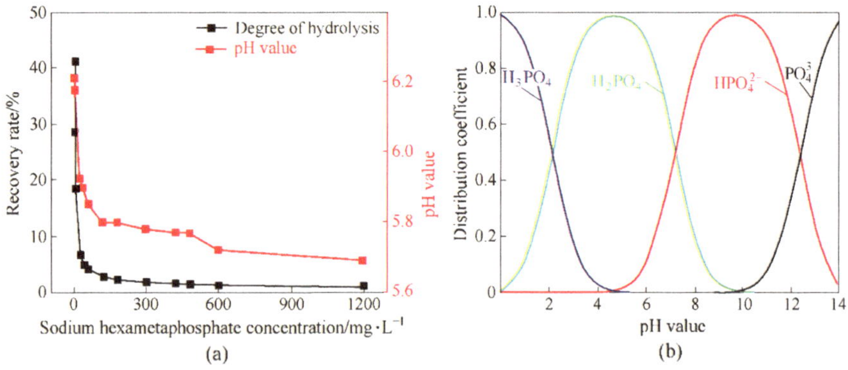

Fig. 4.30 Hydrolysis of sodium hexametaphosphate and distribution of components in the solution [43]

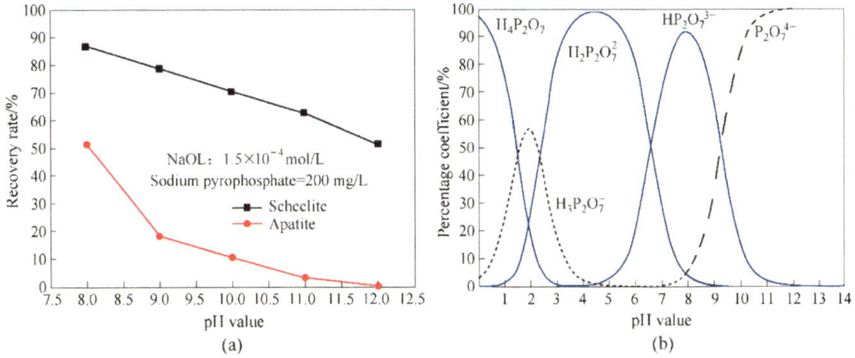

Fig. 4.31 Flotation recovery of scheelite and apatite and distribution of components in sodium pyrophosphate solution under the action of sodium pyrophosphate [45]

Table 4.11 Calculation results of electronegativity of ionic groups [45]

Ionic groups	PO_4^{3-}	$P_2O_4^{4-}$	WO_3^{2-}
X_g	4.95	5.17	5.24

Fig. 4.31, when sodium pyrophosphate exists, the flotation recovery rates of scheelite and apatite decrease with the increase of pulp pH value. When pH > 9.0, sodium pyrophosphate has obvious selective inhibition effect on apatite, scheelite and apatite recovery rates are significantly different, and the flotation separation of scheelite and apatite can be realized. When pH = 8.0–9.0, the main solution groups of sodium pyrophosphate are $HP_2O_7^{3-}$. When pH > 9.0, sodium pyrophosphate mainly exists in the $P_2O_7^{4-}$ form, and its content increases gradually with the increase of pH, so $P_2O_7^{4-}$ is the main component of sodium pyrophosphate inhibition of apatite.

In addition, as shown in Table 4.11, the electronegativity of the three groups can be calculated as follows: phosphate < pyrophosphate < tungstate, the larger the group electronegativity of ions and calcium ions when the ionic bond is stronger, so the order of the three ions and calcium ions is: phosphate < pyrophosphate < tungstate. The group electronegativity of pyrophosphate ion is between phosphate ion and tungstate ion, so sodium pyrophosphate can strongly inhibit apatite, and the inhibition effect of scheelite is weak.

4.4.2 *Organic Regulator and Its Adsorption at the Solid–Liquid Interface*

Citric acid is often used as an inhibitor to inhibit gangue minerals such as fluorite and calcite. Citric acid is a tricarboxylic acid, which is ionized by steps in solution. Its ionization reaction and reaction equilibrium are listed in Table 4.12. According

Table 4.12 Dissolution reaction and reaction constants of citric acid in aqueous solution [46]

Reaction formula	lg*K*	Serial number
$H^+ + L^{3-} \rightleftharpoons HL^{2-}$	6.396	(4.63)
$H^+ + HL^{2-} \rightleftharpoons H_2L^-$	4.761	(4.64)
$H^+ + HL^{2-} \rightleftharpoons H_3L$	3.31	(4.65)

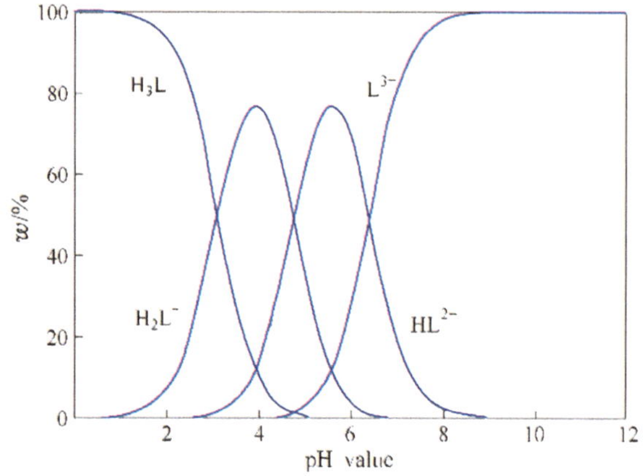

Fig. 4.32 Composition diagram of citric acid solution [47]

to Eqs. (4.63)–(4.65), the composition of citric acid solution is drawn, as shown in Fig. 4.32. The composition of citric acid solution is obviously affected by pH. When pH < 3.1, the dominant group in the solution is HL_3 citric acid molecule. When pH > 3.1, the main component in the solution is citric acid anion. When pH > 6.4, the dominant group is L^{3-} when pH > 9, almost all citric acid in the solution exists in the form of L^{3-}. Therefore, when citric acid is used as an inhibitor in the flotation separation of scheelite and fluorite under alkaline conditions, citric acid mainly plays an inhibitory role in the form of L^{3-}.

Tables 4.13 and 4.14 show the reaction and equilibrium constants of citric acid on scheelite and fluorite surfaces respectively, and Fig. 4.33 shows the Gibbs free energy of citric acid to generate calcium citrate on scheelite and fluorite surfaces respectively. As can be seen from the figure, for scheelite, the Gibbs free energies of formation, CaL^-, CaHL and CaH_2L^+ are all positive in the whole pH range and cannot be generated spontaneously, so the adsorption of citric acid on scheelite surface is weak. F or fluorite, when pH > 9, the Gibbs free energies of citric acid to produce several different products on the surface of fluorite are: $CaL^- < 0 <$ CaHL $<$ CaH_2L^+ which is the most easily generated product on the surface of fluorite. Therefore, citric acid CaL^- selectively inhibits flotation of fluorite by spontaneously generating hydrophilic properties on the surface of fluorite.

Polyaspartic acid (PASP) is a new type of flotation inhibitor of calcium minerals, which is non-toxic, high solubility and easy to degrade. There are many carboxyl groups in the molecular mechanism of poly aspartic acid, which is easy to interact with calcium ions, and can effectively inhibit gangue minerals such as calcite in

Table 4.13 Reaction of citric acid at solid–liquid interface of scheelite and reaction constants [46]

Reaction formula	lg$K$	Serial number
$CaWO_4(s) + L^{3-} \rightleftharpoons CaL^-(s) + +WO_4^{2-}$	−4.68	(4.66)
$CaWO_4(s) + HL^{2-} \rightleftharpoons CaHL(s) + +WO_4^{2-}$	−6.21	(4.67)
$CaWO_4(s) + H_2L^- \rightleftharpoons CaH_2L^+(s) + WO_4^{2-}$	−8.20	(4.68)

Table 4.14 Reaction of citric acid at fluorite solid–liquid interface and reaction constants [46]

Reaction formula	lg$K$	Serial number
$CaF_2(s) + L^{3-} \rightleftharpoons CaL^-(s) + 2F^-$	0.98	(4.69)
$CaF_2(s) + HL^{2-} \rightleftharpoons CaHL(s) + 2F^-$	−0.61	(4.70)
$CaF_2(s) + H_2L^- \rightleftharpoons CaH_2L^+(s) + 2F^-$	−2.60	(4.71)

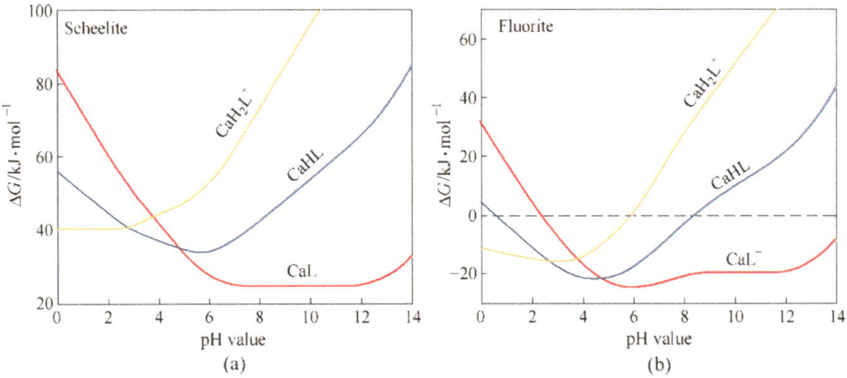

Fig. 4.33 Adsorption products and Gibbs free energy of citric acid on scheelite and fluorite surfaces, reprinted from ref. [46], copyright 2024, with permission from Minerals Engineering

schetungsten flotation. Step reaction and reaction equilibrium of polyaspartic acid in solution are listed in Table 4.15. According to Eqs. (4.72)–(4.75), components of polyaspartic acid solution are drawn, as shown in Fig. 4.34. The composition of polyaspartic acid solution is obviously affected by pH. When pH < 2, the dominant component in the solution exists in the form of polyaspartic acid molecules. When pH > 2, the main components in the solution can be divided into polyaspartic acid anion. When pH > 6, the dominant group was divided into L^{4-}; When pH > 8.5, almost all the polyaspartic acid in the solution existed in the form of L^{4-}. Therefore, when using polyaspartic acid as an inhibitor in the flotation separation of scheelite and calcite under alkaline conditions, polyaspartic acid mainly plays an inhibitory role in the form of L^{4-}.

As a typical organic polymer, carboxymethyl cellulose (CMC) is an anionic polysaccharide obtained from cellulose treated with chloroacetic acid treatment, which can be used as an inhibitor to selectively inhibit calcite flotation in scheelite

Table 4.15 Ionization reaction and reaction constants of polyaspartic acid in solution [48]

Reaction formula	lg$K$	Serial number
$H^+ + L^{4-} \rightleftharpoons HL^{3-}$	5.30	(4.72)
$H^+ + HL^{3-} \rightleftharpoons H_2L^{2-}$	4.22	(4.73)
$H^+ + H_2L^{2-} \rightleftharpoons H_3L^-$	3.72	(4.74)
$H^+ + H_3L^- \rightleftharpoons H_4L$	2.40	(4.75)

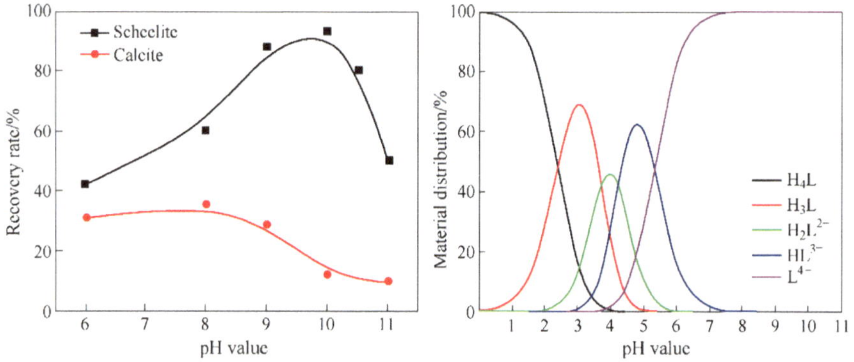

Fig. 4.34 Flotation recovery rates of scheelite and calcite and their solution components under polyaspartic acid ($c = 4$ mg/L) [48]

flotation [49]. In flotation solution, carboxymethyl cellulose carboxylic acid can be ionized to form anion -RCOO$^-$.

$$-CH_2 - COOH \rightleftharpoons -CH_2 - COO^- + H^+ \quad K = 10^{3.8} \tag{4.76}$$

Plot the solution component of carboxymethyl cellulose in solution according to Eq. (4.73), as shown in Fig. 4.35. When pH < 3, the solution is dominated by unionized carboxymethyl cellulose molecules; When pH > 3, the solution is dominated by ionized carboxymethyl cellulose anions [50]. In flotation, carboxymethyl cellulose anion is adsorbed on the surface of calcite with positive potential by electrostatic action, but less adsorbed on scheelite surface [49].

4.4.3 *Collector and Its Adsorption at the Solid–Liquid Interface*

4.4.3.1 Anionic Collector Class

Fatty acids are the most commonly used anionic collectors in the flotation process of tungsten ore. Table 4.16 lists the solubility and dissociation constants of different fatty

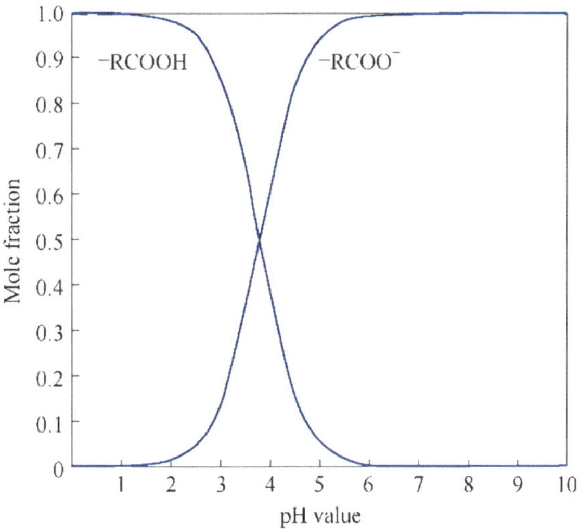

Fig. 4.35 Composition of carboxymethyl cellulose solution, reprinted from ref. [50], copyright 2024, with permission from Minerals Engineering

acids in aqueous solutions. The dissociation constant of saturated fatty acids in water decreases with the increase of relative molecular mass, and the dissociation constant of higher fatty acids above C_{10} is similar, about $10^{-4.92}$. According to the hydrolysis reaction of fatty acids in solution, the distribution of concentration components of different fatty acids in solution was plotted, as shown in Figs. 4.36, 4.37, 4.38, 4.39, 4.40 and 4.41.

By contrast, the activity of fatty acids on minerals is closely related to pH. Fatty acids exist mainly in the form of molecules in the strong acidic environment and in the form of ions in the strong alkaline environment. In weak acid, neutral or weak alkaline environment, there are more fatty acid molecules and fatty acid anions, and in the mutual attraction between hydrocarbon chains to form ion-molecular association, the ion-molecular association is closely related to mineral flotation behavior. Ion-molecular association can be formed in the solution first, and then adsorbed on the mineral surface, or the fatty acid root anion adsorbed on the mineral surface through the association between hydrocarbon chains so that fatty acid molecules adsorbed on the mineral surface, the co-adsorption of ion-molecular association is beneficial to improve the adsorption density of fatty acids on the mineral surface. On the contrary,

Table 4.16 Solubility and dissociation equilibrium constants of different saturated fatty acids [51]

Fatty acids	Dissociation constant	Solubility
Lauric acid	−4.92	0.0055
Myristic acid	−4.90	0.0020
Palmitic acid	−4.92	0.00072
Stearic acid	−4.92	0.00029
Oleic acid	−4.95	$10^{-7.2}$
Linoleic acid	−1.72	0.0139

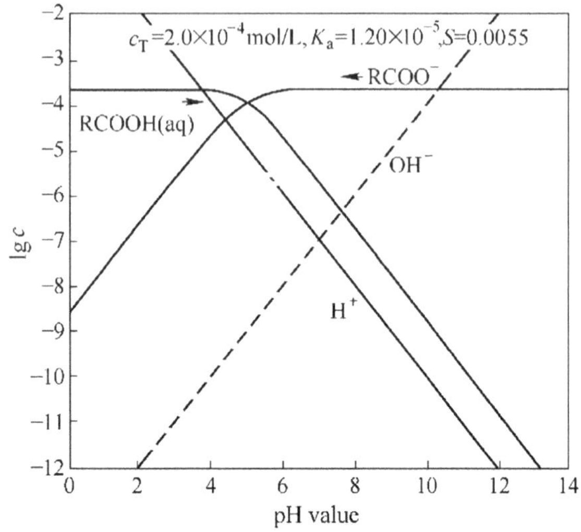

Fig. 4.36 Diagram of dissolved lauric acid components [51]

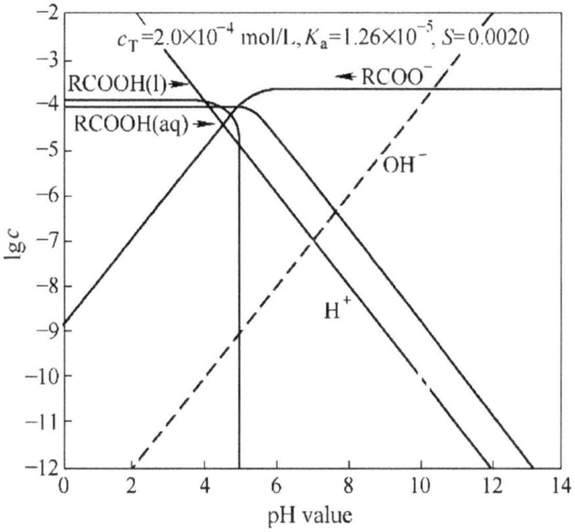

Fig. 4.37 Myristic acid dissolved component diagram [51]

in the strong acidic environment, the fatty acid is mainly in the molecular state, the activity of collecting is not strong. In a strong alkaline environment, due to the high concentration of OH^-, it is easy to have competitive adsorption with fatty acid anions or ion-molecular associations, compete with each other for the mineral surface or remove the absorbed fatty acid components from the surface, thus reducing the collection effect of fatty acids, and even make minerals inhibited.

Figure 4.42 shows the relationship between $\Delta G^{\ominus}$ and pH of different fatty acids reacting with calcium ions. It can be seen from the figure that within the range of 1, there is a large negative value, and the reaction of fatty acid forming precipitation

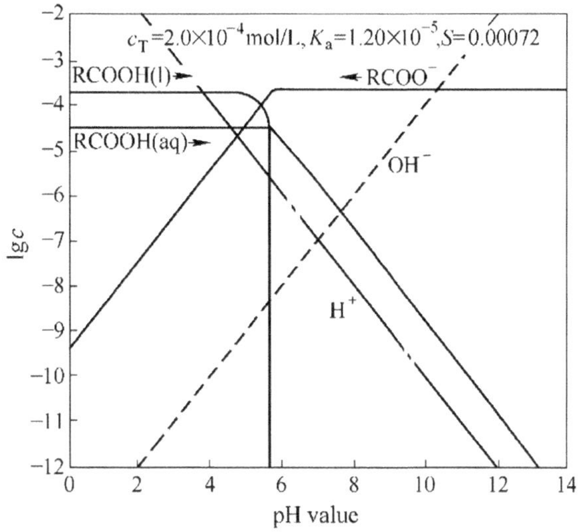

Fig. 4.38 Dissolved components of palmitic acid [51]

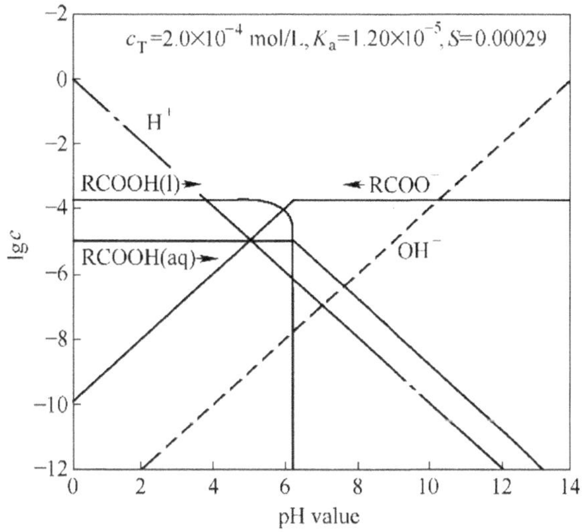

Fig. 4.39 Diagram of dissolved constituents of stearic acid [51]

with calcium ions is easy to proceed, and the adsorption capacity of the collector on the mineral surface is large, and the mineral has good floatability, which is consistent with the conclusion of the adsorption capacity of the agent on the mineral surface. Under the same conditions, with the increase of the carbon atom number of saturated fatty acids, the value of $\Delta G^{\ominus}$ is more negative, indicating that the reaction is easier to proceed. The longer the carbon chain is, the easier the saturated fatty acid is to precipitate at the mineral interface and thus have a better collecting capacity for the mineral. This is consistent with the results of single mineral flotation test: under the same conditions as the dosage of reagent, the longer the carbon chain of the fatty

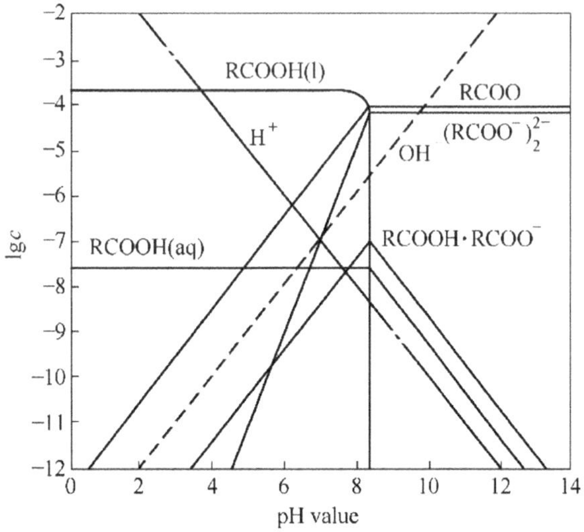

Fig. 4.40 Diagram of dissolved components of oleic acid [51]

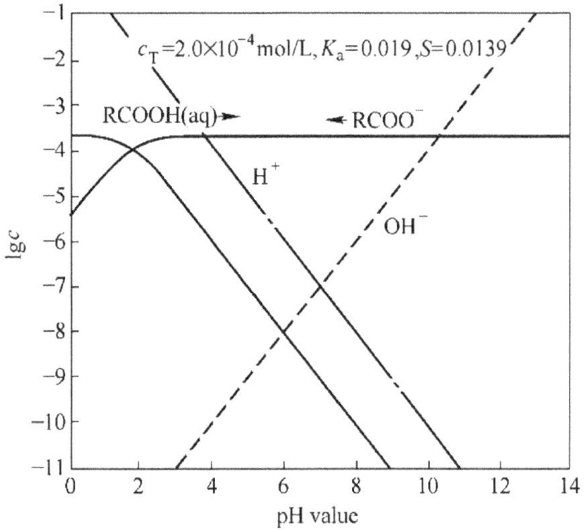

Fig. 4.41 Diagram of dissolved components of linoleic acid [51]

acid, the higher the recovery rate of the mineral. The rise and fall of $\Delta G^{\ominus}$ adsorption capacity and flotation recovery have consistent relationship with the increase or decrease.

4.4.3.2 Cationic Collector

Alkyl primary amine is a typical cationic collector commonly used in the flotation of tungsten ore. In the flotation solution, the positively charged amine ions produced by

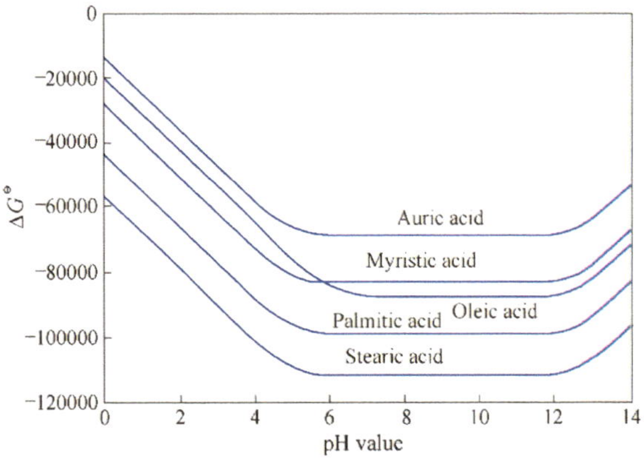

Fig. 4.42 Relationship between pH and reaction of different fatty acids with calcium ions $\Delta G^{\ominus}$ [51, 52]

ionization can interact with the mineral surface and make the mineral hydrophobic floating. Alkyl primary amine is a weak electrolyte. In aqueous solution, the primary amine will be hydrolyzed and dissociated. Therefore, there are both a part of ionic amine ions and a part of molecular amine molecules in the solution. The ratio between the amine molecule and the amine ion in the solution is governed by the medium value. The dissociation properties of amines in aqueous solution are mainly determined by their dissociation constant and the value of the medium. Alkyl amines exist in the following equilibrium in solution:

Dissolution equilibrium:

$$RNH_2(s) \rightleftharpoons RNH_2(aq) \tag{4.77}$$

Hydrolysis balance:

$$RNH_2 + H_2O \rightleftharpoons RNH_3^+ + OH^- \tag{4.78}$$

Dissociation equilibrium:

$$RNH_3^+ \rightleftharpoons RNH_2 + H^+ \tag{4.79}$$

Table 4.17 shows the molecular solubility (S), hydrolysis constant (pK) and dissociation constant (pK_b) of dodecylamine, tedecylamine and octadecylamine. According to Table 4.17 and Eqs. (4.77)–(4.79), the critical pHs values for the precipitation of amine molecules generated by alkyl amines at different concentrations can be calculated, as shown in Table 4.18.

It can be seen from Table 4.18 that, at a certain concentration, the values of molecular precipitation of amine generated by different carbon chain alkyls differ greatly. With the increase of the number of carbon atoms in the alkyl chain, the pHs value of molecular precipitation of amine generated decreases, which is consistent with the scheelite flotation rule in Fig. 4.43 when alkyl primary amine with different

Table 4.17 Molecular solubility, hydrolysis constants and dissociation constants of dodecylamine, tedecylamine and octadecylamine [1, 18, 19]

Alkyl primary amines	S/mol·L^{-1}	pK_b	pK_a
Dodecylamine	$10^{-4.7}$	−3.37	10.63
Tetradecylamine	$10^{-6.0}$	−3.38	10.60
Octadecamide	$10^{-8.3}$	−3.40	10.60

Table 4.18 pHs values of alkyl amines generating amine molecular precipitates at different concentrations [1, 18, 19]

Alkyl primary amines	10^{-2}	10^{-3}	5×10^{-4}	2×10^{-4}	1.5×10^{-4}	5×10^{-5}
Dodecylamine	7.93	8.93	9.25	9.68	9.82	10.45
Tetradecylamine	6.60	7.60	7.90	8.30	8.43	8.91
Octadecamide	4.30	5.30	5.60	6.00	6.12	6.60

carbon chain lengths is the collector. When dodecylamine salt, tetradecylamine salt and octadecylamine salt are used as collectors, the pH values of flotation recovery of scheelite starting to decline under alkaline conditions are 9.7, 8.3 and 6.1, respectively, which correspond to the precipitated values of dodecylamine salt, tetradecylamine salt and octadecylamine salt, respectively. The main effective components of cationic amine collectors are RNH_3^+ which can be obtained when pH > pHs. Most amines precipitate RNH_2 (s), and the concentration of RNH_3^+ is greatly reduced, and the flotation effect will be weakened [19].

In addition, hydrophobic association occurs between hydrocarbon chains of long alkyl primary amines through normal form forces. Under a certain concentration, association also occurs between ions and molecules:

$$\text{Ion association equilibrium}: \ 2RNH_3^+ \rightleftharpoons (RNH_3^+)_2^{2+} \tag{4.80}$$

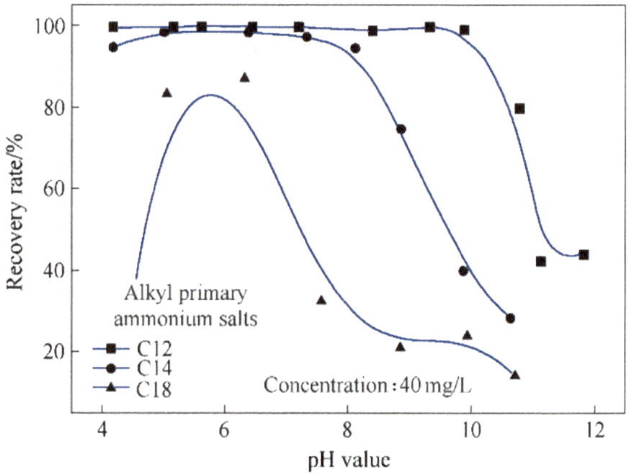

Fig. 4.43 Flotation recovery of scheelite under alkyl primary amines with different carbon chain lengths [19]

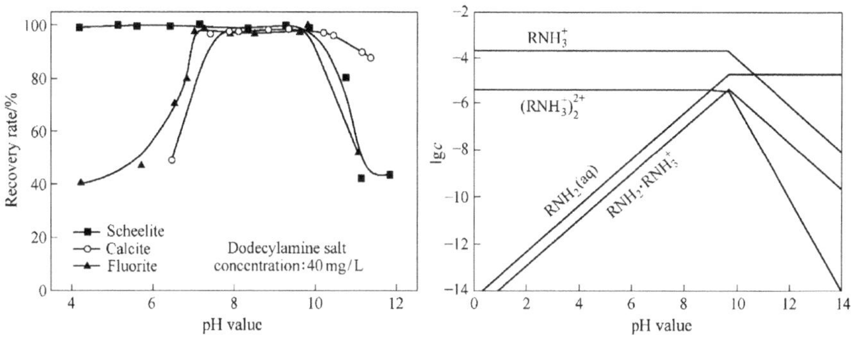

Fig. 4.44 Flotation recovery of three calcium-containing minerals and logarithm of concentration of each component in dodecylamine salt solution when dodecylamine salt is the collector ($c = 2 \times 10^{-4}$ mol/L) [19]

$$\text{Ion - molecular association balance}: RNH_3^+ + RNH_2(aq) \rightleftharpoons \left(RNH_3^+ \rightleftharpoons RNH_2\right)(aq) \quad (4.81)$$

Taking dodecylamine as an example, the relationship between the concentration of each component of dodecylamine in aqueous solution can be calculated according to its reaction and equilibrium constants in solution, as shown in Fig. 4.44.

As shown in Fig. 4.44, the main components of the interaction between dodecylamine and scheelite are mainly ion-molecular associations, and the flotation law of scheelite corresponds to the distribution law of $RNH_3^+ \cdot RNH_2(aq)$ However, the interaction of dodecylamine with calcite and fluorite should be ion-molecular $RNH_3^+ \cdot RNH_2(aq)$ and $RNH_2(aq)$ molecules. The flotation recovery of mineral increases first with the increase, and begins to decline after a certain value, and the flotation recovery of calcite declines slowly.

Table 4.19 shows CMC and HLB values for several typical cationic collectors. In homologues, CMC and HLB decreased with the increase of carbon chain length, and the hydrophobicity increased, but the solubility decreased, which also affected the dissolution and dispersion properties. It can be seen from the HLB value of 1231 that its collection ability is very weak and it has good foaming performance. Therefore, under the action of 1231, the flotation recovery rate of the three calcium-containing minerals is not high. The CMC values of 1231, dodecylamine, 1227 and 1221 were in the order of 1231 > C12 > 1227 > 1221, which corresponded to the order of their influence on the surface potential of the three calcium-containing minerals. From the value of 1227, it can be seen that 1227 is a kind of agent with strong foaming ability, and the better floatability of the three calcium-containing minerals under its action may be attributed to the entrainment effect of foam.

Table 4.19 CMC and HLB values of cationic collector [19]

Reagent		CMC/mol·L^{-1}	HLB
Dodecylamine	$C_{12}H_{25}NH_2$	1.3×10^{-2}	2.9
Tetradecylamine	$C_{14}H_{29}NH_2$	3.27×10^{-3}	1.95
Octadecamide	$C_{18}H_{37}NH_2$	4.0×10^{-4}	0.05
Dodecyl trimethyl ammonium chloride (1231)	$C_{12}H_{25}(CH_3)_3NCl$	1.61×10^{-2}	9.275
Octadecyl trimethyl ammonium chloride (1831)	$C_{18}H_{37}(CH_3)_3NCl$	3.1×10^{-4}	6.425
Didodecyl dimethyl ammonium chloride (1221)	$(C_{12}H_{25})_2(CH_3)_2NCl$	1.8×10^{-4}	4.05
Dodecyldimethylbenzylammonium chloride (1227)	$C_{21}H_{38}NCl$	7.8×10^{-3}	7.425

4.5 Interaction Between Metal Ions and Flotation Reagents and Its Influence

Scheelite, fluorite, calcite and other calcium-containing minerals have certain solubility, which will release a large number of calcium ions to the pulp in the flotation process because of their own dissolution. These calcium ions in the flotation solution will inevitably react with the fatty acid anions to form metal oleic acid complexes, which will affect the flotation process of scheelite.

4.5.1 Coordination Reactions Between Metal Ions and Fatty Acids and Their Effects

4.5.1.1 Solution Chemical Study of Ca–OL

The dominant component diagram can clearly indicate the types of components in the saturated solution where the value of an independent variable in precipitation equilibrium, complexation equilibrium and proton transfer equilibrium is fixed. In the flotation solution of semi-soluble calcium-containing minerals by oleic acid flotation, calcium ions extracted from minerals can react with oleic acid anions to form calcium oleate colloid, which has an impact on flotation. The relationship between the formation conditions of calcium oleate and pH value and oleic acid concentration is discussed with the help of the dominant component diagram [53–56].

In the calcium-oleic acid system, the following equilibrium exists.

$$Ca^{2+} + OH^- \rightleftharpoons CaOH^+ \quad K = 10^{1.40} \tag{4.82}$$

$$CaOH^{+} + OH^{-} \rightleftharpoons Ca(OH)_2(aq) \quad K = 10^{1.36} \tag{4.83}$$

$$HOL(l) \rightleftharpoons HOL(aq) \quad K = 10^{-7.6} \tag{4.84}$$

$$HOL(aq) \rightleftharpoons OL^{-} + H^{+} \quad K = 10^{-4.95} \tag{4.85}$$

$$2OL^{-} + H^{+} \rightleftharpoons HOL \cdot OL^{-} \quad K = 10^{9.9} \tag{4.86}$$

$$2OL^{-} \rightleftharpoons OL_2^{2-} \quad K = 10^{4} \tag{4.87}$$

$$Ca^{2+} + 2OL^{-} \rightleftharpoons CaOL_2(s) \quad K = 10^{15.52} \tag{4.88}$$

Conservation of mass:

$$c(Ca) = c(Ca^{2+}) + c(CaOH^{+}) + c(Ca(OH)_2(aq)) + c(Ca(OL)_2(s)) \tag{4.89}$$

$$\begin{aligned} c(HOL) = {} & c(HOL(l)) + c(HOL(aq)) + c(OL^{-}) + c(2HOL \cdot OL^{-}) \\ & + c\left(2OL_2^{2-}\right) + c(2Ca(OL)_2(s)) \end{aligned} \tag{4.90}$$

According to Eqs. (4.82)–(4.90), the dividing line between the components in the calcium ion-oleic acid system can be calculated and the dominant component diagram of the calcium ion-oleic acid system can be drawn. Figure 4.45 shows the dominant components when the calcium ion concentration is 1×10^{-4}mol/L, 1×10^{-5}mol/L and 1×10^{-4}mol/L.

The dominant range of calcium oleate is determined by calcium ion concentration, oleic acid concentration and pH. At 1×10^{-6} mol/L calcium ion concentration, no matter how oleic acid concentration and pH value change, calcium oleate will not be the dominant component in the system. When the concentration of calcium ion is $10^{-4} - 10$ mol/L, calcium oleate is the dominant component in the solution system in a certain oleic acid concentration and pH range. With the increase of calcium ion concentration, the critical pH and critical oleic acid concentration of calcium oleate gradually decrease, and the interval of calcium oleate as the dominant component gradually increases. When pH = 8–11, the concentration of calcium ion in the saturated solution of fluorite and scheelite is greater than 1×10^{-6} mol/L. At the usual concentration of $10^{-5} - 10^{-4}$ mol/L oleic acid, calcium oleate is stable and exists as the dominant component. When pH > 10, the calcium ion concentration in calcite saturated solution is less than 1×10^{-6} mol/L, at this time, calcium oleate component exists in trace, the main oleic acid group is divided into oleate ion. It can be inferred that in addition to oleic acid molecules and oleic acid anions, calcium oleate also plays an important role in the process of oleic acid flotation of fluorite and scheelite.

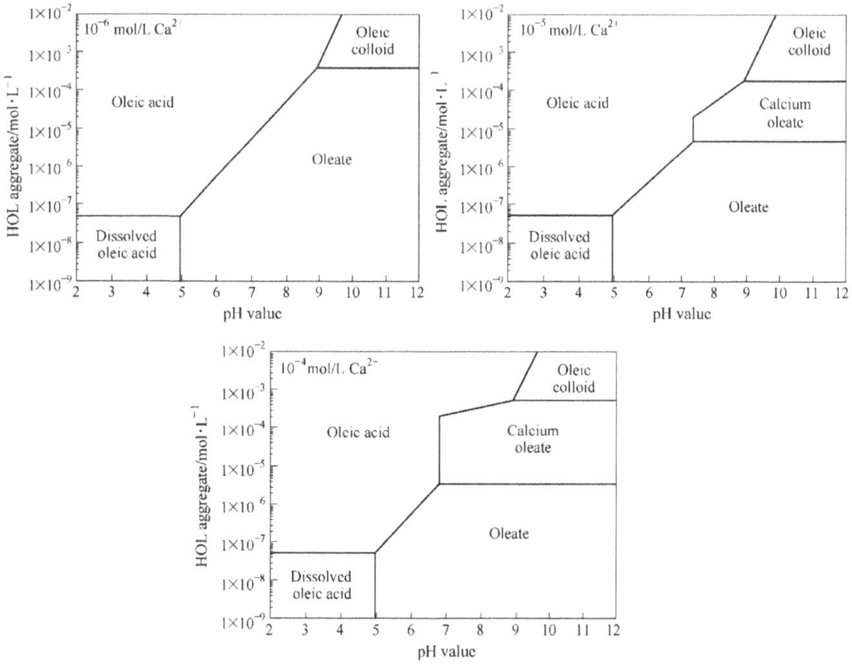

Fig. 4.45 Dominant components in oleic acid solution when Ca^{2+} is 10^{-6} mol/L to 10^{-4} mol/L, reprinted from ref. [57] copyright 2022, with permission from Minerals Engineering

4.5.1.2 Characteristic Functional Groups of Ca–OL

Figure 4.46 shows the infrared spectral analysis results of sodium oleate and Ca–OL colloid. The infrared spectral analysis results of sodium oleate and Ca–OL colloid were similar, and the main functional groups were CH_3, C–H, $-CH_2$, = C–H, –COO–, etc. Tables 4.20 and 4.21 show the absorption peaks of various groups of Ca–OL colloid and sodium oleate. The main difference between the two is that the stretching vibration absorption peak of the free carboxyl dimer (C = O) appears at 1710 cm^{-1} of Ca–OL colloid, and the in-plane bending vibration absorption peak of C–H appears at 1100 cm^{-1}.

4.5.1.3 Colloid Property Analysis of Ca–OL

1. Grain size distribution

Figure 4.47 shows the volume diameter distribution of Ca–OL colloid at different proportions. The bulk diameter of Ca–OL colloids produced by mixing calcium chloride with sodium oleate is mostly distributed in the range of 0.1–1 μm, and

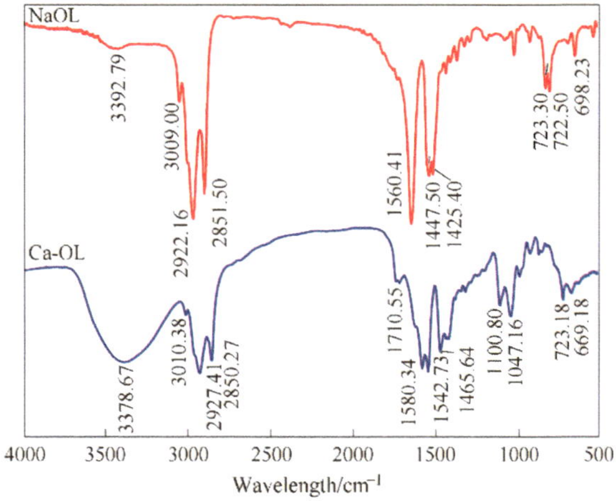

Fig. 4.46 Infrared spectrum of sodium oleate and Ca–OL colloid [58]

Table 4.20 Infrared absorption peaks of sodium oleate functional groups [58]

Infrared absorption peak/cm	Representative group	Type of vibration
3392.79	O–H	Stretching vibration absorption peak
3009.00	= C–H	Stretch vibration absorption peak
2922.16, 2851.50	C–H(–CH_3, –CH_2)	Symmetric vibration absorption peak
1560.41	–COO–	Asymmetric stretching vibration absorption peak
1447.50, 1425.40	–COO–	Symmetric vibration absorption peak
722.50, 698.23	O–H	Out-of-plane bending vibration absorption peak

more than 60% of them are in the range of 190.1–531.2 nm. Only a very few Ca–OL aggregates into large micels with a diameter greater than 1 μm. The volume distribution of Ca–OL generated by mixing calcium chloride and sodium oleate was different. In the three experimental proportions, the colloidal volume diameter of Ca–OL gradually increased with the increase of calcium chloride proportion.

2. Surface active properties

The surface tension results of sodium oleate and Ca–OL colloid measured at different concentrations are shown in Fig. 4.48. Using the pure water used in the test, the results were 71–73 mN/m. As one of the high surfactants, the surface tension of sodium oleate decreased with the increase of concentration, and finally maintained at about 18 mN/m. The critical micellar concentration of sodium oleate is about 1×10^{-3} mol/L, which is basically consistent with the results reported in the literature [59]. The surface tension of Ca–OL solution produced by mixing sodium oleate and

Table 4.21 Infrared absorption peaks of Ca–OL functional groups [58]

Infrared absorption peak/cm	Representative group	Type of vibration
3378.67	O–H	Stretching vibration absorption peak
3010.38	= C–H	Stretch vibration absorption peak
2927.41, 2850.27	C–H(–CH_3, –CH_2)	Symmetric vibration absorption peak
1710.55	C = O	Stretching vibration absorption peak
1580.34, 1542.73	–COO–	Asymmetric stretching vibration absorption peak
1465.64	–COO–	Symmetric vibration absorption peak
1100.80, 1047.16	C–H	In-plane bending vibration absorption peak
723.18, 669.18	O–H	Out-of-plane bending vibration absorption peak

Fig. 4.47 Colloid radius distribution of Ca–OL (NaOL:1 × 10^{-4}mol/L) [58]

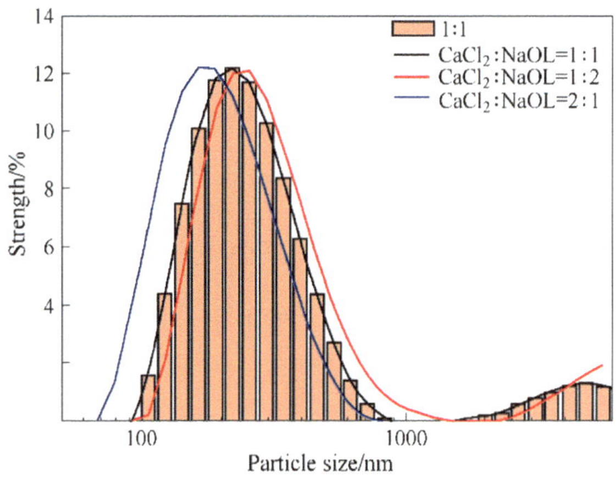

calcium chloride was always higher than that of sodium oleate, but also decreased with the increase of concentration until the visible suspended micelles appeared at 1 × 10^{-3} mol/L. This indicated that a large number of Ca–OL micels were formed in the mixed solution, and the free oleic acid anion in the solution was reduced, leading to the increase of the surface tension of the solution. In addition, the free calcium ions in the solution can electrically neutralize the hydrophilic groups of oleic acid anions, so the surface tension of Ca–OL solution does not rise much compared with that of sodium oleate solution.

3. Foaming properties

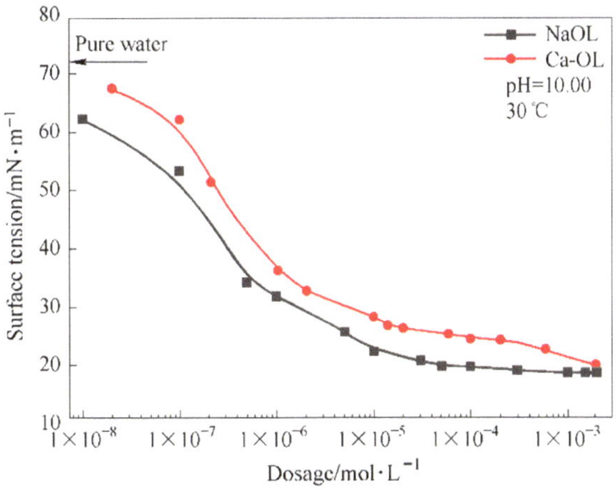

Fig. 4.48 Surface tension curve of sodium oleate and Ca–OL solution [58]

Metal ions have a great influence on the foaming properties of surfactants. Figure 4.49 shows the foam volume of sodium oleate solution with calcium oleate. Sodium oleate itself has the effervescent property. When the concentration is greater than 1×10^{-4} mol/L, the foam quantity begins to show an increasing trend. Once the concentration is higher than 1×10^{-3} mol/L, the amount of foam is too large and leads to the occurrence of overflow trough. In the foam formed by sodium oleate solution, the bottom bubble is small, and the top bubble is mainly large, and the viscosity is low. When calcium ions are introduced into sodium oleate solution, the foaming property of the solution is significantly worse, especially when the concentration of sodium oleate and calcium ions is 2:1, the solution basically does not produce bubbles. This indicates that Ca–OL colloid is formed in the mixed solution, which is not conducive to the formation of foams. When sodium oleate floats semi-soluble calcium-containing minerals, foam is mainly generated by oleate anion.

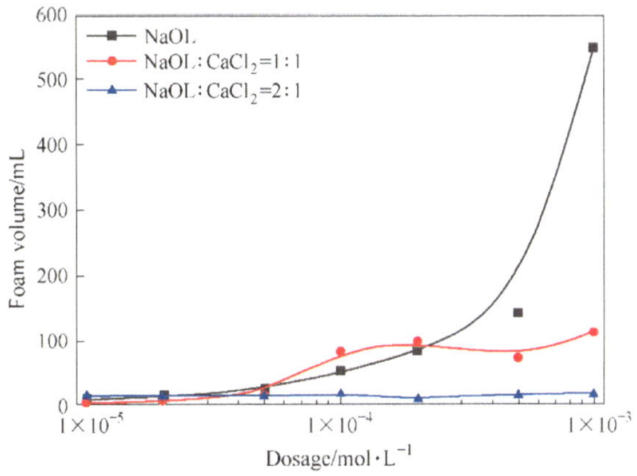

Fig. 4.49 Foaming properties of sodium oleate and Ca–OL [58]

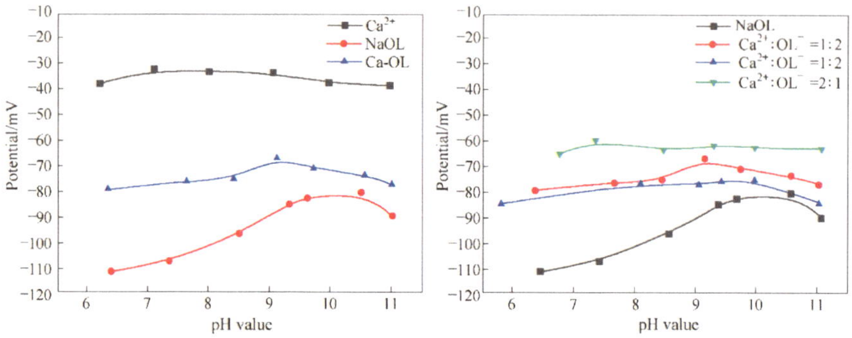

Fig. 4.50 Relationship between Ca–OL colloid electrokinetic potential and pH [58]

4. Surface potential

The electrodynamic potential of calcium ion is always negative, and its electrodynamic potential is almost not affected by pH in the range of pH tested (Fig. 4.50). Sodium oleate electrokinetic potential increased gradually with the increase of pH value, and the highest value appeared at pH = 10.5. In the range of pH tested, the potential curve of Ca–OL colloid is similar to that of Ca^{2+}, which is between Ca^{2+} and NaOL. With the increase of pH, the difference between Ca–OL colloid surface potential and sodium oleate surface potential decreased gradually. With the increase of calcium ion concentration, Ca–OL colloid in solution gradually increased, and the electrokinetic potential also gradually shifted positive.

5. Flotation mechanism

Based on the solution chemistry study of calcium ions and fatty acids, Sun et al. [58] speculated the existence of two modes of action models in fatty acid flotation system, as shown in Fig. 4.51. One is that fatty acid anions interact with calcium dots on mineral surfaces to form fatty acid calcium precipitates, which are then adsorbed on mineral surfaces. The other is that calcium ions dissolved in the solution and oleic acid anions form stable Ca–OL colloid, which then migrates to the mineral surface for adsorption.

4.5.2 *Coordination Reactions of Metal Ions with Hydroxamic Acids and Their Effects*

Chelate collector is an important flotation reagent for tungsten ore flotation, among which benzohydroxamic acid is widely used in industry, but chelate collector flotation tungsten ore often requires activation of metal ions, such as lead ions, iron ions, etc. The interaction between metal ions and flotation collector in the activated flotation process has significant influence on the flotation process.

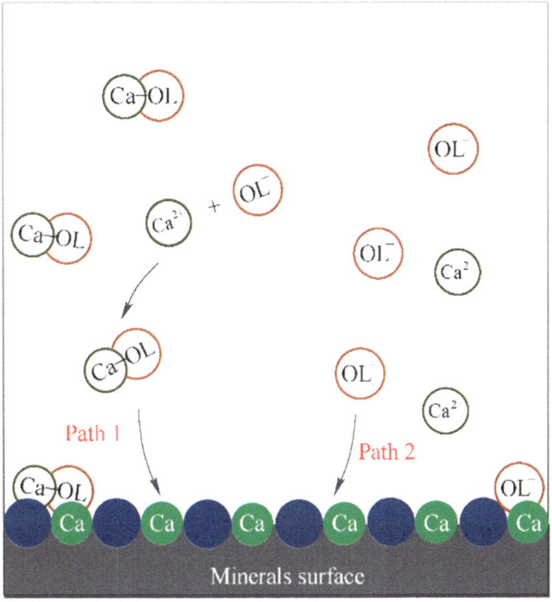

Fig. 4.51 Mechanism of action of sodium oleate and Ca–OL colloid on semi-soluble calcium-containing minerals [58]

4.5.2.1 Stability Constant of Pb–BHA Metal–organic Complex

Many studies believe that lead ion reacts with hydroxamic acid to produce stable PbB_2 complex, but its specific structure has not been confirmed for a long time. Assuming that benzohydroxamic acid reacted with lead nitrate to produce PbB_2, and considering the protonation effect of benzohydroxamic acid, the stability constants at all levels were determined by potentiometric titration.

The formation function $\overline{n}$ represents the mean value of the number of ligands that have been coordinated with the central ion M

$$\overline{n} = \frac{\text{The total concentration of the ligand that has been paired}}{\text{The total concentration of the ion in}} \tag{4.91}$$

For M to form a mononuclear ligand ion system with ligand L, the total concentration of ligands that have coordinated with the central ion is:

$c(\mathrm{ML}) + 2c(\mathrm{ML_2}) + 3c(\mathrm{ML_3}) + \ldots + nc(\mathrm{ML}_n)$ where the total concentration of the central ion is $c(\mathrm{M})_\mathrm{T}$, then,

$$\overline{n} = \frac{c(\mathrm{ML_1}) + 2c(\mathrm{ML_2}) + \ldots + nc(\mathrm{ML}_n)}{c(\mathrm{M}) + c(\mathrm{ML_1}) + \ldots + c(\mathrm{ML}_n)} \tag{4.92}$$

Plug in the expression for the stability constant, then,

$$\overline{n} = \frac{\beta_1 c(\mathrm{L}) + 2\beta_2 c^2(\mathrm{L}) + \ldots + n\beta_n c^n(\mathrm{L})}{1 + \beta_1 c(\mathrm{L}) + \ldots + \beta_n c^n(\mathrm{L})} \tag{4.93}$$

The generating function $\overline{n}$ is just a function of $c(\mathrm{L})$.

$$c(\mathrm{L}) = \frac{c(\mathrm{H})_{\mathrm{T}} - c(\mathrm{H}) + c(\mathrm{OH})}{\sum_{n=1}^{N} n\beta_n^{\mathrm{H}} c^n(\mathrm{H})} \tag{4.94}$$

$$\overline{n} = \frac{c(\mathrm{L})_{\mathrm{T}} - c(\mathrm{L})\left[1 + \sum_{n=1}^{N} \beta_n^{\mathrm{H}} c^n(\mathrm{H})\right]}{c(\mathrm{M})_{\mathrm{T}}} \tag{4.95}$$

where $c(\mathrm{L})$ is the concentration of the ligand; $c(\mathrm{H})_{\mathrm{T}}$ is the total acidity of the system; β_{n}^{H} is the added proton constant for the accumulation of L; $c(\mathrm{L})_{\mathrm{T}}$ is the total concentration of benzohydroxamic acid; $c(\mathrm{M})_{\mathrm{T}}$ is the total concentration of metal ions.

$c_{(\mathrm{BHA})} = 0.01$ mol/L, $c_{(\mathrm{Pb}^{2+})} = 0.01$ mol/L, take 10 mL of BHA and 4 mL of lead nitrate, add 0.5 mL of nitric acid with concentration of 0.068 mol/L, and use potassium nitrate solution with concentration of 0.1 mol/L to volume to 100 mL. In the system, $c_{(\mathrm{BHA})} = 0.001$ mol/L, $c_{(\mathrm{Pb}^{2+})} = 0.0004$ mol/L, titrated with the concentration of 1 mol/L NaOH solution, as shown in Fig. 4.52.

Titration curve of Pb(II)-BHA system Under the titration curves of lead nitrate and BHA, it indicates that lead ions react with BHA to form new products and release H^+ of BHA. The generation curve of Pb(II)-BHA system is drawn according to the formation function formula, as shown in Fig. 4.53.

The stepwise stability constants determined as shown in the figure above are:

$$-\lg c(\mathrm{L})_{n=\frac{1}{2}} = 7.62 \approx \lg K_1 \tag{4.96}$$

$$-\lg c(\mathrm{L})_{n=\frac{3}{2}} = 5.08 \approx \lg K_2 \tag{4.97}$$

K_1 and K_2 are the approximate values of the stepped-stability constants of $[\mathrm{PbB}_2]^+$ and PbB, respectively, and the cumulative stability constant $\lg\beta = 12.70$. It is different

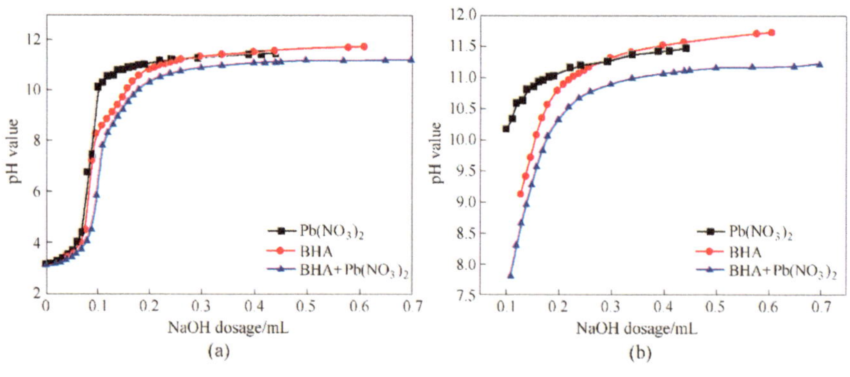

Fig. 4.52 Titration curve of Pb–BHA system [60]

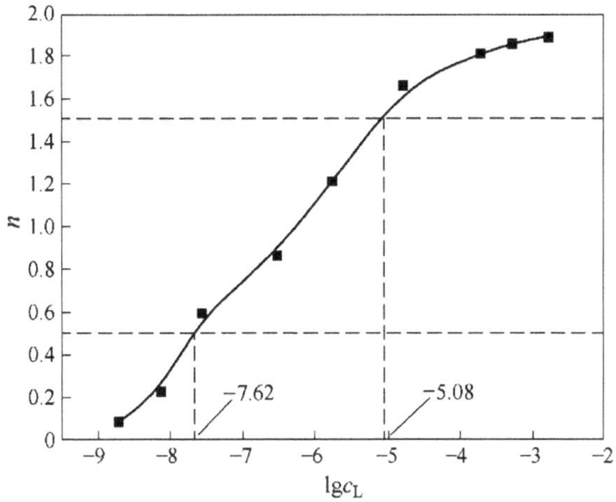

Fig. 4.53 Pb–BHA complex formation curve [60]

from the stability constants of the complex of lead hydroxamic acid (step stability constant $\lg K_1 = 6.7$, $\lg K_2 = 4.0$, cumulative stability constant $\lg \beta = 10.7$), especially the step stability constant. This may be due to the hydroxyl or proton addition reaction in the coordination process of benzohydroxamic acid and lead ion, and the coordination form may not be a single PbB_2 form. Based on the above equilibrium constants, the distribution of lead complexes in the Pb–BHA system as a function of B^- ion concentration was calculated, as shown in Fig. 4.54.

$$Pb^{2+} + B^- \rightarrow (PbB)^+ \quad \beta_1 = \frac{c\left((PbB)^+\right)}{c\left(Pb^{2+}\right)c\left(B^-\right)} = 10^{7.62} \tag{4.98}$$

$$(PbB)^+ + B^- \rightarrow PbB_2 \beta_2 = \frac{c(PbB_2)}{c((PbB)^+)c\left(B^-\right)} = 10^{5.08} \tag{4.99}$$

$$c\left((PbB)^+\right) = \beta_1 c\left(Pb^{2+}\right)c\left(B^-\right) \tag{4.100}$$

$$c(PbB_2) = \beta_2 c\left((PbB)^+\right)c\left(B^-\right) \tag{4.101}$$

$$\begin{aligned} c(Pb)_T &= c\left(Pb^{2+}\right) + c\left((PbB)^+\right) + c(PbB_2) \\ &= c\left(Pb^{2+}\right) + \beta_1 c\left(Pb^{2+}\right)c\left(B^-\right) + \beta_2 c\left((PbB)^+\right)c\left(B^-\right) \\ &= c\left(Pb^{2+}\right) + \beta_1 c\left(Pb^{2+}\right)c\left(B^-\right) + \beta_2\beta_1 c\left(Pb^{2+}\right)c^2\left(B^-\right) \end{aligned} \tag{4.102}$$

$$c(PbB_2) = \beta_2 c\left((PbB)^+\right)c\left(B^-\right) = \beta_2\beta_1 c\left(Pb^{2+}\right)c^2\left(B^-\right) \tag{4.103}$$

$$\phi_{Pb^{2+}} = \frac{c\left(Pb^{2+}\right)}{c\left(Pb^{2+}\right) + \beta_1 c\left(Pb^{2+}\right)c\left(B^-\right) + \beta_2\beta_1 c\left(Pb^{2+}\right)c^2\left(B^-\right)}$$

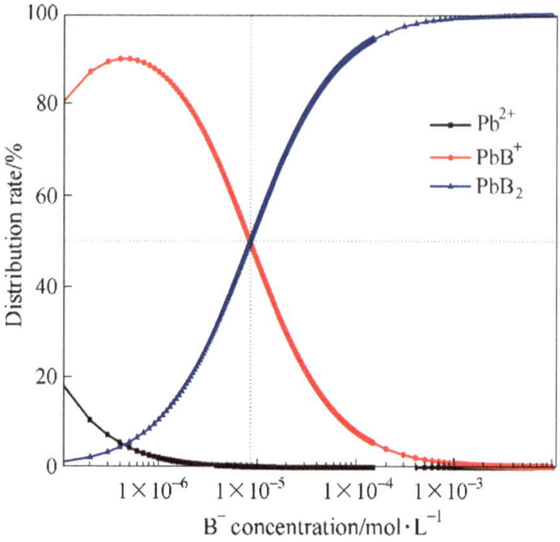

Fig. 4.54 Distribution of coordinated ions at all levels of complex system [60]

$$= \frac{1}{1 + \beta_1 c(B^-) + \beta_2 \beta_1 c^2(B^-)} \tag{4.104}$$

$$\phi_{PbB} = \frac{c((PbB)^+)}{c(Pb^{2+}) + \beta_1 c(Pb^{2+})c(B^-) + \beta_2 \beta_1 c(Pb^{2+})c^2(B^-)} = \phi_{Pb^{2+}} \beta_1 c(B^-) \tag{4.105}$$

$$\phi_{PbB_2} = \frac{c(PbB_2)}{c(Pb^{2+}) + \beta_1 c(Pb^{2+})c(B^-) + \beta_2 \beta_1 c(Pb^{2+})c^2(B^-)} = \phi_{Pb^{2+}} \beta_1 \beta_2 c^2(B^-) \tag{4.106}$$

As can be seen from the chemical composition diagram of benzohydroxamic acid solution, at pH = 9, the ionization of benzohydroxamic acid mainly exists in the form of B^- ion. If the hydroxyl addition of the complex is not considered, when the concentration of B^- ion is greater than 10^{-7} mol/L, lead ion mainly exists in the form of complex. Meanwhile, the single mineral test confirms that at relatively low concentration of B^-, Benzohydroxamic acid on wolframite and scheelite collection ability is very poor, so in the flotation process is very likely to play the role of collection Pb-BHA complex.

It can be seen that no matter what kind of complex form exists in the reaction of Pb^{2+} and BHA, when pH = 9 is around, lead ions mainly exist in the form of complex, free Pb^{2+} or $Pb(OH)^+$ is very little, and some or several complex components may play a major role in the flotation process. However, the coordination reaction between lead ion and benzohydroxamic acid is complex, and its stable complex is not only PbB_2, so it is necessary to consider the influence of hydroxyl, hydrogen ions and other coordination ions.

4.5.2.2 Composition Status of Pb-BHA Metal–organic Complexes in Solution System

Figures 4.55 and 4.56 respectively show the UV/visible spectrum of benzohydroxamic acid and lead nitrate at different concentrations, and their characteristic peaks have slight shifts at different concentrations. The characteristic peaks of benzohydroxamic acid are about 233–237 nm, and that of lead nitrate are 233 nm, 260–263 nm and 297–298 nm.

1. **The continuous change method was used to determine the coordination number of the complex**

c(BHA) = 0.001 mol/L, c(Pb) = 0.001 mol/L, mixed in different proportions, the total volume of the solution was 10 mL, and ultraviolet spectra of different proportions were determined by ultraviolet spectrophotometer, as shown in Fig. 4.57. The results

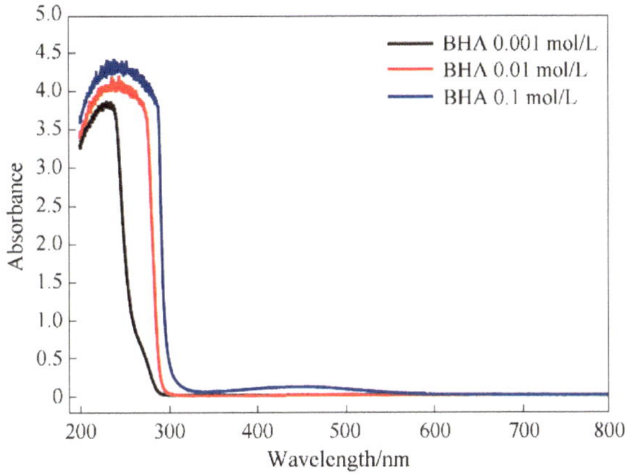

Fig. 4.55 UV/visible spectrum of benzohydroxamic acid (c(BHA) = 0.001 mol/L characteristic absorption peak wavelength 233.5 nm, c(BHA) = 0.01 mol/L characteristic absorption peak wavelength 235 nm, c(BHA) = 0.1 mol/L characteristic absorption peak wavelength 237 nm) [60]

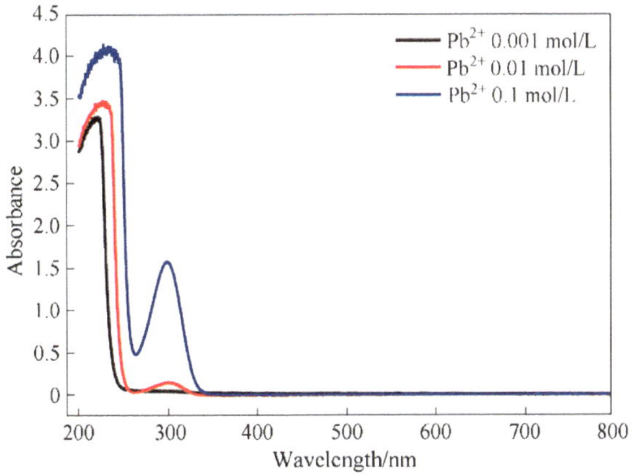

Fig. 4.56 UV/visible spectrum of $Pb(NO_3)_2$ (c(Pb) = 0.001 mol/L characteristic absorption peak wavelength 221.5 nm, c(Pb) = 0.01 mol/L characteristic absorption peak wavelength 232.5/261/297 nm, c(Pb) = 0.1 mol/L characteristic absorption peak wavelength 233/263/298 nm) [60]

showed that the characteristic peak of benzohydroxamic acid and lead nitrate was shifted greatly after mixing, indicating that new products were generated by the reaction of the two. And with the change of the ratio of the two, a series of changes occurred in the peak absorbance of the solution, indicating that the change of the ratio of the two affected the composition and distribution of the complex in the solution system.

Figure 4.58 shows the curve of absorbance variation with the ratio of benzohydroxamic acid and lead nitrate at specific wavelength. Two peaks appear in the curve, indicating that benzohydroxamic acid reacts with lead nitrate to produce a variety of complexes, whose coordination numbers are difficult to be determined by continuous change method, but it can be inferred that there are at least two stable complexes.

2. **Determination of complex coordination number by concentration change method**

(1) Fixed metal ion—transform ligand concentration

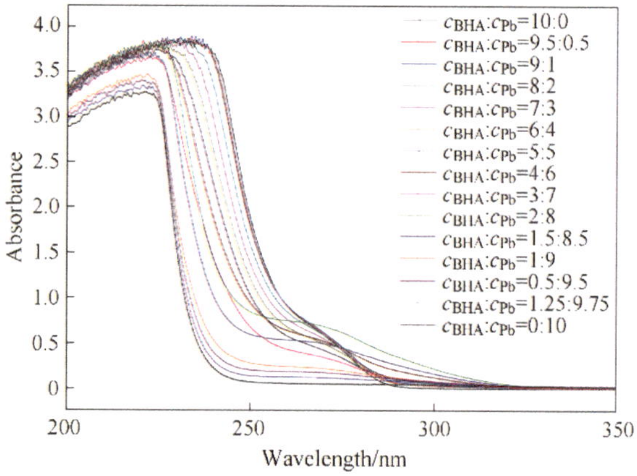

Fig. 4.57 Ultraviolet/visible light spectra of complexes of different proportions [60]

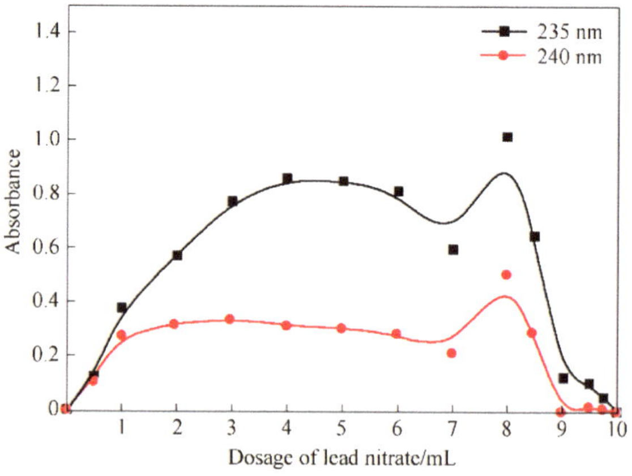

Fig. 4.58 Absorbance-ratio curve at specific wavelength [60]

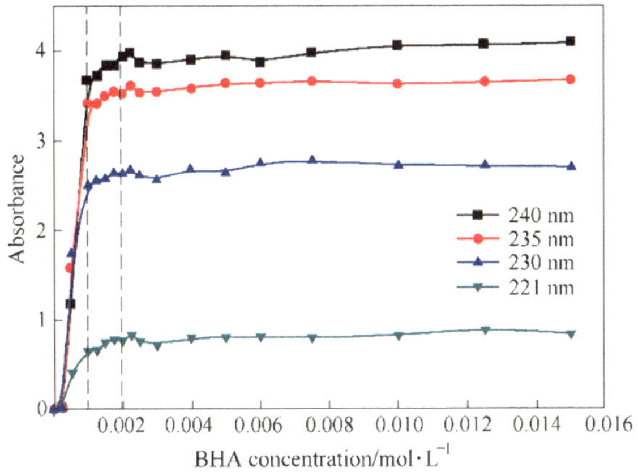

Fig. 4.59 Absorbance-BHA concentration curve [60]

c(Pb) = 0.001 mol/L was fixed, and the concentration of benzohydroxamic acid was gradually changed. The absorbance was measured at 221 nm, 230 nm, 235 nm, and 240 nm respectively, and the absorbence-BHA concentration change curve was drawn, as shown in Fig. 4.59. The results show that the inflection point occurs when c(BHA) is about 0.001–0.002 mol/L, and the molar ratio of lead ion to BHA is about (1–2):2, which indicates that the coordination between lead ion and BHA is not simple to generate PbB_2, but other anions must participate in the reaction.

(2) Fixed ligand—transformed metal ion concentration

c(BHA) = 0.001 mol/L was fixed, and the concentration of lead ion was gradually changed. The absorbance was measured at 235 nm and 240 nm respectively, and the absorbance-Pb^{2+} concentration change curve was drawn, as shown in Fig. 4.60. The results show that two platform regions appear in the absorbance-Pb^{2+} curve within the measurement range. The first platform region is estimated to be the complex generated under the condition of low Pb/BHA ratio, while the second platform region is the complex generated under the condition of high Pb/BHA ratio, and the complex is more stable under the condition of low ratio.

Therefore, it is basically determined by UV/visible light spectrophotometry that benzohydroxamic acid reacts with Pb^{2+} to generate at least two relatively stable complexes. Considering that the reaction of Pb^{2+} with OH^- ion may lead to the formation of hydroxylated ions, PbB_2 or $Pb_m(OH)_{2m-1}B$ is speculated, and the former is more stable.

4.5.2.3 Solution Chemistry of Lead Benzohydroxamic Acid Complexes in the Interfacial Region

The above experimental studies show that Pb-BHA complexes have similar properties to Pb^{2+}: they are easy to combine with hydroxyl group and precipitate in a certain

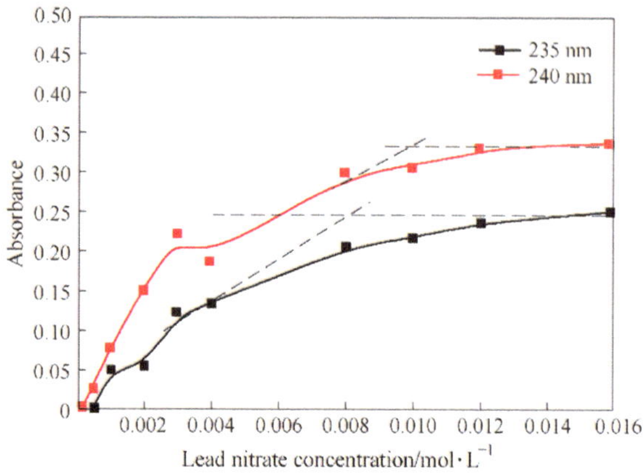

Fig. 4.60 Absorbance-Pb^{2+} concentration curve [60]

pH range. Therefore, it can be speculated that the behavior of Pb-BHA complexes on mineral surfaces may be similar to that of lead ions on mineral surfaces.

When the adsorption of Pb-BHA complexes at the interface reaches equilibrium, a certain component in the process of ionic hydrolysis also reaches equilibrium in the solution and the interface region, and its potential chemical phase is equal. According to solution chemistry, the concentration of component i in the ionic solution in the mineral/water interface region is larger than the concentration of component i in solution. The pH required for the complex to precipitate in the mineral/water interface region is lower than the pH required to produce a precipitate in solution. When the positioning ion on the surface of the mineral is H^+ or OH^-, the adsorption of H^+ is approximately treated as electrostatic force only, and the relationship between the pH of the interface region and the pH of the solution is

$$\mathrm{pH}' = \mathrm{pH} + \frac{\mathrm{e}\psi_d}{2.3RT} \tag{4.107}$$

When $\psi_d < 0$, the potential on the mineral surface is negative, $\mathrm{pH}' < \mathrm{pH}$; and vice versa, $\mathrm{pH}' > \mathrm{pH}$.

In the optimal flotation pH range of scheelite and wolscheelite (8–10), the mineral surface is negatively charged ($\psi_d < 0$), the pH in the mineral interface region is lower than that of the solution, but the concentration of the complex in the interface region is higher than that of the solution system, and the dielectric constant in the interface region is far smaller than that of the solution. Therefore, in the interface region, Pb-BHA complexes may generate hydroxyl complex precipitation on the surface of minerals, so as to realize the collection of minerals. The adsorption mechanism of Pb-BHA complexes on tungsten ore surface will be discussed in detail in Chap. 5.

4.5.3 *Coordination Reactions of Metal Ions with Silicic Acid Colloid and Their Effects*

Sodium silicate is a commonly used inhibitor in tungsten flotation, and the introduction of different metal ions (Fe^{2+}, Fe^{3+}, Ca^{2+}, Mg^{2+}, Pb^{2+}, Al^{3+}) has a great impact on its flotation effect. Among them, Al^{3+} modified sodium silicate is the most commonly used in industry, but its specific action mechanism is controversial. This section discusses the mechanism of action from the point of view of metal ion coordination assembly.

4.5.3.1 Infrared Spectroscopic Analysis of Al–Na_2SiO_3 Polymerization Structure

In order to study the self-assembly mechanism of Al–Na_2SiO_3 polymer, it is necessary to first characterize its structure. The infrared spectrum characterization results of Na_2SiO_3 and Al–$NaSiO_3$ polymer are shown in Fig. 4.61.

According to literature reports [63, 64], the absorption spectrum peaks in the infrared spectrum of standard sodium silicate are shown in Table 4.22. According to Fig. 4.61 and Table 4.22, the absorption bands 3324.73 cm^{-1} and 3396.63 cm^{-1} are the vibration absorption peaks of intermolecular associated hydroxyl group. The peak at 3396.63 cm^{-1} is the stretching vibration absorption peak of the associated hydroxyl group between sodium silicate molecules, 1646.20 cm^{-1} is the bending vibration absorption peak of water molecule hydroxyl group, 990.71 cm^{-1} and 460.92 cm are the stretching vibration peak of $SiO_4{}^{4-}$ ion and the bending vibration peak of Si–O–Si in $SiO_4{}^{4-}$ ion, respectively.

Fig. 4.61 Infrared spectra of Na_2SiO_3 and Al–Na_2SiO_3 [40, 61, 62]

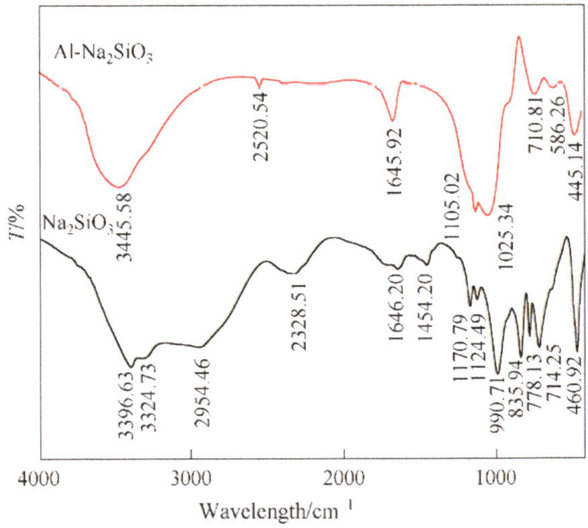

Table 4.22 Absorption peaks of water glass infrared spectrum [61]

Group name	Infrared absorption peak/cm^{-1}	Absorption peak shape
Free hydroxyl group	3670–3580, 1500–1250	The former is pointed, while the latter is strong and wide
Intermolecular association hydroxyl group	3550–3200	–
Dimer hydroxyl	3550–3450	Pointier
Polymeric hydroxyl group	3400–3200	Strong and wide
Crystallised hydroxy group	3600–3100 1645–1615	The former is weak and pointed, while the latter is wide and weak
Si–OH	3680 ± 10	–
Si–O–Si	1093–1020, 500–400	The former is strong and sharp

By comparing the IR spectra of Na_2SiO_3 and Al–Na_2SiO_3 polymer, two new absorption peaks of 2520.54 cm^{-1} and 586.26 cm^{-1} appeared in the spectra of Al–Na_2SiO_3. At the same time, the absorption peaks at 2954. 46 cm^{-1}, 2328. 51 cm^{-1}, 1454.20 cm^{-1} and 835.94 cm^{-1} disappeared, and the other absorption peaks shifted to different degrees, which indicated that the self-assembly reaction between Al ion and Na_2SiO_3 resulted in a certain degree of change in the structure of sodium silicate. The absorption peak at 3445.58 cm^{-1} was generated by the stretching vibration of the free hydroxyl group, and the absorption peak at 2520.54 cm^{-1} was mainly the result of O–Al bonding. The angular deformation frequency of water molecules was within the region of 1645.92 cm^{-1}. The absorption peaks at 1105.02 cm^{-1} and 445.14 cm^{-1} were generated by Si–O–Al vibration formed between sodium silicate, aluminum ion and Al–Na_2SiO_3 polymer, and the absorption peaks at 586.26 cm^{-1} could be attributed to the bending vibration of Al–OH super imposed on the absorption peaks of water molecules [65]. Therefore, through the change of peak strength at 1104.11 cm^{-1}, 445 cm^{-1} and 586.26 cm^{-1}, it can be seen that after the reaction of aluminum ion with sodium silicate, aluminum ion formed a bond with Si–O bond, and thus Si–O–Al bond appeared [66].

By comparing and analyzing the infrared spectra of Na_2SiO_3 and Al–Na_2SiO_3 polymers, it is shown that the self-assembly reaction between aluminum ion and NaSiO changes the structure of sodium silicate. The appearance of Si–O–Al vibration peak indicates that aluminum ion and sodium silicate are not simply combined, but interact with each other. The two are linked by chemical bonds. The self-assembly results in a more stable Al–Na_2SiO_3 polymer.

4.5.3.2 Self-Assembly of Al^{3+} with Na_2SiO_3 in Aqueous Solution

As a strong base and weak acid, Na_2SiO_3 will undergo a series of hydrolysis reactions in aqueous solution to generate hydrolytic products of various components. The hydrolytic reaction products and the distribution of each component have been discussed in detail in Sect. 4.3.1 and will not be repeated here. In addition to the hydrolysis reaction, the $Si(OH)_4$ molecules in Na_2SiO_3 solution will undergo further polymerization reaction to produce polymeric silicic acid with a larger molecular weight. $Si(OH)_4$ has different polymerization behavior under different pH conditions. $Si(OH)_4$ is combined with H^+ in acidic solution to form $H_5SiO_4^+$, and the concentration increases with decreasing pH value. Therefore, in acidic solution, $Si(OH)_4$ and $H_5SiO_4^+$ undergo polymerization reaction to form polysilicic acid, and according to this reaction to form polysomes, the reaction formula is as follows.

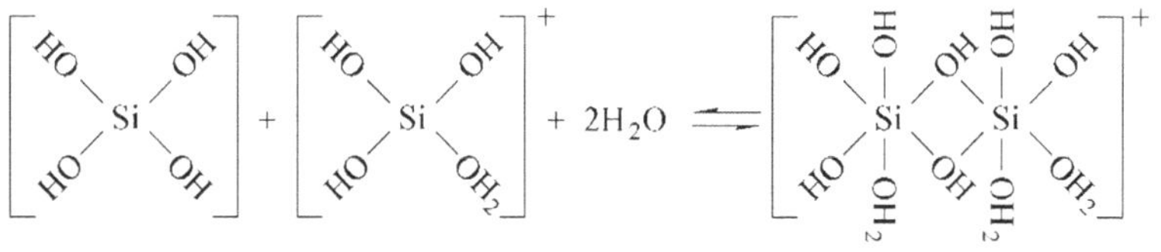

$Si(OH)_4$ is formed by the loss of hydrogen ions in neutral or alkaline solution to form $H_3SiO_4^-$, so under alkaline or neutral conditions, $Si(OH)_4$ and $H_3SiO_4^-$ polymerization reaction to form dipolysilicic acid, the reaction equation is as follows:

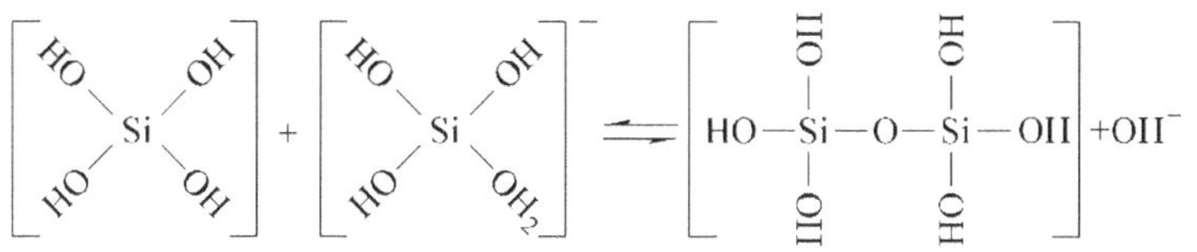

Neutral molecules of dipolysilicate can interact with OH^- to form $H_5Si_2O_7^-$ negative ions, and neutral molecules of dipolysilicate and negative ions can be further polymerized with $H_5Si_2O_7^-$ to form silicic acid molecules with higher degree of polymerization. When the polymerized polysilicate continues to polymerize, the reaction can occur not only at the end of the chain, but also in the middle of the chain, so spherical or ellipsoidal silica colloidal particles will be formed. The internal structure of polysilicate acid is a network structure formed by siloxane bond (–Si–O–Si–) connection, and the surface contains a large number of hydroxyl groups (–OH) [67]. The structure diagram is shown in Fig. 4.62.

After aluminum salt was added to the aqueous solution, Al^{3+} underwent step-by-step hydrolysis reaction. In aqueous solution, aluminum ions will coordinate with 6 water molecules to form the octahedral ligand structure hydrate of $[Al(H_2O)_6]^{3+}$and $[Al(H_2O)_5OH]^{2+}$ etc. Under the alkaline environment, Al^{3+} will undergo the following hydrolysis reaction in aqueous solution [40, 61, 69].

$$Al^{3+} + H_2O = Al(OH)^{2+} + H^+ \quad K = 1 \times 10^{-4.99} \tag{4.108}$$

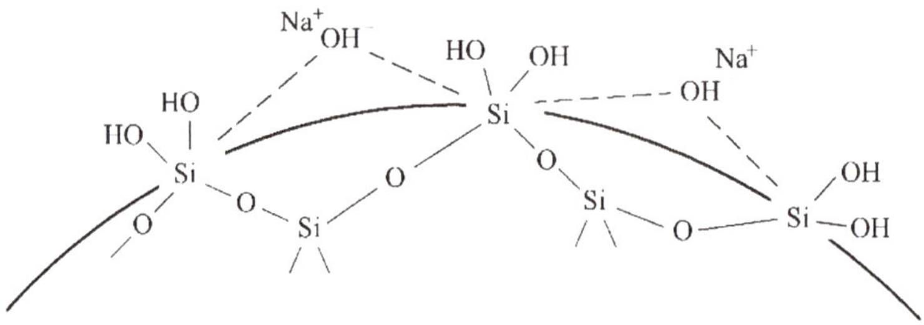

Fig. 4.62 Schematic diagram of the colloidal structure of polysilicate [67, 68]

$$Al^{3+} + 2H_2O = Al(OH)_2^+ + 2H^+ \quad K = 1 \times 10^{-10.0} \tag{4.109}$$

$$Al^{3+} + 3H_2O = Al(OH)_3(aq) + 3H^+ \quad K = 1 \times 10^{-9.35} \tag{4.110}$$

$$Al^{3+} + 4H_2O = Al(OH)_4^- + 4H^+ \quad K = 1 \times 10^{-23.0} \tag{4.111}$$

The contents of hydrolyzed components of Al^{3+} under different pH conditions are plotted by the above formula, as shown in Fig. 4.63. When pH is in the range of 3–5, Al^{3+}, $Al(OH)^{2+}$, $Al(OH)_2{}^+$ and other monomer hydroxyl complex ionic components are dominant. When pH is in the range of 7–8, $Al(OH)_3$ is the main component, and when pH > 9, $Al(OH)_4{}^-$ is the main component.

The resulting mononuclear hydrolyzed component will also undergo hydroxyl bridging polymerization to form dimer aluminum hydrate, such as.

Fig. 4.63 Component contents of monomer aluminum solution under different pH conditions [40]

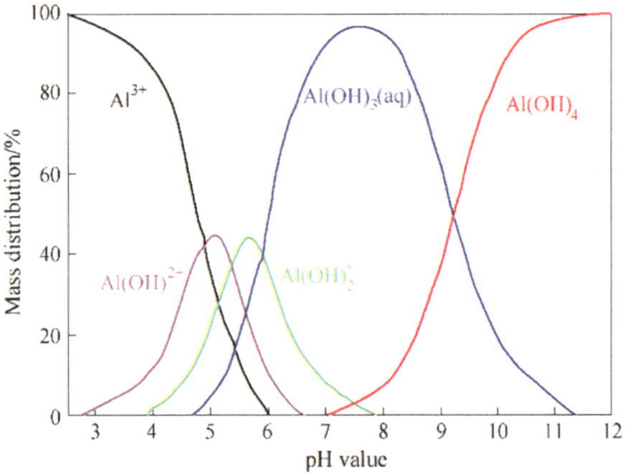

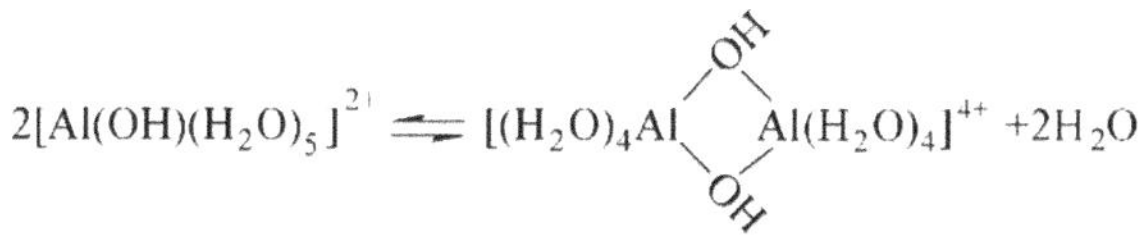

The dimers will continue to polymerize into polymers, resulting in the formation$_2$ of highly charged complex ions with different degrees of polymerization such as $[Al_2(OH)_2(H_2O)_8]^{4+}$, $[Al_3(OH)_4(H_2O)_{10}]^{5+}$, $[Al_6(OH)_{15}]^{3+}$, $[Al_7(OH)_{17}]^{4+}$. When Al^{3+} is added into the sodium silicate solution, on the one hand, the hydrolyzed product of monomer aluminum may use the hydroxyl group on the surface or structure of polysilicate acid as a bridge polymerization reaction to generate poly aluminum hydrate. On the other hand, the complex ions of the monomer aluminum hydroxyl group and the poly aluminum hydrate will undergo dehydration and polycondensation reaction with the hydroxyl group on the surface or structure of the polysilicate to form a new composite aluminum silicate polymer [59, 67]. Taking the hydrolyzed components of monomer silicic acid and the hydrolyzed products of monomer aluminum as examples, the dominant components $Si(OH)_4$ and $Al(OH)_3$ in the solution undergo dehydration and polycondensation under acidic and neutral conditions as follows [40, 61].

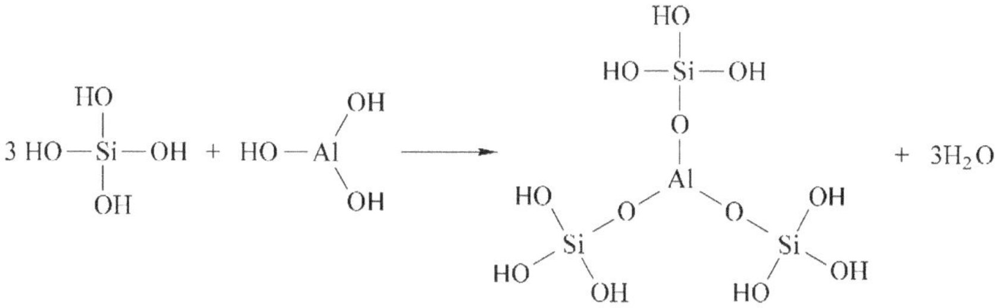

Under alkaline conditions, the main groups in the solution can be divided into $H_3SiO_4^-$ and $Al(OH)_4^-$ for the following dehydration polycondensation reaction [40, 61].

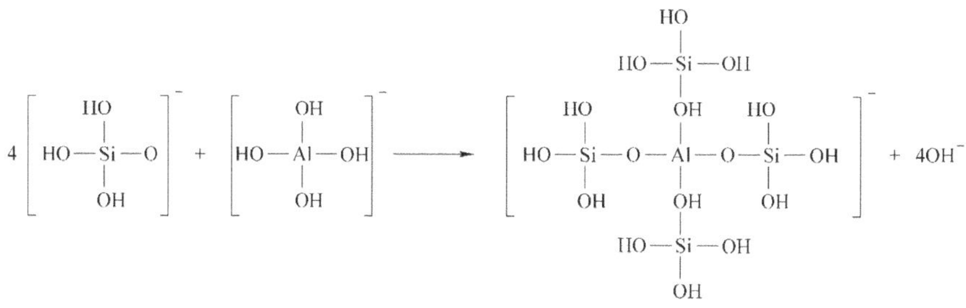

As a result of the comprehensive action between the monomer aluminum hydroxyl complex ion and the silicic acid component, there are Si–O–Al bonds besides Si–O–Si bonds and O–Al–O bonds in the system. The silico-oxygen tetrahedron and al-oxygen tetrahedron form a three-dimensional network structure of the composite aluminosilicic acid polymer with oxygen atoms as the bridge, which is consistent with the results of infrared spectrum detection [40, 61]. The above dehydration condensation reaction will also occur between the polymerization of aluminum hydrolyzed products and the polymerization product of silicic acid polysilicic acid, resulting in the formation of more complex structure and larger groups of aluminosilicic acid polymer. Solution chemical reaction description: aluminum hydrolyzed products can react with silicate acid through chemical bonding to form Al–Na_2SiO_3, Al–Na_2SiO_3 polymer, and the poly aluminum hydrate can react with polysilic acid to form Al–Na_2SiO_3 polymer with more complex structure. Compared with sodium silicate, the structure of Al–Na_2SiO_3 has changed, the composition of the solution is more complex, the molecular weight of the polymer micelles is larger, and the physical and chemical properties have also changed to a certain extent.

4.5.3.3 Influence of Al^{3+} on the Potential Properties of Na_2SiO_3 Colloidal Particles

In mineral flotation, the electrical properties of reagents have a great impact on the adsorption of reagents on mineral surfaces. In order to study the adsorption of Na_2SiO_3 and Al–Na_2SiO_3 on mineral surfaces, it is necessary to first characterize the electrical properties of both. The Zeta potential of 100 mg/L Na_2SiO_3 and Al–Na_2SiO_3 in pure water solution was determined, as shown in Fig. 4.64. The results show that in the range of pH = 4–12, the colloidal potential of sodium silicate is negative, and with the increase of pH value, the potential is negative. In the range of pH 8–10, the potential is about –20 mV. For Al–Na_2SiO_3, its potential changes more dramatically in the range of pH = 4–12, its potential is positive when pH < 5, as pH increases, the potential is constantly negative, in the range of pH 8–10, the potential is about −40 mV.

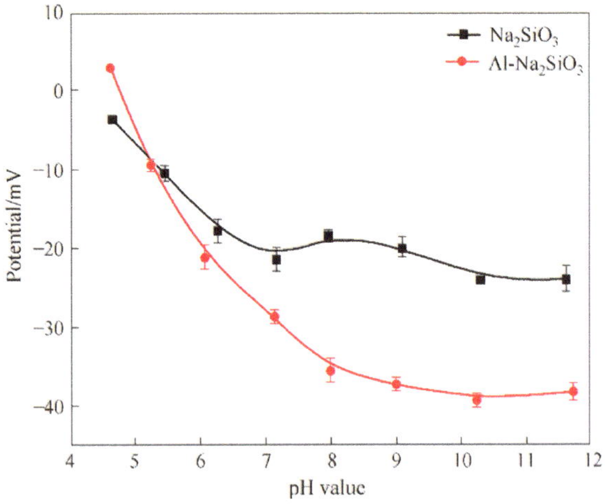

Fig. 4.64 Potentiodynamic difference between Na_2SiO_3 and Al–Na_2SiO_3 [62]

In the range of pH = 4–12, the potential of Al–Na_2SiO_3 is more negative, which may be due to the self-assembly reaction between aluminum ion and sodium silicate, which leads to the polymerization of sodium silicate into a polymer with higher molecular weight. The hydroxyl groups on the surface of Al–Na_2SiO_3 polymer become more negative. On the other hand, due to the hydrolysis of aluminum ions in aqueous solution, the hydrolysis reaction of sodium silicate was promoted, resulting in more silica colloidal particles in the solution, which made the potential of the whole system more negative. Compared with Na_2SiO_3, Al–Na_2SiO_3 has a more negative potential in the flotation range of pH = 8–10, which makes it more difficult to adsorb Al–Na_2SiO_3 on the surface of negatively charged scheelite, and easier to adsorb al-Nasio on the surface of positively charged minerals such as calcite, which will facilitate its adsorption on the surface of calcite. But not conducive to adsorption on scheelite surface.

4.6 Summary

Tungsten solution chemistry plays an important role in the theoretical research and production practice of tungsten flotation. Scheelite and its associated minerals belong to salt minerals, which have dissolution reaction and a series of side reactions in the pulp system. The ions released by the dissolution of minerals and metal ions in the solution will react with the surface of the associated minerals, complete different degrees of surface transformation, change the solution components, mineral surface potential, surface topography, surface active sites and other properties, thus affecting the flotation behavior of minerals and the role of agents on the surface of minerals; on the other hand, flotation reagents occur dissociation and chemical reactions in the solution system, which determines their adsorption behavior on the surface of minerals, and then affects the flotation behavior of minerals. Therefore, the in-depth understanding of tungsten flotation solution chemistry is conducive to promoting the progress of tungsten flotation technology.

Acknowledgements This work was supported by National key research center and development Program of the 14th Five-Year Plan (No. 2022YFC2905105 and No. 2022YFC2905104), National Natural Science Foundation of China (NSFC) (No. 52122406and No. 52374291), Hunan High-tech Industry Technology Innovation Leading plan (No. 2022GK4056), Hunan Innovative Province construction special Project (No. 2020RC3001).

References

1. Wang D, Hu Yuehua (1988) Flotation solution chemistry
2. QI M (2015) Study on the relationship between surface characteristics and floatability of wolframite. Jiangxi University of Science and Technology

3. Zhang H (2017) Study on the effect of metal ions on wolframite flotation behavior and its mechanism. Kunming University of Science and Technology
4. Luo L (2013) Flotation performance evaluation of chelated collector for wolfrelite. Jiangxi University of Science and Technology
5. Hu W (2013) Study on mechanism and application of combined collector in flotation of fine wolframite. Jiangxi University of Science and Technology
6. Meng Q, Yuan Z, Ma L et al (2018) Journal of Northeastern University (Natural Science Edition), 39(04):599–603
7. Meng QY, Feng QM, Ou LM (2016) Flotation behavior and adsorption mechanism of fine wolframite with octyl hydroxamic acid. J Central South University 23(6):1339–1344
8. Meng QY, Feng QM, Shi Q et al (2015) Studies on interaction mechanism of fine wolframite with octyl hydroxamic acid. Miner Eng 79:133–138
9. Marinakis KI, Kelsall GH (1985) Adsorption of dodecyl sulfate and decyl phosphonate on wolframite, (Fe, Mn)WO_4, and their use in the two-liquid flotation of fine wolframite particles. J Colloid Interface Sci 106(2):517–531
10. Yuehua H, Chi R, Xu Z (2003) Solution chemistry study of salt-type mineral flotation systems: role of inorganic dispersants. Ind Eng Chem Res 42(8):1641–1647
11. Hu Y-H, Wang D-Z (1996) Study on flotation behavior of scheelite/fluorite in solution chemistry. Min Metall 5(1):7
12. Wang R, Han H, Sun W et al (2021) Surface transformation behavior of calcium-bearing minerals and its application in flotation. Metal Mine 06:52–59
13. Zeng QJ, Lin RX, Zhang XH et al (2006) Study on mineral processing technology of scheelite in north-east China. J Guangdong Non-Ferrous Metals
14. Qin Q, Hu Y (1999) Chemical behavior of semi-soluble salt mineral solution and its effect on mineral floatability. Min Metall Eng 19(2):4
15. Zou Y (2016) Study on synthesis. Jiangxi University of Science and Technology, Surface solubility and adsorption properties of scheelite/calcite/fluorite
16. Yang YU, Sun CY, Shuo-Shi LU (2013) Study of floatability and crystal chemistry analysis of scheelite and calcium minerals. J China Univ Min Technol 42(2):278–283
17. Liu H (2010) Research on new flotation process of refractory scheelite at room temperature. Central South University
18. Hu Y, Wang D (1990) Study on the performance of alkylamine on the collection of salt minerals in solution chemistry. J Central South University Min Metall 21(1):8
19. Li S (2010) Experimental study on separation of scheelite from calcium-bearing gangue minerals by flotation with cationic collector. Central South University
20. Gao Y, Qiu X, Feng Q (2003) Study on chemistry of flotation solution of scheelite by benzohydroxamic acid. Nonferrous Metals (Min Process Sect) 000(004):28–31
21. Cheng X (2000) Study on the new separation method of tungsten minerals and calcium-bearing minerals and the action mechanism of chemical agents. Metal Ore Dressing Abroad
22. Sun W, Tang H, Chen C (2013) Solution chemical behavior of sodium silicate in flotation separation system of fluorite and scheelite. Chinese J Nonferrous Met 23(08):2274–2283
23. Zhang Z, Wang B (1995) Effect of mutual transformation of mineral surface on flotation of co-existing system of salt minerals. Metal Mine 4:5
24. Zhang Z, Wang B (1994) Mutual transformation of mineral surface in coexistence system. Western Explorat Eng 6:9–11
25. Xiao L, Chen J (1992) Flotation solution chemistry of salt minerals. Chinese J Nonferrous Metals 000(003):19–24
26. Hu Y, Wang D (1992) Dissolution, surface property change and flotation separation control design of salt minerals. J Central South University: Sci Technol 023(003):273–279
27. Zhang Z, Wang B (1997) Dissolution and transformation of minerals in flotation of salt minerals. Nonferrous Metals Sci Eng 011(003):20–23
28. Ruolin, Wang, Zhao, et al (2019) Fluorite particles as a novel calcite recovery depressant in scheelite flotation using Pb-BHA complexes as collectors—ScienceDirect. Miner Eng 132:84–91

29. Huang W, Xu T, Han H et al (2019) Fatty acid flotation versus BHA flotation of tungsten minerals and their performance in flotation practice. Metal Mine
30. Han HS, Liu WL, Hu YH et al (2017) A novel flotation scheme: selective flotation of tungsten minerals from calcium minerals using Pb–BHA complexes in Shizhuyuan. Rare Met 36(06):533–540
31. Hu Y, Han H, Tian M et al (2018) Action mechanism and application of lead benzohydroxamic acid metal-organic complex in flotation of oxidized ore. Conserv Utilization Miner Resour 1:7
32. Wang R, Sun W, Han H et al (2021) Fluorite particles as a novel barite depressant in terms of surface transformation. Miner Eng 166(6):106877
33. Zuo Q, Zhang Q, Deng B et al (2011) Effect of three kinds of adjuster on dispersion behavior of fine-grained hematite. Metal Mine 2:3
34. Wang Y, Chen X, Hu Y (2007) Effect of sodium carbonate on dispersion behavior of fine aluminum silicate minerals. J China Univ Min Technol 36(3):6
35. Feng B, Lu Y, Weng C (2016) Dispersion mechanism of carbonate on serpentine/pyrite flotation system. J Central South Univ Sci Technol 47(4):7
36. Tang J, Zhang S, Zhang J et al (2020) Dispersion mechanism of sodium carbonate on fine fluorite and quartz. Non-metallic Ores
37. Zhu C (1990) Effect of pH regulator on separation of scheelite from calcite and fluorite. Min Metall Eng 10(1):5
38. Chen C (2011) Effect of inorganic anions on flotation behavior of three typical calcium-containing salts and its mechanism. Central South University
39. Marinakis KI, Shergold HL (1985) Influence of sodium silicate addition on the adsorption of oleic acid by fluorite, calcite and barite. Int J Miner Process 14(3):177–193
40. Wei Z, Hu Y, Han H et al (2019) Selective ZSSEMBLY of H_2SiO_3 Polymer Using Al_3^+ in Pb-BHA Flotation. Minerals 9(1)
41. Separation of scheelite from fluorite, calcite and quartz by Flotation. Central South University
42. Dong LY, Jiao F, Qin WQ et al (2019) Selective depressive effect of sodium fluorosilicate on calcite during scheelite flotation. Miner Eng 131:262–271
43. Zhang H, Xu X, Chen G et al. Application and mechanism of sodium hempetaphosphate in phosphate ore flotation. Conserv Utilization Miner Resour 20,40(6):6
44. Feng Q, Zhou Q, Zhang G et al (2011) Inhibition mechanism of sodium hmetaphosphate on calcite. Chinese J Nonferrous Metals 21(02):436–441
45. Wang J, Yin W, Sun Z (2019) Study on selective inhibition and mechanism of scheelite and apatite flotation. Nonferrous Metals Eng 9(02):66–69
46. Dong L, Jiao F, Qin W et al (2021) New insights into the depressive mechanism of citric acid in the selective flotation of scheelite from fluorite. Miner Eng 171(107):117
47. Yao J, Li D, Yin W et al (2017) Dispersion mechanism of citric acid in flotation system of Hematite containing carbonate. J Northeastern University (Nat Sci Edn) 38(05):720–724
48. Wei Z, Fu J, Han H et al (2020) A highly selective reagent scheme for scheelite flotation: polyaspartic acid and Pb–BHA complexes. Minerals 10
49. Dong LY, Jiao F, Qin WQ et al (2019) New insights into the carboxymethyl cellulose adsorption on scheelite and calcite: adsorption mechanism, AFM imaging and adsorption model. Appl Surf Sci 463:105–114
50. Lopez-Valdivieso A, Lozano-Ledesma LA, Robledo-Cabrera A et al (2017) Carboxymethyl-cellulose (CMC) as PbS depressant in the processing of Pb–Cu bulk concentrates. Adsorption and studies on floatability. Miner Eng 112:77–83
51. Yang Y (2010) Mixing effect of fatty acid collector in scheelite flotation. Central South University
52. Jiang Q-M (2009) Effect of mixed fatty acids on separation of schetungsten from fluorite and calcite. Central South University (in Chinese)
53. Fa K, Nguyen AV, Miller JD (2005) Hydrophobic attraction as revealed by AFM force measurements and molecular dynamics simulation. J Phys Chem B 109(27):13112–13118
54. Fa K, Nguyen AV, Miller JD (2006) Interaction of calcium dioleate collector colloids with calcite and fluorite surfaces as revealed by AFM force measurements and molecular dynamics simulation. Int J Miner Process 81(3):166–177

55. Free ML, Miller JD (1996) The significance of collector colloid adsorption phenomena in the fluorite/oleate flotation system as revealed by FTIR/IRS and solution chemistry analysis. Int J Miner Process 48(3):197–216
56. Fa KQ, Tao JA, Nalaskowski J et al (2003) Interaction forces between a calcium dioleate sphere and calcite/fluorite surfaces and their significance in flotation. Langmuir 19(25):10523–10530
57. Sun W, Han H, Sun W et al (2022) Novel insights into the role of colloidal calcium dioleate in the flotation of calcium minerals. Miner Eng 175
58. Sun W (2017) Research on mechanism and application of Ca-oleic acid colloid collector in flotation of calcium-containing minerals. Central South University, Changsha (in Chinese)
59. Fang Y, Noihiko O (1995) Study on the Relationship between composition and structure of aluminum-modified silica sol Colloidal particles. J Chinese Ceramic Soc 23(2):6
60. Han H (2017) Study on application and mechanism of new metal complex collector in flotation of tungsten ore. Central South University, Changsha (in Chinese)
61. Wei Z (2018) Self-assembly behavior of Al–SiO_3 colloidal particles and its application in flotation separation of scheelite and calcite. Central South University, Changsha (in Chinese)
62. Wei Z, Sun W, Han H et al. Effect mechanism of Al–Na_2SiO_3 polymer inhibitor on flotation separation of schetungsten and calcite. Chinese J Non-Ferrous Metals 20,30(12):3006–3017
63. Xie J, Chang J, Wang X (2001) Application of infrared spectroscopy in organic chemistry and pharmaceutical chemistry
64. Yang N, Yue W (2000) Handbook of atlas of inorganic nonmetallic materials. handbook of atlas of inorganic nonmetallic materials
65. Yu H, Gao B, Yue Q et al (1999) Study on structure characteristics of polysilicon aluminum chloride coagulant by infrared spectroscopy. J Shandong University
66. Chen W, Zhong H, Yang J et al (2002) Study on infrared spectra and crystal morphology of polysilical chloride by different preparation processes. Chem Res Appl 14(1):3
67. Ma J (2005) Study on preparation and application mechanism of modified microparticle silica sol. J Nanjing Forestry Univy: Nat Sci Edn 29(1):1
68. Lu Q (2017) Preparation of large particle size silica Sol particles from sodium silicate. Shandong University
69. Luan Z (1987) Hydrochemical reaction and morphological composition of aluminum. Chinese J Environ Eng 02:3–12 (in Chinese)

Chapter 5
Flotation Reagents of Tungsten Ore

Haisheng Han, Jianjun Wang, Jianhua Kang, Zhao Wei, and Wei Sun

Abstract Tungsten is the importantly rare and strategic metal and plays an irreplaceable role in the military, civil, industrial, and other fields. Scheelite and wolframite account for approximately 70% and 30% of the world's total tungsten resources, respectively. Tungsten minerals are mainly enriched through froth flotation, especially scheelite. However, the selective separation of scheelite from the other Ca-containing minerals is still a worldwide problem due to their similar surface properties. Achieving precise regulation of interfacial physicochemical properties constitutes the scientific foundation for successful scheelite flotation. It is the key to develop excellent flotation reagents with good collecting ability and high selecting. Therefore, tungsten flotation reagent and its molecular design, synergistic effect and interfacial assembly of flotation reagents and metal ion coordination regulation of molecular assembly are introduced in detail.

Keywords Tungsten ore · Collector · Depressant · Coordination assembly

5.1 Introduction

Flotation reagent, equipment and process are troika of flotation technology, among which flotation reagent is the scientific basis and key of this technology, and it is also the root of the strengthening of the flotation process. In the past hundred years, a lot of researches have been carried out in the field of flotation reagents for tungsten ore, especially in the field of improving the selectivity of flotation reagents and

H. Han (✉) · J. Wang · J. Kang · Z. Wei · W. Sun
Central South University, Changsha, China

Engineering Research Center of Ministry of Education for Carbon Emission Reduction in Metal Resource Exploitation and Utilization, Central South University, Changsha, China

State Key Laboratory of Mineral Processing, Changsha, China

H. Han
Technology Research Center of Hunan Province for Comprehensive Utilization of Associated Fluorite and Fluorine Chemical Engineering, Chenzhou, China

W. Sun et al., *Flotation Chemistry of Tungsten Minerals and Its Application*,
https://doi.org/10.1007/978-981-96-5077-4_5

strengthening the interaction between flotation reagents and mineral surfaces, and a series of achievements have been made, which support the efficient development and utilization of complex tungsten resources.

Although there are a lot of reports on flotation reagents of tungsten ore, the systematic summary and analysis of these literatures are still lacking, especially the design principles and molecular assembly of collectors. Therefore, this chapter summarizes the tungsten flotation reagents and their research progress, including different collectors, depressants, pH regulators, dispersants, activators and their combined use. Moreover, the design and assembly principle of tungsten flotation reagents and the related interaction mechanisms are also revealed, so as to provide reference for the design and development of new tungsten flotation reagents.

5.2 Tungsten Flotation Reagent and Its Molecular Design

5.2.1 Flotation Collectors for Tungsten Ore

There are many kinds of collectors for tungsten ore flotation, including anionic type, cationic type, amphoteric type, and complex type. The common classification of collectors for tungsten flotation is shown in Table 5.1.

Table 5.1 Commonly used classification of scheelite flotation collectors

Collector types	Representative reagents	Characteristics
Anionic collector	Fatty acids: oleic acid, sodium oleate, oxidized paraffin soap, tall oil, etc.	Strong collecting, poor selectivity
	Chelate: benzohydroxamic acid, octyl hydroxamic acid, salicylhydroxamic acid, cupric iron and so on	Strong selectivity, poor collecting
	Sulfonic acid: sodium dodecyl sulfate, sodium dodecyl benzene sulfonate	Strong collecting, strong foaming
	Phosphoric acid: styrenephosphonic acid	Strong selectivity, general collecting
Cationic collector	Primary amine: DDA; Quaternary ammonium: didecyl dimethyl ammonium chloride, dodecyl trimethyl ammonium chloride	Strong collecting and selectivity
Amphoteric collector	Alpha-benzoamino-benzyl phosphate, carbamic acid	High selectivity and cost
Complex collector	Pb-BHA (lead benzohydroxamic acid complex)	Strong selectivity, weak collecting

5.2.1.1 Anionic Collector

Tungsten anionic collectors can be divided into fatty acids, chelates, sulfonic acids and phosphonic acids, among which fatty acids and chelates are the most widely used. Fatty acid collectors have low cost, strong collecting ability, but poor selectivity, and usually need to be assisted by a large number of inhibitors, such as water glass, sodium hexacaphosphate, to inhibit other gangue minerals. Chelate collectors have good selectivity, but high cost. Common anionic collector types are described below.

1. Fatty acid collector

Fatty acid collector is the most widely used in anionic collectors for tungsten ore, mainly including various fatty acids and their soaps, such as oleic acid, sodium oleate, linoleic acid, linolenic acid, castor oleic acid, tall oil, palmitic acid, naphthenic acid, oxidized paraffin soap, etc. Among them, oleic acid and its sodium salt and oxidized paraffin soap are commonly used as fatty acid collector. Oxidized paraffin soap contains long-chain fatty acids that can form complexes with metal ions on the surface of tungsten ore, greatly improving the surface hydrophobicity of tungsten ore. It also has foaming properties and is inexpensive. It is mostly used in the flotation of scheelite, less in wolframite. In addition, the amount of fatty acid content of oxidized paraffin soap can be divided into different models, the common model is 731 and 733. 731 oxidized paraffin soap has strong collecting ability and foaming ability, but poor selectivity, and is usually mixed with crude tar oil [1]. 733 is more saponification than 731 and has good selectivity but is more expensive than 731.

Studies have shown that the main mechanism of action of fatty acid collectors is to use carboxyl group to coordinate with metal active sites on the surface of tungsten ore to form collector bilayer film or oleic acid-metal ion precipitation [2], as shown in Fig. 5.1. The Ca^{2+} active site on the surface of tungsten ore is the main action site of the fatty acid collector, and the carboxyl group of the fatty acid collector interacts with the Ca^{2+} on the surface of minerals to generate fatty acid calcium salt precipitation, which makes the surface of scheelite hydrophobic, so as to realize the scheelite collection.

Although fatty acid collectors share common carboxyl groups and have similar interaction mechanisms with minerals, the effect of fatty acid collectors with different carbon chain lengths and number of branch chains on tungsten and other calcium-bearing gangue minerals is different. Yaohui Yang et al. studied the effects of different chain lengths, number of branch chains, saturation degree and whether there are hydroxyl groups in them on the collection capacity of fatty acids (Fig. 5.2) [4]. Moreover, they summarized the "construction-activity" relationship of fatty acid collectors, and screened out fatty acids with excellent selectivity for the separation of scheelite and other calcium-containing minerals, achieving remarkable separation effect. The main conclusions were as follows:

(1) When fatty acids are used alone, the longer the carbon chain and the greater the degree of unsaturated, the better the collecting capacity of fatty acids for scheelite, fluorite and calcite is.

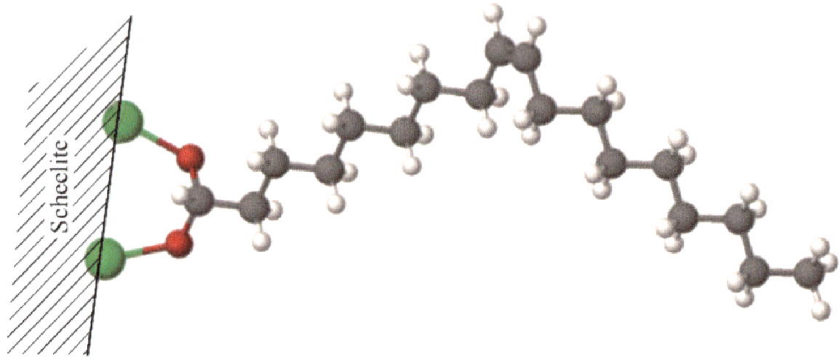

Fig. 5.1 Adsorption mechanism of fatty acids on scheelite surface, reprinted from Ref. [3], copyright 2018, with permission from International Journal of Mining Science and Technology

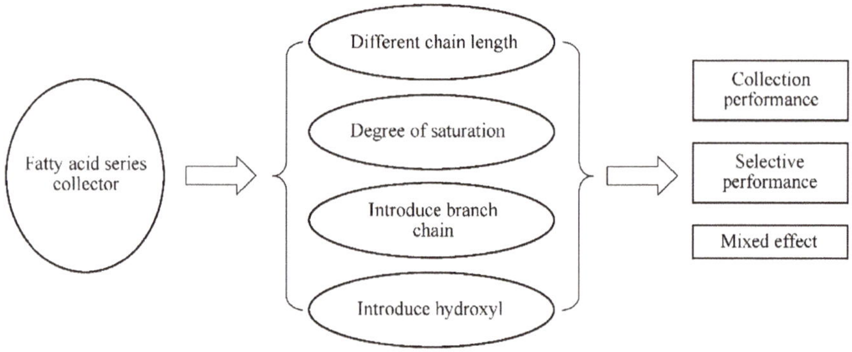

Fig. 5.2 Basic ideas for designing different fatty acids [4]

(2) The flotation effect of branched fatty acids is better than that of normal hydrocarbon chain fatty acids.
(3) The flotation effect of fatty acid molecules with hydroxyl group becomes worse.
(4) When fatty acids are mixed, there is a synergistic effect, and the mixed collector has a better flotation effect on calcium minerals than the single collector.
(5) Strengthening the collector (oleic acid) first or adding two kinds of collectors at the same time can produce positive synergistic effect, adding weak collector (lauric acid) first can produce no synergistic effect, and even produce negative synergistic effect.
(6) The mixing effect of oleic acid and lauric acid is better, indicating that the mixing of fatty acids with large differences in long and short chains is conducive to collecting the calcium minerals.

The results of this study provide a basis for the design and assembly of fatty acid flotation reagents.

In addition to the mixing of different fatty acids, the mixing of fatty acids with other agents can also improve the flotation effect. Therefore, in recent years, fatty acid research mainly focuses on the modification of traditional fatty acids to improve their low temperature resistance, water hardness and other properties, so as to improve the selectivity of collector to tungsten ore, such as GY series, TAB series, BK series, EA series, K series, etc. [5]. In addition, the fatty acids of C_{17-20} can also be modified by introducing halogen atoms, such as chlorine, bromine and other atoms, and the obtained modified fatty acids can improve the recovery of scheelite. Taking brominated fatty acids as an example, brominated fatty acid collectors DHT-1($C_{13}H_{25}O_2Br$) and DHT-3($C_{18}H_{33}O_2Br_3$) have better harvesting capacity for wolframite and scheelite than conventional fatty acids [6]. The mechanism is that under the weak acid condition (pH = 5–6), the bromo-fatty acid collector is adsorbed on the surface of wolframite through the synergistic action of Br-H·· O and carboxyl group (chemical bonding and hydrogen bonding), and the adsorption capacity is stronger. Under alkaline conditions (pH = 7.5–8.5), a large number of hydroxide ions will cover the surface of scheelite and play a bridge role to form hydrogen bonds with bromo-fatty acid collector Br–H·· O, improving scheelite flotation.

2. Chelating collector

Chelating collector refers to surfactants that can form stable chelates with metal ions on the surface of a mineral. Common hydroxamic acids and arsenate collectors include benzohydroxamic acid, salicylohydroxamic acid, naphthohydroxamic acid, benzyl arsenate, toluene arsenate, etc. In addition, 8-hydroxyquinoline, α-nitroso-β-naphthol, cupferron are also commonly used chelating agents. Because of the strong toxicity of arsenic chelating agents, the impact on the ecological environment is great. Up to now, this kind of chelating agents is less used. Other chelating agents with less toxic are widely used, such as GY series (hydroxamic acids), CF series (nitrobenzene hydroxylammonium salt), etc.

Hydroxamic acid collector is a representative chelating agent due to its strong chelating ability with metal ions such as Cu^{2+}, Co^{2+}, Ni^{2+}, Fe^{3+}, Zn^{2+} and rare earth metal ions to form stable chelate [7]. However, compared with fatty acids with the same alkyl structure, hydroxamic acids have weak dissociation and small solubility. Therefore, hydroxamic acids generally need to be dissolved in weakly alkaline solutions for use. Taking benzohydroxamic acid as an example [7–9], in the range of pH value 4.7–13.7, the O atom in the -CONHOH group of hydroxamic acid can have strong chelation with metal ions on the surface of tungsten, forming a stable "O, O" five-membered cyclochelate structure. Hydroxamic acid is also adsorbed on wolframite surface in the form of "O, O" five-membered cyclochelate. But the active sites of the action are controversial. One view thinks that the main active site on wolframite surface is Mn^{2+} [10], and another view thinks that Fe^{2+} on wolframite surface can form a more stable product with hydroxamic acid [11].

In order to improve the collecting ability and selectivity of hydroxamic acid collector, many new hydroxamic acid agents have appeared in recent years, such as cyclohexyl hydroxamic acid, amide hydroxamic acid, octyl hydroxamic acid, capric hydroxamic acid, phosphonic hydroxamic acid, alkyl hydroxamic acid and cinnamic hydroxamic acid [12]. In fact, the modification of hydroxamic acid mostly follows the classical hydroxamic method, which synthesizes different hydroxamic

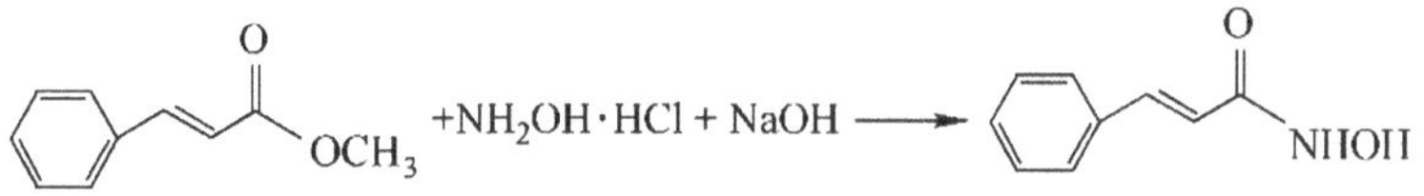

Fig. 5.3 Preparation process of cinnamic hydroxamic acid [13]

acids by changing the ingredients of raw materials. Taking cinnamomic hydroxamic acid as an example [13], methyl cinnamate and hydroxylamine are used as raw materials to generate a new hydroxamic acid-cinnamomic hydroxamic acid (CIHA) by hydroxylamine method. The preparation process is shown in Fig. 5.3.

In addition, amido-hydroxamic acid is also a relatively novel chelating agent [14]. Such as (N-(6-(hydroxyl amino)-6-carbonyl) benzamide (NHOB), N-(6-(hydroxyl amino)-6-carbonyl) simba amide (NHOO), N-(6-(hydroxyl amino)-6-carbonyl) capric amide (NHOD) and N-(4-(hydroxyl amino)-4-carbonyl) simba amide (NOBO), etc. The flotation results show that, compared with other amido-hydroxamic acids, NHOD has better collecting ability of scheelite. NHOD can be chemically adsorbed on scheelite by forming surface complex in addition to electrostatic adsorption. Moreover, NHOD has unique bond types and hydrophobic groups, which can form intermolecular hydrogen bonds on scheelite surface. These characteristics make NHOD an efficient collector for scheelite/calcite flotation separation.

3. Sulfonic acid collector

Sulfonic acid collector refers to a class of anionic surfactants with sulfonic acid groups, which are rarely used in industry. Compared with fatty acid collectors, sulfonic acid collectors have better selectivity and more foam, but their collecting ability is weaker than fatty acids. Sodium alkyl sulfonates are representative of sulfonic acid collectors, such as the common sodium dodecyl sulfonate. Sulfonic acid collector is often used as an auxiliary component of fatty acid collector, used to improve foam structure and selectivity. Oleic acid and other fatty acids can also be treated by sulfuric acid for sulfonation, in order to improve the selectivity of traditional fatty acids with similar collecting properties [15]. Sulfonic acid ions and active metal sites on the surface of a tungsten mineral form insoluble substances to make tungsten mineral surface hydrophobic, realizing the separation of tungsten ore and gangue minerals. Compared with fatty acid collectors, sulfonic acid collectors have lower collecting ability, but their resistance to hard water is strong because of their weak binding ability with calcium. Therefore, compared with fatty acid collectors, sulfonic acid collectors have better selectivity and strong foaming ability. If necessary, other agents with strong collecting ability can also be added to enhance the sorting effect [16]. In addition to sodium dodecyl sulfonate, sodium dodecyl benzene sulfonate and sodium petroleum sulfonate are also widely used in tungsten ore flotation.

4. Phosphonic acid collector

Phosphonic acid collector refers to a class of anionic surfactants with phosphonic acid groups, can form a relatively stable complex with metal ions through its own phosphonic acid group. This kind of collector is widely used in wolframite flotation, such as p-ethylbenzene phosphonic acid, hexylphosphonic acid, heptyphosphonic acid, octylphosphonic acid, styrene phosphonic acid and other alkyl phosphonate esters. Phosphonic acid agents have good selectivity, and belong to low toxicity or non-toxic chemical products, which is conducive to the protection of ecological environment. However, the synthesis process of this class of agents is complicated, and the production rate of the products is low, resulting in high cost. In most cases, it is difficult to use alone, so it needs to be combined with fatty acid collector [17].

Phosphonic acid collector can produce insoluble salt with Fe^{2+}, Fe^{3+}, Ca^{2+}, etc., and has good collecting capacity for wolframite and scheelite [18]. If styrene phosphonic acid is used as the collector of wolframite [19], for the wolframite with a grade of WO_3 0.41%, the wolframite concentrate with a grade of WO_3 53.08–55.95% and a recovery of 81.31–84.77% can be obtained by flotation-gravity concentration-flotation process. Using LP-08 (isopropyl alkyl phosphinic acid) for scheelite flotation [20], it has selective chemical adsorption on scheelite surface, and less adsorption on the surface of gangue minerals such as fluorite and garnet, so as to achieve a good separation of scheelite and gangue minerals. In addition, when FXL-14 (tetradecane amino dimethylphosphonic acid) is used as collector [21], whether with or without the use of Pb^{2+} as activator, wolframite has a good flotation result, showing that FXL-14 has a strong collecting capacity for wolframite. In addition, alkyl phosphonate is also an important phosphonic acid collector, its selectivity is better than conventional collector [22–24].

5.2.1.2 Cationic Collector

The above anionic collectors enrich tungsten ore mainly by interacting with metal active sites on the surface of tungsten ores. However, due to the similar surface structure and solubility of scheelite ($CaWO_4$) with the calcium-containing minerals calcite ($CaCO_3$) and fluorite (CaF_2), flotation separation of scheelite from other calcium-containing minerals has lacked effective and highly selective collectors. At present, although the most widely used fatty acid collectors, such as oleic acid and oxidized paraffin soap, have the same mechanism of action with calcium-containing minerals. Namely such collectors realize mineral flotation through chemical adsorption or calcium salt precipitation caused by the interaction of carboxyl group of fatty acid and Ca^{2+} on mineral surface. As a result, the separation of fatty acid collectors for scheelite and other calcium-containing minerals is poor. In order to achieve good separation of scheelite and other calcium-containing minerals, when the fatty acid collector is used, it is usually necessary to add a large number of inhibitors such as water glass, and the tungsten roughing concentrate can be further cleaned by "Petrov" method to obtain high-quality tungsten concentrate. However, this process has high energy consumption and high pollution. Therefore, it is an important topic in the field of tungsten flotation to develop highly selective collector and realize scheelite flotation at room temperature.

In aqueous solution, scheelite and other calcium-containing minerals are subject to the action of water dipole and solute, and the surface will be charged. Studies have shown that due to the unbalanced dissolution of mineral surface components, the surface charges of different minerals are different. For example, the WO_4^{2-} ion and Ca^{2+} ion dissolution imbalance in the scheelite lattice makes scheelite surface generally negatively charged, while the F^- ion and Ca^{2+} ion dissolution imbalance on the fluorite surface makes the fluorite surface positively charged [25–28]. In addition, Ca^{2+} ion and Mg^{2+} ion can make scheelite surface potential positive shift, and silicate, phosphate and polyphosphate can increase the scheelite surface electronegativity. Therefore, the potential negative shift [28–30], that is, by regulating pH value and adding foreign metal ions activator, can expand the difference in surface charge among scheelite and calcite, fluorite. Therefore, according to the principle of electrostatic action, it is feasible to use cationic collector to realize the flotation separation of scheelite from other calcium-containing minerals in theory.

Most cationic collectors are nitrogen-containing organic surfactants (amine compounds), and a few contain phosphorus, sulfur or silicon, etc., which are often used for silicate, carbonate, phosphate, potassium salt, and metal oxide minerals flotation. As an effective collector of scheelite, cationic collector mainly refers to amine collectors, such as fatty amine and aromatic amine, etc., that is, common dodecylamine (DDA), dodecyltrimethyl ammonium chloride and trioctyl methyl ammonium chloride. Aliphatic amines are first discovered in 1850 and are not used as flotation collectors until 1920. However, its shortcomings such as high synthetic cost, low solubility and poor selectivity hinder the popularization and application of cationic collector. After decades of research and development, the type of amine collector has been developed from a single fatty amine to a variety of amines, such as ether amine, amide, polyamine, condensed amine, aromatic amine, morpholine and its quaternary ammonium salt, and promoted the development of cationic flotation process. The functional group of these amine collectors is the positively charged amine group $R\text{-}NH_3^+$, so the amine collector can be selectively adsorbed on the surface of negatively charged scheelite by electrostatic action, so as to achieve effective separation of scheelite and other common calcium bearing gangue minerals.

Among many amine collectors, DDA is the most common amine collector. However, DDA is insoluble in water and needs to be acidified into salts before being used and then dissolved in water. Among them, the commonly used acids are acetic acid, hydrochloric acid and sulfuric acid, with acetic acid being the most common. The adsorption mechanism of DDA on scheelite surface is shown in Fig. 5.4. DDA forms $(RNH_3^+)_2WO_4^{2-}$ through electrostatic adsorption between NH_3^+ and WO_4^{2-} on mineral surface, thus enhancing the hydrophobicity of scheelite surface. It is generally believed that the adsorption of alkyl amine salts on mineral surfaces is physical adsorption [32], but the adsorption strength of DDA acetate on scheelite is greater than that on calcite. This adsorption difference comes from the cation RNH_3^+ of DDA, which is easy to adsorption on tungstate site, but limited on carbonate site, thus strengthening the flotation separation of scheelite from calcite.

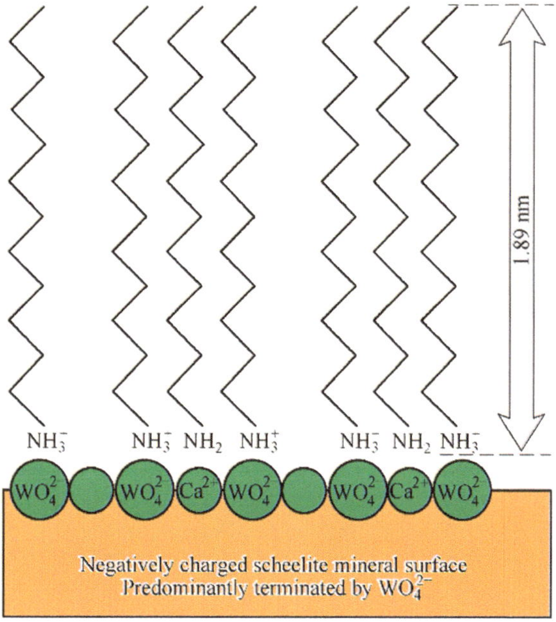

Fig. 5.4 Schematic diagram of different adsorption behavior of DDA species on scheelite surface, reprinted from Ref. [31], copyright 2015, with permission from Minerals Engineering

In addition to DDA, quaternary ammonium salts are also widely studied amine collectors. They have positively charged halides with NR_4^+ structure, namely quaternary ammonium cations, in which R is alkyl or aryl. Therefore, the main difference of quaternary ammonium salts is the four substituents on the central nitrogen, and the structural differences of these substituents lead to different flotation properties. On the other hand, fatty amines, such as primary, secondary and tertiary amines, do not carry a positive charge on their own, but only carry a charge when they bind H^+ in solution, so these three amine collectors are sensitive to solution pH. They only show strong activity in weak acidic environments, and in strong alkaline and acidic environments usually do not have the ability to collect. However, quaternary ammonium salts are quaternary ammonium ion in acidic, neutral, and alkaline environments. Their morphology in solution will not change under the influence of pH, and they always maintain the surface-active part with positive charge. Therefore, quaternary ammonium salt collectors are more adaptable to the flotation solution environment and have broader application potential.

Didecyl dimethyl ammonium chloride (DDAC), trioctylmethyl chloride (TOAC) and dodecyl trimethyl ammonium chloride (DTAC) are three typical quaternary ammonium cationic collectors, and their structural formulas are shownin Fig. 5.5a [33]. They contain two, three and one long-chain alkyl substituents, respectively, and these structural differences determine their different collecting properties. The flotation results show that the collecting and selectivity of DDAC and TOAC are much higher than that of DTAC, while the collecting capacity of DDAC is slightly higher than that of TOAC, but the selectivity is weaker than that of TOAC. Compared with the traditional collector sodium oleate, DDAC and TOAC have better collecting and selectivity for scheelite. TOAC is used as collector to further cleaning for a scheelite roughing concentrate from Shizhuyuan at normal temperature. The WO_3 grade of tungsten concentrate is 52.01%, and the working recovery is 51.54%. The normal temperature flotation separation of tungsten ore and calcium-bearing gangue minerals is basically realized. Under the flotation condition, the central nitrogen atom of quaternary ammonium salt can neither provide valence electrons nor empty orbitals to form covalent bonds with the atoms on the surface of calcium minerals, indicating that quaternary ammonium salt will not be chemically adsorbed on the surface of scheelite and calcite. In addition, the central nitrogen atom of quaternary ammonium salt is surrounded by four alkyl substituent groups, and the steric hindrance of interaction with the mineral surface is large, which is also not favorable for chemical adsorption of quaternary ammonium salt on the mineral surface. Therefore, physical action is likely to be the main mechanism of quaternary ammonium salt to capture scheelite. Zeta potential measurements and infrared spectrum analysis also further confirmed that the physical adsorption of quaternary ammonium salts on the surface of scheelite may have occurred, but no chemical adsorption [34]. In addition, the adsorption configuration of TOAC on the common exposed surfaces of scheelite and calcite was analyzed by molecular dynamics simulation, as shown in Fig. 5.5b, c. The results show that the TOAC molecule is initially adhered to the scheelite surface, and then shifted upward under the action of water molecules. When calcite reached equilibrium in 1.2 ns system, TOAC showed a more vertical configuration compared with its configuration on scheelite. The acting group is still the head group, but the acting area is smaller than that on the scheelite surface.

In addition, quaternary ammonium salt dioctadecyl dimethyl ammonium bromide (BDDA) is a new type of amine collector [35]. The flotation results show that the flotation effect of BDDA is better than that of oleic acid, which can realize the selective separation of scheelite from calcite. At the same time, the mechanism study shows that BDDA is also adsorbed on the surface of scheelite by electrostatic action.

Therefore, the cationic collector has good selectivity and low temperature resistance, showing great potential in the flotation separation of scheelite and calcium-bearing gangue minerals, especially quaternary ammonium salt and its derivatives. However, limited by the source, cost and other factors, the problem of its large-scale industrial application remains to be solved.

5.2.1.3 Amphoteric Collector

Amphoteric collector, namely amino acid collector, can be used in almost the whole pH range and have strong adaptability to pulp pH. Common amphoteric collector mainly refer to amino acids, such as α-benzoamino-benzyl phosphate, α-amino-aryl phosphate, β-amino-alkyl phosphonic acid, β-amino-alkylphosphonate and so on. As for the mode of action of amphoteric collector, studies have shown that under the condition of acidic pulp, this type of collector presents the characteristics of cationic collector, the acting group is $–NH^{3+}$. At this time, the surface of scheelite

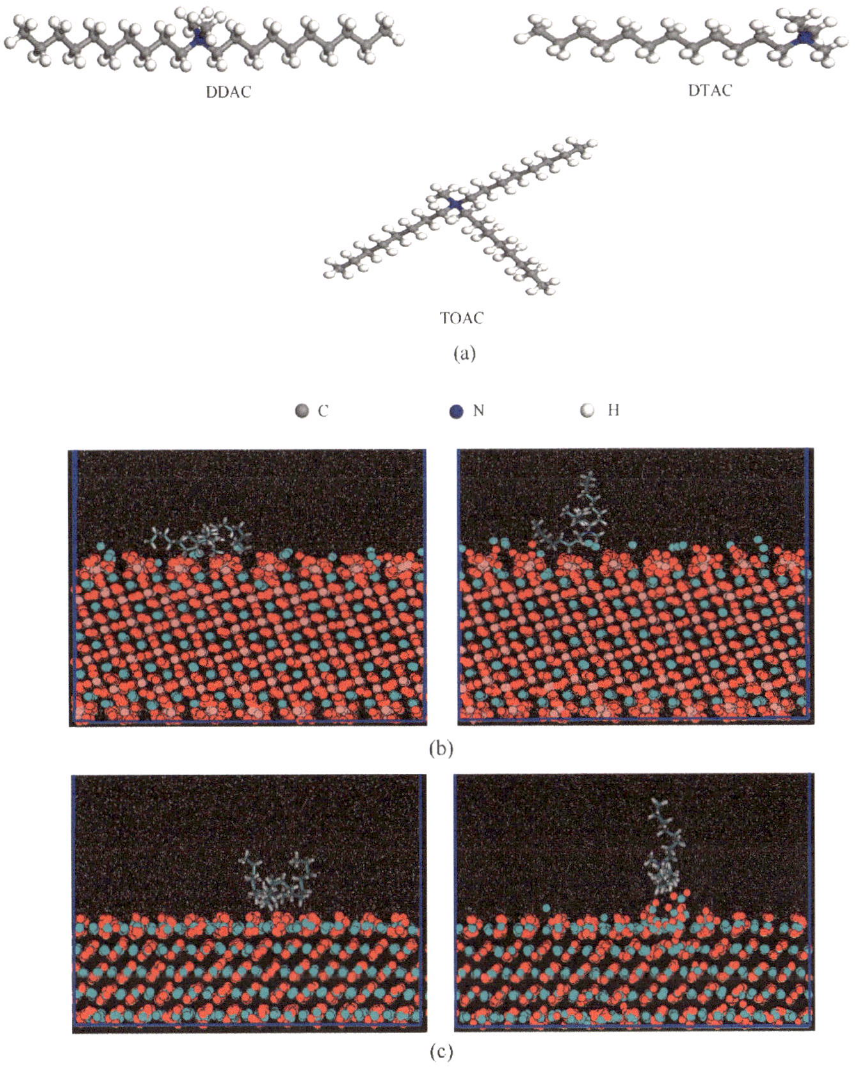

Fig. 5.5 Cationic structure of quaternary ammonium collector **a** and adsorption configuration of TOAC at scheelite (101) **b** and calcite (104) **c** in aqueous solution (left) and at 1.2 ns (right) [33]

is negatively charged so that the collector mainly relies on electrostatic force to be adsorbed on the surface of scheelite [36]. When the pulp is alkaline, some researchers through zeta potential measurement, collector adsorption determination and infrared spectrum characterization, find that amphoteric amino acid collector presents anionic collector characteristics, among which the univalent collector anion is an active conjugate, in the way of chemical adsorption on the surface of calcium minerals. Thus, it can change the mineral surface hydrophilic/hydrophobic, and improve the flotation behavior of minerals [12]. Because amphoteric collector has two charges and exhibits different ionic properties at different pulp pH values, they can be applied in a wide pH range from strong acid to strong base, and can also be used well in combination with many other collectors and foaming agents to strengthen foam, improve hydrophobic effects, and adapt to the flotation of various minerals [37]. In addition, amphoteric collector has better resistance to hard water, so amphoteric collector has better selectivity than fatty acid collector [16].

The effects of different amphoteric collectors on the floatability of tungsten ore and gangue minerals are different. By studying the collection performance of β-aminylalkylphosphonic acid and β-aminylalkylphosphonate on fluorite, barite and scheelite, it can be seen that β-aminylalkylphosphonic acid has strong collection ability for fluorite in a wide range of pH, and the separation of fluorite and scheelite can be realized by adjusting the pH of pulp to alkaline [38]. Amphoteric collectors such as carbamic acid and diphosphonic acid have good selectivity for scheelite, but the mechanism of action has not been studied deeply [12]. α-benzamino-benzylphosphate is a new amphoteric collector [39]. The flotation results show that fluorite has a good floatability in the pH range of 6–10, calcite has a good floatability in the pH range of 8–10, and scheelite shows a limited floatability in the pH value of 8. Therefore, the amphoteric collector α-benzamino-benzylphosphate is used. Selective flotation separation of scheelite from other calcium-containing minerals can be realized by controlling the pH of pulp. In addition, amphoteric collectors such as Medearam, HostponT and Flotble AM20 (n-alkyl-sodium chole-aminopropionate) and AM21 (sodium oleate aminosulfonate) have selective collecting capacity for wolframite [40]. Despite the excellent performance of amphoteric collectors, in order to obtain high-grade tungsten concentrate with WO_3 content of over 65%, the cleaning process still needs to refer to the "Petrov method" heating flotation process [41].

Compared with single anionic or cationic collector, amphoteric collector will show different anionic or cationic collector characteristics at different pulp pH values, which helps to capture more target minerals, and the intermediate hydrophobic section is like a "bridge". This feature of amphoteric collector is very significant, helping to improve the flotation of micro-fine minerals. Similar to the agglomeration phenomenon, this "bridge" can bridge more micro-fine minerals together, thus forming hydrophobic particles with larger apparent size. However, it should be noted that, because the amphoteric characteristics can only be displayed at different pH values, it is necessary to adjust to different pH values to capture more target minerals. At this time, the possibility of desorption of functional groups adsorbed on the surface of target minerals should be considered. In addition, many amphoteric collectors have strong bubbling. Therefore, for the design and assembly of new amphoteric collectors, in addition to the design of both ends of the amphoteric functional groups and the middle of the hydrophobic chain, its bubbling property also need to be focused. Otherwise, in the actual application process, the amphoteric collector dosage will be greatly reduced, which will weaken the amphoteric collector on adsorption of the target mineral. To this end, we can improve the application range and selective collection ability of amphoteric collector by optimizing the synthetic pathway and combining with other agents (including other types of collectors, foaming agents and defoamers, etc.).

5.2.2 *Flotation Regulators for Tungsten Ore*

The flotation process of tungsten ore is a complex system with great difficulty in separation. The floatability difference between tungsten ore and gangue minerals is often achieved through a series of regulators, so as to achieve efficient flotation separation. Tungsten flotation regulators mainly include pH regulator, inhibitor, dispersant and activator.

5.2.2.1 pH Regulator

In the flotation process, pulp pH is of great significance, and all kinds of minerals can be effectively flotation only under their appropriate pH conditions. The optimal pH value of mineral flotation not only depends on the interface properties of minerals, but also depends on the flotation reagent system and other conditions. For the same reagent, the optimal flotation pH value of different minerals is not necessarily the same; for the same mineral floated by different reagents, the optimal flotation pH value is not necessarily the same. Therefore, adjusting pulp pH is one of the most important parameters to control the flotation process. Wolframite and scheelite are generally floated under alkaline conditions, and the common pH regulator is mainly inorganic, such as sodium hydroxide (NaOH), water glass (Na_2SiO_3), sodium carbonate (Na_2CO_3) and calcium oxide (CaO) so on [42, 43].

In scheelite flotation practice, sodium carbonate can not only eliminate the influence of unavoidable ions in pulp, but also adjust the pH value of pulp, which is the most common application. In the flotation of skarn scheelite with more calcite content, Na_2CO_3 is often used as the pH regulator. In the operation of requiring higher flotation pH, sodium carbonate is generally used to adjust to weak alkalinity, and then sodium hydroxide to strong alkali. In addition, another characteristic of sodium carbonate is that it can keep the pH value of pulp in a relatively stable range, because it is a strong base and weak acid, the ionization and hydrolysis reaction process is as follows:

$$\text{Ionized: } Na_2CO_3 = 2Na^+ + {CO_3}^{2-} \tag{5.1}$$

$$\text{Hydrolysis: } {CO_3}^{2-} + H_2O \rightleftharpoons {HCO_3}^- + OH^- \quad K_1 = 2.26 \times 10^{-4} \tag{5.2}$$

$${HCO_3}^- + H_2O \rightleftharpoons H_2CO_3 + OH^- \quad K_2 = 2.95 \times 10^{-8} \tag{5.3}$$

where K_1 and K_2 are the hydrolysis constants of step 1 and step 2, respectively. It can be seen from the above equations that OH^-, HCO_3^- and CO_3^{2-} ions obtained after hydrolysis, and a small amount of foreign H^+ or OH^- ions have little effect on the pulp pH. Therefore, the pH adjusted by sodium carbonate is relatively stable. It can keep the pH of pulp between 8 and 10, which is conducive to the flotation separation of minerals with harsh pH requirements. There are also views that the use of sodium carbonate to adjust the pH value [44], CO_3^{2-} and Ca ions in the pulp generate calcium carbonate to cover on scheelite surface, CO_3^{2-} ion can be selectively adsorbed on scheelite. There is competitive adsorption between CO_3^{2-} and SiO_3^{2-} ions, so the presence of CO_3^{2-} ion reduces the adsorption of SiO_3^{2-} on scheelite surface, and then improves the scheelite flotation.

In addition to acting as a pH regulator, sodium carbonate also has different inhibitory abilities to different calcium-containing minerals: fluorite > calcite > calcium-containing silicate > scheelite and apatite [45]. Therefore, the role of sodium carbonate is not only limited to precipitation or limiting calcium and magnesium ions in pulp, buffering pH and improving mineral flotation performance, but also has a selective inhibition effect on calcium-containing minerals, especially calcite. However, for ores that require higher alkalinity to be flotation, sodium carbonate cannot adjust the pH of the pulp to a high enough level, in which case sodium hydroxide works better. In scheelite flotation with high fluorite and low calcite, NaOH is often used as pH regulator. The practice of tungsten flotation at home and abroad has proved that the choice of sodium carbonate and sodium hydroxide is concluded: generally, for ores containing more soluble or slightly soluble minerals, sodium carbonate is the best, and sodium hydroxide is the other way around [46].

5.2.2.2 Inhibitor

There are many types of tungsten flotation inhibitors, which can be divided into inorganic and organic inhibitors [3]. Among them, inorganic inhibitors mainly include water glass, sodium phosphate, sodium fluosilicate, etc.; organic inhibitors mainly carboxymethyl cellulose, citric acid and polycarboxylic acid. Among them, cheap inorganic inhibitors are widely used.

1. Inorganic inhibitor

Water glass is the most widely used inhibitor in scheelite flotation. The aqueous solution of water glass, commonly known as water glass, is a kind of viscous aqueous solution with high concentration and strong alkalinity. It is produced by melting quartz sand and sodium carbonate at high temperature, and its color is light yellow or bluish gray. Sodium silicate type water glass is represented by $Na_2O \cdot mSiO_2$, where m is the "modulus" of water glass, and the acid–base properties of water glass are usually divided by $m = 3$. When $m < 3$, it is alkaline, and when $m \geq 3$, it is neutral. Sodium silicate is a strong base and weak acid salt, and a series of hydrolysis and ionization reactions will occur in aqueous solution to generate hydrolytic products of

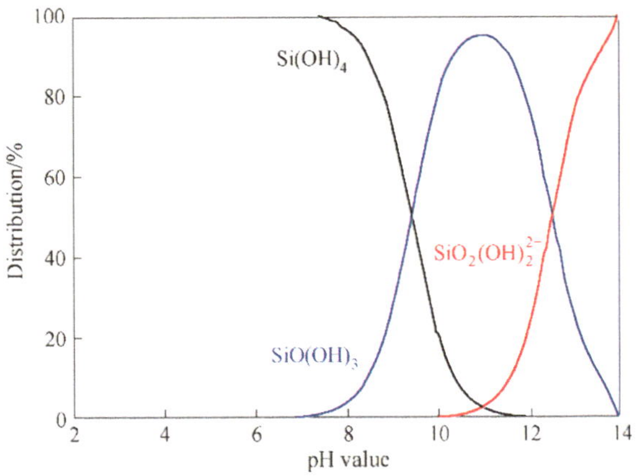

Fig. 5.6 Distribution of solution components of sodium silicate under different pH conditions [47]

various components. The solution component distribution of sodium silicate under different pH conditions is shown in Fig. 5.6.

Water glass has strong inhibition on gangue minerals and silicate minerals, and the effect is significant in scheelite flotation at room temperature. Water glass has a strong ability to resolve the collector films of calcite, fluorite and other calcium-bearing gangue minerals, but the collector on the surface of scheelite is almost not affected, so the efficient flotation separation of scheelite and gangue minerals can be realized. The adsorption of silicate ion on calcite surface is easier than that of oleate ion, which indicates that the use of water glass as an inhibitor can separate calcite from scheelite. Therefore, when the scheelite ore composition type is relatively simple, single water glass can achieve satisfactory separation index. When the mineral composition is complex, the selectivity of single water glass is often poor. While inhibiting gangue minerals such as calcite and fluorite, scheelite also has a certain degree of influence, restricting the improvement of scheelite recovery [48, 49].

In addition to water glass, phosphate inhibitors and sodium fluosilicate have strong inhibition ability for silicate and some calcium-containing minerals, so they are also used in scheelite flotation. Phosphate will be ionized and hydrolyzed in solution to form a series of phosphate anions, which will react with Ca^{2+} ions on the surface of calcite and generate $Ca_3(PO_4)_2$, so that the surface of calcite is strongly hydrophilic and inhibited [50]. Sodium fluosilicate also has selective inhibition on silicate minerals such as quartz and feldspar [51]. It should be noted that sodium fluosilicate is a water-soluble compound, and it needs to be diluted to very low concentration or directly interact with minerals in the grinding process to strengthen the adsorption effect. Sodium hexametaphosphate [52] ($(NaPO_3)_6$) is ionized in water, and the anion can form a complex with the Ca^{2+} on the surface of the mineral, which is then converted into a soluble complex, thus achieving the purpose of inhibiting calcite or fluorite. At the same time, calcium oleate molecules on the mineral surface can be dissolved in sodium hexametaphosphate, producing soluble calcium complex into the liquid phase, desorption of the collector on the surface of calcium minerals to achieve the purpose of inhibition. Compared with sodium hexametaphosphate, the adsorption of sodium hexametaphosphate makes the negative potential shift of calcite greater, thus reducing the adsorption of anionic collector on the surface of calcite, and thus reducing the floating ability of calcite. It should be noted that the dosage of sodium hexametaphosphate and pH value have a great impact on the floatability of scheelite and other gangue minerals [53]. The optimal pH value of flotation separation is about 10. When the dosage of sodium hexametaphosphate is greater than 20 mg/L, it can inhibit the flotation of fluorite; when the dosage of sodium hexametaphosphate is greater than 80 mg/L, it can inhibit the flotation of calcite.

2. Organic inhibitor

At present, in terms of industrial application, water glass and modified water glass are still the most used inhibitors with the best inhibitory effect, while the application of organic inhibitors is less. However, when water glass is used in large quantities, it will cause difficulties in slurry settlement, resulting in low recycling efficiency of wastewater. Due to the wide source and variety of organic inhibitors, more and more organic inhibitors have been studied in scheelite flotation in recent years. The development of non-toxic and pollution-free environmentally friendly organic inhibitors has become an important research direction of scheelite flotation inhibitors.

Scheelite flotation organic inhibitors according to molecular weight can be divided into small molecule and large molecule inhibitors. Small molecule inhibitors mainly include oxalic acid, citric acid and lactic acid. Macromolecular inhibitors are mainly starch, cellulose, tannins, glycans, EDTA (ethylene diamine tetraacetic acid) and so on. Organic inhibitor molecules generally contain more than two polar bases, these polar bases have very high activity, adsorption on the mineral surface, making the mineral surface strongly hydrophilic and thus inhibited. According to the different functional groups and mineral surface action, inhibitors can be divided into carboxylic acid, sulfonic acid/sulfuric acid, hydroxyl and phosphoric acid, among which carboxylic acid mainly includes polyaspartate, sodium humate, sodium alginate, xanthan gum, carboxymethyl cellulose and citric acid [54–58]. Sulfonic acid/sulfuric acid group mainly includes calcium lignosulfonate and sodium glucan sulfate, etc. [59, 60]. The hydroxyl group is mainly tannic acid and glucose [61], and the phosphoric acid is mainly phytic acid [62].

For small molecule organic inhibitors, citric acid is a typical representative, it is a natural acid, and exists in many animals and plants, commonly used as a preservative, food additives due to its environmentally friendly. The molecular structure of citric acid contains 3 carboxyl groups and 1 hydroxyl group, so it is easily soluble in water and has strong acidity. It can strongly chelate with calcium ions. Hu Yuehua et al. [63] used citric acid as an inhibitor of fluorite and F305 as a collector of scheelite. In the range of pH = 6–12, citric acid could strongly inhibit the flotation of fluorite, but had no effect on scheelite flotation. The flotation test of artificial mixed ore showed that scheelite concentrate with WO_3 grade of 65.2% and recovery of 88.3% could be obtained by using citric acid. In addition, for the tungsten ore containing scheelite and wolframite (WO_3 0.40%, CaF_2 16.46%) from polymetal flotation plant of Hunan Shizhuyuan Company, a scheelite concentrate with WO_3 grade of 3.24% and recovery of 88.45% can be obtained by roughing flotation with this reagent system in one stage, which further confirms the selective inhibition ability of citric acid on fluorite. Zeta potential measurements showed that citric acid was more adsorbed on the surface of fluorite, but less adsorbed on scheelite. Solution chemistry calculation further showed that the Gibbs free energy of calcium citrate produced by the reaction on the surface of fluorite was negative, but positive on the surface of scheelite, indicating that citric acid can spontaneously react with calcium ions on the surface of fluorite, but it is difficult to react on the surface of scheelite.

For macromolecular organic inhibitors, common polycarboxylic acid, natural gum, carboxymethyl cellulose (CMC) and humic acid and their sodium salts, etc. The following is a brief analysis of the effects of these macromolecular organic inhibitors on the flotation behavior of tungsten ore and gangue minerals.

Polycarboxylic acid is the representative of environmentally friendly pharmaceutical. It has a number of carboxylic acid functional groups, is a good scale inhibitor with good chelating ability with metal ions and is often used in industrial water treatment with green and efficient characteristics. Therefore, polycarboxylic acids have the potential to act as inhibitors of calcium-containing minerals based on their molecular structure and good calcium chelating ability. Polyaspartic acid (PASP), polyacrylic acid (PAA), polymaleic acid (HPMA), and polyepoxy succinic acid (PESA) are four common polycarboxylic acids. When sodium oleate is used as collector, the flotation results show that: except PESA, PAA, PASP and HPMA can realize the flotation separation of scheelite from fluorite and calcite under certain conditions, and PASP has the best selective inhibition effect [64]. Compared with traditional water glass inhibitors, PASP has the characteristics of environmental protection, high efficiency, small dosage, good selectivity and so on. The adsorption mechanism showed that the surface potential of the three calcium-containing minerals shifted negatively after interaction with polycarboxylic acid inhibitors, indicating that the inhibitors were adsorbed on their surfaces. After the interaction with sodium oleate, the surface potential of fluorite and calcite was basically unchanged, while the surface potential of scheelite continued to shift negatively. This finding indicated that the adsorption of inhibitor hindered the adsorption of the collector sodium oleate on the surface of fluorite and calcite, but had little effect on the adsorption of sodium oleate on scheelite surface. Further analysis showed that polycarboxylic acid had strong chemical adsorption on the surface of fluorite and calcite, but relatively weak adsorption on the surface of scheelite.

In addition to polycarboxylic acids, organic carboxylic acid agents containing carboxylic acid and phosphonic acid groups are also important scale inhibitors, and also have good calcium chelating ability. Phosphonyl butane tricarboxylic acid (PBTCA) can be selectively adsorbed on the surface of calcite, which prevents the subsequent adsorption of sodium oleate, and thus is well inhibited [65]. However, less adsorption on scheelite surface has less impact on the adsorption of sodium oleate and the floatability of scheelite, so as to achieve the selective separation of scheelite and calcite, but the dosage should not be too large, otherwise scheelite will be inhibited.

In addition, natural gum is also an important macromolecular inhibitor, it belongs to the chitosan macromolecular agents with a high degree of stability, safety, and easy modification. At the same time, it has the flotation adjustment, fine mineral flocculation, and dispersion function. This makes the natural glue show a broad application prospect in mineral processing, especially for the mineral engineering field faced with fine and micro-fine mineral processing problems. The flotation results show that acacia gum, guar gum, xanthan gum have inhibitory effects on calcium minerals such as scheelite [66]. When sodium oleate was used as collector, the recovery rate of scheelite decreased significantly with the increase of the amount of natural gum, while the recovery rate of calcite and fluorite decreased slowly, indicating that the natural gum had stronger inhibition effect on scheelite. In addition, the combination of water glass and natural glue can improve the selective inhibition of natural glue for scheelite and fluorite. Metal ions have a better effect on the inhibition of calcite by xanthan gum. Stirring time has little effect on the inhibition effect of natural gum. The experiment of artificial mixed mineral shows that natural glue has good separation effect for scheelite and calcite, but bad separation effect for fluorite and calcite. The mechanism study shows that different natural adhesives have different adsorption modes on the surface of calcium-containing minerals. The adsorption of acacia gum and xanthan gum on the surface of minerals is mainly caused by electrostatic force, hydrogen bonding and other physical and chemical effects. While, guar gum mainly shows hydrogen bonding adsorption. The calculation of molecular structure and properties of natural gum shows that the branched chain structure, hydroxyl group and relative molecular weight of natural gum have great influence on the inhibition of natural gum.

Carboxymethyl cellulose (CMC) and humic acid and their sodium salts are also a kind of polymer organic inhibitors widely used in flotation. Researchers have found that carboxymethyl cellulose has a very significant inhibitory effect on calcite and fluorite at pH = 7.5–9.5. Under good dosage and pH conditions, carboxymethyl cellulose can effectively inhibit fluorite and calcite in scheelite [53]. Its inhibition mechanism is carboxyl group and mineral surface Ca^{2+} combination. At the same time, The polar groups that do not function play a hydrophilic role [67]. Humic acid is a kind of natural polymer polyelectrolyte widely existing in soil and coal measure minerals. It is easy to form colloids and disperse in solution. Its basic structure is composed of aromatic ring and lipid ring. The active functional groups are mainly carboxyl, hydroxyl, carbonyl, quinone, methoxy, etc. As a natural metal chelating agent, it forms insoluble compounds after complexing with calcium and magnesium ions. The addition of humic acid inhibitors in the flotation process can significantly reduce the solubility of fluorite and calcite, reduce the amount of Ca^{2+} adsorption by collector, so as to achieve the purpose of selectively inhibiting the gangue containing calcium and efficiently recovering scheelite [68].

For other organic inhibitors, the inhibition ability of different functional groups on tungsten ore and gangue minerals is different. Studies have shown that when 731 oxidized paraffin soap is used as collector, the pH of pulp suitable for separating scheelite from calcite and fluorite is about 10, regardless of tannin extract or citric acid inhibitor. Among them, tannin extract has the strongest inhibition ability on calcite.

Therefore, tannin extract is an effective inhibitor for separating scheelite when the calcium-containing gangue co-existing with scheelite is only calcite. Citric acid has the strongest inhibition ability on fluorite, which is stronger than malic acid, gallic acid, etc. Therefore, when the calcium-containing gangue co-existing with scheelite is only fluorite, citric acid is an effective inhibitor of scheelite [69]. In addition, in the flotation of scheelite and calcite [70], using sodium oleate as collector, carboxylated chitosan, chitosan, xanthan gum, yellow potato gum, calcium llignin sulfonate, and sodium alginate depressants have better selective inhibition effect on calcite. And sodium carboxymethyl starch, methyl cellulose, pectin can inhibit calcite but also inhibit scheelite, which have poor selectivity. Especially, the inhibition effect of sodium carboxymethyl starch on calcite is weaker than that on scheelite; Locust bean gum and hydroxyethyl cellulose had no significant inhibition effect on calcite. Artificial mixed mineral and actual ore tests confirmed that carboxylated chitosan, chitosan, calcium lentisulfonate, sodium alginate, yellow potato gum five polymer inhibitors can play an effective role for inhibiting calcite in the flotation process of scheelite.

The adsorption mechanism of these organic inhibitors on mineral surface is generally believed to be the combination of oxygen atoms in oxygen-containing functional groups of organic inhibitors and calcium points on mineral surface, in addition, it may also include hydrogen bonding with mineral surface [56, 71]. As can be seen from the adsorption model in Fig. 5.7, functional groups of organic inhibitors mainly bond with calcium points on mineral surfaces through oxygen atoms and have some hydrogen bonding with oxygen atoms on mineral surfaces. The adsorption of organic inhibitors on the surfaces of scheelite and gangue minerals is selective. Studies show that organic inhibitors are more inclined to adsorption on gangue minerals such as calcite and fluorite, but less adsorption on scheelite surface. The selectivity of organic inhibitors is closely related to the calcium point density and surface electrical properties on the mineral surface. Calcite and fluorite have higher calcite density than scheelite, which is conducive to the adsorption of organic inhibitors on their surfaces. Calcite and fluorite surfaces tend to be positively charged in alkaline environments with $pH = 8–11$, while scheelite is negatively charged, which makes it easier for organic inhibitors with negative charges in solution to be adsorbed on calcite and fluorite surfaces.

5.2.2.3 Dispersant

In general, agents that disperse fine particles or micro-fine particles are dispersants. Dispersants are generally divided into two categories: inorganic dispersants and organic dispersants. Scheelite and wolframite flotation commonly used inorganic dispersants are water glass, sodium carbonate, sodium hexametaphosphate, etc. The organic dispersants commonly used are carboxymethyl cellulose, tannic acid and so on.

Water glass Na_2SiO_3 has a dispersive effect on the slime, by adding water glass to weaken the harmful effect of slime on flotation, but the dosage should not be

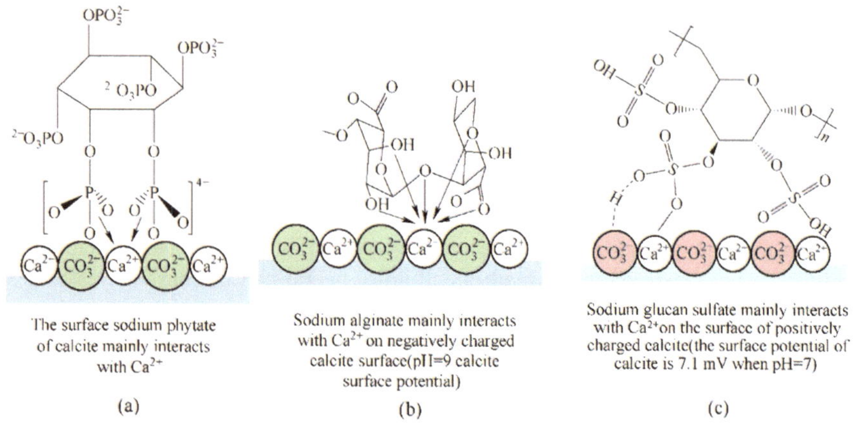

Fig. 5.7 Adsorption model of organic inhibitor on calcite surface: **a** sodium phytate, reprinted from Ref. [62], copyright 2018, with permission from Minerals Engineering, **b** sodium alginate, reprinted from Ref. [56], copyright 2017, with permission from Minerals Engineering **c** sodium glucan sulfate, reprinted from Ref. [60], copyright 2017, with permission from International Journal of Mineral Processing

too large. In the flotation of scheelite, water glass is widely used as an inhibitor, and it is also commonly used as a slime dispersant. The main mechanism of action is that the colloidal silicic acid, $HSiO_3^-$ and SiO_3^{2-} ions dissociated from water glass are adsorbed on the slime surface, forming a layer of strong hydrophilic and negatively charged "anti-condensation" covering. On the one hand, it enhances the strength and hydrophilicity of the hydration layer on the surface of the slime. More importantly, the absolute value of the negative potential on the surface of the slime is greatly increased. And the electrostatic repulsion of the same charge between the micro-particles is enhanced, so that they are difficult to approach each other. Water glass is also often used in combination with sodium hydroxide to achieve the best dispersion of microparticle groups.

Sodium carbonate is a widely used pH regulator in the flotation of wolframite and scheelite. Its pH range is 8–10, and it also has a certain dispersion effect on fine mud. Therefore, when the pH required in the flotation process is not high, and the fine mud is required to disperse, then the use of Na_2CO_3 as the regulator can have both these two effects. Sometimes in order to strengthen its dispersion, a small amount of water glass and sodium carbonate is often used in combination. Some scholars have found that the combination of $Na_2SiO_3 + Na_2CO_3$ can improve the selective inhibition ability of fluorite and apatite [72].

5.2.2.4 Activator

Activator usually refers to the activation of target minerals with metal ions, such as Cu^{2+}, Fe^{2+}, Pb^{2+}, Ca^{2+}, etc. In tungsten flotation, whether using chelating agents such as benzohydroxamic acid or conventional fatty acid collector such as sodium oleate, Pb^{2+} ion is the most common and relatively most effective activator in tungsten flotation [5]. According to the traditional activation theory, when the lead nitrate is used as the activator, the lead nitrate in the solution will hydrolyze into Pb^{2+}, $Pb(OH)^+$ and $Pb(OH)_2$ with active components at different pH values. These active components can be selectively adsorbed on the surface of tungsten minerals, and then chelating agents or fatty acids can interact with these active components to achieve selective separation of tungsten minerals from gangue minerals.

5.3 Synergistic Effect and Interfacial Assembly of Flotation Reagents for Tungsten Ore

In recent years, domestic and foreign research institutes have carried out a lot of research work on the design and development of flotation reagents. The assembled reagent has better surface activity and synergistic effect than the single reagent. Through the combination of different agents, the selectivity and activity of different agents can complement each other, playing a synergistic effect, so that the pulp environment adaptability is better, the reagent consumption is lower, and the harvesting ability is stronger [73]. On the basis of the classic reagent mix, domestic and foreign researchers have carried out in-depth research on the combination of reagents and their synergistic mechanism, and developed the interfacial molecular assembly theory of flotation reagents. It includes anion collector assembly, the same type of ion collector assembly, anion and neutral molecule assembly, anion and macromolecule assembly and metal ion coordination regulation molecule assembly [74–76]. And a series of new flotation agents have been developed, which have been widely used in tungsten ore flotation.

5.3.1 Synergistic Effect of Anion-Anion Collector

Anionic collector mainly through anionic functional groups, such as carboxyl, sulfonic and phosphonic acid groups, interact with Ca, Fe and Mn and other active sites on the surface of tungsten ore to form coordination compounds such as single, double and bridge ring coordination [77], and then make the surface of tungsten ore hydrophobic through its own different hydrophobic carbon chains. For complex tungsten resources, it is often difficult to achieve effective separation of tungsten ore and gangue minerals (fluorite, calcite, barite, etc.) by using single anionic collectors such as carboxylic acids, sulfonic acids, hydroxamic acids and chelate. The collaborative assembly of anionic collectors with different structures can improve the performance of anionic collectors and make up for the defects of activity or selectivity of a single type of agent, as shown in Fig. 5.8.

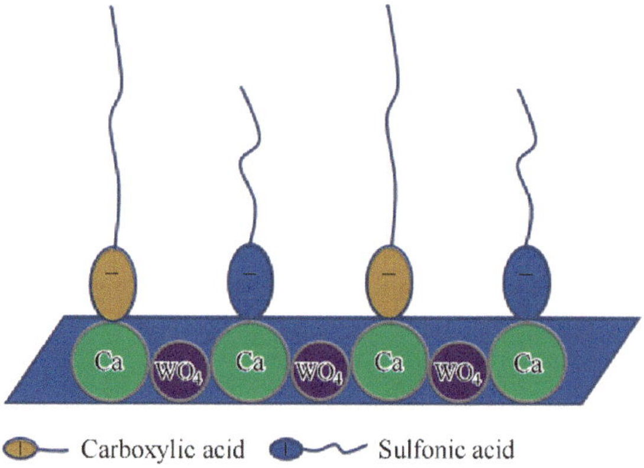

Fig. 5.8 Synergistic mechanism model of anion-anion assembly collector on scheelite surface

Oleic acid anionic collectors have long carbon chain structure and strong bonding ability with metal ions, especially Ca^{2+} and Mg^{2+} ions, resulting in low temperature resistance and poor adaptability to hard water. The floatability of scheelite and fluorite is generally improved after the combination of sodium oleate and sodium alkyl fatty acid, while that of calcite is generally worse. The better effect is the combination of sodium oleate, sodium laurate and sodium stearate. Jiang [78] used sodium oleate and its homologue sodium laurate, sodium stearate to separate scheelite from calcite and fluorite by flotation. The results of flotation experiment are shown in Fig. 5.9. When sodium oleate and sodium n-caprylate are used in combination, the floatability of calcite is poor when the dosage of reagent is less than 8.22×10^{-6} mol/L. When sodium oleate is combined with sodium laurate and sodium stearate, the floatability of scheelite is better than that of fluorite and calcite under the same dosage of reagent. Under the same pH value and dosage condition, if scheelite and fluorite reach the same recovery, the dosage of the combined agent is lower than that of sodium oleate. The results indicated that there is a positive synergistic effect between sodium oleate and homologous fatty acids.

Hydroxamic acid is a surfactant agent with high selectivity for metal ions, with certain harvesting capacity and good selectivity. When sodium oleate and hydroxamic acid surfactants are assembled and mixed, the collecting ability and selectivity are significantly improved. Huang [79] investigated the effects of cyclohexyl methyl hydroxamic acid (CHA), p-tert-butylbenzohydroxamic acid (THA), benzohydroxamic acid (BHA), salicylhydroxamic acid (SHA), octylhydroxamic acid (OHA) on the flotation behavior of scheelite and wolframite. The results showed that cyclohexyl hydroxamic acid had stronger harvesting ability for scheelite and wolframite than salicylhydroxamic acid, benzohydroxamic acid. The harvesting performance of five hydroxamic acids was as follows: cyclohexyl hydroxamic acid > p-tert-butylbenzohydroxamic acid > octylhydroxamic acid > salicylhydroxamic acid > benzohydroxamic acid. Then, taking benzohydroxamic acid and cyclohexyl hydroxamic acid as representatives, the flotation performance of the collector mixed with hydroxamic acid and sodium oleate was investigated. As shown in Figs. 5.10 and

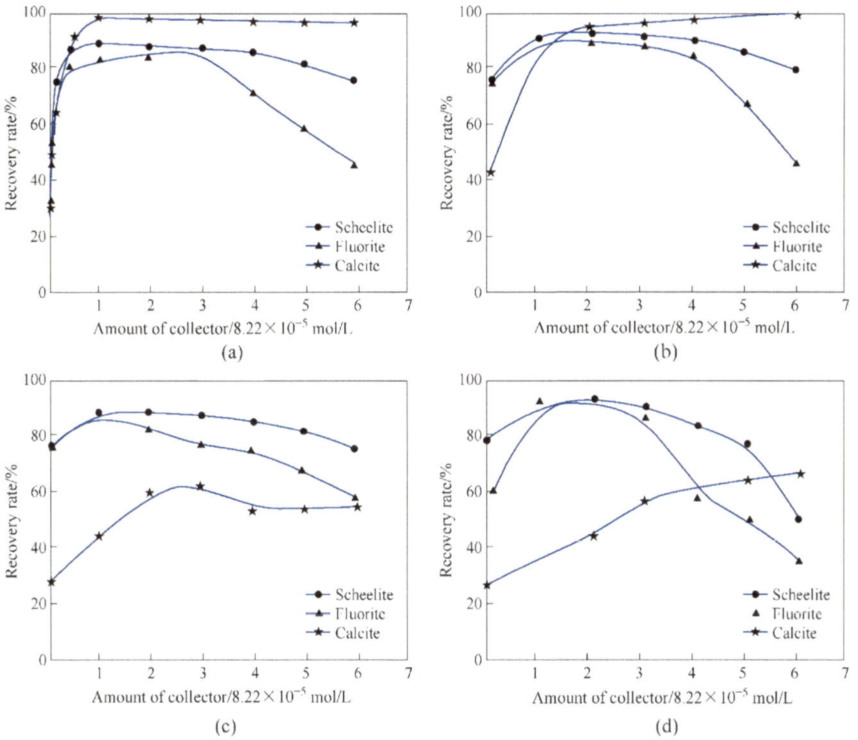

Fig. 5.9 Effect of collector dosage on mineral floatability **a** sodium oleate; **b** sodium oleate + sodium n-caprylate; **c** sodium oleate + sodium laurate; **d** sodium oleate + sodium stearate, reprinted from Ref. [78], copyright 2012, with permission from International Journal of Mineral Processing Mining and Metallurgy Engineering

5.11, when the total amount of collector is the same, compared with sodium oleate alone, the addition of hydroxamic acid collector significantly improves the flotation recovery of tungsten ore. Mechanism analysis showed that sodium oleate or hydroxamic acid preferently adsorbed on tungsten surface may change the physical and chemical properties of tungsten surface and promote the secondary adsorption of collector. Orhydroxamic acid and sodium oleate co-adsorption, produce positive synergistic effect, and improve the flotation recovery of scheelite and wolframite.

Hu [73] studied the effects of fatty acid collector 731 combined with benzohydroxamic acid and salicylaldoxime respectively on the floatability of wolframite, fluorite and quartz, as shown in Figs. 5.12 and 5.13. When 731 and benzohydroxamic acid were used alone, the recoveries of wolframite were 69.38% and 75.60%, respectively. When benzohydroxamic acid accounted for 20%, 30% and 40% in the combined collector, the recovery rate of wolframite was higher than 69.38%. When the combined collector was dominated by 731, there was a synergistic effect on wolframite, among which the proportion of benzohydroxamic acid in the combined collector was 30%. Namely, when the dosage ratio of 731: benzohydroxamic acid

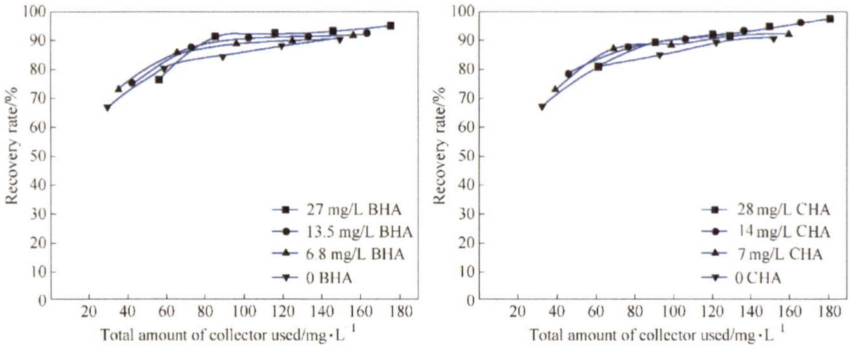

Fig. 5.10 Effect of BHA and CHA on sodium oleate flotation of scheelite [79]

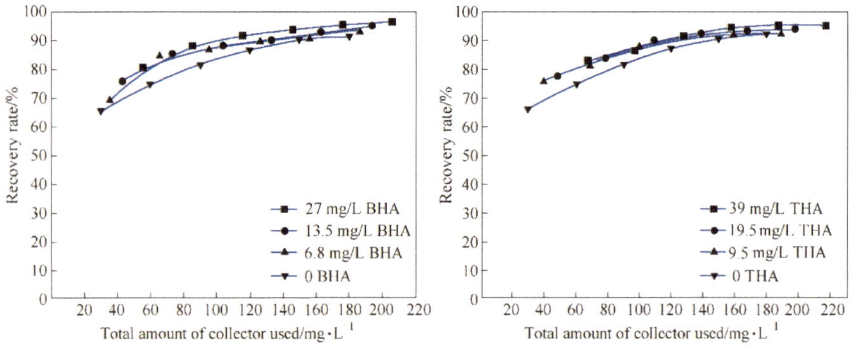

Fig. 5.11 Effect of BHA and CHA on wolframite flotation by sodium oleate [79]

was 7:3, the synergistic effect was the most obvious. The recovery rate of wolframite reached 79.35%. The collecting ability of fluorite and quartz was not enhanced before and after the combination of the collectors. When 731 and salicylaldoxime were used alone, the recovery of wolframite was 65.43% and 69.60%, respectively. When salicylaldoxime accounted for 10%, 20% and 40% of the combined collector, the wolframite recovery was higher than 72.60%, that is, when 731 was the main collector, there was a synergistic effect on wolframite. When benzohydroxamic acid accounted for 10% of the combined collector, the synergistic effect was the most obvious, and the wolframite recovery reached 81.83%. There was no obvious synergistic effect on fluorite and quartz of the combined collector with each combination ratio.

Gao et al. [80] used water glass (WG) as an inhibitor to study the effects of 733 and MES collector on the flotation behavior of scheelite, fluorite and calcite under different mass ratios, and the results are shown in Figs. 5.14 and 5.15. The mixture of 733 and MES has higher selectivity for the separation of scheelite from calcite and fluorite than the single 733 or MES. When the mass ratio of 733 to MES is 4:1, the concentrate product with a grade of 65.76% and recovery of 66.04% can be

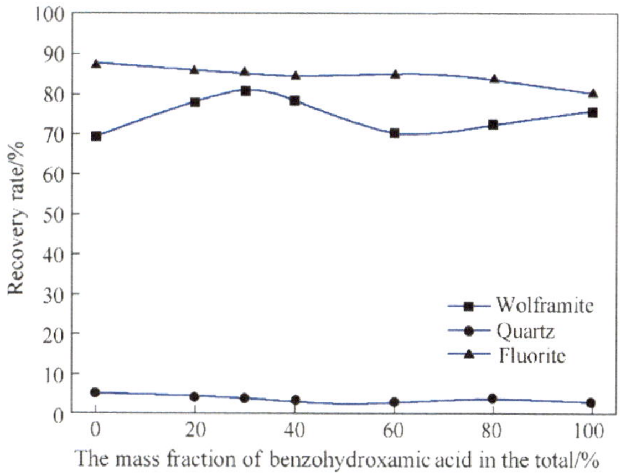

Fig. 5.12 Effect of ratio of 731 and benzohydroxamic acid on flotation of single mineral (lead nitrate 400 mg/L, 731 + BHA 200 mg/L) [73]

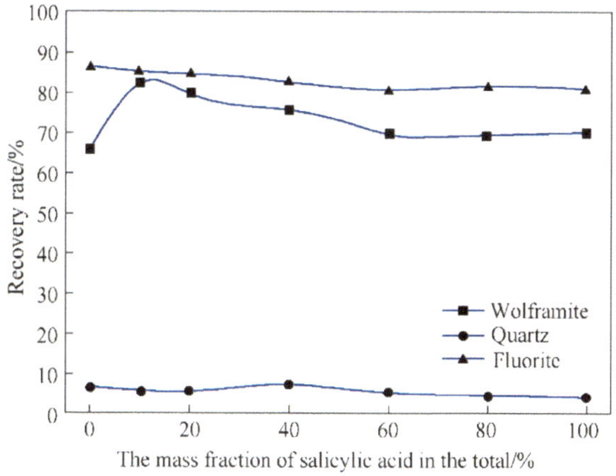

Fig. 5.13 Effect of ratio of 731 and salicylaldoxime on flotation of single mineral (lead nitrate 400 mg/L, 731 + Salicylaldoxime 200 mg/L) [73]

obtained. In the collector system assembled by MES and 733 (mass ratio 1:4), the influence of Ca^{2+} or Mg^{2+} concentration on the flotation behavior of scheelite (pH = 10, total collector dosage: 50 mg/L) is shown in Fig. 5.16. The results show that in the assembled collector system, with the increase of calcium and magnesium ion concentration, the reduction range of scheelite recovery was significantly weakened, which proves that the collection capacity of the assembled collector scheelite was less affected by Ca^{2+} or Mg^{2+}, and had a high tolerance to hard water. Through the combination of fatty acid methyl ester sodium sulfonate (MES) and oxidized paraffin soap (733), the adaptation ability of the agent to hard water was effectively enhanced, and the selectivity of the agent was improved.

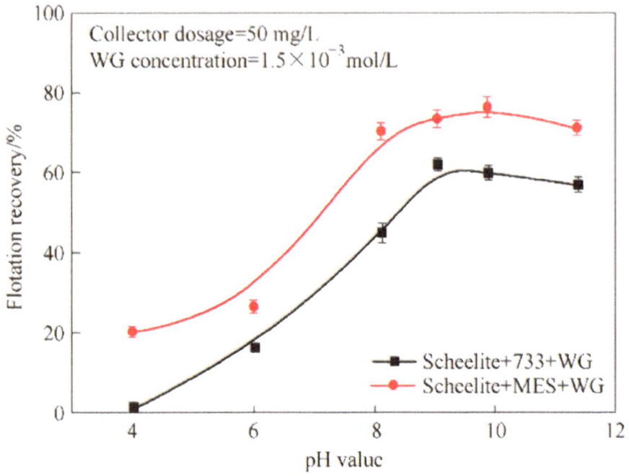

Fig. 5.14 Effect of pulp pH on scheelite flotation using 733 or MES as collector and WG as depressant, reprinted from Ref. [80], copyright 2015, with permission from Minerals Engineering

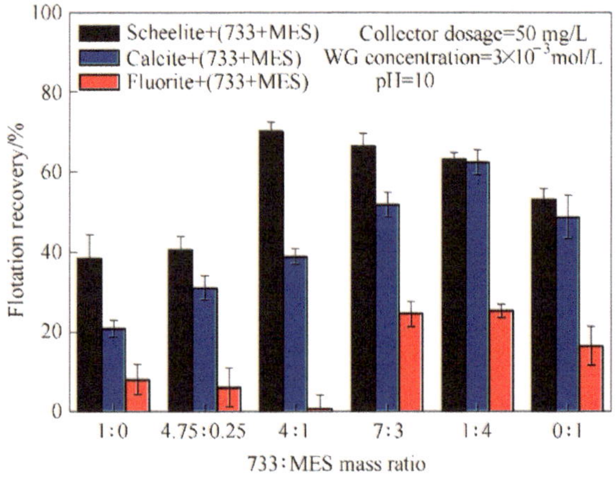

Fig. 5.15 Effect of mass ratio of 733 + MES on flotation behavior of scheelite, fluorite and calcite, reprinted from Ref. [80], copyright 2015, with permission from Minerals Engineering

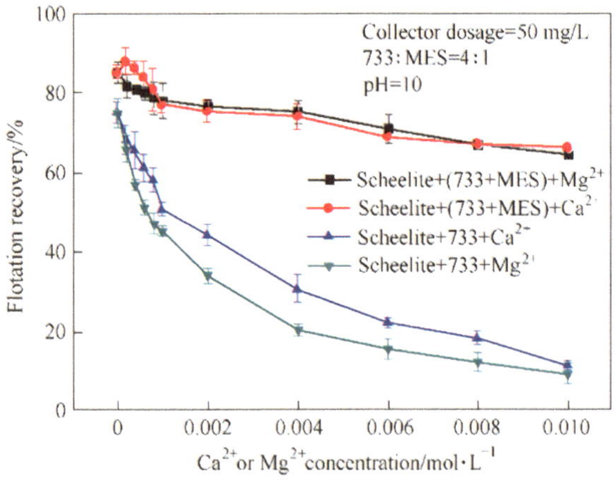

Fig. 5.16 Effect of Ca^{2+} or Mg^{2+} concentration on the flotation behavior of scheelite with 733 or 733 + MES as collector, reprinted from Ref. [80], copyright 2015, with permission from Minerals Engineering

5.3.2 *Synergistic Effect of Anion-Cation Collector*

The main mode of action of cationic collector on mineral surface is electrostatic adsorption, which is used in combination with anionic surfactant. Through charge compensation, the cationic collector and anionic surfactant are co-adsorbed on mineral surface, so as to improve reagent performance, strengthen adaptability and improve flotation effect. The mechanism model is shown in Fig. 5.17.

The combination of sodium oleate (NaOL) and DDA (DDA) can improve the foam viscosity and hydrophobicity of DDA and improve the separation efficiency. Dong Liuyang et al. [81] studied the role and mechanism of cationic collector DDA (DDA) and anionic collector sodium oleate (NaOL) and their combined collectors in the flotation separation of scheelite and calcite. As shown in Figs. 5.18 and 5.19, when pH = 7 and the combined ratio of DDA and NaOL was 9:1, the recovery rate of scheelite reached 95%, which was significantly higher than that of DDA and sodium oleate alone. When DDA and NaOL were combined at 9:1, the recovery of scheelite and calcite reached the maximum when the total amount of combined collector was 1.5×10^{-4} mol/L, which were 97% and 76%, respectively. The flotation recovery of scheelite was higher than that of calcite, which was conducive to the flotation separation of scheelite and calcite. As can be seen from Fig. 5.20, after the addition of inhibitor acidizing water glass, the floatability of scheelite was slightly reduced, but it could still reach 90%, while the recovery of calcite was reduced from about 80% to about 40%, creating a good condition for the flotation separation of scheelite and calcite.

Yang [33] used pre-oxidized paraffin soap (733) + benzohydroxamic acid (BHA) as mixed collector for roughing flotation of scheelite ore from Shizhuyuan, and then strongly stirred scheelite roughing concentrate under acidized water glass condition. After a blank cleaning, quaternary ammonium collector (decadecyl dimethyl ammonium chloride (DDAC), octylmethyl ammonium chloride (TOAC) or dodecyl trimethyl ammonium chloride (DTAC)) was added for normal temperature cleaning.

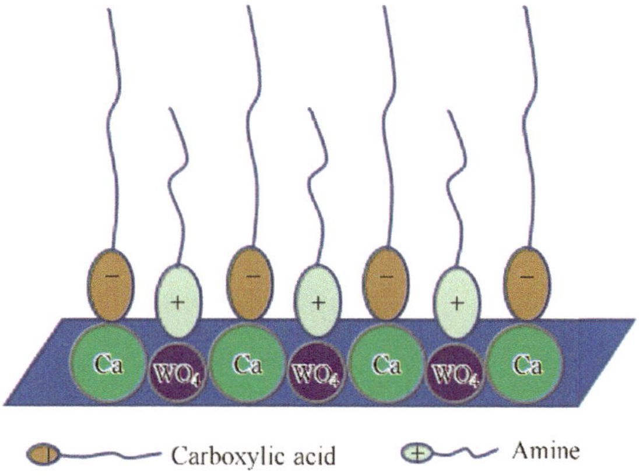

Fig. 5.17 Synergistic mechanism model of anion-cation assembly collector on scheelite surface

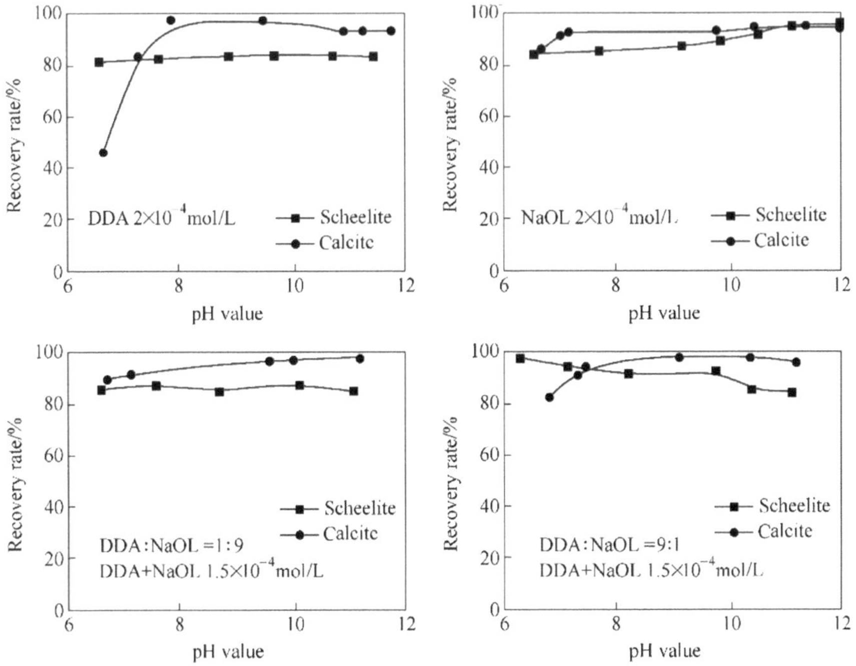

Fig. 5.18 Effect of pH values on recoveries of scheelite and calcite, reprinted from Ref. [81], copyright 2018, with permission from Mining and Metallurgical Engineering

Fluorite could be separated well in roughing flotation, but the separation effect of scheelite and calcite was poor. In the process of normal temperature separation by adding quaternary ammonium collector, the efficiency of separation was significantly improved, and the final WO_3 grade of concentrate product was increased from 37.54% to more than 50%. The principle of its action was that a large number of anionic collectors were adsorbed on scheelite in the roughing separation process, and cationic collectors were further adsorbed on scheelite in the cleaning process, realizing the co-adsorption of anionic collectors and cationic collectors, and greatly improving the floatability of scheelite. Therefore, the assembly and combination of sodium oleate and quaternary ammonium salts can significantly improve the separation efficiency of scheelite and calcite.

5.3.3 Synergistic Effect of Anion-Nonionic Collector

Nonionic surfactants act on mineral surfaces mainly in the form of lenticular adsorption, co-adsorption and bridging, with emulsification, dispersion, anti-ion interference and synergistic abilities [14, 82]. Therefore, adding a certain amount of nonionic surfactant to the anionic surfactant can generate mixed micelles due to the

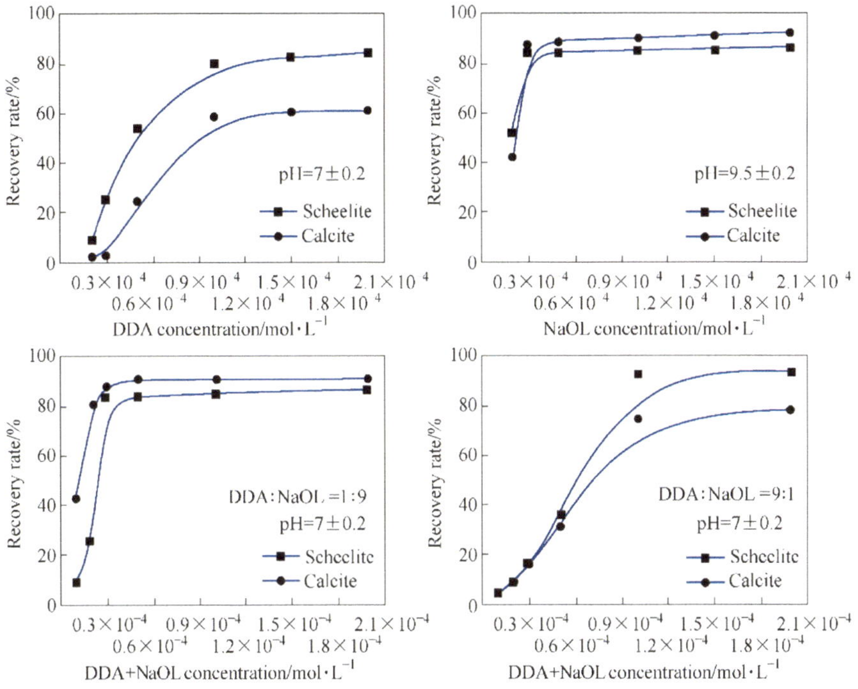

Fig. 5.19 Effect of collector concentration on recoveries of scheelite and calcite, reprinted from Ref. [81], copyright 2018, with permission from Mining and Metallurgical Engineering

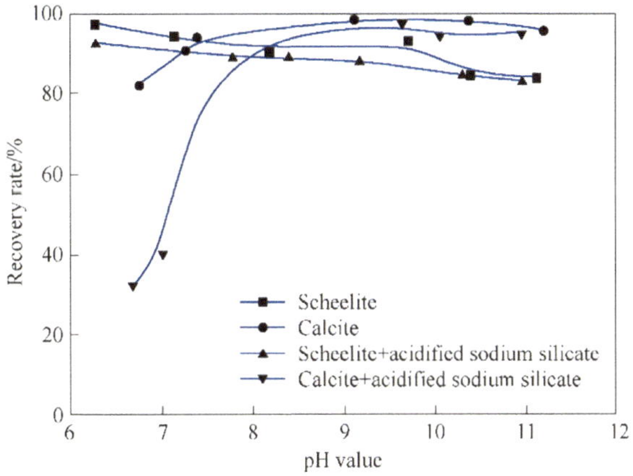

Fig. 5.20 The effect of pH on the recovery of scheelite and calcite before and after the addition of inhibitor, reprinted from Ref. [81], copyright 2018, with permission from Mining and Metallurgical Engineering

interaction between surfactant molecules, which significantly reduces the surface tension and critical micelle concentration of the composite system. And it has better dispersibility and adaptability, which is beneficial for resisting the effects of calcium and magnesium ions, temperature, and pH on mineral flotation [83]. At present,

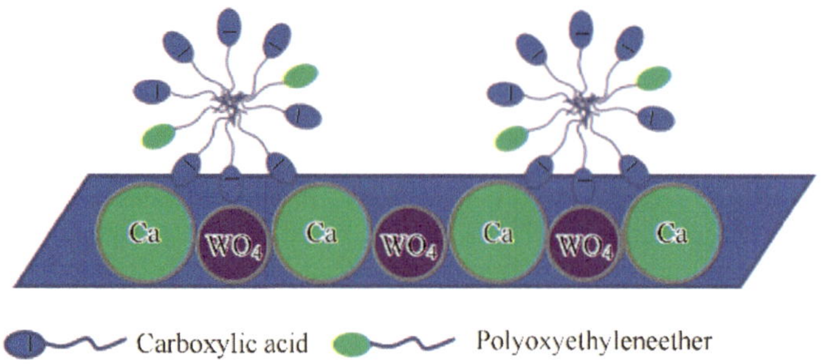

Fig. 5.21 Synergistic mechanism model of anion-nonionic assembly collector on scheelite surface

nonionic surfactants widely used in the flotation process of tungsten ore mainly include polyoxyethylene ethers and fatty acid amides, and the mechanism model is shown in Fig. 5.21.

At low temperature, sodium oleate has poor collection ability for scheelite, while the assembly and combination of sodium oleate (NaOL) and polyoxyethylene ether (MOA-9) can improve the collection ability of scheelite to varying degrees. Zhu et al. [84] studied the influence of fatty alcohol polyoxyethylene ether with different structures in sodium oleate system on the low-temperature flotation behavior of scheelite through single mineral flotation test, and applied it in the practice of scheelite low-temperature flotation. As shown in Figs. 5.22 and 5.23, when the pH value was 10 and the dosage of sodium oleate was 60 mg/L, the flotation recovery rate of scheelite at 10 °C was only 20%. When NaOL and MOA-9 were mixed according to the mass ratio of 5:1, the flotation recovery of scheelite could reach 85% at the same concentration of NaOL. The floatability of scheelite was significantly improved, the amount of collector was greatly reduced, and the influence of the MOA reagents with different structures on scheelite flotation behavior had the same rule. Namely, with the increase of pH value, the flotation recovery of scheelite first increased and then decreased, and the floatability of scheelite was the best when pH was 8–10. In addition, the flotation recovery rate of scheelite using different reagents was ranked in descending order from high to low, including isomeric alcohol polyoxyethylene ether, linear dodecanol polyoxyethylene ether, and linear hexadecanol polyoxyethylene ether.

The combination of oleophilic sulfonic acid collector and hydrophilic nonionic surfactant can obtain a complex system with moderate hydrophilic and oleophilic properties. The complex system has additive hydrophilic and oleophilic properties, which can significantly reduce interfacial tension and achieve synergistic effect. Sodium alkyl benzene sulfonate is a nonionic sulfonate surfactant, hydrolysis and ionization in water, with high stability, good performance in hard water, good compatibility with other types of surfactants and good solubility in water, etc. Wang et al. [85] studied the synergism between three kinds of sodium branched alkyl benzene sulfonate (3C10ΦS, 7C14ΦS, 8C18ΦS) and three kinds of polyoxyethylene ether nonionic surfactants (LAP-9, OP 10, TW-20) on the interfacial tension of oil and

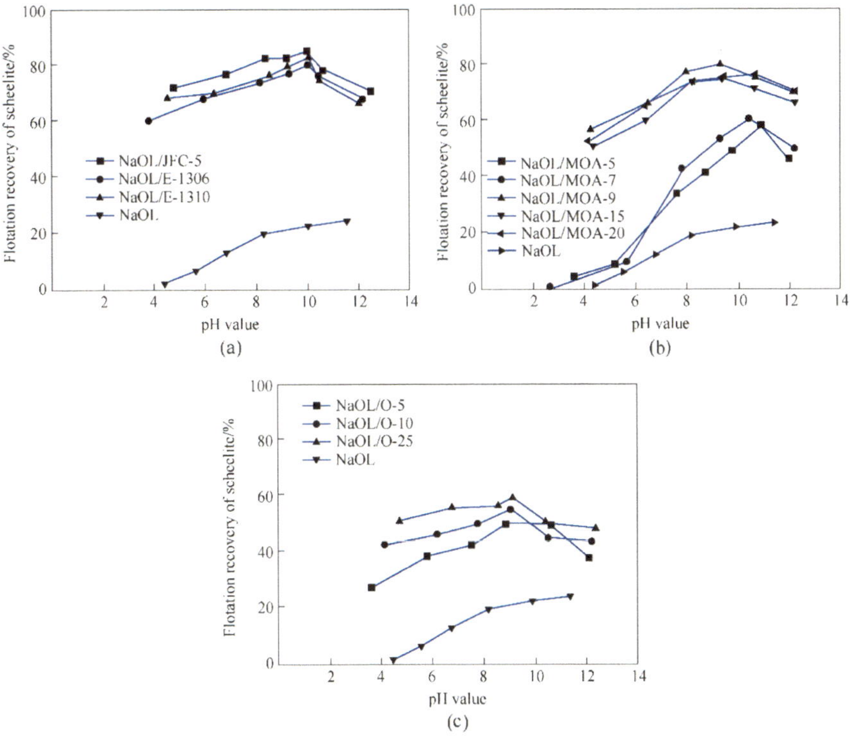

Fig. 5.22 Influence of pH value on flotation behavior of scheelite at 10 °C [84]

water. The results showed that: 3C10ΦS, 7C14ΦS and LAP-9 are hydrophilic, and the order of hydrophilicity is 3C10ΦS > LAP-9 > 7C14ΦS. And 8C18ΦS is oleophilic, in the high carbon number alkane hydrophilic oil performance is close to moderate, to achieve ultra-low interfacial tension (Fig. 5.24); at the same time, the combination of oily sulfonate 8C18ΦS and hydrophilic LAP-9 nonionic surfactant can obtain a moderate mixture system of hydrophilic and oilophilic, so that the interfacial tension is reduced, to achieve synergistic effect; and the compound system of the hydrophilic sulfonate 3C10ΦS, 7C14ΦS and the hydrophilic LAP-9 nonionic surfactant is still hydrophilic, and the interfacial activity is low. The above results indicate that the hydrophilicity and lipophilicity of the complex system of sulfonates and non-ionic surfactants have additive properties (Fig. 5.25).

Fatty acid amide is a kind of nonionic surfactant with strong surface activity, which has good bubble stability and wetting and emulsifying ability. It can be used as a synergist to improve the collection performance of fatty acid. Based on the interface assembly method of sodium oleate and oleate amide, the anion-nonionic association collector was designed and developed. Compared with single sodium oleate, the association collector showed better selectivity and weakened the collection ability for calcite and fluorite, which was conducive to the flotation separation

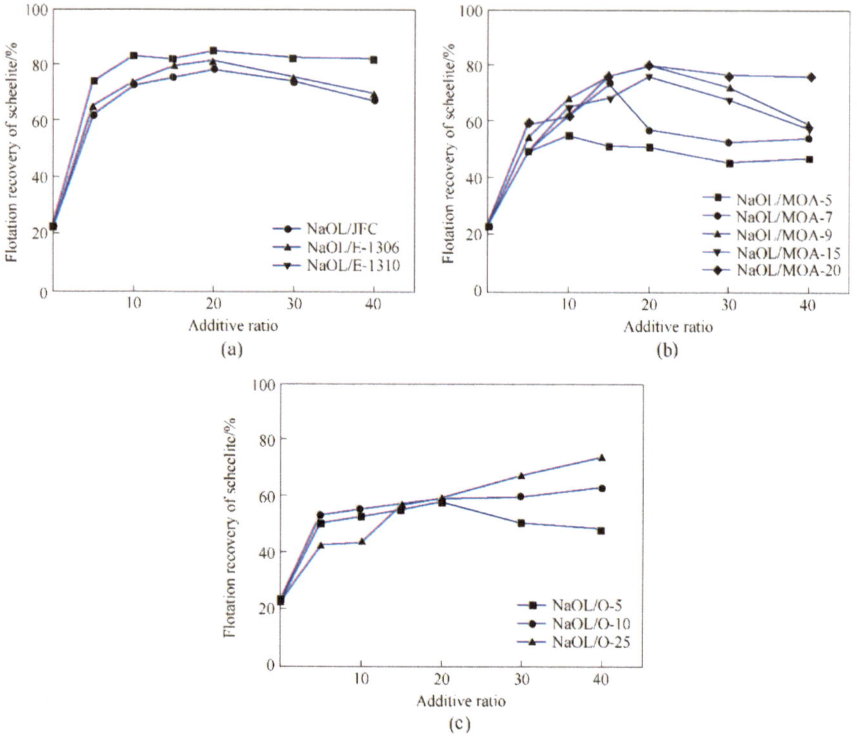

Fig. 5.23 Effect of polyoxyethylene ether addition ratio on flotation behavior of sodium oleate at low temperature (10 °C) [84]

Fig. 5.24 Interfacial tension between a single surfactant and a n-alkanes, reprinted from Ref. [85], copyright 2012, with permission from Oilfield Chemistry

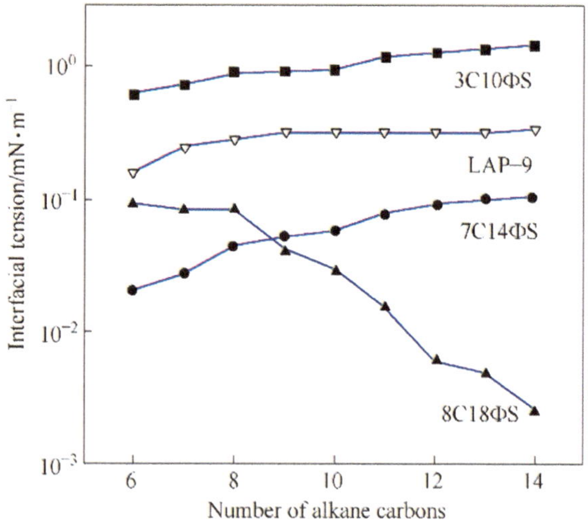

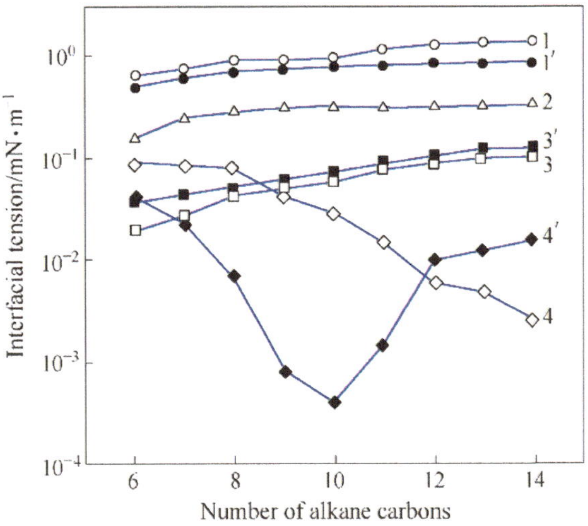

Fig. 5.25 Interfacial tension between sulfonate-Lap-9 compound system and n-alkanes (compound volume ratio 2:1), reprinted from Ref. [85], copyright 2012, with permission from Oilfield Chemistry 1—3c10ΦS; 1′—3C10ΦS and LAP-9 compound system; 2—LAP-9; 3—7C14ΦS; 3′—7C14ΦS, LAP-9 compound system; 4-8—c18ΦS; 4′—8C18ΦS and LAP-9 compound system

of scheelite from fluorite and calcite. Oleate amide has excellent emulsification, dispersibility, resistance to ion interference and synergistic ability. The addition of oleamide changes the micellar morphology of single sodium oleate and forms a stable configuration of mixed micelles through hydrocarbon chain association with sodium oleate, which is uniformly adsorbed on the surface of fluorite mineral in the form of massive, forming dense multilayer adsorption. While, the sparse mixed micellar adsorption on the surface of calcite results in poor hydrophobicity on the surface of calcite, significantly enhancing the difference in flotation behavior between fluorite and calcite. It provides the possibility for efficient flotation separation of fluorite and calcite.

5.3.4 Interface Assembly Mechanism Between Flotation Reagents

The adsorption of different reagents on mineral surfaces is a complex process, and the description of the action mechanism of the combined collector is only a hypothesis and inference to a large extent. The adsorption mechanism can be summarized as co-adsorption, lengthening of hydrophobic end, charge compensation, promoting adsorption and so on. In the field of interfacial chemistry, a large number of experts and scholars have studied the micellar morphology of mixed surfactants in solution and the adsorption at the water/gas interface and proposed the interfacial assembly mechanism of mixed surfactants in solution. Surfactant interfacial assembly is defined as the spontaneous and ordered assembly of surfactant molecules to form a thermodynamically stable and ordered structure due to non-covalent bonding such as electrostatic interaction, hydrogen bonding and hydrophobic interaction. The adsorption of these special colloid structures at the solid/liquid and gas/liquid interfaces is highly selective. Due to the complexity of tungsten ore (single crystal purity, optical properties, solubility, etc.), the assembly and adsorption tests of flotation reagents at the tungsten ore/water interface are difficult, and the microscopic mechanism of flotation reagents has not been studied deeply. However, researchers at home and abroad have carried out a lot of research work on the interface assembly of flotation reagents with silicate minerals as the object, trying to reveal the mechanism of the interface action of flotation reagents.

Xu et al. [86] introduced the synergistic index of mixed surfactants into the field of combined flotation reagents, systematically studied and revealed the synergistic assembly behavior of combined flotation reagent molecules at the interface of aqueous solution and mineral. By means of surface tension test, combined with colloid interface chemistry theory, solution theory and Rubingh theory, the synergistic index of anionic/cationic assembly collector was deduced and calculated systematically. At the same time, according to the theory of solution chemistry, the relationship between the active component of the anionic/cationic assembly collector and the pH value of the solution was constructed, which could be used to predict the flotation performance of the anionic/cationic assembly collector. Wang and Jiang et al. [87, 88] studied the flotation enhancement mechanism of the anionic/cationic assembly collector for mica minerals such as white mica, biotite and lematolite in an in-depth and systematic way. They found that the cationic collector not only promoted the adsorption of the anionic collector on the surface of mica, but also increased its adsorption capacity on the surface of mica, which enhanced the hydrophobicity of mica and its selective separation from quartz. Through contact angle measurement, molecular dynamics simulation, adsorption capacity measurement, infrared spectroscopy, X-ray photoelectron spectroscopy (XPS) and other interface analysis methods, the microscopic level of the co-assembly mechanism of anionic/cationic assembly collector on the surface of mica was revealed. It was found that the assembly collector could achieve a "1 + 1 > 2" synergistic effect through proper assembly. The assembly adsorption of anionic/cationic collector not only reduces the electrostatic repulsion between the same molecules, but also enhances the electrostatic attraction between the two kinds of heterogeneous molecules, so that the distance between the collectors is shortened, and the number of pharmaceutical molecules arranged on the unit area is increased. Guided by multi-scale calculation simulation, Wang et al. [89–91] conducted experimental verification and feedback optimization at the same time, and studied the assembly behavior of anionic/cationic assembly collector in aqueous solution and the secondary assembly mechanism of pharmaceutical association from aqueous solution to mineral/water interface from macro, mesoscopic and microscopic levels. Effective regulation of adsorption capacity, adsorption morphology and adsorption strength of pharmaceutical molecules on mineral/water interface was realized, as shown in Fig. 5.26. Through molecular dynamics simulation and experimental research, it was found that the adsorption and association structures of two types of drugs with different ratios at the gas/liquid interface were different. Further analysis was conducted on the cluster size, molecular composition, interaction between surfactant molecules and water molecules in the complex, and the mode of interaction between surfactants. Through the simulation calculation of mineral/water interface, it was found that compared with a single agent, the combination agent was co-adsorbed on the surface of mica, and the hydrophobic carbon chain tilted to the surface to form a close hydrophobic association molecular layer. On the surface of quartz, very individual cations were adsorbed on the quartz surface, and the remaining agents were distributed near the quartz surface through the hydrophobic association of carbon chains, forming a loose hydrophilic association structure. Based on the above studies, the mechanism of interfacial assembly of composite collector can be understood from a new perspective: the adsorption of combined collectors on mineral surfaces is different from the single-layer adsorption of a single agent, but rather the adsorption of the connective structural units composed of multiple agent molecules on the active sites of mineral surfaces. The adsorption of combined collector on mineral/water interface forms association structure with different morphologies, leading to the enhancement of mineral selective hydrophobic phenomenon.

The assembly of ionic surfactants and neutral molecules is an important class of flotation reagent mixtures. They have strong synergistic effect, lower CMC value, more effectively reduce the surface tension of water than the corresponding single component, and have better wettability, solubilization and foaming properties. Studies have shown [84, 85] that when a certain concentration is reached, the presence of neutral molecular surfactants can weaken the electrostatic mutual repulsion between anions, thus increasing the adsorption of collector on the mineral surface. The long hydrophobic hydrocarbon chains contained by neutral surfactants can also enhance their absorbability to the hydrophobic surface and the hydrophobicity of the agent. In combination of reagents, as the concentration of neutral molecules increases, the neutral surfactant may replace the adsorption of anions on the strong hydrophobic surface.

Wang et al. [92] used the combination of cationic surfactant cetyltrimethyl ammonium bromide (CTAB) and neutral molecular surfactant octylphenoxyethylene ether (TX-100) to change the surface wettability of oil-wet sandstone, and built an adsorption model, as shown in Fig. 5.27. The results show that the adsorption of the combined surfactant on the quartz surface is related to its concentration. At low concentration, the two surfactants compete to adsorb on the quartz surface. The addition of TX-100 under medium concentration conditions, reducing the electrostatic repulsion between CTAB, the formation of a mixture has a lower CMC valve than a single surfactant. CTAB will be adsorbed on the quartz surface before TX-100, and then form aggregates with TX-100, thereby increasing the adsorption of cationic surfactants. When the concentration of the surfactant is greater than CMC value, CTAB will form ion pairs with the carboxylic acids and be desorbed from the quartz surface. Jiang et al. [93] found that the mixed cationic collector cetyltrimethylammonium chloride (CTAC) and the non-ionic collector octanol (OCT) had a synergistic effect on the flotation and adsorption of muscovite. The mixing ratio of collector has a significant effect on the CMC value. The CMC value firstly decreases with the

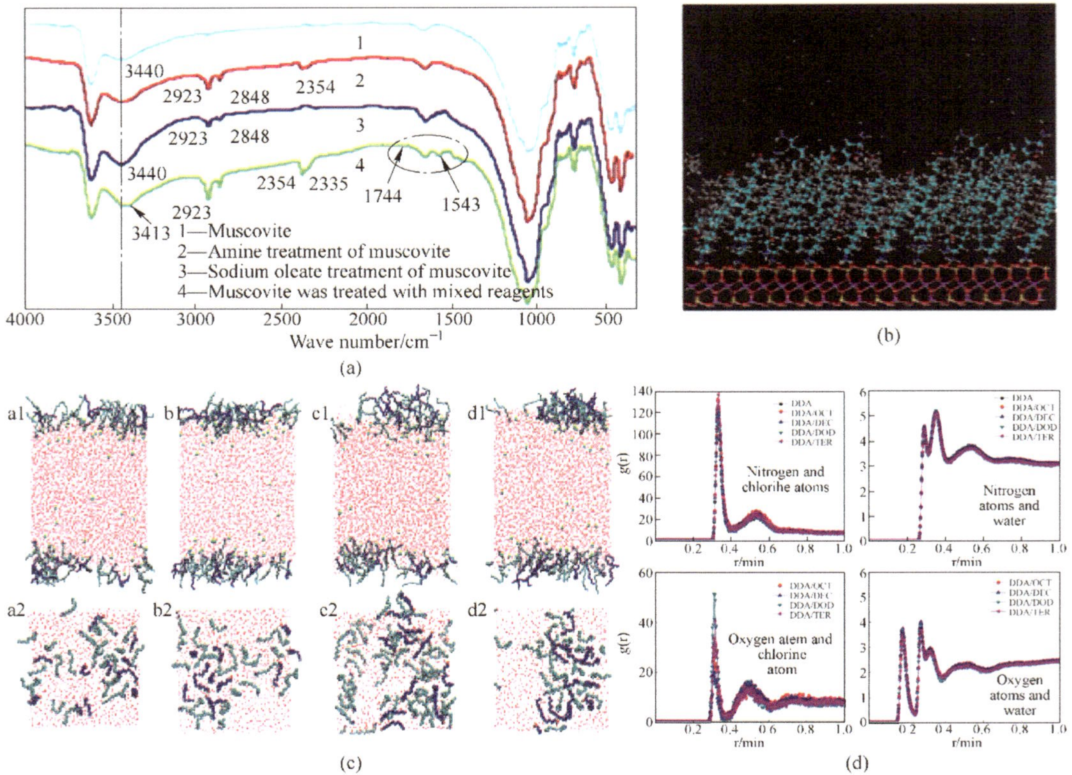

Fig. 5.26 Interface mechanism of the combined reagents, **a** reprinted from Ref. [89], copyright 2014, with permission from Minerals engineering, **b** reprinted from Ref. [90], copyright 2015, with permission from Powder Technology, **c**, **d** reprinted from Ref. [91], copyright 2017, with permission from Journal of Molecular Liquids

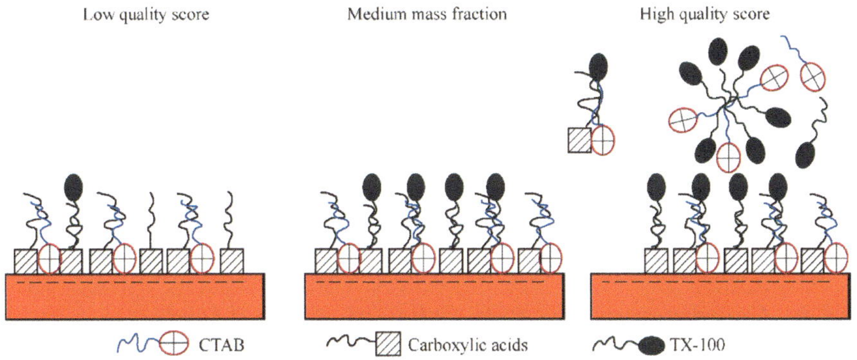

Fig. 5.27 Adsorption model of cationic/nonionic composite surfactant (CTAB/TX-100) changing the surface wettability of oil-wet sandstone [92]

increase of CTAC ratio. The CMC value reaches the minimum when the molar ratio of CTAC/OCT reaches $\alpha_{CTAC} = 0.67$, and all surfactants show synergistic effect. The results show that in the first stage, except for the cation exchange adsorption between the collector and the potassium ion on the surface of muscovite, the electrostatic adsorption between the combined collector and the negatively charged (001) surface is the main driving force for the increase of the adsorption capacity. In the second stage, the non-polar carbon chains of the collector are highly interwoven with each other and form a hydrophobic layer on the surface of muscovite, making it appear hydrophobic.

In fact, the flotation application and mechanism of flotation reagent interface assembly are still in the initial stage. The pulp system is complex and there are many interference factors. Previous research methods are relatively simple, generally only rely on the adsorption capacity of agents on mineral surface, infrared spectrum testing and other means, the useful information obtained is very limited. There are few studies on the interaction and adsorption mechanism of the combined collector in solution and at the mineral interface, and the description of the mechanism of the combined collector is only a hypothesis and deduction to a large extent. The mechanism of reagent assembly and adsorption at the molecular level is still lack of accurate analysis, which restricts the efficiency of flotation reagent development. In recent years, quantum chemical calculation, molecular dynamics simulation, interfacial spectroscopy and other rapid development, for the precise analysis of the interfacial assembly mechanism of flotation reagents to provide good support, is an important direction of future development. For example, Nguyen et al. [94] studied in situ the adsorption behavior of cationic surfactants at the gas/liquid interface by sum-frequency vibration spectroscopy. It was found that the water substructure and hydrophobic hydrocarbon chain structure of the gas/liquid interface in the mixed system were greatly different from that in the single reagent system, and strong synergistic interaction occurred between the anionic and cationic groups. The molecular configuration of cationic surfactants at the gas/liquid interface was deduced by the analysis of the sum frequency vibration spectra, as shown in Fig. 5.28. This work provides a basis for the interface regulation of flotation reagents and a new means for the analysis of the interface assembly mechanism.

5.4 Metal Ion Coordination Regulation of Molecular Assembly in Tungsten Ore Flotation Reagents

5.4.1 Coordination Assembly Behavior of Metal Ions in Flotation Process

5.4.1.1 Classical Theory of Activated Flotation of Metal Ions

Metal ions play a very important role in mineral flotation process. In the practice of mineral flotation, metal ions can be used as activators to improve the recovery rate of mineral flotation and can also be used as inhibitors to inhibit mineral flotation, and play a role in directional regulation of mineral flotation behavior in flotation. For example, Ca^{2+} and other metal ions can be used for the flotation of quartz and other silicate minerals; Pb^{2+} can activate the flotation of scheelite, cassiterite and ilmenite oxide minerals; Ca^{2+}, Mg^{2+} and Fe^{3+} and other metal ions can strengthen the inhibition of inhibitors on gangue minerals such as dolomite [75]. However, for a long time, the understanding of the regulatory mechanism of metal ions in flotation system is not comprehensive, and the explanation of the mechanism of metal ions in flotation also exists diversity. In 1965, Fuerstenau et al. [75, 95] studied the activation of metal ions in the flotation process of quartz and beryl, and found that the optimal activation pH value was consistent with the pH with the largest amount of metal ion monohydroxyl complex, and proposed the hypothesis that the main component of metal ion activation was monohydroxyl complex (Fig. 5.29).

According to the classical activated flotation theoretical model, metal ion hydrate is first adsorbed with the hydroxyl group on the mineral surface through dehydration condensation to form metal ion active site on the mineral surface, and then collector ion forms metal-collector complex by complexing reaction with the metal ion active site on the mineral surface, thus adsorbed on the mineral surface. The classical activated flotation model can explain the role of metal ions in activated flotation to a certain extent, but this hypothesis ignores the following three problems: (1) The pH with the largest amount of mono-hydroxyl complex is often corresponding to the pH generated by hydroxide precipitation; (2) The range of pH generated by monohydroxyl complex is narrow and the concentration is very low, and the concentration is usually less than 1%-10% of the total concentration, which is obviously less than the effective acting concentration; (3) the precipitated solubility product of metal hydroxide at the mineral-aqueous interface is smaller than in solution [97]. Based on the above understanding, in 1987, Hu and Wang et al. [98] discussed the activation of metal ions in terms of the effects of adsorption of metal ions on the surface of oxidized

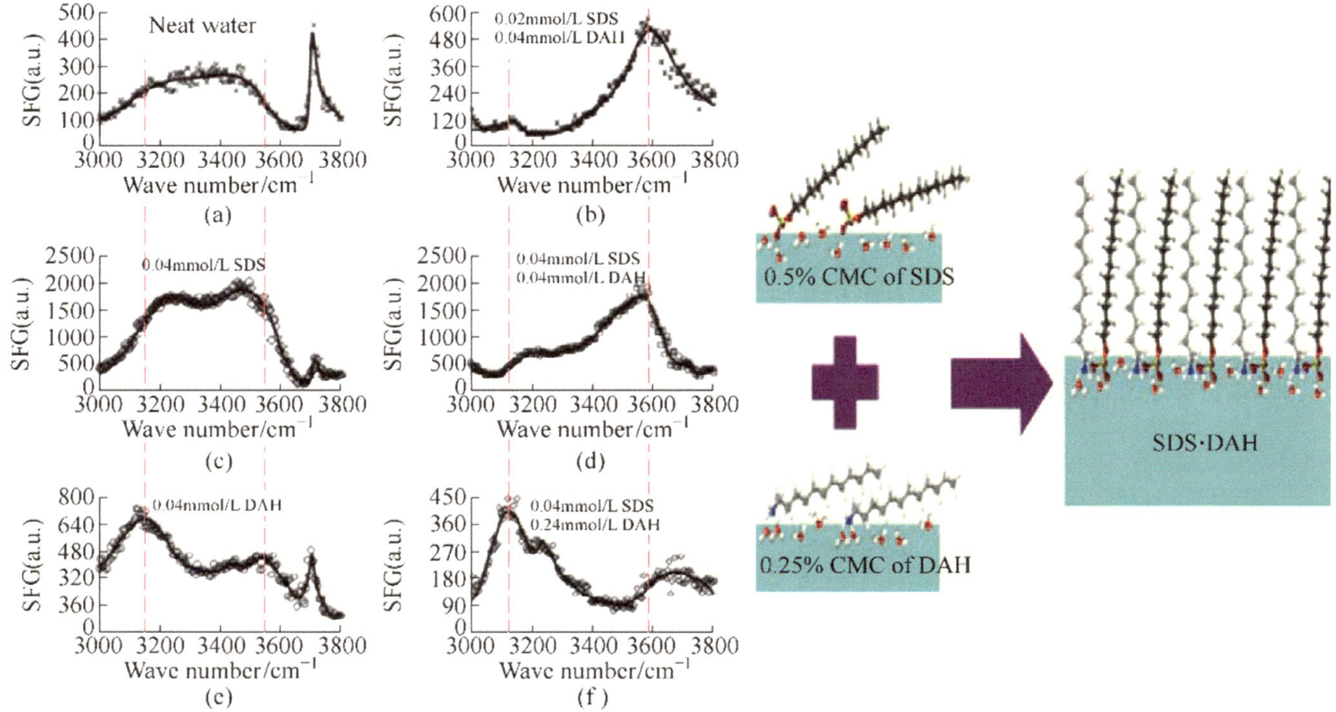

Fig. 5.28 SFG signals of water molecules at the gas/liquid interface under different reagent systems, reprinted from Ref. [94], copyright 2014, with permission from Journal of Langmuir

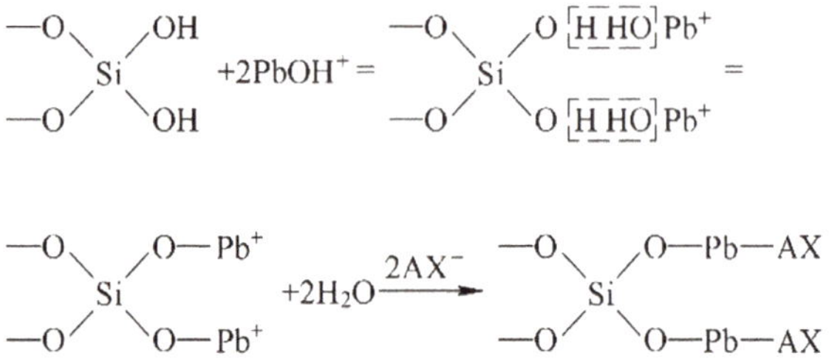

Fig. 5.29 Classical activation flotation theoretical model, reprinted from Ref. [96], copyright 2016, with permission from International Journal of Mineral Processing

ore on electrical properties and floatability and other experimental results, and put forward another activation mechanism: the surface precipitates of metal hydroxide may be the effective components of adsorption and activation of metal ions on the surface of oxidized ore. The formation of surface precipitation is controlled by the activity, pH and solubility product of metal ions in the interfacial region [97].

In fact, metal ions exist in the form of hydrated metal ions in the solution, and the adsorption of metal ions on the mineral surface will lead to the change of the structure of the hydration layer on the mineral surface, thus affecting the adsorption of the collector. On the other hand, hydrated metal ions and collectors in the solution system will inevitably interact to form certain complexes, and the role of these complexes in the flotation process is often ignored. Taking Benzohydroxamic acid (BHA) collector and Pb^{2+} ion as an example, at pH 9, BHA ionization mainly exists in the form of B^- ion (Fig. 5.30), and lead nitrate mainly exists in the form of $Pb(OH)^+$ ($10^{-4.3}$ mol/L), $Pb(OH)_2$(aq) ($10^{-4.7}$ mol/L) and Pb^{2+} ($10^{-5.5}$ mol/L) (Fig. 5.31). In the liquid phase system, BHA and lead ion components will be complexed to form a certain complex, and the role of this component in mineral flotation has not been paid attention to for a long time. Therefore, the classical activated flotation theory is a simplified adsorption model, and the explanation of the adsorption process of metal ions and collectors is not perfect, which needs to be further expanded.

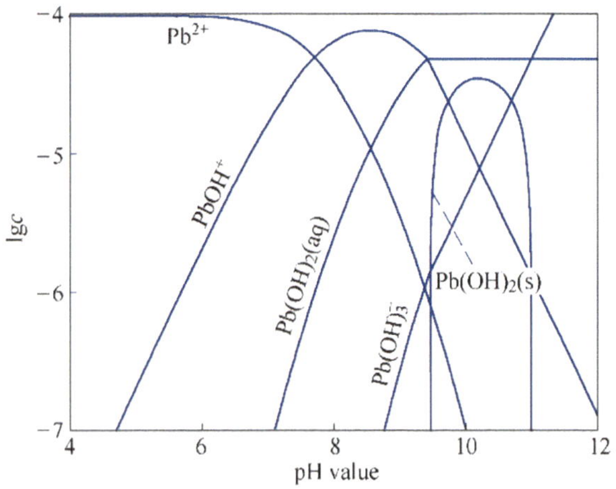

Fig. 5.30 Logarithmic diagram of concentration of hydrolyzed component of Pb^{2+} ($c_{Pb} = 1 \times 10^{-4}$ mol/L), reprinted from Ref. [99], copyright 2021, with permission from Chemical Engineering Science

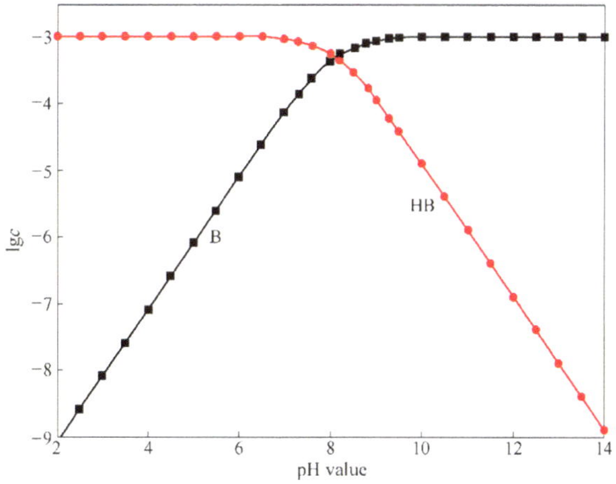

Fig. 5.31 Logarithmic diagram of concentration of each component of benzohydroxamic acid solution ($c_{BHA} = 1 \times 10^{-3}$ mol/L), reprinted from Ref. [100], copyright 2018, with permission from Minerals Engineering

5.4.1.2 Coordination Assembly Behavior of Calcium Ions in Fatty Acid System

In 1996, Miller et al. [101] studied the adsorption mechanism of oleic acid on mineral surfaces by in-situ FTIR/IRS (Fig. 5.32). By studying the adsorption of oleate and calcium oleate precipitates in the fluorite/oleic acid system, the following conclusions were obtained: (1) At low oleate concentration, calcium oleate formed a single layer film on the surface of fluorite due to its unique chemisorption reaction, while at high oleate concentration, multilayer surface calcium oleate precipitates on the surface of fluorite; (2) When the oleate concentration was 4×10^{-6}–4×10^{-5}, the precipitation and adsorption of calcium oleate had a significant effect on the flotation process; (3) The precipitation and adsorption of calcium oleate only occurred in the alkaline pH range; (4) Calcium oleate was formed in solution and then transported to the surface, rather than nucleation and growth on the surface as surface precipitation. From this, it is speculated that oleic acid adsorption on mineral surfaces may be: oleic acid components first formed calcium oleate colloid in solution, and then calcium oleate colloid was adsorbed on the mineral surface, rather than directly adsorbed on the mineral surface.

Later, Fa et al. [102–104] conducted a detailed study on the mechanism of oleate in the flotation of calcite and fluorite through infrared spectroscopy, flotation experiments, atomic force microscopy (AFM), molecular dynamics simulation and DLVO theory (Fig. 5.33). Infrared spectrum analysis found the existence of asymmetric carboxylic acid-coo-tensile coupling at 1575 and 1538 cm^{-1} nearby, indicating that calcium oleate precipitates were formed on the surface of calcite and fluorite. The flotation experiments showed that the colloid of calcium oleate collector has better performance as a collector for fluorite than sodium oleate, and has almost no ability to collect calcite in a certain concentration range, which can realize the flotation separation of fluorite and calcite. By modifying the spherical calcium oleate colloid on the AFM probe, the force of calcium oleate on the surface of calcite and fluorite

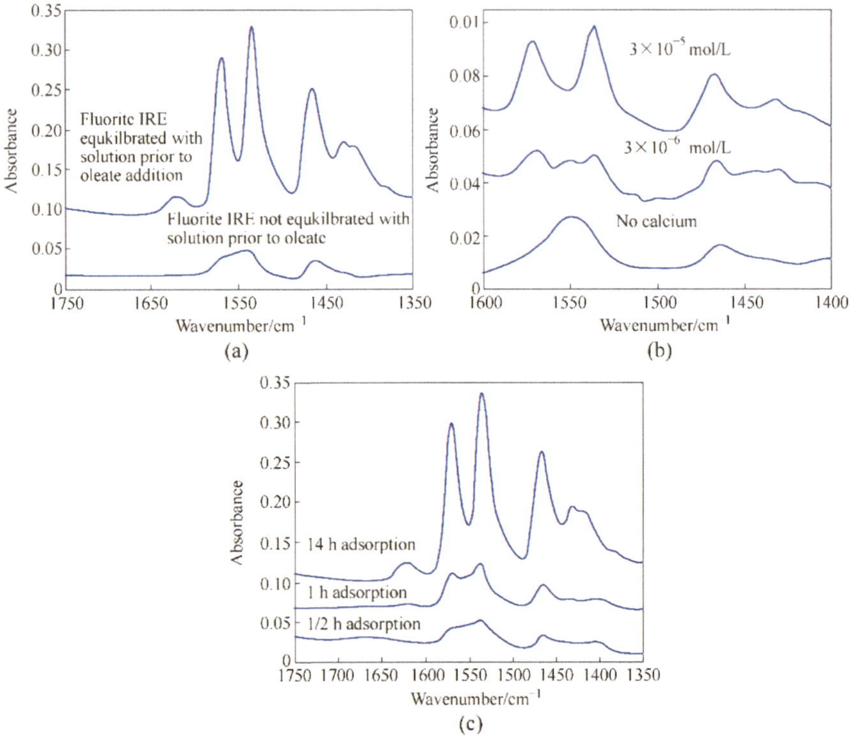

Fig. 5.32 Adsorption of oleate and calcium oleate precipitates in fluorite/oleic acid system **a** In situ FTIR/IRS spectra of adsorbents on the fluorite surface at different calcium concentrations; **b** In situ FTIR/IRS spectra of oleate adsorbents at dissolution equilibrium and non-equilibrium of fluorite; and **c** comparison of adsorption time to relative adsorption capacity, reprinted from Ref. [101], copyright 1996, with permission from International Journal of Mineral Processing

was measured, and compared with the classical DLVO gel force. Combined with the molecular dynamics simulation, the different interfacial water structures of the fluorite—calcium water interface were revealed, and it was found that the surface of fluorite had certain hydrophobicity. There is a long range hydrophobic force between calcium oleate colloidal particles and the hydrophobic surface of fluorite, so calcium oleate colloidal particles are more easily adsorbed on the surface of fluorite.

Sun et al. [105] studied the flotation behavior of three typical semi-soluble calcium-containing minerals, fluorite, scheelite and calcite, in sodium oleate system and pre-mixed calcium oleate colloid system, as shown in Fig. 5.34. The results showed that at pH = 10.00, the flotation effect of sodium oleate and calcium oleate was similar to that of scheelite and fluorite at low concentration. With the increase of collector concentration, the gap between calcium oleate and sodium oleate increases, indicating that the flotation recovery of scheelite and fluorite was higher under the action of calcium oleate. But the flotation condition of calcite was different from scheelite and fluorite, and the flotation recovery of calcium oleate was significantly

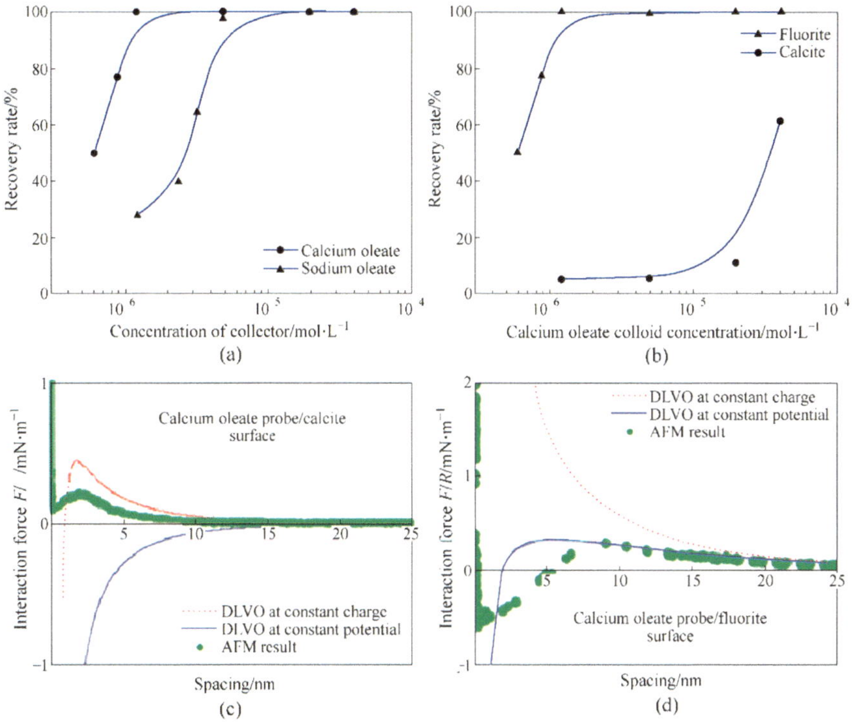

Fig. 5.33 Curves of selective harvesting capacity and interaction forces of calcium oleate colloidal particles on fluorite and calcite **a** the effect of sodium oleate and calcium oleate concentrations on the flotation recovery of fluorite, **b** the effect of calcium oleate concentration on the flotation recovery of fluorite and calcite, **c** the force curve and DLVO force comparison of the AFM colloid probe on the surface of calcite, **d** the force curve and DLVO force comparison of the AFM colloid probe on the surface of fluorite, **a**, **b** reprinted from Ref. [102], copyright 1996, with permission from Langmuir, **c**, **d** reprinted from Ref. [103], copyright 2005, with permission from Journal of Physical Chemistry B

lower than sodium oleate. Solution chemistry calculation showed that the concentration of dissolved calcium ion in the saturated solution of scheelite and fluorite was 10^{-5}–10^{-6} mol/L, under the test of sodium oleate concentration, calcium oleate existed as the dominant component in the flotation solution, while the concentration of dissolved calcium ion in the saturated solution of calcite was less than 10 mol/L, calcium oleate could only exist in the solution. This further indicated that calcium oleate played an important role in the flotation system of sodium oleate, and calcium oleate had a stronger collecting ability of scheelite and fluorite.

The study on calcium oleate colloid collector shows that the complexes formed by metal ions and flotation reagents are likely to be the real components in activated flotation, which is a new understanding of the role of metal ions in mineral flotation. This research result shows that for semi-soluble minerals, the role of colloid formed by metal cations and collectors must be considered in the flotation process, and the

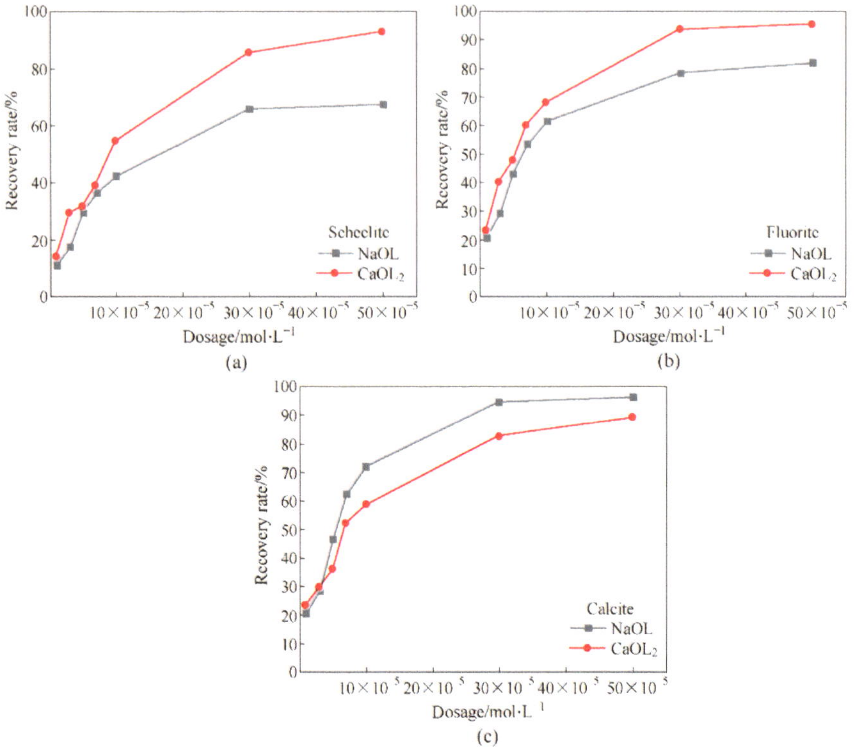

Fig. 5.34 Effect of sodium oleate and calcium oleate dosage on flotation recovery rates of **a** scheelite, **b** fluorite and **c** calcite, reprinted from Ref. [105], copyright 2022, with permission from Minerals Engineering

colloidal complexes formed by metal ions and collectors are likely to be the effective components.

5.4.1.3 Coordination Assembly Behavior of Lead Ions in Benzohydroxamic Acid System

Lead ion and benzohydroxamic acid are widely used in the flotation of scheelite, wolframite, cassiterite, ilmenite and other oxide ores in industry. In the traditional activated flotation practice, lead ions are usually added preferentially to activate the target minerals, and then hydroxamic acid collector is added to collect the target minerals. In 2017, Han et al. [106, 107] systematically investigated the influence of the addition of lead nitrate and benzohydroxamic acid on the flotation of scheelite and cassitite, as shown in Fig. 5.35. The results showed that the pre-mixing of benzohydroxamic acid and lead nitrate was significantly better than the sequential dosing, so it was speculated that the complex formed by the reaction of benzohydroxamic

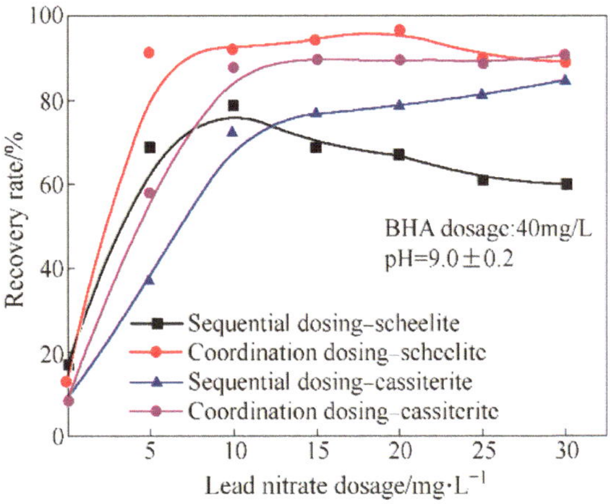

Fig. 5.35 Effect of Pb ion dosage on the flotation of scheelite and cassiterite by benzohydroxamic acid [108]

acid and lead ion might be the effective component in activated flotation, and this component had stronger collecting ability.

Tian et al. [109, 110] used solution chemistry calculation method to analyze the main active components of Pb-BHA complex in aqueous solutions with different pH values. Based on density functional theory and first principle calculation, they studied the difference of adsorption energy of reagent on cassitar surface in two models of activated flotation and complex flotation, as shown in Fig. 5.36a. The results showed that at the optimum flotation pH = 8–9, the adsorption energy of Ho-Pb-BHA complex on cassiterite surface was − 201.38 kJ/mol, while the adsorption energy of $Pb(OH)(H_2O)_5^+$ and BHA anion was − 55.63 kJ/mol. Therefore, from the perspective of thermodynamics, The Pb-BHA complex was more easily adsorbed on the surface of cassiterite. Hu Wenjihao et al. measured the surface force curve of BHA/Pb-BHA and spodumene minerals in aqueous solution by using the colloidal probe method of AFM, and achieved the modification of BHA molecules on the probe by synthesizing n-hydroxy-4-mercapto benzenamide, as shown in Fig. 5.36b [111]. The results showed that the adhesion energy of BHA and Pb^{2+}-activated spodumene (110) surface was 1.99 mJ/m^2, while the adhesion energy of Pb-BHA complex and spodumene (110) surface was 3.77 mJ/m^2, which proved from the perspective of kinetics that the adsorption capacity of Pb-BHA complex on the mineral surface was stronger. Therefore, thermodynamic and kinetic studies proved that the flotation model of the complex was superior to the activated flotation model.

On the basis of experimental testing and theoretical analysis, Hu et al. [112] proposed a metal ion interface preassembly model, as shown in Fig. 5.37: Model 1: activated flotation model, the colloids of Pb^{2+}, $PbOH^+$ and $Pb(OH)_2$ in the solution are adsorbed on the surface of scheelite and wolframite by electrostatic action, and hydroxybridge dehydration reaction occurs on the surface to form precipitation. The BHA anion reacts with the lead particles on the mineral surface to form "O, O" five-member ring chelate to achieve the purpose of flotation collection. Model 2: metal ion interface preassembly model, in the solution of Pb^{2+}, $PbOH^+$ and $Pb(OH)_2$ colloid

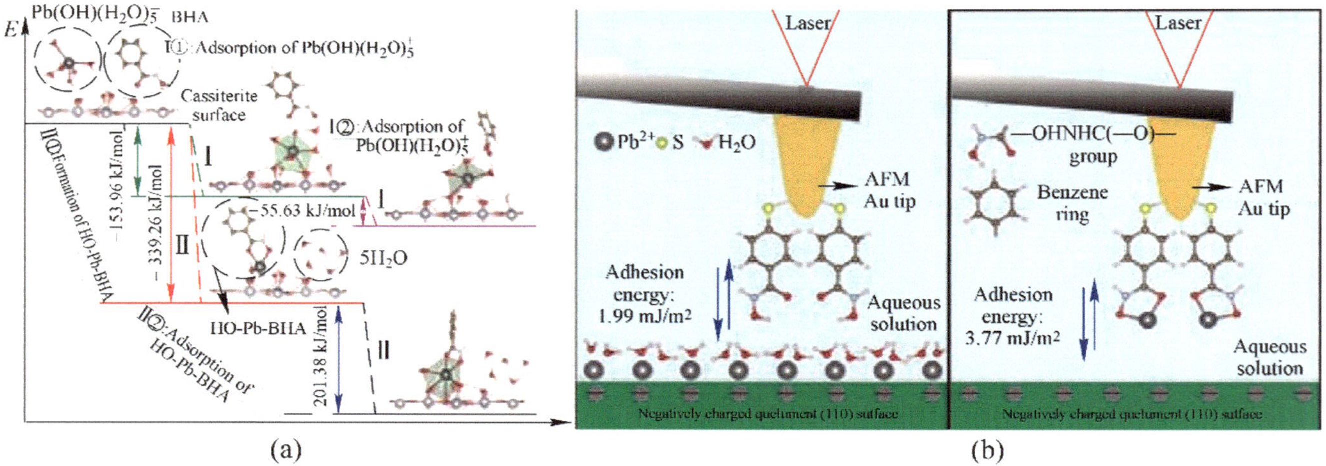

Fig. 5.36 Differences in adsorption behaviors of reagents on mineral surfaces between active flotation and complex flotation models: **a** differences in adsorption energy of reagents on cassinette surface under the two models, reprinted from Ref. [109], copyright 2018, with permission from Journal of Colloid Interface Science; **b** differences in adhesion energy of reagents on spodumene surface under the two models, reprinted from Ref. [111], copyright 2020, with permission from Langmuir

and BHA anion ligand reaction form some or some complexes, which has similar properties to lead ions, colloidal structure charged positively, through electrostatic adsorption on the surface of the mineral, hydroxy-bridge shrinkage reaction on the surface of the mineral, strengthening the adsorption of the collector on the mineral surface. In the traditional activated flotation process, the two mechanisms coexist, and the mechanism shown in Model 1 accounts for a considerable proportion, while the complex system is mainly based on the mechanism shown in Model 2. Although the two mechanisms seem to lead to the same destination, both of which use lead ions as active particles to achieve the adsorption of BHA on the mineral surface, there may be some differences in the component state and structure of BHA on the mineral surface, such as the structure and spatial arrangement of BHA, and these differences will lead to different flotation effects under the two reaction systems. The experimental results show that, compared with adding lead nitrate activator and benzohydroxamic acid (BHA) collector in the pulp, Pb-BHA metal–organic complex has stronger collecting ability and selectivity in the flotation separation of scheelite and calcite under the condition of the same dosage of agent. In the solution of pH = 8–10, this kind of new collector has strong positive charge, easy to selectively adsorb on the surface of scheelite under negative charge by electrostatic force and chemical action, and the main gangue minerals of scheelite flotation (fluorite, calcite) have positive charge in this pH range, which is not conducive to the adsorption of the new collector, so that it shows high selectivity of flotation. Based on this research finding, Hu Yuehua and Sun Wei et al. put forward the concept of metal ion coordination regulation molecular assembly, which provided a new idea and direction for the design and development of new flotation reagents.

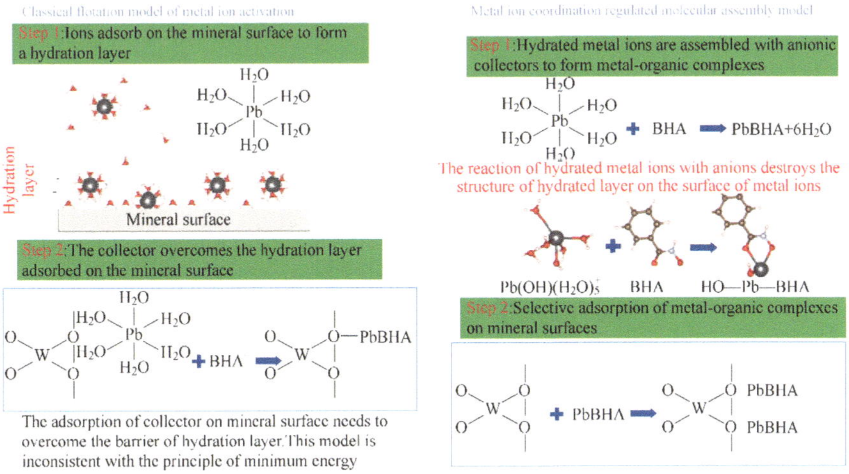

Fig. 5.37 Assembly model of metal ion coordination regulatory molecules, reprinted from Ref. [112], copyright 2018, with permission from Conservation and Utilization of Mineral Resources

5.4.2 Collector of Metal Ion-Hydroxamic Acid Organic Complex

The study of scheelite crystal chemistry and solution chemistry revealed the microscopic differences between scheelite and fluorite in weakly alkaline solution systems, that is, scheelite has a negatively charged surface rich in oxygen atoms, while fluorite has a positively charged surface rich in calcium atoms. This difference provides a new scheme for the design and development of new flotation reagents. Many researchers at home and abroad have carried out a lot of research work around the cationic collector (such as DDA, quaternary ammonium salt, etc.), and made certain progress in research. It is worth noting that the assembly of metal ion coordination regulatory molecules can take advantage of the high selectivity of metal groups to achieve selective adsorption, that is, the metal group as a functional group and anionic groups on the surface of minerals, rather than bonding with calcium atoms. Therefore, the template effect of metal ions is used to control the assembly of organic ligand, forming metal–organic complex collector with metal base as polar group and organic ligand as hydrophobic group. It is possible to realize the targeted adsorption of collector on scheelite surface, so as to fundamentally solve the flotation separation problem of scheelite and calcium-bearing gangue minerals (Fig. 5.38).

5.4.2.1 Coordination Assembly Behavior of Lead Ions with Benzohydroxamic Acid

1. **Quantum chemical study on the coordination assembly of lead ions with benzohydroxamic acid**

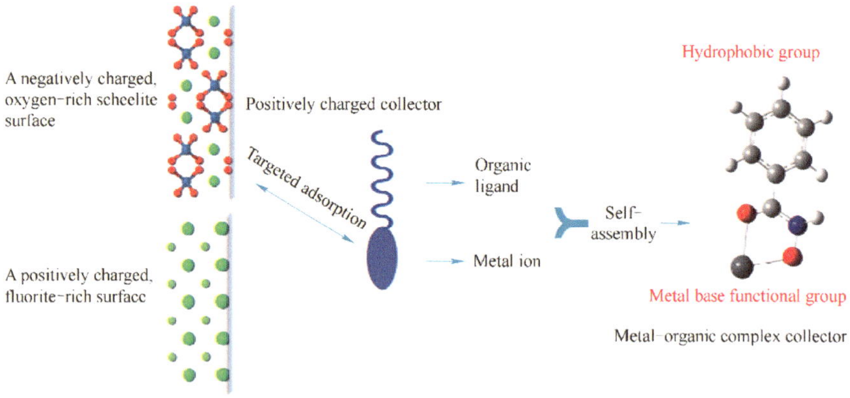

Fig. 5.38 Design of metal–organic complex collector, reprinted from Ref. [113], copyright 2021, with permission from Chemical Engineering Journal

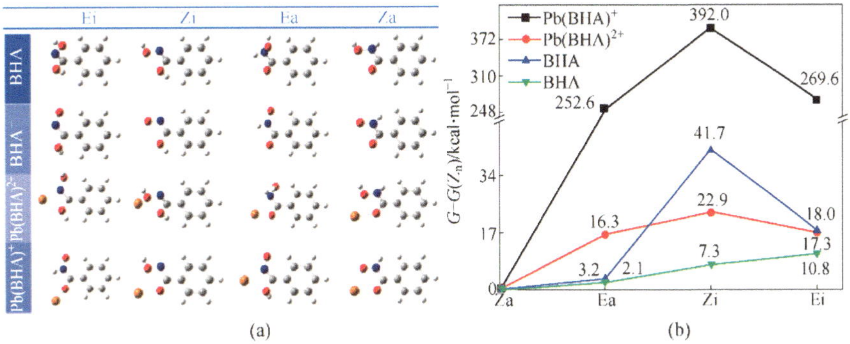

Fig. 5.39 **a** Optimal structure of BHA, BHA^-, $Pb(BHA)^{2+}$, and $Pb(BHA)^+$; **b** Gibbs free energy of BHA, BHA^-, $Pb(BHA)^{2+}$, and $Pb(BHA)^+$ isomers (in terms of zero Gibbs free energy of Za type) [108]

Quantum chemical calculation combined with detection technology is an important means to study the structure of Pb-BHA complex. He et al. [108] studied the coordination assembly of lead ion and benzohydroxamic acid by quantum chemical calculation, and optimized the structure of BHA^-, BHA, $Pb(BHA)^{2+}$, and $Pb(BHA)^+$ under the theoretical level of B3LYP/aug-cc-pVDZ. Figure 5.39a shows all the optimized structures and their complexes with Pb(II) ions. These optimized structures were divided into four categories: Ei, Zi, Ea and Za. The calculation results of Fig. 5.39b show that Za type was the most stable configuration of BHA^-, BHA, $Pb(BHA)^{2+}$, and $Pb(BHA)^+$, and that BHA was more likely to cooperate with Pb(II) to form "O, O" five-membered rings than "O, N" four-membered rings.

He et al. [114] calculated the optimal structure of BHA, $Pb(BHA)^+$ and $Pb(BHA)_2$ and their HOMO and LUMO frontier orbitals under the solvent model by using quantum chemistry calculation (Fig. 5.40a). The results showed that LUMO of $Pb(BHA)^+$ was mainly concentrated on lead ions. The LUMO of $Pb(BHA)_2$ was distributed throughout the complex, and the gap between HOMO and LUMO of $Pb(BHA)^+$ was small, indicating that $Pb(BHA)^+$ was more active than $Pb(BHA)_2$, and it was an electrophilic molecule with lead ion as its potential affinity point. He et al. [108] used quantum chemistry calculation to study the structural stability of the complexes generated by Pb-BHA complexes in solution under different Pb/BHA ratios. The study believed that lead ions and BHA existed in various forms of complexes in solution, and first of all, a single coordination $Pb(BHA)^+$ was formed. Then form complexes with high coordination number such as $Pb(BHA)_2$, $Pb(BHA)_2$ + BHA^- (Fig. 5.40b). Pb-BHA^- was a favorable morphology thermodynamically, and the free energy of BHA^- and Pb^{2+} (− 1492.80 kJ/mol) was much higher than that of water molecules and Pb (− 211.35 kJ/mol), indicating that the coordination efficiency of BHA^- and Pb^{2+} was higher than that of water molecules and Pb^{2+}. The positive change of Gibbs free energy produced by the coordination reaction of the third BHA^- was 0.13 kJ/mol, which indicated that the binding of the third water molecule with the previous $Pb(BHA)_2$ was thermodynamically unfavorable.

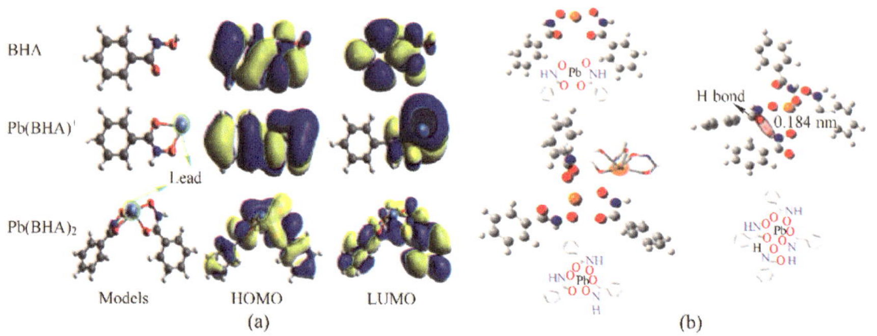

Fig. 5.40 **a** Optimized structures of BHA, $Pb(BHA)^+$ and $Pb(BHA)_2$ and their HOMO and LUMO frontier orbitals, reprinted from Ref. [114], copyright 2018, with permission from Applied Surface Science. **b** Optimized structures of Pb-BHA with different ligands [108]

In addition, when the number of BHA ligands reached 4, intramolecular aggregation occurred due to hydrogen bonding, and the hydrogen bond length between the added BHA and the adjacent BHA was 0.184 nm, and N–H··O bond angle was 153.9°. The structure of the complex was unstable obtained when the number of ligands was greater than 3, but Pb(II)-BHA could form complexes with three or more BHA ligands through intermolecular interactions such as van der Waals forces, hydrogen bonding, etc.

2. Molecular structure analysis of lead ion and benzohydroxamic acid complex

Quantum chemistry studies have preliminarily revealed the assembly behavior of lead ion and benzohydroxamic acid, while the exact analysis of Pb-BHA complex structure requires strong technical means and more comprehensive research methods. Single crystal X-ray structure analysis is the most direct and convincing characterization method to study the structure of metal–organic complexes [115]. It can provide the exact spatial position of all the atoms of a compound in the solid state, so as to obtain the material structure, arrangement, atomic connection, accurate bond length and bond angle data [116]. The structure of Pb-BHA complex collector is affected by the ratio of Pb to BHA and the pH of the solution. The X-ray crystal structure analysis of Pb-BHA metal complex shows that the chemical formula of Pb^{2+} and BHA is $C_{56}H_{47}N_{12}O_{28}Pb_6$, the molecular formula is $Pb_6L_8(NO_3)_4$ (HL = BHA), the molecular weight is 2579.19, the monoclinic system, the space group is $P2_1/c$, and the cell parameters are: $a = 1.78428$ (3) nm, $b = 1.80279$ (3) nm, $c = 2.10876$ (4) nm, $\alpha = 90.00°$, $\beta = 107.992$ (2), $\gamma = 90.00°$, $z = 4$. In the complex molecule, 6 lead ions are coordinated with 8 ligands and 4 nitrates to form the complex structure of $Pb_6L_8(NO_3)_4$ (Fig. 5.41a). Pb(II) ions coordinate with the oxygen atom of the L ligand to form a non-planar "O, O" five-membered ring complex. The coordination environment of Pb(II) ions in the complex is shown in Fig. 5.41b. There are 6 Pb(II) ion centers in the complex molecule, and all Pb(II) ions are coordinated with oxygen atoms. The length range of Pb–O bond is 0.2377 (4)–0.2789 (5) nm. The Pb1, Pb2,

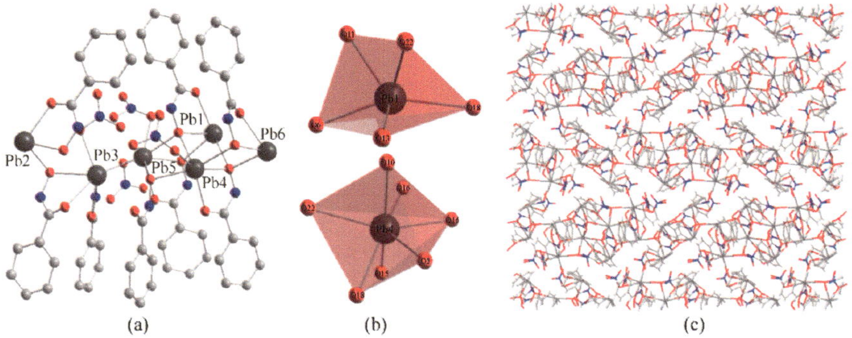

Fig. 5.41 **a** Molecular structure of $Pb_6L_8(NO_3)_4$, **b** coordination environment of Pb1 and Pb4, and **c** three-dimensional$_4$ extended frame structure of $Pb_6L_8(NO_3)$, reprinted from Ref. [117], copyright 2023, with permission from Journal of Molecular Liquids

Pb3 and Pb6 ions have a five-coordination structure, while the Pb4 and Pb5 ions have a seven-coordination structure. The Pb1, Pb2, and Pb6 ions are coordinated with the oxygen atoms of the L ligand, while the Pb3, Pb4, and Pb5 ions are coordinated with the oxygen atoms of the L ligand and NO_3^- with an O-Pb-O angle range of 58.58 (15)°–158.02 (15)°. Adjacent complex molecules are combined by Pb–O bond connections to form three-dimensional extended frame structure polymers with repetitive structures (Fig. 5.41c). The L ligand provides N–H bonds, while the L or NO_3^- hydroxyl oxygen provides oxygen atoms to form intramolecular and intermolecular N–H··O hydrogen bonds.

Pb^{2+} and BHA generate complexes with the molecular formula $[Pb_6L_8(NO)_3]_3NO_3$ under the condition of 2:1 molar ratio. $[Pb_6L_8(NO)_3]_3NO_3$ crystallizes in the monoclinic system with space group $P2_1/c$ and lattice constants a = 1.79246 (11) nm, b = 1.81038 (11) nm, c = 2.14467 (13) nm. In the molecular structure of the complex, six Pb^{2+} ions form the inner boundary of the complex with eight L ligands and three NO_3^- anions, and there is a free NO_3^- as the outside of the complex (Fig. 5.42a). Pb coordinates with the oxygen atom of the L ligand to form a non-planar five-membered ring complex. Figure 5.42b shows the coordination environment of Pb^{2+} in the complex. There are 6 Pb^{2+} ion centers present in the complex molecule, all of which are coordinated with oxygen atoms, and the Pb–O bond spacing is 0.2364 (5)–0.2751 (5) nm. In these Pb^{2+}, Pb1, Pb2, Pb3 and Pb6 are pentacoordinated, while Pb4 and Pb5 are hexacoordinated. Pb1, Pb2 and Pb5 ions are coordinated to the oxygen atom of the L-ligand, while Pb3, Pb4 and Pb6 ions are coordinated to the oxygen atom of the L-ligand or NO_3^- ligand. The Pb^{2+} and oxygen atoms in the ligand bond to adjacent Pb^{2+} complex molecules to form a three-dimensional extended frame structure polymer with a repetitive structure (Fig. 5.42c). There are N–H··O hydrogen bonds in the three-dimensional extended frame structure, where the L ligand provides the N–H bond, while The hydroxyl oxygen of L or NO_3^- provides oxygen atoms to form intramolecular and intermolecular hydrogen bonds.

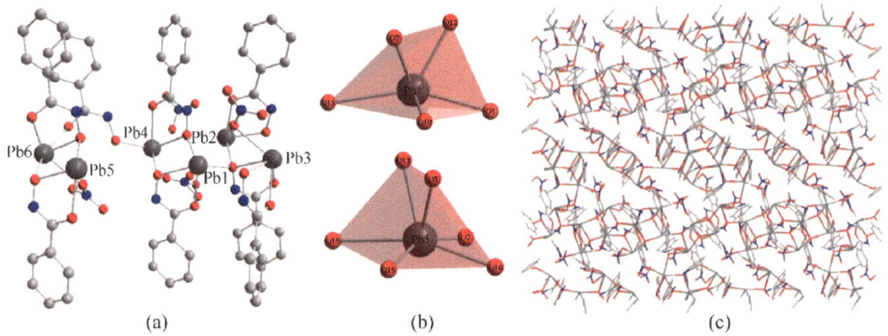

Fig. 5.42 **a** molecular structure of $[Pb_6L_{88}(NO)_3]_3NO_3$, **b** coordination environment of Pb1 and Pb5, and **c** three-dimensional$_3$ extended frame structure of $[Pb_6L_8(NO_3)_3]NO_3$, reprinted from Ref. [113], copyright 2021, with permission from Chemical Engineering Journal

The morphology of complexes will evolve during crystal, liquid phase and interfacial adsorption. The structural evolution of the crystal–liquid phase–solid/liquid interface where Pb^{2+} and BHA generate complexes under the condition of 1:1 molar ratio is shown in Fig. 5.43. The molecular formula of the single crystal structure of the complex is $Pb_6L_8(NO_3)_4$ and the structure of the liquid precipitate is $Pb_4L_5(OH)_3$. Lead ions not involved in the composition of the complex generate $Pb(OH)_2$ precip$_2$itates, forming the suspension of $Pb_4L_5(OH)_3$ and $Pb(OH)_2$. The structure of the adsorbent produced by the complex on the surface of the mineral is $Pb_4L_5(OH)_3$. The complex uses lead ions to bond with the oxygen atom on the surface of scheelite to form the Pb–O bond with the bond length of 0.2215nm. In addition, $Pb(OH)_2$ is also adsorbed on the surface of scheelite.

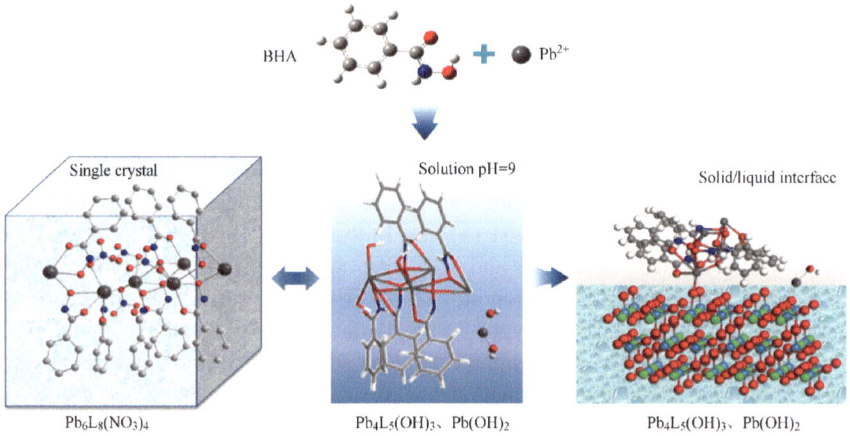

Fig. 5.43 Evolution of the crystal–liquid phase–solid/liquid interface of the complex, reprinted from Ref. [117], copyright 2023, with permission from Journal of Molecular Liquids

5.4.2.2 Mineral Flotation Behavior of Lead Ion-Benzohydroxamic Acid Complex Collector System

Han et al. [107] studied the flotation behavior of Pb-BHA complex collector, and explored in detail the effects of the ratio of Pb and BHA and solution pH on the flotation of scheelite and wolframite by the complex, as shown in Fig. 5.44. The results showed that the collection capacity of the complexes generated under different molar ratios of lead ions and BHA for different minerals was different. When $c(Pb^{2+})$:$c(_{BHA})$ is 1:2, 1:1, 2:1, the complexes formed have strong collecting ability of scheelite, and the common effective pH value range of the three ratios is 8.5–10.2. When $c(Pb^{2+})$:$c(_{BHA})$ was 1:2, 1:1, 2:1, the formed complexes had strong collecting capacity for wolframite and calcite, and the effective pH range of the complexes formed under different proportions was different. When $c(Pb^{2+})$:$c(_{BHA})$ is 1:1, the complexes formed have a certain collecting ability for fluorite, but when $c(Pb^{2+})$:$c(_{BHA})$ is 2:1, 1:2, 4:1, the complexes formed have no collecting ability or very weak collecting ability for fluorite. This indicates that the ratio of Pb and BHA affects the complexes formed by the two and the state of their components, which reflects the differences in the collection performance of different minerals.

The results show that the ratio of Pb to BHA in the complex is 1:2–2:1, scheelite and wolframite have stronger harvesting capacity, while the scheelite has better floatability in the range of pH = 8–10. Therefore, the Pb-BHA complex collector has a strong selective capturing ability of scheelite, while fluorite does not float (Fig. 5.45) [48, 49], indicating that the complex has a good selectivity, which is conducive to the efficient flotation separation of tungsten/fluorite associated resources. By adjusting the ratio of Pb/BHA and pH of pulp reasonably, the floatability of the complex to tungstite and gangue minerals can be changed, so as to realize the efficient separation of tungsten-bearing minerals from calcium-bearing gangue minerals.

Tian et al. [110] studied the behavior of Pb-BHA collector in cassite flotation, as shown in Fig. 5.46a. When the concentration of BHA is 80 mg/L, the recovery rate of cassiterite reaches a maximum of 46%, while under the condition of using Pb-BHA, the recovery rate of cassiterite is significantly increased to 86.15%. Compared with using BHA alone, the maximum recovery rate of cassiterite is increased by 40%, indicating that Pb-BHA collector has a strong collecting ability for cassiterite flotation. The behavior of Pb-BHA collector in spodumene flotation is shown in Fig. 5.46b. In the absence of lead nitrate (LN), the recovery rate of spodumene is only about 2%, and in the case of 40 mg/L LN, the recovery rate of spodumene is about 40%. With the concentration of BHA increasing to 120 mg/L, the recovery rate of spodumene increases rapidly. It reaches the maximum of 81.61%, indicating that Pb-BHA has a strong collecting ability for spodumene flotation [118].

5.4.2.3 Adsorption Mechanism of Lead Ion-Benzohydroxamic Acid Complex Collector at the Solid/liquid Interface

1. Zeta potential analysis

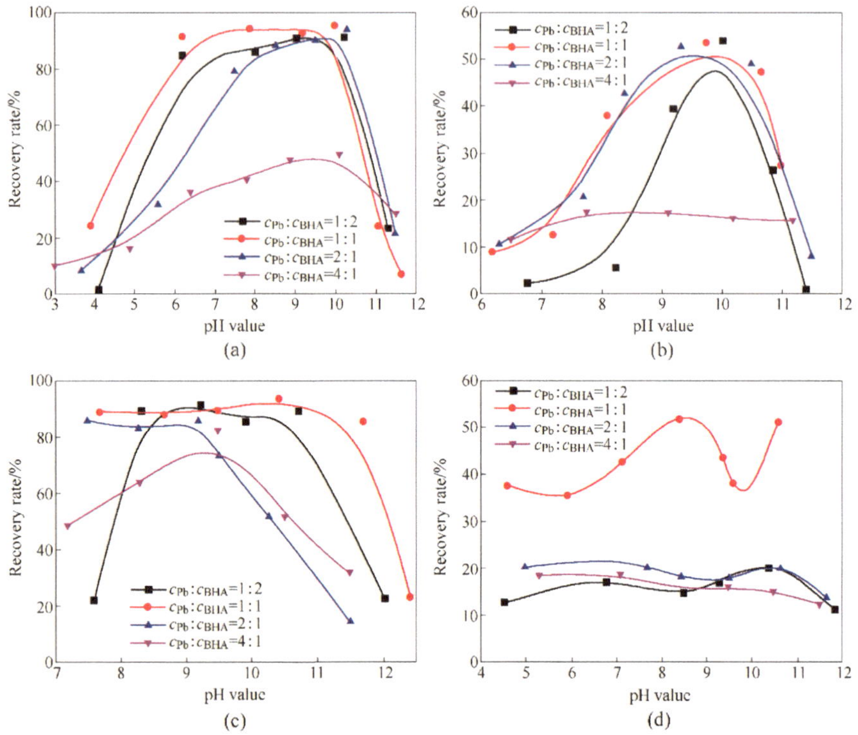

Fig. 5.44 The change of mineral collection capacity of Pb benzohydroxamic acid complex with pH value: **a** scheelite, **b** wolframite, **c** calcite, **d** fluorite. ($c_{BHA} = 1.5 \times 10^{-4}$ mol/L, $c_{terpilenol} = 12.5$ μL/L), reprinted from Ref. [107] copyright 2017, with permission from International Journal of Mineral Processing

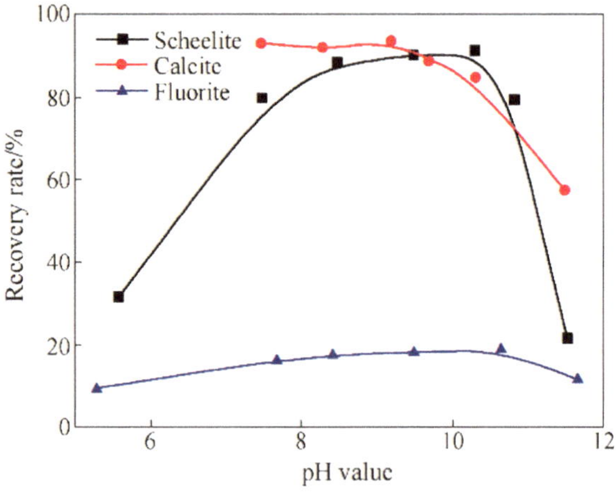

Fig. 5.45 Buoyancy of calcium-bearing minerals in the third medium of Pb-BHA collector system, reprinted from Ref. [49], copyright 2018, with permission from Minerals Engineering

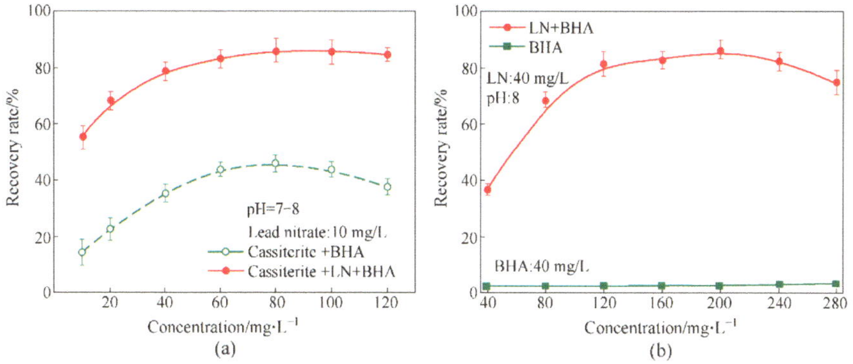

Fig. 5.46 Pb-BHA collector for **a** cassiterite, reprinted from Ref. [110], copyright 2018, with permission from Minerals Engineering, and **b** spodumene flotation capacity reprinted from Ref. [118], copyright 2019, with permission from Minerals Engineering

The surface electrokinetic potential of the colloid/precipitate of the lead ion, Pb-BHA complex is shown in Fig. 5.47. Curves 2 and 4 show the zeta potential curve of the complex formed by adding BHA on the basis of the dosage of reagent in curves 1 and 3, respectively. It can be seen that the surface electrokinetic potential of the complex is positive under weakly alkaline and alkaline conditions, and it is smaller than that of the same concentration of lead ionic colloid. But under the condition of strong alkali, it is the opposite: the complex surface galvanic potential is higher than that of Pb ionic colloid. The complex of curve 2 is the potential curve of colloid or precipitate formed by 0.75×10^{-4} mol/L $Pb(NO_3)_2$ and 1.5×10 mol/L BHA; when $pH < 11$, the surface potential of the complex is positive, and is smaller than that of the same concentration of lead ionic colloidal solution, the highest potential appeared at $pH = 8$. When $pH > 11$, the complex surface potential is negative. The complex of curve 4 is a potential curve of colloid or precipitate formed by matching 1.5×10^{-4} mol/L $Pb(NO_3)_2$ and 1.5×10 mol/L BHA. Its rule is similar to that of curve 2, but in the range of pH tested, it is higher than the potential of curve 2, and the highest value of curve 4 appears at $pH = 7.2$. The above results show that the complex and the lead hydroxide colloid or precipitate have similar charging laws, and the potential of the complex under different proportions is different to some degree.

Figure 5.48 shows the relationship between scheelite surface potential and solution pH, as well as the effect of a single agent on mineral surface potential. When BHA was added to the solution, the scheelite surface galvanic potential shifted negatively. Both the scheelite surface and the BHA ions are negatively charged, so it can be inferred that the BHA has non-electrostatic adsorption on the scheelite surface: the BHA may have chemically bonded with the Ca^{2+} and is adsorbed on the scheelite surface. Under acidic conditions, the negative potential shift on scheelite surface is small, because the degree of BHA dissociation is small under acidic conditions, and it is difficult to react with calcium points. When a certain amount of lead nitrate is added into the solution, lead ions, hydroxyl complexes and hydroxide colloid are positively

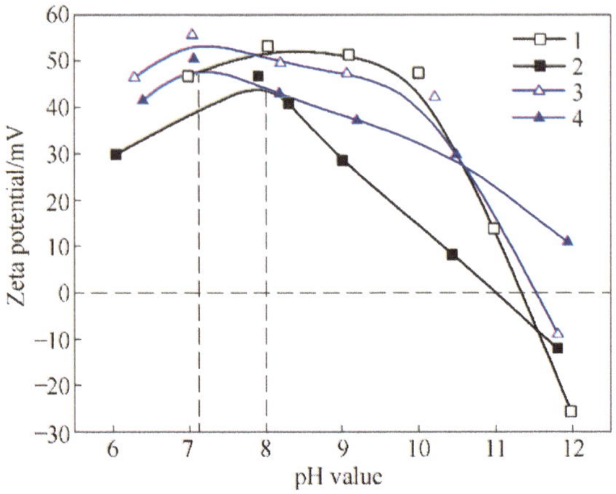

Fig. 5.47 Relation between colloid potential and pH of Pb-BHA complexes 1—$Pb(NO_3)_2$ 0.75 × 10^{-4} mol/L solution; 2—complex solution: $Pb(NO_3)_2$ 0.75 × 10^{-4} mol/L + BHA 1.5 × 10^{-4} mol/L; 3—$Pb(NO_3)_2$ 1.5 × 10^{-4} mol/L solution; 4—complex solution: $Pb(NO_3)_2$ 1.5 × 10^{-4} mol/L + BHA 1.5 × 10^{-4} mol/L, reprinted from Ref. [119], copyright 2018, with permission from Minerals Engineering

charged and easy to adsorb on the surface of scheelite under negative charge. When the concentration of lead nitrate in solution is 0.75 × 10^{-4} mol/L, scheelite surface potential is significantly positive shift. With the increase of pH, scheelite surface potential rapidly increased. Scheelite surface potential is the highest around pH = 8. When pH > 10, scheelite surface potential rapidly decreased, and even when the pH is high enough scheelite surface potential turned negative.

The results of surface zeta potential measurement of the Pb-BHA complex and scheelite show that the colloid or precipitate surface of the complex has a positive charge, while the scheelite surface has a negative charge, so the complex is easy

Fig. 5.48 Influence of single agent addition on scheelite surface potential, reprinted from Ref. [119], copyright 2018, with permission from Minerals Engineering

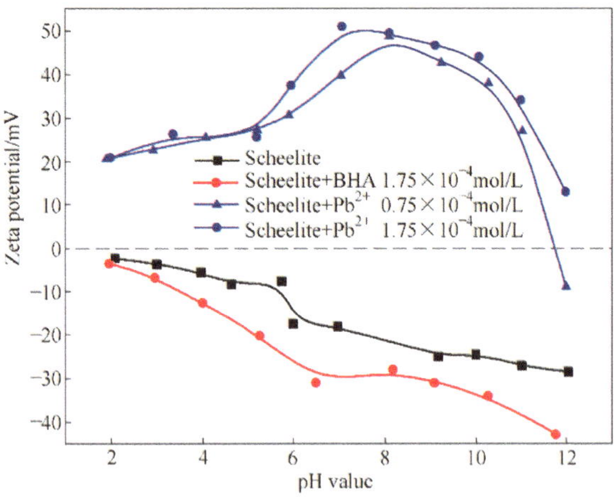

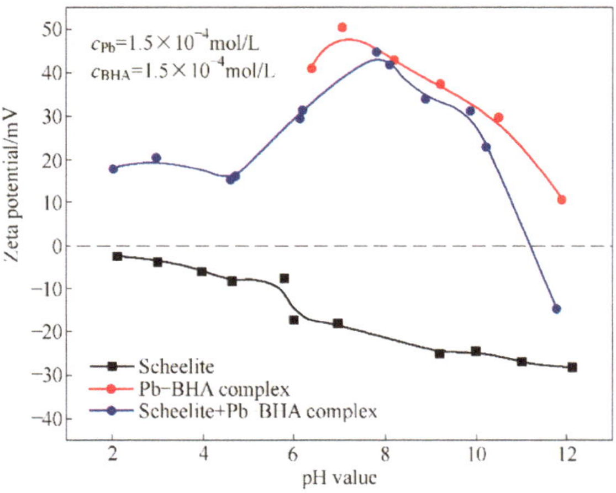

Fig. 5.49 Zeta potential changes before and after the interaction of scheelite with Pb-BHA complex, reprinted from Ref. [120], copyright 2020, with permission from Minerals Engineering

to adsorb on the mineral surface by electrostatic force. The electrokinetic potential before and after the interaction between Pb-BHA complex and scheelite is shown in Fig. 5.49. After the addition of Pb-BHA complex, the galvanic potential of scheelite surface has a positive shift. The galvanic potential first increases with the increase of pH, and there is a small concave point in the galvanic potential curve when pH = 4.7. According to the diagram of dissolved components in the scheelite saturated solution, this may be due to the potential change caused by the surface dissolution of scheelite. In the range of pH = 1.3–4.7, the concentration of HWO_4^- and WO_4^{2-} in the solution increased rapidly, and the mineral surface potential shifted rapidly negatively, and the negative shift reaches the maximum when pH = 4.7. The scheelite surface potential reaches the maximum at pH = 8, and then decreases sharply with the increase of pH, and the isoelectric point shifts to the right at pH_{IEP} = 11. Therefore, it can be speculated that the colloid of Pb-BHA complex has similar properties to that of lead hydroxide and can be absorbed on the surface of scheelite by electrostatic force.

2. Infrared spectrum analysis

The infrared spectrum of Pb-BHA collector with different lead ion and BHA ratio is shown in Fig. 5.50. For the infrared spectrum of BHA, due to the superposition of O–H stretching vibration and N–H stretching vibration of BHA, a strong wide absorption peak is formed at 3295.33 cm^{-1}, which is the characteristic peak of hydroxamic acid. 3060.17 cm^{-1} is the stretching vibration of N–H bond, and 3025.81 cm^{-1} is the peak of C-H stretching vibration. It is generally believed that the C=N double bond stretching vibration of oxime class is within 1690–1650 cm^{-1}, and for the oxime with aromatic ring such as BHA, C=N stretching vibration is about 40 cm^{-1} lower, and the C=O double bond of oxime class also has a similar rule, hydroxamic acid isomerization form of these two bonds exist at the same time, Therefore, 1645.50 cm^{-1} is the C=N double bond or C=O double bond of BHA, which is reflected as a wide strong peak, which is the characteristic peak of aromatic hydroxamic acids. The peak at 1567 cm^{-1}, 1488 cm^{-1} and 1456.02 cm^{-1} is sharper, which is the benzene ring skeleton deformation vibration of aromatic oxime, or respiratory vibration, and also the characteristic peak of BHA. 1163.29 cm^{-1} is the stretching vibration of C–N. Due to the vibration splitting of N–O, the splitting peak is formed at 1076.64 cm^{-1}, 1042.02 cm^{-1} and 1019.91 cm^{-1}. The peaks formed at 689.99 cm^{-1} and 521.94 cm^{-1} may be caused by the out-of-plane bending vibration of C–H on the benzene ring. The infrared spectrum of the complex changed greatly compared with that of BHA, indicating that the lead nitrate and BHA had a chemical reaction to form a new product. With the increase of the molar ratio of lead nitrate to BHA, the BHA peak of 3295.33 cm^{-1} (–OH stretching vibration peak) gradually weakened, and the BHA peak of 1645.50 cm^{-1} (C=N and C=O superposition vibration peak) shifted and gradually weakened, especially the C=O 1760–1770 cm^{-1} peak changed significantly. At the same time, a new peak (possibly O-Me) was formed in the range of 3500–3700 cm^{-1}, which indicated that lead ion reacted with benzohydroxamic acid and the C=O double bond was opened. The peaks at 1567 cm^{-1}, 1488 cm^{-1} and 1456.02 cm^{-1} were greatly shifted after the introduction of lead ion, indicating that the benzene ring skeleton of aromatic oxime was deformed. The widening of peaks at 1000–1200 cm^{-1} may be caused by the superposition of C–N and N–O bonds after the introduction of lead ion.

Figure 5.51 shows the infrared spectra of scheelite and its interaction with Pb-BHA. In the infrared spectrum of scheelite, 808.54 cm^{-1} and 440.08 cm^{-1} are the characteristic peaks of tungstate, while 3425.01 cm^{-1} is the asymmetric stretching vibration of O–H bond of water molecules on the surface of scheelite. After the scheelite was treated with the complex, the characteristic peaks appeared at 3424.94 cm^{-1}, 1598.59 cm^{-1}, 1561.73 cm^{-1}, 1403.35 cm^{-1}, 1150.82 cm^{-1}, 806.70 cm^{-1} and 439.80 cm^{-1}. Among them, 3424.94 cm^{-1}, 806.70 cm^{-1} and 439.80 cm^{-1} have a certain degree of displacement relative to the characteristic peaks of scheelite treated by the complex. The characteristic peaks of 1598.59 $cm^{-1}$and 1561.73 cm $^{-1}$ are benzene skeleton bending vibrations, 1403.35 cm^{-1} are O–H

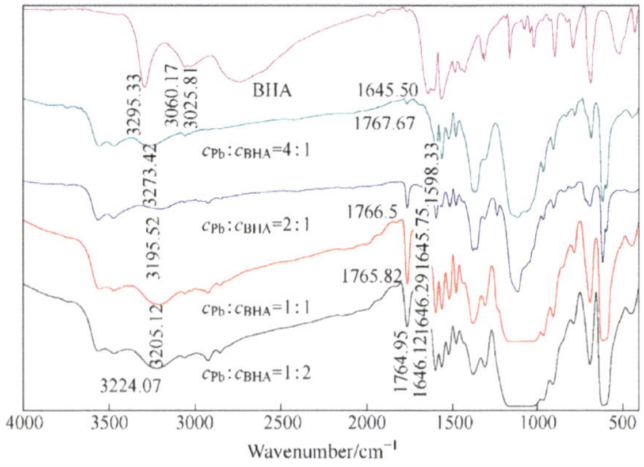

Fig. 5.50 Infrared spectra of Pb-BHA complexes at different proportions

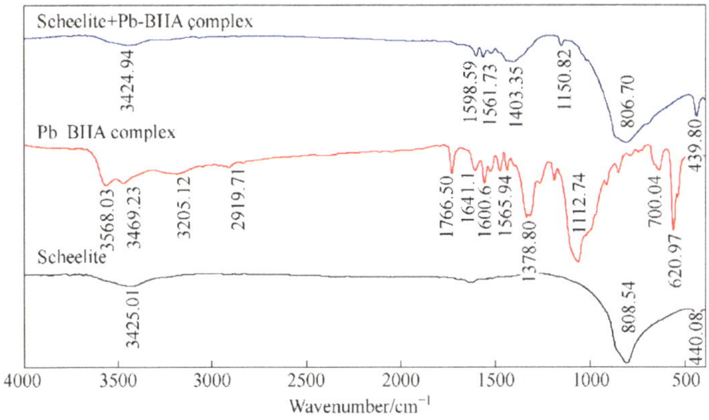

Fig. 5.51 Infrared spectra of the Pb-BHA complex before and after its action on scheelite surface, reprinted from Ref. [120], copyright 2020, with permission from Minerals Engineering

bending vibrations, and 1150.82 cm^{-1} are C-N stretching vibrations. The occurrence of characteristic peaks of these complexes indicates that the main functional groups of the complexes appear on the scheelite surface.

3. X-ray photoelectron spectroscopy analysis

The photoelectron spectroscopy analysis of BHA and Pb-BHA complexes is shown in Table 5.2. After BHA interacts with Pb^{2+} ion, the binding energies of O1*s*, N1*s* and Pb 4*f* atoms shift greatly, indicating that both of them may bond with Pb. The N element and its electron binding energy displacement can be used as an indicator to characterize the adsorption of Pb-BHA metal–organic complex on mineral surface.

The full spectrum of XPS before and after Pb-BHA action on scheelite surface is shown in Fig. 5.52, and the binding energy of XPS element is shown in Table 5.3. Full spectrum analysis shows that Pb 4*f* and N1*s* peaks appear on the surface of scheelite after it is treated by the complex, which indicates the adsorption of the complex on the surface of scheelite. The binding energy data analysis shows that the binding energy changes (ΔE) of O 1*s*, Ca 2*p*, W 4*f*, Pb 4*f* and N 1*s* on scheelite surface before and after being treated by Pb-BHA are + 0.55 eV, + 0.19 eV, + 0.12 eV, + 0.22 eV and -0.14 eV, respectively. The shift of Pb 4*f* and O 1*s* is the most significant. ΔIt indicates that oxygen atom is the active site of complex adsorption on scheelite surface. The displacement of Ca 2*p* also appears to some extent, which may be due to the hydrogen bonding between the complex and the mineral surface.

Table 5.2 Surface element binding energy of BHA and Pb-BHA complexes (eV)

Elements	BHA	$Pb(OH)_2$	Pb-BHA	ΔE
O1*s*	532.37		531.18	1.19
N1*s*	400.21		399.16	1.05
Pb 4*f*		138.40	138.70	+ 0.30

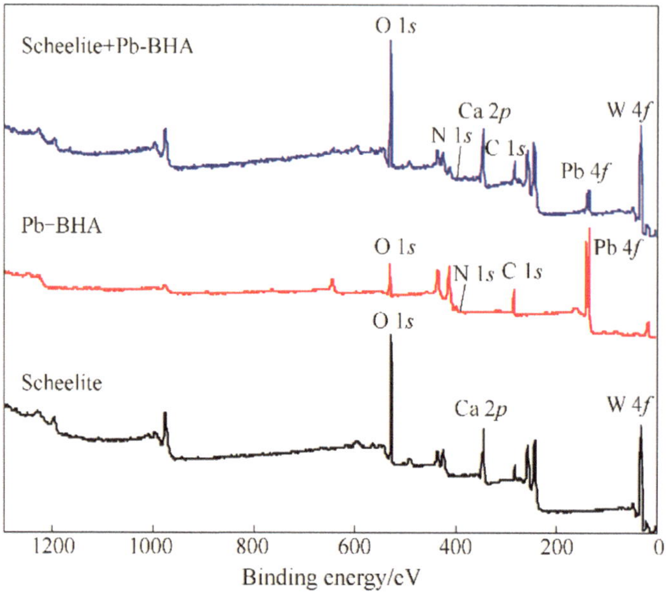

Fig. 5.52 Full spectrum of XPS before and after Pb-BHA interaction on scheelite surface, reprinted from Ref. [120], copyright 2020, with permission from Minerals Engineering

Table 5.3 Binding energy of Pb-BHA before and after action on scheelite surface (eV)

Elements	Scheelite	Pb-BHA	Scheelite + Pb-BHA	ΔE
C 1*s*	284.8	284.8	284.8	0
O 1*s*	530.31	531.26	530.86	+ 0.55
Ca 2*p*	346.86		347.05	+ 0.19
W 4*f*	35.31		35.43	+ 0.12
Pb 4*f*		138.50	138.72	+ 0.22
N 1*s*		398.99	398.85	− 0.14

The O 1*s*, N 1*s* and Pb 4*f* maps before and after the adsorption complex on the scheelite surface were analyzed, as shown in Fig. 5.53. Figure 5.53a shows that before scheelite is treated by the complex, two chemical states of oxygen elements appear on the surface of scheelite. The peaks at positions 531.79 eV and 530.29 eV correspond to the Ca–O bond and W–O bond on scheelite surface. Figure 5.53b shows that after scheelite is treated by the complex, five oxygen chemical states appear on the surface, and the peaks at 531.82 eV and 530.71 eV correspond to the Ca–O bond and W–O bond on the scheelite surface. The peaks at 532.21 eV, 531.25 eV and 529.70 eV correspond to N–O bond, C–O bond and Pb–O bond formed by the complexes adsorbed on scheelite surface. Figure 5.53c shows that after scheelite is treated by the complex, two nitrogen chemical states appear on the surface. The peaks at 400.47 eV and 398.84 eV correspond to the C–N bond and N–O bond (or C=N bond, the binding energy of the two is very close, so they are not distinguished) two nitrogen chemical states. Figure 5.53d shows that after the scheelite surface is treated by the complex, two Pb splitting peaks, Pb $4f_{5/2}$ and Pb $4f_{7/2}$, appear at 143.60 eV and 138.72 eV, which shift by + 0.22 eV compared with the two Pb 4*f* peaks in the complex. The presence of N–O bonds, C–O bonds, C–N bonds and Pb–O bonds indicates that the functional groups of the complex appear on the scheelite surface, while the presence of Pb–O bonds indicates that the complex acts on the oxygen particles on the mineral surface through lead ions and thus chemical adsorption occurs.

4. Adsorption model of Pb-BHA complexes on scheelite surface

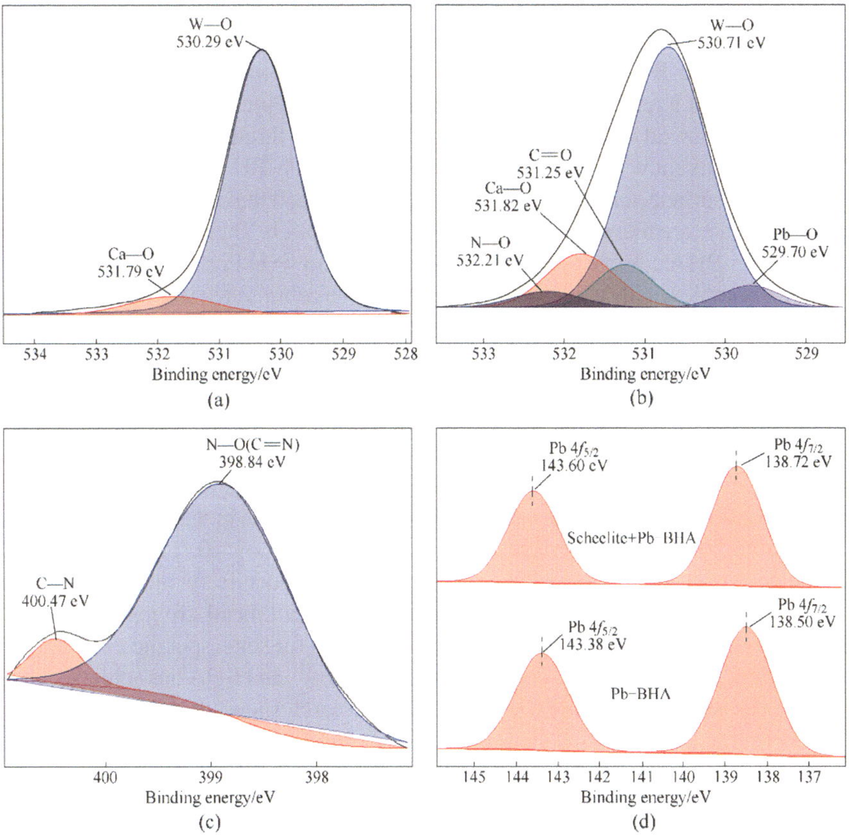

Fig. 5.53 Binding energy spectrum analysis of XPS elements on scheelite surface before and after complex treatment: **a–b** O 1*s*, **c** N 1*s*, **d** Pb 4*f*, reprinted from Ref. [113], copyright 2021, with permission from Chemical Engineering Journal

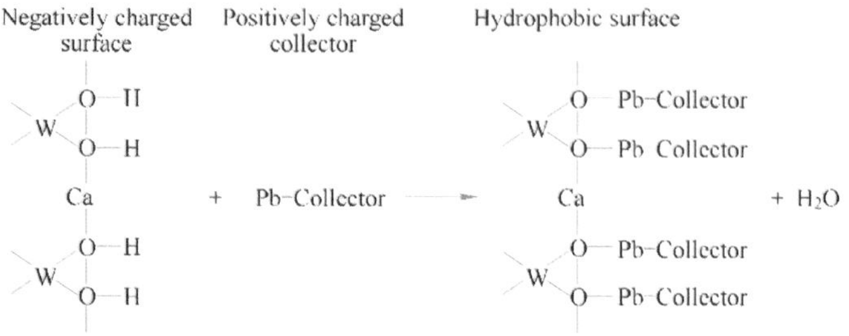

Fig. 5.54 Self-assembly structure of Pb-BHA complex and its adsorption mechanism on mineral surface, reprinted from Ref. [120], copyright 2020, with permission from Chemical Engineering Journal

The adsorption mechanism of the complex at the solid–liquid interface is also significantly different from that of traditional collectors. Zeta potential and XPS adsorption mechanism of Pb-BHA complex on scheelite surface showed that Pb-BHA complex colloid in solution reflects the electrical property of lead ions, is positive in a certain pH range, can be adsorbed on the surface of minerals through electrostatic action. The XPS results show that the chemical adsorption of Pb-BHA complexes occurs on the scheelite surface, and the lead ions in the complex and the oxygen particles on the mineral surface are the active sites of chemical adsorption. Therefore, the adsorption model of Pb-BHA complex on scheelite surface is shown in Fig. 5.54. The colloid of Pb-BHA complex with strong positive surface charge migrates to scheelite surface with negative surface charge under the attraction of electrostatic force, and then the lead ionic groups in the complex have chemical adsorption with the oxygen particles on the mineral surface to adsorb on the mineral surface.

Metal–organic complex collector uses metal ions as its functional group, which is fundamentally different from traditional collectors. Metal ion groups have good selective adsorption ability, making the complex have high selectivity. Pb-BHA complex collector has been proved to obtain good indicators in the flotation of cassiterite, rutile, ilmenite, spodopene and other oxidized ores [110, 121, 122]. Due to the numerous types of metal ions and ligands, the structure of the complexes is diverse. Through the selection and assembly of different metal groups and ligands, the complex collector with targeted adsorption ability for some specific minerals can be designed. For example, the complex collector of Fe^{3+} and BHA has achieved good results in cassiterite flotation [100]. The above research shows that some complexes formed by metal ions and organic ligand assembly have certain characteristics in terms of selectivity and collection ability of flotation, and this finding has a certain universality. From the perspective of metal ion coordination regulation molecular assembly, the metal–organic collector has a variety of structure due to the numerous types of metal ions and ligands, so the development and application of complex collector has great potential, which provides a new direction for the design and development of new flotation reagents.

5.4.3 *Structure–Activity Relationship of Metal Ion-Hydroxamic Acid Complex Collector*

As a new collector of metal–organic complexes, Pb-BHA complexes have been widely used in the flotation of scheelite, wolframite, cassiterite, ilmenite, rutile and other oxidized ores, and have obtained good flotation indexes [77, 121, 123]. However, the amount of Pb-BHA in mineral flotation is relatively large and the bubbling ability is insufficient. Taking scheelite flotation as an example, the dosage of Pb-BHA is usually 2–3 times that of fatty acids, and auxiliary foaming agent is needed [120, 124]. Due to the high production cost of hydroxamic acid, the reagent cost of flotation scheelite with Pb-BHA is high. Therefore, how to construct the "structure–activity" relationship of metal ion-hydroxamic acid complex collector, and develop more efficient and low-cost new flotation reagents is of great significance.

5.4.3.1 "Structure–Activity" Relationship of Hydroxamic Acid Molecules Based on Electron Effect of Substituents in Conjugated System

There are two direct factors affecting the flotation ability of the complex: (1) the adsorption ability of the functional groups of the complex on the mineral surface; (2) the hydrophobic ability of the non-polar groups of the complex. Ligand is an important component of the complex collector. The hydrophobicity and foaming properties of BHA benzene ring are much worse than those of long chain hydrocarbon groups. Therefore, improving the binding ability of BHA to the active site on the mineral surface and enhancing the hydrophobicity of non-polar groups are effective ways to improve the flotation efficiency of the complex. The molecular modification method with hydroxamic acid as the core framework has been reported in the literature, and new hydroxamic acid molecules have been designed, including cyclohexyl hydroxamic acid [125], p-tert-butylbenzohydroxamic acid [79], N-(6-(hydroxyamino)-6-oxy-hexyl) decamamide [14], alkyl hydroxamic acid [126–128], etc. These studies showed that the introduction of different groups had a great impact on the performance of hydroxamic acid [83, 128]. When these hydroxamic acids were used as ligands of metal ion-organic complex collector, the selectivity and collecting ability of the agent changed significantly. However, systematic regular summary has been lacking. Based on the collector of lead benzohydroxamic acid metal ion-organic complex, the research idea of "structure–activity" relationship between hydroxamic acid ligand molecules and their metal ion complexes was put forward. The electron effect was used to improve (or reduce) the electron density of hydroxamic acid group, and then the influence of modified hydroxamic acid molecules and its metal ion-organic complex on the flotation performance of scheelite was investigated. As shown in Fig. 5.55, two new collector molecules, p-methyl benzohydroxamic acid (MBHA) and p-bromohydroxamic acid (BBHA), were synthesized by introducing

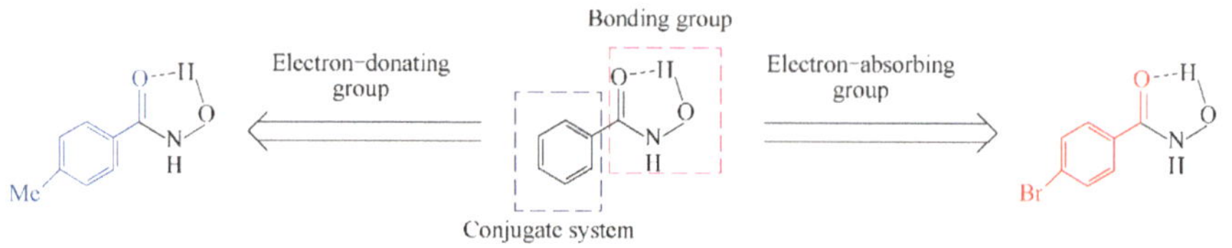

Fig. 5.55 Molecular modification of benzohydroxamic acid collector based on electronic effects, reprinted from Ref. [129], copyright 2020, with permission from Minerals Engineering

the electron-giving group methyl and the electron-absorbing group bromohydroxamic acid in the benzene ring of BHA molecule, thereby increasing or decreasing the electron density of the hydroxamic acid group in the molecule respectively. The carbonyl group and benzene ring are in the conjugated position, and the electron giving/sucking substituents on the benzene ring will cause the change of the electron density of the carbonyl group. Oxime N in the carbonyl α position, the change of the carbonyl electron density can cause the change of oxime N electron density through the hyperconjugation effect; Hydroxyl group is closely associated with carbonyl group through intramolecular hydrogen bonding, so the change of carbonyl group electron density can cause the change of hydroxy-oxygen electron density through intramolecular hydrogen bonding. Therefore, conjugation, hyperconjugation and intramolecular hydrogen bonding can enhance or weaken the electron density of bonding groups. Due to the similar size of methyl group and bromine group, MBHA and BBHA collector can exclude the influence of the size of substituent group on the interface reaction when adsorbing mineral surface and interacting with bubbles.

1. Synthesis of novel hydroxamic acid molecules

Synthesize three hydroxamic acids according to the method shown in Fig. 5.56.

(1) Benzohydroxamic acid (BHA). Yield: 91.5%. Element content (%) obtained by elemental analysis: C, 61.33; H, 5.12; N, 10.38. Calculated as $C_7H_7NO_2$ (%): C, 61.31; H, 5.11; N, 10.22. ^{1}H NMR (500 MHz, DMSO-d_6) δ *11.23* (s, 1H), 9.05 (s, 1H), 7.75 (d, J = 7.3Hz, 2H), 7.50 (q, J = 7.1Hz, 1H), 7.44 (t, J = 7.5Hz, 2H). ^{13}C NMR (126 MHz, DMSO-d_6) δ 164.33, 132.89, 131.23, 128.47, 126.96.

(2) P-methylbenzohydroxamic acid (MBHA). Yield: 90.1%. Element content (%) obtained by elemental analysis: C, 63.60; H, 5.98; N, 9.26. Calculated as $C_8H_9NO_2$ (%): C, 63.58; H, 5.96; N, 9.27. ^{1}H NMR (500 MHz, DMSO-d_6)

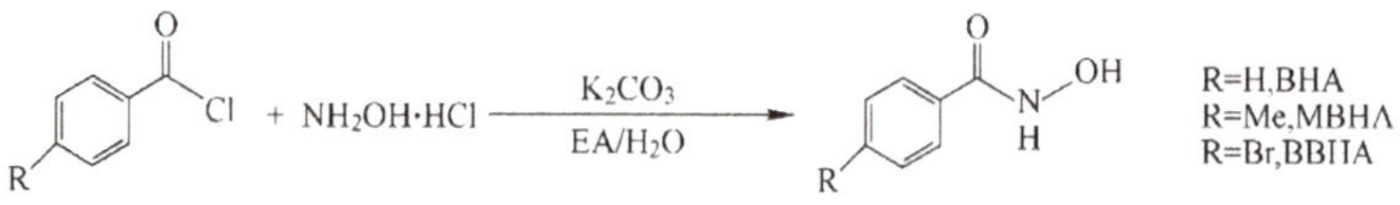

Fig. 5.56 Synthesis methods of three hydroxamic acids, reprinted from Ref. [129], copyright 2020, with permission from Minerals Engineering

δ 11.15 (s, 1H), 8.97 (s, 1H), 7.65 (d, J = 8.0Hz, 2H), 7.24 (d, J = 8.0Hz, 2H), 2.33 (s, 3H). ^{13}C NMR (126 MHz, DMSO-d_6) δ: 164.39, 141.13, 130.09, 129.01, 127.00, 21.07.

(3) P-bromo-benzohydroxamic acid (BBHA). Yield: 88.5%. Element content (%) obtained by elemental analysis: C, 38.92; H, 2.81; N, 6.47. Calculated as $C_7H_6NO_2Br$ (%): C, 38.89; H, 2.78; N, 6.48. ^{1}H NMR (500 MHz, DMSO-d_{66}) δ 11.31 (s, 1H), 9.11 (s, 1H), 7.71–7.65 (m, 4H). ^{13}C NMR (126 MHz, DMSO-d) δ 163.81, 132.53, 131.29, 128.81, 126.89.

2. Flotation performance of BHA, MBHA and BBHA scheelite

The harvesting capacity of MBHA, BBHA and BHA for scheelite at the same dosage is shown in Fig. 5.57a. The results show that the flotation recovery of scheelite increases rapidly with the increase of the amount of collector, and then gradually flattens out until the highest recovery is obtained. When the amount of collector is 1.5×10^{-4} mol/L, the recovery rate of MBHA is 79.6%, while that of BHA and BBHA are 63.6% and 58.4%, respectively. It shows that MBHA has a stronger collection capacity of scheelite, and the optimal amount is 1.5×10^{-4} mol/L. Under the optimal dosage of agent, the effects of different pulp pH values on the flotation of scheelite by three hydroxamic acids were studied, as shown in Fig. 5.57b. The results show that the three reagents are more suitable for flotation under alkaline conditions, and the optimal pH value is 9. At this pH value, the recovery rates of MBHA, BHA and BBHA were 80.6%, 59.43% and 57.7%, respectively. It can be seen that MBHA has the strongest collection capacity. Therefore, the results of Fig. 5.57 show that the collecting capacity of the three hydroxamic acids is ranked as MBHA > BHA > BBHA, which means that the electron-donating substituent group (methyl) enhances the collecting capacity of the collector, while the electron-absorbing group (bromine) is on the contrary.

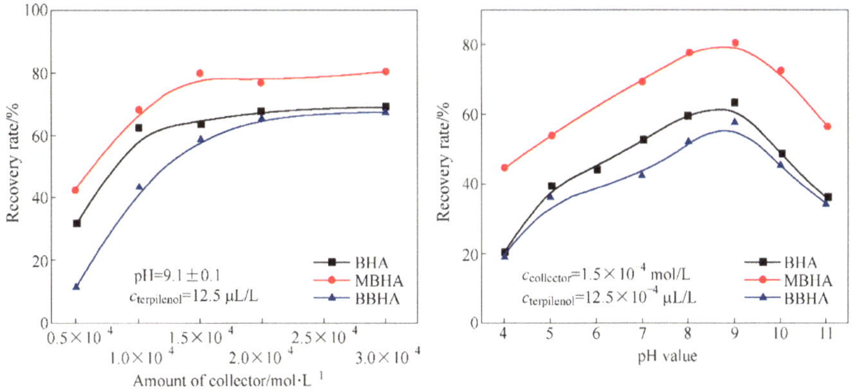

(a) (b)

Fig. 5.57 **a** Effect of collector dosage and **b** pH value on the flotation recovery of scheelite, reprinted from Ref. [129], copyright 2020, with permission from Minerals Engineering

3. Quantum chemical study of the effect of electron effect on the molecular structure and properties of BH

The influence of electron effect on the structure and properties of BHA molecules was studied by quantum chemical calculation based on DFT. The molecular structure, charge distribution, MEP, HOMO and LUMO energies of BHA, MBHA and BBHA are shown in Fig. 5.58. The Mulliken charge distribution shows that the negative charge is mainly distributed on the carbonyl oxygen and hydroxyl oxygen, which is the bonding site (electron donor site) of the three collector molecules on the scheelite surface. The Mulliken charges of carbonyl oxygen and hydroxyl oxygen in BHA and BBHA are − 0.718/− 0.468 and − 0.714/− 0.470, respectively, while the Mulliken charges of MBHA are − 0.756/− 0.47. In addition, the Mulliken charges of the nitrogen atoms of BHA, MBHA and BBHA are 0.317, 0.277 and 0.323, respectively. The Mulliken charge distribution results show that the introduction of methyl groups increases the electron density of carbonyl oxygen and hydroxyl oxygen, and decreases the electron density of nitrogen atoms. Figure 5.58 shows that the region with the largest negative electrostatic potential in the BHA is the oxygen atom, especially the carbonyl oxygen. The oxygen potential of the MBHA is significantly lower than that of the BHA, while the oxygen potential of the BBHA is significantly higher. The bond lengths of C=O and N–O are 0.1241/0.1400 nm for BHA, 0.1240/0.1398 nm for BBHA and 0.1241/0.1401 nm for MBHA, indicating that the introduction of methyl group will lead to further polarization of the two covalent bonds, C=O and N–O. C=O and N–O bond lengths increase. The E value is widely regarded as an evaluation criterion of the electron giving ability of the electron group. It can be seen E_{HOMO} from Fig. 5.58 that MBHA has the highest E value and E_{HOMO} BBHA has the lowest E value, which indicates that MBHA has a stronger electron giving ability. The comprehensive analysis of charge distribution, electrostatic potential, bond length, and E value shows that the introduction of a methyl group into the benzene ring increases the electron density of the carbonyl oxygen through conjugation effect, then the carbonyl oxygen reduces the electron density of the secondary amine through hyperconjugation effect, and the hydroxyl oxygen increases the electron density through intramolecular hydrogen bonding. Therefore, the secondary amine electron density of the MBHA molecule is reduced, while the electron density of the two bonded atoms of carbonyl oxygen and hydroxy-oxygen is increased. The bonding site of MBHA is more active than that of BHA and BBHA, making MBHA a stronger electron donor, which means that MBHA will exhibit greater activity than BHA in binding to Ca^{2+}.

5.4.3.2 Pb-MBHA Harvesting Performance of Scheelite and Its Action Mechanism

1. Harvesting performance of Pb-MBHA scheelite

Figure 5.59a shows the effect of dosage of MBHA and BHA on scheelite flotation when lead ion is used as activator. The results show that with the increase of the

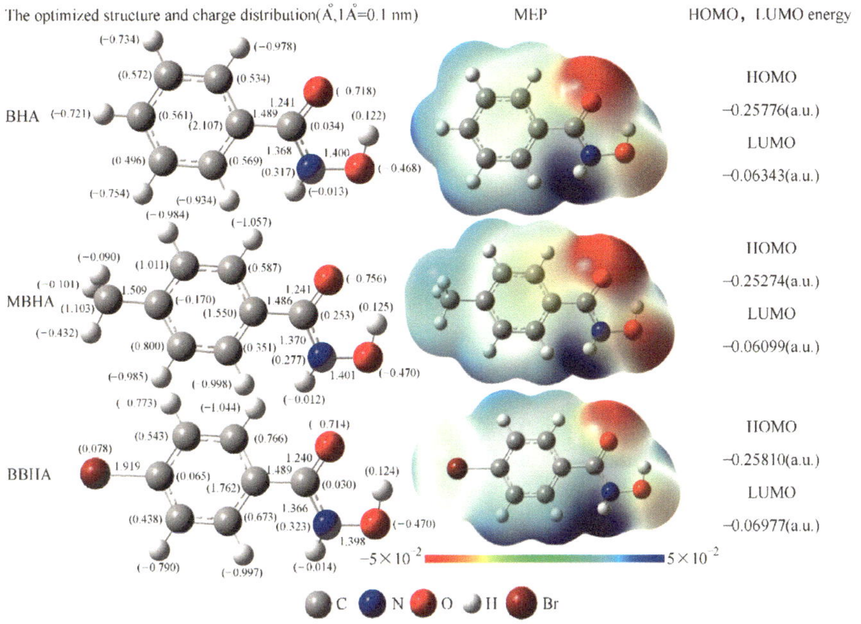

Fig. 5.58 Structure and charge distribution of BHA, MBHA and BBHA after optimization (left), MEP (middle), HOMO and LUMO energy (right), reprinted from Ref. [129], copyright 2020, with permission from Minerals Engineering

amount of collector, the recovery rate of scheelite increases rapidly, and the recovery rate of MBHA is always higher than that of BHA. Therefore, in the presence of lead ions, the harvesting capacity of scheelite by MBHA is still stronger than that by BHA. Figure 5.59b shows the effect of the addition sequence of lead ions and MBHA on scheelite flotation. The results show that the recovery rate of scheelite flotation by mixing lead ions and MBHA is higher than that by sequential addition. When the dosage of lead ions is 1×10^{-4} mol/L, the recovery rate of scheelite as complex collector is the highest, reaching 93.25%. Therefore, the Pb-MBHA complex collector formed by the assembly of lead ions and MBHA has a stronger collection capacity of scheelite.

The effect of pH on the flotation of scheelite and fluorite by Pb-MBHA complexes is shown in Fig. 5.60. The results show that compared with Pb-BHA, Pb-MBHA is almost unaffected by pH over the whole pH = 5–11 range. At pH = 9.0, the recovery rates of Pb-MBHA and Pb-BHA are 94.05% and 80.6%, respectively, indicating that Pb-MBHA has better harvesting capacity and pH adaptability. At the same time, the recoveries of fluorite were all lower than 20%, indicating that Pb-MBHA has a poor ability to collect fluorite, which indicates that Pb-MBHA complex has good selectivity, which is conducive to efficient flotation separation of scheelite and fluorite. Therefore, Pb-MBHA has stronger harvesting capacity and pH adaptability

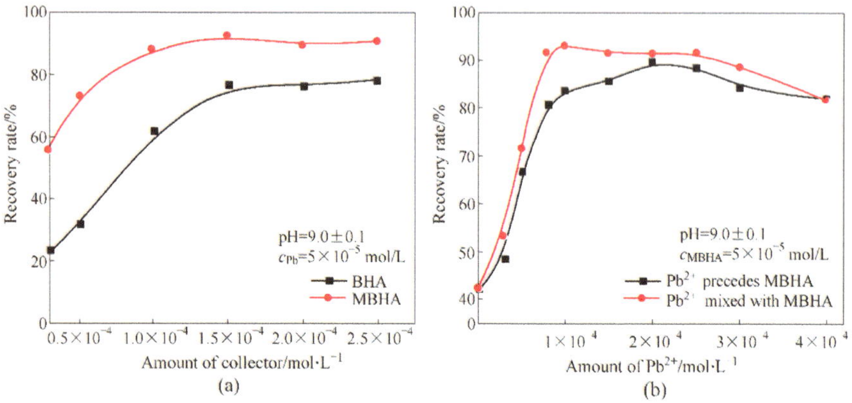

Fig. 5.59 **a** Effects of dosage of BHA and MBHA on scheelite flotation in the presence of lead ions, and **b** effects of addition sequence of lead ions and MBHA on scheelite flotation, reprinted from Ref. [99], copyright 2021, with permission from Chemical Engineering Science

than Pb-BHA, and maintains good selectivity, which indicates that Pb-MBHA is a more efficient collector than Pb-BHA, and can be used in natural pulp pH conditions.

2. Mechanism of Pb-MBHA in scheelite flotation

The zeta potentials of BHA, MBHA, Pb-MBHA and Pb-BHA are shown in Fig. 5.61a. The results show that the zeta potential of BHA and MBHA decreases with increasing pH value, and the MBHA carries more negative charge. However, Pb-BHA and Pb-MBHA are positively charged, and Pb-MBHA has a higher potential than Pb-BHA, which indicates that MBHA has a stronger binding capacity with lead ions. The zeta potential of scheelite treated with both collectors is shown in Fig. 5.61b. At a pH of 6–12, the scheelite surface is negatively charged. After collector treatment, the

Fig. 5.60 Effect of pH value on flotation of scheelite and fluorite with Pb-MBHA complex collector, reprinted from Ref. [99], copyright 2021, with permission from Chemical Engineering Science

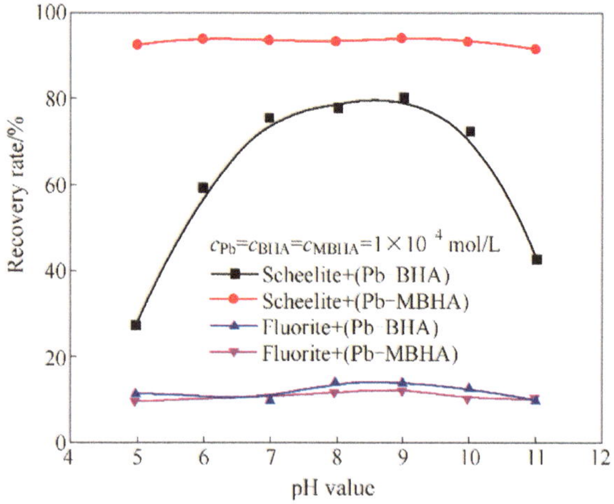

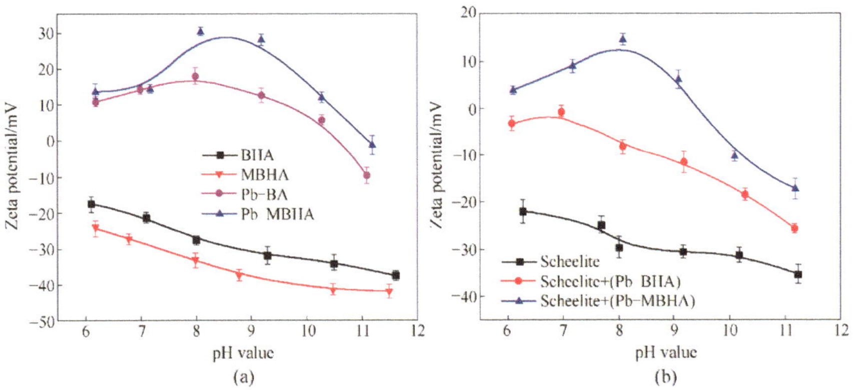

Fig. 5.61 **a** Zeta potential of BHA, MBHA, Pb-MBHA and Pb-BHA, and **b** Zeta potential of scheelite before and after being treated with Pb-BHA and Pb-MBHA reprinted from Ref. [99], copyright 2021, with permission from Chemical Engineering Science

potential of scheelite showed a positive shift, indicating that Pb-BHA and Pb-MBHA were adsorbed on scheelite surface, while the potential shift after adsorption of Pb-MBHA on scheelite surface was larger, indicating that the adsorption of Pb-MBHA on scheelite surface was stronger than that of Pb-BHA.

The flotation pH of scheelite is usually weakly alkaline, while Pb(OH)MBHA, $Pb(MBHA)^+$ and $Pb(MBHA)_2$ are the dominant components of PB-MBHA in weakly alkaline solutions. Pb-MBHA complex collector has more active bonding groups, has a strong positive charge in the solution, and is more likely to adsorb on the surface of negatively charged scheelite. At the solid–liquid interface, the lead ionic groups of $Pb(MBHA)^+{}_2$ and Pb(MBHA) are selectively bonded with the oxygen atoms of $WO_4{}^{2-}$ groups on the scheelite surface, the hydroxyl group in Pb(OH)MBHA can be bonded with the hydroxyl group on the scheelite surface through dehydration reaction, and the chemical adsorption of Pb-MBHA component is stronger than PB-BHA. At the same time, the introduction of methyl group improves the hydrophobic ability of the non-polar groups of the complex, and the adsorption of Pb-MBHA makes the surface of scheelite more hydrophobic, and it is easier to form a stable foam at the gas–liquid interface.

Therefore, in order to introduce electron group substituents to BHA molecules, the electron density of BHA functional oxygen atoms can be enhanced through the conjugate system substituent electron effect, so as to enhance the adsorption of Pb-MBHA components on scheelite surface, which provides a new method and idea for the study of "structure–activity" relationship between metal ion-organic complex flotation reagents.

5.4.4 Design of Metal Ion-Polyligand Complex Collector and Its Action Mechanism in Tungsten Flotation

In the field of chemistry and materials research, multi-ligand metal–organic complexes (multi-ligand metal–organic frame materials) are complexes formed by the coordination of multiple types of ligand and metal ions. Compared with single ligand complexes, multi-ligand complexes have varied spatial structure and more excellent physical and chemical properties [130–132]. Lead ions in Pb-BHA complexes have stereochemical activity and flexible coordination environment, which can form complexes with hydroxamic acid group, carboxyl group, sulfuric acid/sulfonic acid group, phosphonic acid group and other organic functional compounds. Therefore, by introducing a new ligand collector to coordinate with lead ions, a multi-ligand metal-based collector is formed. May be an important way to improve the adsorption capacity of metal groups and the hydrophobic ability of non-polar groups of Pb-BHA complexes. Due to the large number of ligands and the diversity and complexity of the ligand and metal ion assembly modes, the development of multi-ligand metal-based collector requires a systematic design approach.

5.4.4.1 Molecular Design of Multi-ligand Metal-Based Collector

1. Principles for molecular design of multi-ligand metal-based collectors

In the design of multi-ligand metal-based collector, the metal base is still lead ion, but the ligand coordination with lead ion will directly affect the chemical properties of lead ion, and then affect the adsorption ability of lead ion to the active site on the surface of minerals. Therefore, the polar base design of multi-ligand metal-based collector is mainly reflected in the influence of the new ligand on the overall nature of the complex groups or molecules.

The classical methods of polar base design include group electronegativity method and molecular orbital calculation method [133], among which the molecular orbital method through quantum chemical calculation to obtain a variety of molecular orbital index, not only can calculate the physical and chemical properties of pharmaceutical and mineral crystals, but also can combine pharmaceutical and mineral calculation, with high precision, accurate structure, calculation method system and other advantages. For complex collector, due to the complex and oxidized ore action for oxygen atoms to lose electrons and lead ions to gain electrons, so the LUMO energy of the complex can reflect the ability of the complex to gain electrons. The lower E_{LUMO} is, the more easily the metal base of the complex gains electrons. Therefore, reducing the E_{LUMO} of Pb-BHA complex molecules is an important indicator of the polar base design of multi-ligand metal-based collector. The design of the non-polar base is mainly based on the allocation coefficient of the ligand, and increasing the lg$P$ of the Pb-BHA complex molecule is an important index for the design of the non-polar base of the multi-ligand metal-based collector.

Energy criteria are needed to determine and predict the strength of the interaction between collector and mineral. Commonly used energy criteria include the energy equation of agent activity and the calculation of frontier orbit between agent and mineral, etc. The energy band theory of solid states that the energy levels of valence band top (VBM) and conduction band bottom (CBM) are equivalent to the HOMO and LUMO of molecules, respectively [134, 135]. During chemical reactions, electrons are transferred from the HOMO of molecule 1 to the LUMO of molecule 2 and from the HOMO of molecule 2 to the LUMO of molecule 1. According to frontier orbital theory, the smaller the absolute value of the difference between the energy value of one reactant and the energy value of the other reactant, the stronger the interaction between the two molecules and the weaker the interaction between the two molecules. This theory is applied to the calculation of flotation reagent with mineral surface analysis, through the $|E_{HOMO(mineral)} - E_{LUMO(reagent)}|$ and $|E_{HOMO(reagent)} - E_{LUMO(mineral)}|$ calculation, can be on different agents with the same minerals, same agents with different mineral form is analyzed, the smaller the difference, the stronger the reagent and the effect of mineral [136, 137]. Therefore, reduce the $|E_{HOMO(mineral)} - E_{LUMO(reagent)}|$ is more ligand metal matrix and the function of the mineral collector important criterion index.

So much ligand metal matrix collector design principle is: through the introduction of new ligands, decrease the molecular complexes E_{LUMO}, increasing the lg$P$, reduce the $|E_{HOMO(mineral)} - E_{LUMO(reagent)}|$, in raising the ability of collecting complexes at the same time maintain its good selectivity.

2. Ligand metal matrix more collector molecular design method

According to the design principles, the design method of multi-ligand metal-based collector can be established, as shown in Fig. 5.62. The design method consists of the following 4 steps:

(1) Crystal structure and electronic structure analysis of target minerals and gangue minerals. With scheelite, wolframite and cassiterite as target minerals, and calcite and fluorite as gangue minerals, the crystal structure, surface charge, state density (DOS, PDOS), E_{HOMO}, E_{LUMO} and other electronic structure data in the common view of minerals were calculated by first principles.
(2) Calculation of quantum chemical properties of ligands. Taking common oxide ore flotation collector as an alternative ligand, the new ligand was preliminarily screened through quantum chemical calculation of the ligand molecules' E_{LUMO}, E_{HOMO}, molecular electronegativity x_m, octanol water partition coefficient lg$P$ and other data.
(3) Construction of multi-ligand metal-based collector and calculation of quantum chemical properties. By introducing a new ligand into the Pb-BHA complex to construct a multi-ligand metal-based collector molecule, the E_{HOMO}, E_{LUMO}, x_m, lg$P$ and other data were calculated, and all the data of the multi-ligand collector were sorted.
(4) Determination of the strength of the interaction between the multi-ligand metal-based collector and the mineral surface and prediction of flotation performance.

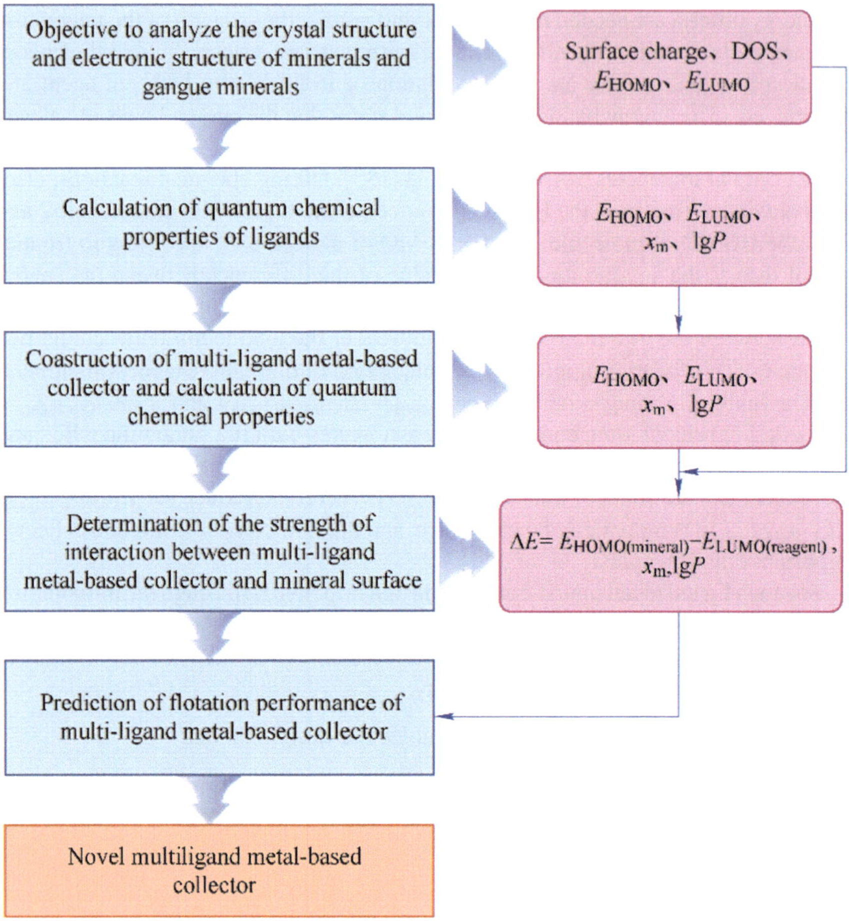

Fig. 5.62 Design method of multi-ligand metal-based collector, reprinted from Ref. [138], copyright 2024, with permission from Separation and Purification Technology

By $\Delta E = |E_{HOMO(mineral)} - E_{LUMO(reagent)}|$ calculation, combined with multiple ligands x_m, lg$P$ data collector, collector of ligand strength judgement to the mineral surface, and the ability of collecting, selective flotation performance prediction, comprehensive collecting ability and selectivity to get more new ligands metal matrix collector.

3. Construction of multi-ligand metal-based collector and prediction of flotation performance

The state densities of scheelite, wolframite, cassiterite, calcite and fluorite are shown in Fig. 5.63. For scheelite, calcite and fluorite, it can be seen that the vicinity of the Fermi level of scheelite and calcite is mainly composed of O 2p orbitals, while fluorite is mainly composed of F 2*p* orbitals. The calcium atomic density composition of scheelite, calcite and fluorite is similar, the valence band near − 40 eV is mainly composed of Ca 3*s* orbitals, and the valence band near − 20 eV is mainly composed of Ca 3*p* orbitals, which makes the properties of calcium atoms on the mineral surface similar. However, the composition of the state density near the Fermi level is the most critical in the chemical reaction. Since scheelite and calcite are mainly composed of O 2*p* orbitals, while fluorite is composed of F 2*p* orbitals, this indicates that scheelite and calcite have strong activity of oxygen atoms, while fluorite has strong activity of fluorine atoms, which becomes an important basis for selective separation of the three. The state density of wolframite, citrine and scheelite is quite different. The Mn 3*p* orbital of wolframite is mainly distributed between − 50 and − 40 eV, the Mn 3*d* orbital is mainly distributed between − 10 eV and − 5 eV valence band and conduction band level, the Sn 4*d* orbital of cassiterite is mainly distributed near − 20 eV, and the Sn 5*p* orbital is mainly distributed in conduction band level. However, the density of states near the Fermi level of the three minerals is mainly composed of O 2*p* orbitals, indicating that the oxygen atomic activity of the three minerals is high. The complex collector is chemically adsorbed with the oxygen atoms on the surface of the three minerals, which provides an important basis for the collection of scheelite, wolframite and cassiterite and the flotation separation from calcite and fluorite.

The front-line orbital energies of scheelite (112) plane, wolframite (010) plane, cassiterite (100) plane, calcite (104) plane and fluorite (111) plane are shown in Table 5.4.

Using $Pb_4(L_1)_5(OH)_3$ as the molecular composition and spatial configuration of the Pb-BHA complex, the multi-ligand metal-based collector molecule is constructed by introducing a new ligand to replace the hydroxyl group at the position of $Pb_4(L_1)_5(OH)_3$ identical lead ion. A multi-ligand metal-based collector with the molecular formula $Pb_4(L_1)_5(L_2)(OH)_2$ (HL_1 = BHA, with HL_2 or NaL_2 as the new ligand) was constructed. In PBE0/def2tzvp//B3LYP/def2svp calculation level and SMD implicit solvent model, get the Pb-BHA, Pb-BHA-OHA, Pb-BHA-SDS, Pb-BHA-OL, Pb-BHA-SPA, optimization of 5 kinds of complexes molecular configuration as shown in Fig. 5.64.

The quantum chemical properties of multi-ligand metal-based collectors are shown in Table 5.5. According to the calculated data, the quantum chemical properties of several complexes can be sorted as follows (L_2 represents the corresponding multi ligand complexes):

(1) E_{HOMO} order: OH > OHA > SPA > OL > SDS
(2) E_{LUMO} order: OH > OHA > SDS > OL > SPA
(3) x_{m} order: SPA > SDS > OL > OHA > OH
(4) lg*P* order: OL > SDS > OHA > SPA > OH

E_{LUMO} is a scale to measure the ability of the complex to obtain electrons. The lower E_{LUMO}, the easier the metal group of the complex to obtain electrons. Therefore, the order of the ability of several complexes to obtain electrons was SPA > OL >

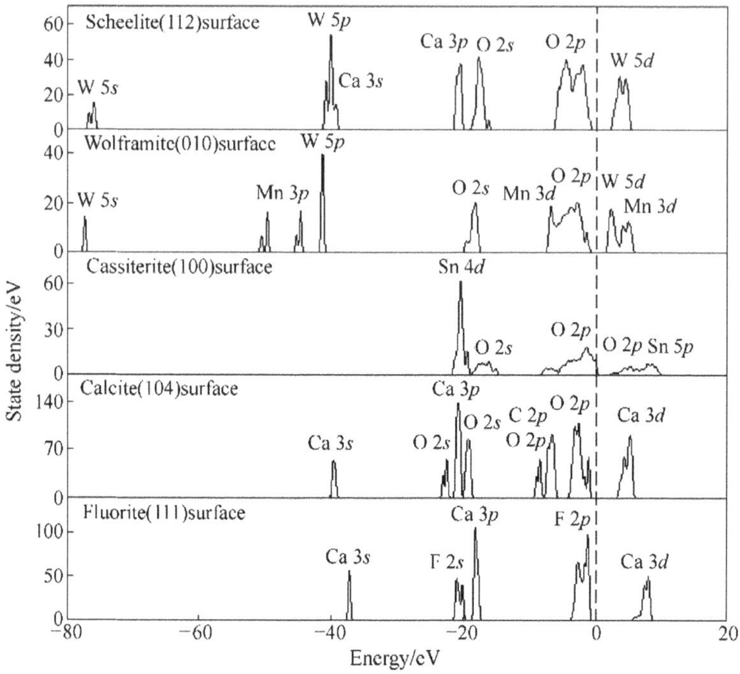

Fig. 5.63 State densities of scheelite, wolframite, cassiterite, calcite and fluorite, reprinted from Ref. [138], copyright 2024, with permission from Separation and Purification Technology

Table 5.4 Front orbit energy of scheelite (112) plane, wolframite (010) plane, cassiterite (100) plane, calcite (104) plane and fluorite (111) plane

Minerals	E_{HOMO} (a.u.)	E_{LUMO} (a.u.)
Scheelite (112) face	0.24133	0.12277
$MnWO_4$ (010) face	0.23253	0.13254
Cassiterite (100) face	0.22352	0.19229
Calcite (104) face	0.20718	0.0356
Fluorite (111) face	0.28241	0.0385

SDS > OHA > OH, indicating that the introduction of new ligands improves the ability of the Pb-BHA complex to obtain electrons. And x_m ordering showed that the introduction of new ligand enhanced the electronegativity of the complex molecules, therefore, combining E_{LUMO} and x_m could predict the collecting ability of several multi-ligand collector, the collecting ability of the order was SPA > OL > SDS > OHA > OH. lg*P* sequencing showed that the introduction of new ligand enhanced the hydrophobic ability of the complexes, and the hydrophobic ability of several multi-ligand collectors was ranked as OL > SDS > OHA > SPA > OH.

Comprehensive minerals and orbital energy calculation result, calculating the $|E_{HOMO(mineral)} - E_{LUMO(reagent)}|$, the results are shown in Table 5.6. According to the

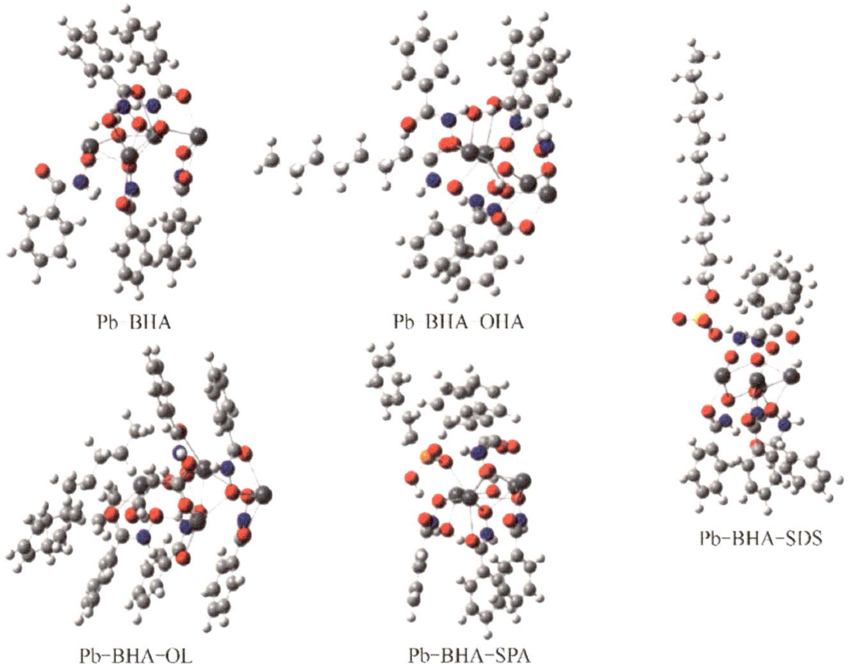

Fig. 5.64 Optimal configuration of multi-ligand metal-based collector, reprinted from Ref. [138], copyright 2024, with permission from Separation and Purification Technology

Table 5.5 Quantum chemical properties of multi-ligand metal-based collectors

L_2	E_{HOMO}/a.u.	E_{LUMO}/a.u.	x_m	lg$P$
OH	0.21582	0.04568	3.56	4.47
OHA	0.21712	0.04706	3.59	6.37
SDS	0.22288	0.04865	3.69	7.57
OL	0.22142	0.04882	3.68	10.69
SPA	0.21755	0.05408	3.70	4.55

ΔE value, the collecting ability of different collectors for the same mineral and the collecting ability and selectivity of the same collector for different minerals can be predicted.

The ΔE ordering of polyligand collectors with scheelite, wolframite and cassiterite is: OH > OHA > SDS > OL > SPA, and the smaller ΔE is, the stronger the effect on mineral surface is. Combining with the x_m (SPA > SDS > OL > OHA > OH) and lg$P$ (OL > SDS > OHA > SPA > OH) of polyligand collector, the collecting ability of the polyligand collector of scheelite, wolframite and cassiterite can be predicted, and the order of collecting ability is OL > SPA > SDS > OHA > OH.

Table 5.6 Multi-ligand metal based collectors and $|E_{HOMO(mineral)} - E_{LUMO(reagent)}|$ calculation results

L_2	Scheelite (a.u.)	Wolframite (a.u.)	Cassiterite (a.u.)	Calcite (a.u.)	Fluorite (a.u.)
OH	0.195647	0.18685	0.177836	0.161498	0.236732
OHA	0.194267	0.18547	0.176456	0.160118	0.235352
SDS	0.192677	0.18388	0.174866	0.158528	0.233762
OL	0.192507	0.18371	0.174696	0.158358	0.233592
SPA	0.187247	0.17845	0.169436	0.153098	0.228332

The order of ΔE of polyligand collector, calcite and fluorite is also OH > OHA > SDS > OL > SPA, which indicates that the introduction of new ligand enhances the effect on target minerals and gangue minerals. The ΔE order of the same polyligand collector and five minerals is fluorite > scheelite > wolframite > cassiterite > calcite, indicating that the introduction of the new ligand maintains the order of the effect of Pb-BHA on the five minerals, but has no significant effect on the selectivity. Therefore, the selectivity of the multi-ligand collector can be predicted in the order of OH > OHA > SDS > SPA > OL.

By synthesizing the collection ability and selectivity of the multiligand collectors, it can be predicted that the PB-BHA-SDS and Pb-BHA-OHA multi-ligand collectors formed by the two new ligands, SDS and OHA, have stronger collection ability than the Pb-BHA multiligand collectors, while maintaining good selectivity. Therefore, the molecular design of multi-ligand metal-based collectors shows that Pb-BHA-SDS and Pb-BHA-OHA are new collectors with good collecting ability and selectivity for the flotation of scheelite, wolframite and cassiterite.

5.4.4.2 Flotation Behavior and Action Mechanism of Pb-BHA-SDS Polyligand Metal-Based Collector

1. Behavior of Pb-BHA-SDS complex in scheelite flotation

The effect of SDS dosage on the flotation of scheelite by Pb-BHA is shown in Fig. 5.65. The results show that when the concentration of BHA and Pb^{2+} is 1.0×10^{-4} mol/L, the recovery of scheelite is only 63.75% without SDS. With the increase of SDS dosage, scheelite recovery gradually increases and then stabilizes. When SDS dosage is 1×10^{-5} mol/L, scheelite recovery reaches 91.45%, which is 27.7 percentage points higher than that without SDS. Therefore, a small amount of SDS can significantly improve the collection capacity of scheelite, which indicates that Pb-BHA-SDS has a stronger collection capacity of scheelite than Pb-BHA under the same dosage condition.

The effect of pH on the flotation of scheelite, calcite and fluorite by Pb-BHA-SDS complex is shown in Fig. 5.66. The results show that the flotation recovery of scheelite is higher in the range of pH = 8–11, indicating that Pb-BHA-SDS has strong collecting ability and pH adaptability of scheelite. The flotation recovery of fluorite

Fig. 5.65 Effect of SDS dosage on the flotation of scheelite by Pb-BHA complex, reprinted from Ref. [139] copyright 2021, with permission from Minerals Engineering

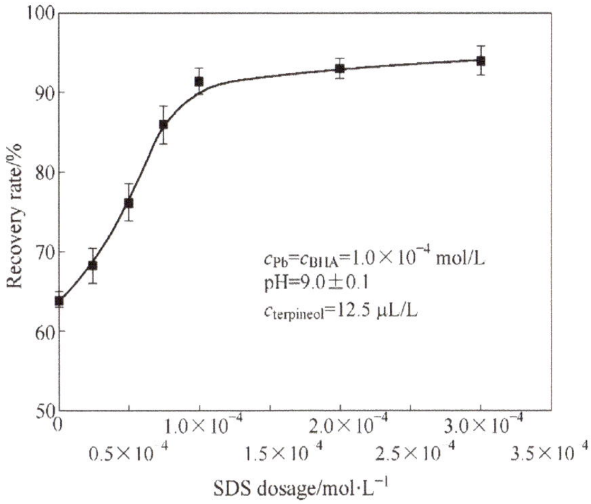

is less than 20%, indicating that the collection ability of the Pb-BHA-SDS complex for fluorite is very poor. But the flotation recovery of calcite is lower than scheelite on the whole. At the optimum flotation pH value of 9.0 scheelite, the flotation recovery rates of scheelite, calcite and fluorite are 91.45%, 79.57% and 12.34%, respectively, indicating that the Pb-BHA-SDS complex has good selectivity for the flotation of calcium minerals.

2. Mechanism of Pb-BHA-SDS multiligand collector in scheelite flotation

The zeta potential of scheelite before and after Pb-BHA and Pb-BHA-SDS treatment is shown in Fig. 5.67. In the pH range of 2–12, scheelite is surface-negatively charged in solution. The zeta potential of Pb-BHA-SDS colloids is positive in the pH range

Fig. 5.66 Effect of pH on the flotation of scheelite, calcite and fluorite by Pb-BHA-SDS complex, reprinted from Ref. [139], copyright 2021, with permission from Minerals Engineering

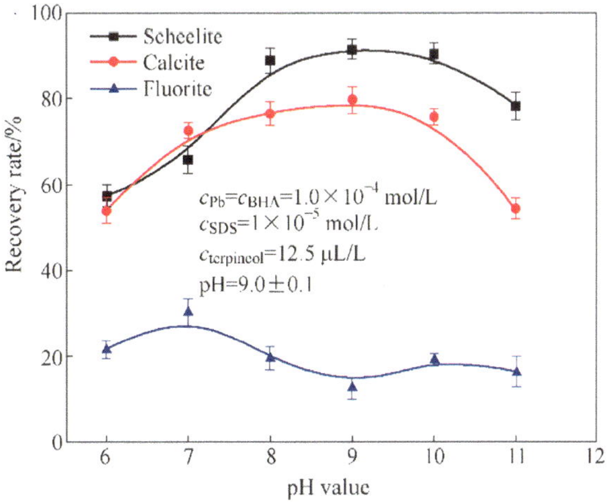

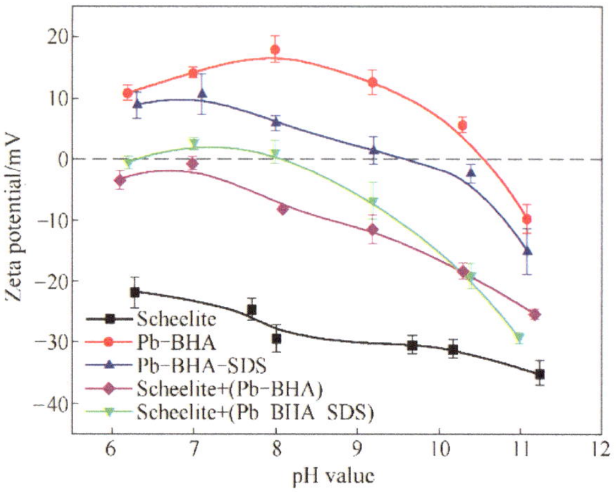

Fig. 5.67 Potential of scheelite treated by Pb-BHA and Pb-BHA-SDS ($c_{Pb} = c_{BHA} = 1.0 \times 10^{-4}$ mol/L, $c_{SDS} = 1 \times 10^{-5}$ mol/L), reprinted from Ref. [139], copyright 2021, with permission from Minerals Engineering

of 6–10, indicating that Pb-BHA-SDS colloids are positively charged in a certain pH range. Compared with Pb-BHA, the potential value of Pb-BHA-SDS is relatively small, indicating that the colloid of Pb-BHA-SDS has relatively less positive charge, which may be because the introduction of SDS in Pb-BHA-SDS weakens the overall positive charge value of the complex. The positive shift of zeta potential of scheelite after Pb-BHA-SDS treatment indicates that the molecules of Pb-BHA-SDS are adsorbed on scheelite surface, and the electrostatic attraction may play an important role in the adsorption process. Compared with Pb-BHA, the zeta potential displacement value of scheelite treated by Pb-BHA-SDS is larger in the range of pH = 6–10, which indicates that the adsorption of Pb-BHA-SDS on scheelite surface is stronger than that of Pb-BHA.

Figure 5.68 shows the contact angle of scheelite before and after being treated with Pb-BHA, SDS and Pb-BHA-SDS. The results show that the contact angle of scheelite before and after treatment is 56.76°. When scheelite is treated with Pb-BHA, the contact angle is 82.94°, and the contact angle increases by 26.18°, indicating that the hydrophobicity of scheelite is significantly enhanced after Pb-BHA treatment. After SDS treatment, the contact angle is 78.48° and the contact angle is increased by 21.72°, indicating that the hydrophobicity of scheelite is also enhanced after SDS treatment, but the increase is less than that of Pb-BHA. When scheelite is treated with Pb-BHA-SDS, the contact angle is 87.95°, and the contact angle is increased by 31.19°, which is higher than Pb-BHA, indicating that the hydrophobicity of scheelite is enhanced most significantly after Pb-BHA-SDS treatment. After the polar group is adsorbed on the mineral surface by the complex collector, the non-polar hydrophobic groups face outward, which leads to the hydrophobic surface of the mineral. The benzene ring in the Pb-BHA structure has hydrophobic ability, but its hydrophobic ability is limited. In addition to benzene ring, there are also SDS long-chain alkyls with strong hydrophobic ability in Pb-BHA-SDS structure. The introduction of these long-chain alkyls significantly enhanced the hydrophobic ability of non-polar groups

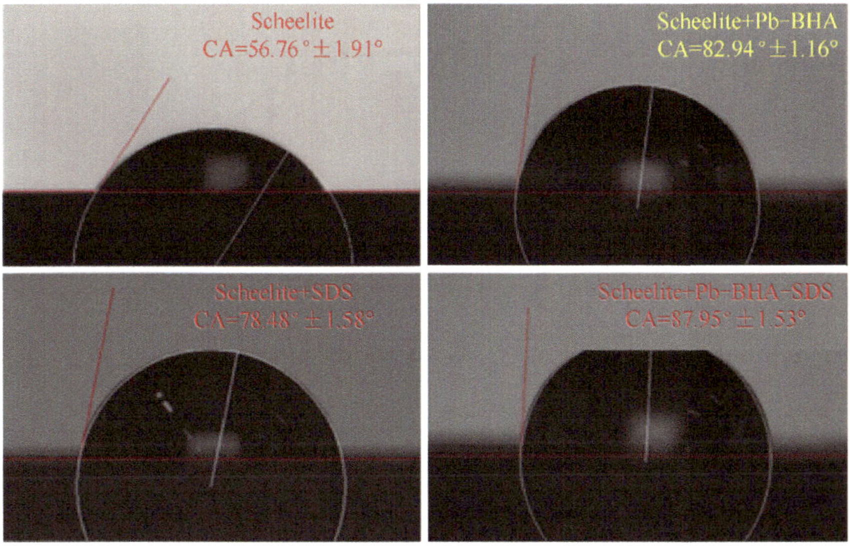

Fig. 5.68 Contact angle of scheelite before and after being treated by Pb-BHA, SDS and Pb-BHA-SDS ($c_{Pb} = c_{SDS} = 1.0 \times 10^{-5}$ mol/L, $c = 1.0 \times 10$ mol/L, pH $= 9.0 \pm 0.1$)

in the collector structure, so that the adsorption of Pb-BHA-SDS on scheelite surface enhanced the hydrophobic ability of scheelite.

Solution surface tension of SDS and Pb-BHA-SDS is shown in Fig. 5.69. The results showed that the surface tension of SDS solution decreased significantly with the increase of SDS concentration, and the surface tension value did not decrease but increased slightly when SDS concentration reached 8×10^{-3} mol/L. The lowest surface tension was 22.89 mN/m when SDS concentration reached 8×10^{-3} mol/L, so the critical micellar concentration (CMC) of SDS solution was 8×10^{-3} mol/L. The surface tension of Pb-BHA-SDS solution decreased more obviously with the increase of SDS concentration, and the surface tension did not decrease when SDS concentration reached 2×10^{-3} mol/L, and the lowest value was 14.19 mN/m, so the corresponding CMC value was 2×10^{-3} mol/L. The results show that SDS, as a typical surfactant, can significantly reduce the surface tension of the solution at a very low dosage, and the Pb-BHA-SDS formed in the PB-BHA structure has a larger molecular weight and colloidal particle size, and its adsorption on the gas–liquid interface has a more significant reduction on the surface tension than SDS. The decrease of CMC value indicates that Pb-BHA-SDS is more likely to form micelles in solution, indicating that the adsorption of Pb-BHA-SDS on the gas–liquid interface reaches saturation faster.

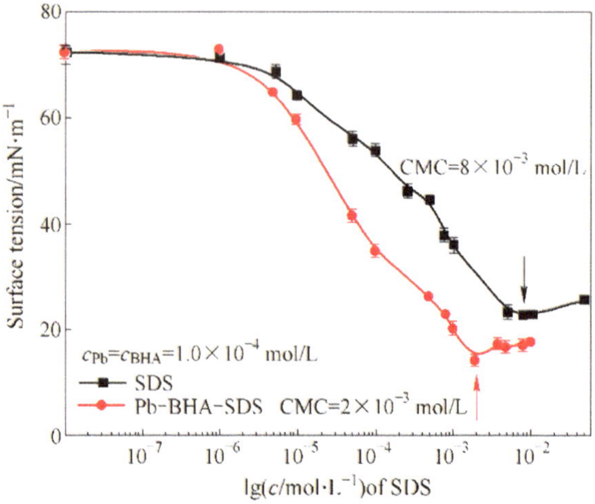

Fig. 5.69 Solution surface tension of SDS and PB-BHA-SDS

5.4.4.3 Mechanism of Pb-BHA-OHA Multiligand Collector in Scheelite Flotation

The behavior of Pb-BHA-OHA complex collector in the flotation of scheelite, calcite and fluorite, is shown in Fig. 5.70. The results show that, in the range of pH = 8–10, the flotation recovery of scheelite is higher, indicating that Pb-BHA- OHA has a strong ability to collect scheelite. The flotation recovery rate of fluorite is less than 20%, indicating that the collection ability of Pb-BHA-OHA complex for fluorite is poor. The flotation recovery of calcite is lower than scheelite on the whole. When the optimum flotation pH value of scheelite is 9.0, the flotation recovery rates of scheelite, calcite and fluorite are 90.10%, 69.30% and 17.70%, respectively. The flotation recovery rates of the three minerals are scheelite > calcite > fluorite, indicating that the Pb-BHA-OHA compound has good selectivity for the flotation of calcium-containing minerals. Single mineral flotation experiment showed that 1×10^{-5} mol/L OHA significantly improved the collection ability of Pb-BHA to scheelite, while maintaining good selectivity of the complex, indicating that the collector of Pb-BHA-OHA complex has a strong selective collection ability of scheelite flotation.

The scheelite zeta potential before and after Pb-BHA and Pb-BHA-OHA treatments is shown in Fig. 5.71a. The zeta potential of Pb-BHA-OHA colloids is positive in the pH range of 6–10, indicating that the Pb-BHA-OHA colloids are positively charged in a certain pH range. The zeta potential of scheelite after Pb-BHA-OHA treatment has a positive shift, indicating that Pb-BHA-OHA molecules adsorbed on scheelite surface, and electrostatic attraction may play an important role in the adsorption process. Compared with Pb-BHA, the zeta potential displacement value of scheelite treated by Pb-BHA-OHA is larger in the range of pH = 6–10, which indicates that the adsorption of Pb-BHA-OHA on scheelite surface is stronger than that of Pb-BHA. The contact angle of scheelite before and after being treated by Pb-BHA, OHA and Pb-BHA-OHA is shown in Fig. 5.71b. The contact angle of

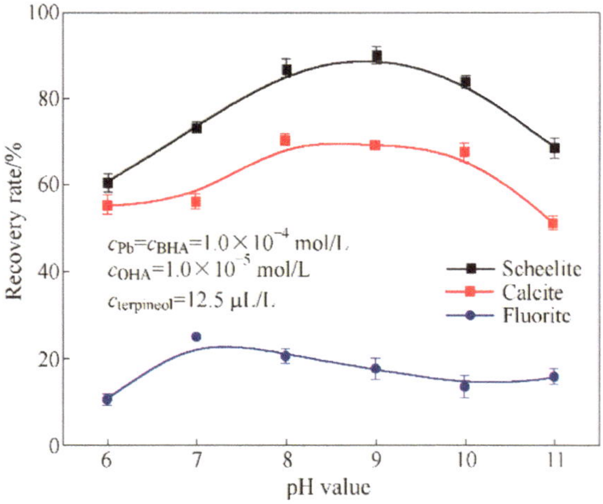

Fig. 5.70 Effect of pH on flotation of scheelite, calcite and fluorite by Pb-BHA-OHA complex

scheelite treated by Pb-BHA is 82.94°, while the contact angle of scheelite treated by Pb-BHA-OHA is 87.56°, indicating that the hydrophobicity of scheelite treated by Pb-BHA-OHA is enhanced most significantly. The solution surface tension of OHA and Pb-BHA-OHA is shown in Fig. 5.71c. The results showed that the surface tension of OHA solution decreased significantly with the increase of OHA concentration, and the CMC value was 3×10^{-2} mol/L. The surface tension of Pb-BHA-OHA solution decreased significantly with the increase of OHA concentration, and the CMC value was 2×10^{-2} mol/L. The results show that when OHA enters the Pb-BHA structure, the adsorption of Pb-BHA-OHA on the gas–liquid interface reduces the surface tension more significantly than that of OHA. The decrease of CMC value indicates that Pb-BHA-OHA is more likely to form micelles in solution, and its adsorption on the gas–liquid interface reaches saturation faster.

5.4.5 *Design and Assembly of Metal Ion-Inorganic/Organic Complex Inhibitors*

5.4.5.1 Metal-Silicate Polymer Inhibitors

The use of metal ions to regulate the assembly behavior of silicate colloids in water glass can strengthen the selectivity of water glass to different minerals, and achieve good application results in practice. The regulation of Fe^{2+} [140, 141], Pb^{2+} [142], Al^{3+} [49, 143], Co^{2+} [144] and other metal ions on the self-assembly of silicate colloidal particles has been reported in the literature, and the flotation separation of scheelite and calcite is realized by using metal ion-silicate polymer as calcite inhibitor (Fig. 5.72). The results showed that the addition of metal ions significantly enhanced the selectivity of water glass, and the modified water glass weakened the

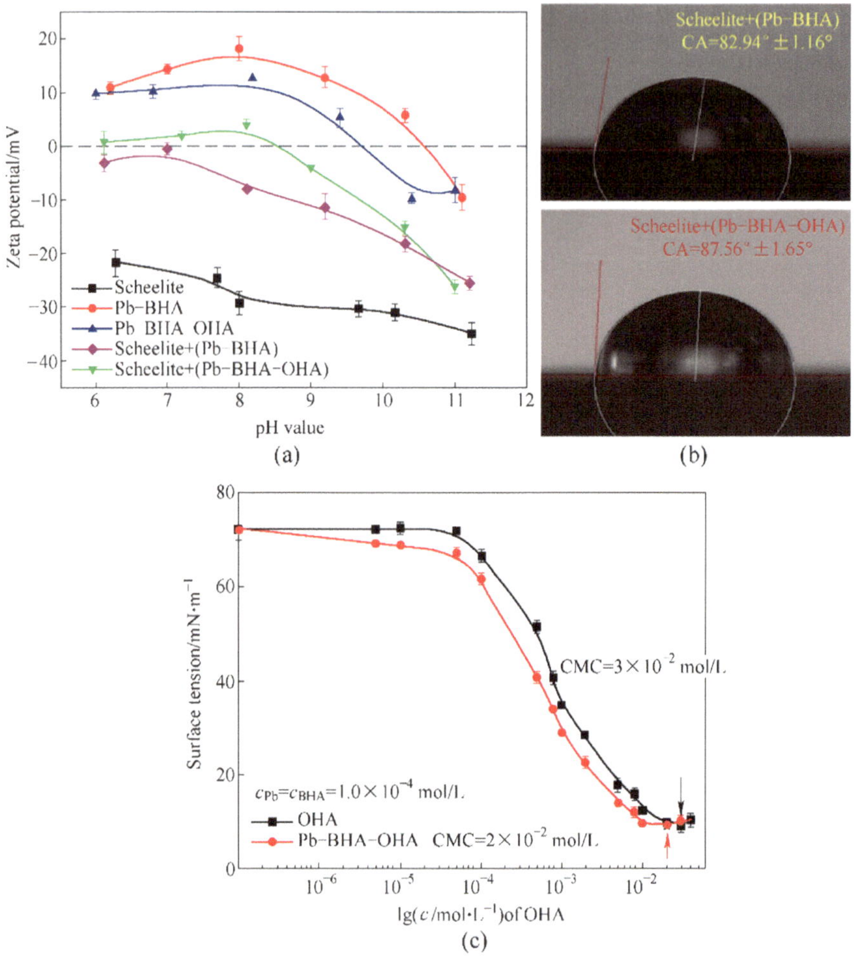

Fig. 5.71 **a** Potential of scheelite before and after being treated by Pb-BHA and Pb-BHA-OHA ($c_{Pb} = c_{BHA} = 1.0 \times 10^{-4}$ mol/L, $c_{OHA} = 1 \times 10^{-5}$ mol/L); **b** contact angle ($c_{Pb} = c_{OHA} = 1.0 \times 10^{-4}$ mol/L, $c_{OHA} = 1.0 \times 10^{-5}$ mol/L); **c** solution surface tension of OHA and Pb-BHA-OHA of scheelite before and after Pb-BHA, OHA and OHA treatment

inhibition effect on target minerals, but strengthened the inhibition effect on gangue minerals. The ratio of metal ions to water glass, the way of adding metal ions, and the pH value of solution can significantly affect the selectivity of the metal ion-silicic polymer to scheelite and calcite.

It is generally believed that there are two mechanisms for metal ions to regulate the assembly of silicic acid colloids: one is that the hydroxyl group in the water glass solution will interact with metal ions, thus promoting the hydrolysis of the system to produce a large number of silicic acid colloids [51]. This hydrolysis theory holds that the hydrolysis of metal ions in aqueous solution will increase the concentration of

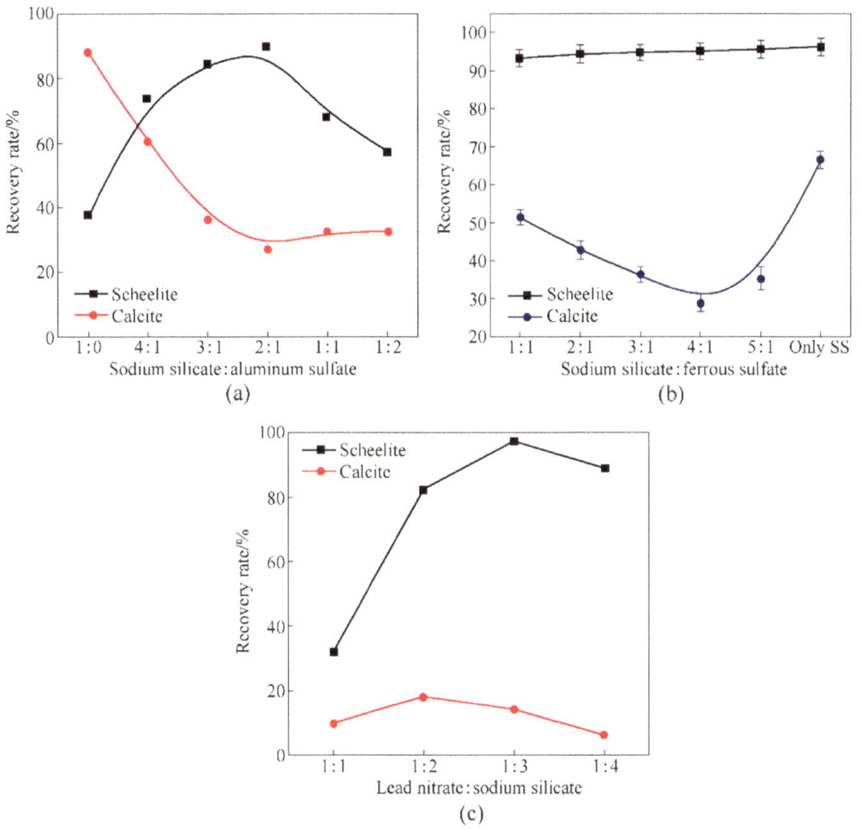

Fig. 5.72 Effect of metal ion to water glass ratio on the flotation of scheelite and calcite. **a** Al^{3+} [47], **b** Fe^{2+}, reprinted from Ref. [141], copyright 2018, with permission from Minerals Engineering, **c** Pb, reprinted from Ref. [142], copyright 2016, with permission from Separation Science and Technology

H^+ in the solution, which promotes the hydrolyzation of water glass in the direction of forming silicic acid colloidal particles, which have high activity and selectivity. Some scholars believe that water glass is a chemical reaction with metal ions to produce a composite silicate colloidal, this silicate colloidal micelle is larger, more hydroxyl groups on the surface, higher activity, the selective inhibition of adsorption on the surface of minerals is stronger [145]. This salt-forming theory believes that metal ions after adding water glass and water glass chemical reaction occurred, the formation of metal-silicate composite salt colloid, this composite silicate colloid selective adsorption ability is very strong, significantly enhance the selectivity of water glass.

In view of the second salting theory, some scholars have carried out research on the structure of metal-silicic acid complex salt. The structure$_3$ of Al–Na_2SiO_3 was studied by infrared spectroscopy, and the detection showed that Si–O–Al bond

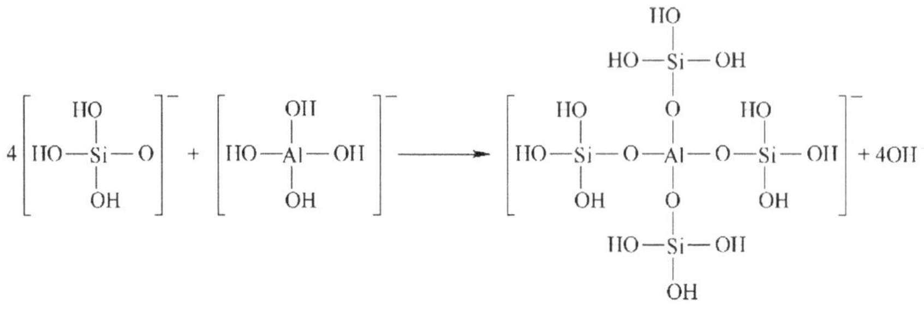

Fig. 5.73 Self-assembly reaction of $H_3SiO_4^-$ and $Al(OH)^-$ in solution [47]

appeared in the infrared map of salinized water glass, which proved that Al–Na_2SiO_3 is a composite aluminosilicate polymer [47, 143]. Solution chemistry studies suggest that the hydrolyzed products of aluminum may undergo dehydration polycondensation reaction with the hydroxyl group on the surface or structure of polysilicic acid to form a new composite aluminosilicate polymer. It is speculated that $H_3SiO_4^-$ and $Al(OH)^-$, both dominant components, can undergo dehydration polycondensation in solution according to the reaction shown in Fig. 5.73 under alkaline environment [47, 49]. The hydrolyzed components of metal ions in solution were assembled by dehydration polycondensation reaction to form a composite aluminosilicate polymer with Si–O–Al bond. However, some scholars believe that there is no conflict between the theory of hydrolysis and the theory of salting, and the formation of composite silicate colloid will also promote the formation of active silicate colloid in the system. Therefore, the modification mechanism of metal ions on water glass is the coexistence of two mechanisms, which should be judged comprehensively according to the reaction balance of the solution system [46].

According to the self-assembly mechanism of silicic acid colloidal particles controlled by metal ions, the team of Professor Hu Yuehua and Sun Wei from Central South University designed and developed Al–Na_2SiO_3 polymer inhibitor [47, 49, 107], which was applied to the flotation separation of scheelite and calcite, and achieved good separation effect in the flotation practice of Shizhuyuan and other typical tungsten ores. The combination of Al-Na_2SiO_3 polymer and Pb-BHA metal–organic complex collector can well realize the flotation separation of scheelite from calcite and fluorite. The adsorption model of Al–Na_2SiO_3 polymer inhibitor and Pb-BHA complex collector on scheelite and calcite surfaces is shown in Fig. 5.74. Scheelite is a negatively charged surface of oxygen-rich atoms, so strongly positively charged Pb-BHA complexes can more easily bind to the oxygen atoms on scheelite surface, while negatively charged Al–Na_2SiO_3 polymers can have a more difficult adsorption of calcium atoms on scheelite surface. Whereas calcite is the surface of positively charged calcium-rich atoms, Pb-BHA is more difficult to adsorb on the surface of calcite, whereas Al–Na_2SiO_3 polymer is easier to adsorb on the surface of calcite. Therefore, when Al–Na_2SiO_3 polymer is used as inhibitor and Pb-BHA complex is used as collector for scheelite flotation, the selectivity advantages of the

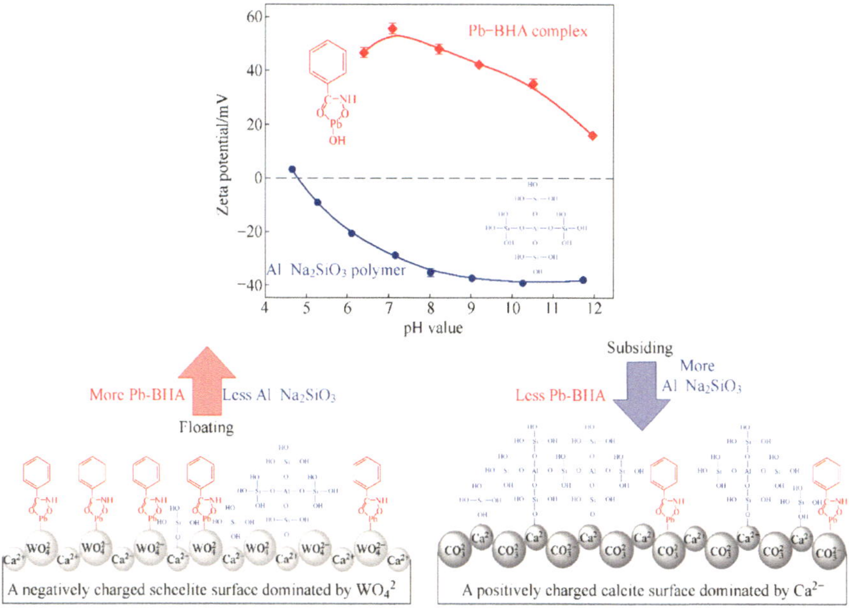

Fig. 5.74 Adsorption model of Al–NaSi2O polymer3 and Pb-BHA complexes on scheelite and calcite surfaces [47]

two will be mutually enhanced, reflecting synergistic selectivity, and thus realizing efficient flotation separation of scheelite and calcite.

5.4.5.2 Metal Ion-Organic Complex Inhibitors

The complexes formed by the assembly of metal ions and organic inhibitors also show significant advantages in terms of selectivity, especially the assembly of metal ions and macromolecular organic inhibitors. Dong et al. [146] studied the inhibition effect of flotation separation of schebellite and calcite by mixing iron ions (Fe^{3+}) with the organic chelating agent tartaric acid (TA). The results show that TA/Fe^{3+} inhibitor has good selectivity in the flotation separation of scheelite and calcite. Mechanism analysis shows that the mixed inhibitor TA/Fe^{3+} is preferentially adsorbed on the surface of calcite in the form of chelate FeL_2^-, and the pre-adsorption of iron ions provides a large number of adsorption sites for the adsorption of TA. The adsorption of TA/Fe^{3+} obviously hindered the adsorption of the collector sodium oleate. Wei et al. [147] studied the effect of the coordination assembly of Fe^{3+} and citric acid (CA) on calcite flotation. The results showed that single citric acid or Fe^{3+} had limited inhibitory effect on calcite in the flotation system of sodium oleate (NaOL), while the addition of Fe^{3+} greatly improved the inhibitory effect of CA on calcite. When pH is 8.5 and Fe^{3+}/CA mixed inhibitor concentration is 1×10^{-4} mol/L, the recovery

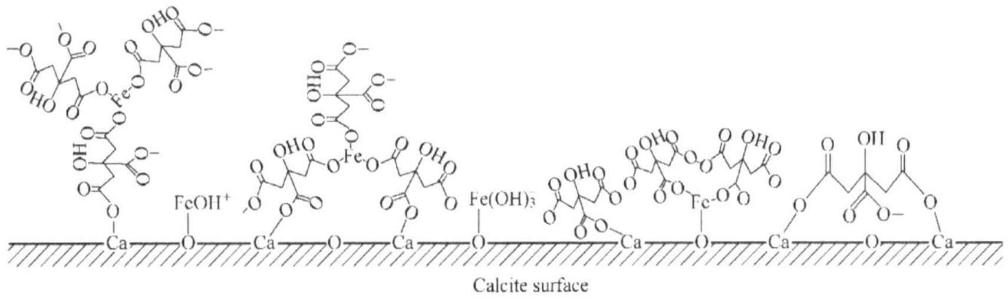

Fig. 5.75 Fe(III)-citric acid inhibitor adsorption model on calcite surface, reprinted from Ref. [147], copyright 2019, with permission from Colloids and surfaces. A, Physicochemical and engineering aspects

rate of calcite is only 3.9%. Mechanism analysis shows that Fe^{3+} in the pulp first reacts with CA and then adsorbs at the calcium atomic site of calcite. In addition, Fe^{3+} can be first adsorbed on the oxygen atomic site of calcite, and then citric acid can be further adsorbed on the Fe^{3+} site. Thus, Fe^{3+}/CA complex inhibitors can be adsorbed to both the calcium and oxygen atomic sites of calcite (Fig. 5.75).

The above studies show that it is feasible to design new flotation reagents through the assembly of metal ion coordination regulated molecules. The key lies in: establishing the matching relationship between mineral surface characteristics and metal–organic complex structure, realizing the directional regulation of metal–organic complex structure, and forming the coordination regulated molecular assembly theory suitable for flotation system. The assembly of metal ion coordination regulatory molecules in the field of coordination polymers will provide certain theoretical guidance for the development of new flotation reagents for metal–organic complexes. However, mineral flotation is different from chemical industry, medicine, materials and other fields. It is a complex system about solid/liquid/gas three-phase interface. The existing metal ion coordination regulatory molecular assembly theory and technology is difficult to be directly applied to the flotation system, and needs to be further improved and expanded, which is mainly reflected in the following aspects:

(1) The real component states of collector and metal ions in aqueous solution system in flotation system are controversial, their assembly behavior and assembly conditions are not clear, and it is difficult to achieve directional regulation;
(2) Different from the traditional anionic collector, the effective functional group of metal–organic complex collector is metal base, and its microscopic adsorption structure and adsorption mechanism at the solid/liquid interface are not perfect.
(3) The crystal chemistry of mineral and the structure of collector are the fundamental reasons that determine the difference of mineral floatability, and the matching relationship between them needs to be further studied.

The rapid development of high performance computing and modern analysis and detection technology has laid a good foundation for the solution of the above problems. Computer simulation, transmission electron microscopy (TEM) and atomic force microscopy (AFM) can directly observe the three-dimensional structure of interfacial aggregates, provide the microscopic behavior and physicochemical properties of aggregates at the molecular level, and help to reveal the self-assembly process of complex system and the interaction mechanism and law of each component. Flotation reagents for metal ion-organic complexes, as a new flotation reagent, show unique properties, but the related theories are not perfect and need to be further enriched and developed.

5.5 Summary

The difference of physical and chemical properties between tungsten ore and gangue minerals and the anisotropy of their surface properties determines the need for targeted design and assembly of flotation reagents, so that tungsten ore can be selectively separated from other minerals. This chapter summarizes and analyzes the design and assembly principle of flotation reagents for tungsten ore and reviews the research progress of the two most important types of agents for tungsten ore (collector and inhibitor) in detail. Among many tungsten collectors, fatty acid agents are commonly used as collectors, but their selectivity is poor, requiring a large amount of water glass to inhibit gangue minerals, and in calcite and fluorite and other calcium minerals content is high, need to use heating method to improve the grade, the process is more complicated, the recovery of associated fluorite, cassiterite and other minerals will also be affected. Metal–organic complex collector has better selectivity for tungsten ore, and has little effect on the subsequent flotation of fluorite, cassiterite and other minerals, so it is gradually replacing the traditional fatty acid method. Although there are many new inhibitors, especially organic inhibitors have been widely reported, but water glass and its modifiers are still indispensable in practical application. How to combine these new inhibitors with water glass and its modifiers to adapt to different ore types is of great significance.

Acknowledgements This work was supported by National key research center and development Program of the 14th Five-Year Plan (No. 2022YFC2905105 and No. 2022YFC2905104), National Natural Science Foundation of China (NSFC) (No. 52122406 and No. 52374291), Hunan High-tech Industry Technology Innovation Leading plan (No. 2022GK4056), Hunan Innovative Province construction special Project (No. 2020RC3001), and Hunan Postgraduate Research and Innovation Project (No. CX20220200).

References

1. Hu Y (2014) Mineral ftation. Central South University Press
2. Rudolph M, Hartmann R (2017) Specific surface free energy component distributions and flotabilities of mineral microparticles in flotation-an inverse gas chromatography study. Colloids Surf A Physicochemical Eng Aspects 513:380–388

3. Kupka N, Rudolph M (2018) Froth flotation of scheelite—a review. Int J Min Sci Technol 28(3):373–384
4. Yang Y (2010) Mixing effect of fatty acid collector in scheelite flotation process
5. Li T, Qiu X, Zhou X (2018) Research Status of scheelite flotation reagents. Mater Res Appl 12(01):8–12
6. Sun H (2015) Study on collecting performance and action mechanism of new collector for tungsten ore
7. Gao Y, Qiu X, Zhong C et al (2012) Properties of hydroxamic acid collectors and flotation characteristics of tungsten ore. China Tungsten Ind 27(02):10–14
8. Gao Y, Qiu X, Feng Q (2003) Study on chemistry of flotation solution of scheelite captured by benzohydroxamic acid. Nonferrous Met (Miner Process Sect) 04:28–31
9. Gao Y, Qiu X, Han Z (2015) Flotation Mechanism of scheelite by hydroxamic acid. Chin J Nonferrous Met 25(05):1339–1344
10. Gao Y, Qiu X, Xia Q et al (2001) Study on the mechanism of interaction between benzohydroxamic acid and wolframite. Guangdong J Nonferrous Met 02:92–95
11. Ye Z (2000) Study on the collecting mechanism of wolframite by benzohydroxamic acid. Nonferrous Met (Miner Process Sect) 05:35–39
12. Yang X (2018) Beneficiation studies of tungsten ores—a review. Miner Eng 125:111–119
13. Wang L (2019) Preparation of new hydroxamic acid collector and flotation characteristics of scheelite and wolframite
14. Deng L, Zhao G, Zhong H et al (2016) Investigation on the selectivity of N-((hydroxyamino)-alkyl) alkylamide surfactants for scheelite/calcite flotation separation. J Ind Eng Chem 33:131–141
15. Zhou Y, Wu Y (2013) Research status of scheelite flotation. China Tungsten Ind 28(01):19–24
16. Wang Q, Zhang X, Li K et al (2015) Research progress of flotation reagents for scheelite. China Tungsten Ind 30(06):21–27
17. Liu L, Fan Z, He W (2012) Study on phosphonic acid compounds. Nonferrous Met (Miner Process) 02:67–70
18. Hu W, Xinyang Y (2013) Research status of flotation of micro-fine wolframite. Nonferrous Met Sci Eng 4(05):102–107
19. Institute JNMR (1976) Study on flotation of wolframite from tungsten slime by styrene phosphonic acid. Nonferrous Met (Smelting Sect) 02:45–55
20. Lu Y, Lin Q, Wang D (1993) A new reagent for flotation separation of fluorite scheelite garnet—LP series collector. Nonferrous Met Min Metall 01:20–25
21. Xiao Y, Zhu Y (1987) Effect mechanism of FXL-14 on wolframite harvesting. Min Metall Eng 02:33–36
22. Bogdanov OS (1974) Hydroxamic acids as collectors in the flotation of wolframite, cassiterite and pyrochlore. Am Chem Soc 553–564
23. Srinivas K, Sreenivas T, Padmanabhan NPH et al (2004) Studies on the application of alkyl phosphoric acid ester in the flotation of wolframite 25(04):253–267
24. Bhagat RP, Pathak PN (1996) The effect of polymeric dispersant on magnetic separation of tungsten ore slimes. Int J Miner Process 47(3):213–217
25. Ozcan O, Bulutcu AN (1993) Electrokinetic, infrared and flotation studies of scheelite and calcite with oxine, alkyl oxine, oleoyl sarcosine and quebracho. Int J Miner Process 39(3):275–290
26. Gao Z, Sun W, Hu Y et al (2013) Surface energies and appearances of commonly exposed surfaces of scheelite crystal. Trans Nonferrous Met Soc China 23(7):2147–2152
27. Marinakis KI, Kelsall GH (1987) The surface chemical properties of scheelite (CaWO4) II. Collector adsorption and recovery of fine scheelite particles at the iso-octane/water interface. Colloids Surf 26:243–255
28. Atademir MR, Kitchener JA, Shergold HL (1979) The surface chemistry and flotation of scheelite. I. Solubility and surface characteristics of precipitated calcium tungstate. J Colloid Interface Sci 71(3):466–476

29. Arnold R, Warren LJ (1974) Electrokinetic properties of scheelite. J Colloid Interface Sci 47(1):134–144
30. Hyung SC, Kook NH (1963) Electrokinetic Property and flotation characteristics of scheelite. J Korean Chem Soc 7(01):17–24
31. Gao Z, Sun W, Hu Y (2015) New insights into the dodecylamine adsorption on scheelite and calcite: an adsorption model. Miner Eng 79:54–61
32. Hiçyìlmaz C, Atalay Ü, Özbayoglu G (1993) Selective flotation of scheelite using amines. Minera Eng 6(3):313–320
33. Yang F (2013) Application of quaternary ammonium collector in scheelite flotation and research on its mechanism of action
34. Yang F, Sun W, Hu Y et al (2015) Cationic flotation of scheelite from calcite using quaternary ammonium salts as collector: adsorption behavior and mechanism. Miner Eng 81:18–28
35. Hu Y, Yang F, Sun W (2011) The flotation separation of scheelite from calcite using a quaternary ammonium salt as collector. Miner Eng 24(1):82–84
36. Wang D (1986) Theory and practice of mineral flotation and flotation reagent. Central South University of Technology Press, Chang Sha
37. Fang H, Ai G, Liu Y (2016) Research status and development trend of scheelite beneficiation process. China Tungsten Ind 31(03):27–31
38. Hu Y, Wang D (1989) Study on flotation of fluorite, barite and scheelite by a new amphoteric collector. Nonferrous Met (Miner Process Sect) 04:10–3+4
39. Hu YH, Xu ZH (2003) Interactions of amphoteric amino phosphoric acids with calcium-containing minerals and selective flotation. Int J Miner Process 72(1–4):87–94
40. Xu H, Li W, Chen W (2019) Research status of flotation collector for tungsten ore and preparation and industrial application of new reagents. China Tungsten Ind 34(01):37–44+57
41. Xu S, Meng S, Guo D (1986) Flotation of scheelite by amphoteric collector. Nonferrous Met (Miner Process Sect) 04:24–28
42. Li S, Li Q (2019) Research status of scheelite flotation. Compr Utilization Miner Resour 03:17–21
43. Li W (2019) Research progress of flotation reagents for scheelite. World Nonferrous Met 14:245–247
44. Zhu C, Feng Q, Zhu J (1990) Effect of pH regulator on separation of scheelite from calcite and fluorite. Min Metall Eng 01:19–23
45. Kupka N, Rudolph M (2018) Role of sodium carbonate in scheelite flotation—a multi-faceted reagents. Miner Eng 129:120–128
46. Sun W, Hu Y, Qin W et al (2000) Research progress of flotation reagents for tungsten ore. Prot Utilization Miner Resour 03:42–46
47. Wei Z, Hu Y, Han H et al (2019) Selective separation of scheelite from calcite by self-assembly of H2SiO3 polymer using Al3+ in Pb-BHA flotation. Minerals 9(1)
48. Han H, Liu W, Hu Y et al (2017) A novel flotation scheme: selective flotation of tungsten minerals from calcium minerals using Pb-BHA complexes in Shizhuyuan. Rare Met 36(6):533–540
49. Wei Z, Hu Y, Han H et al (2018) Selective flotation of scheelite from calcite using Al-Na2SiO3 polymer as depressant and Pb-BHA complexes as collector. Miner Eng 120:29–34
50. Chen H (1998) Study on the inhibition behavior of calcite flotation by several inorganic and organic inhibitors. Sichuan Nonferrous Met 02:42–45
51. Zhu J, Zhu Y (1996) Chemical principle of flotation reagents (Revised edition). Central South University of Technology Press
52. Changgen L, Yongxin L (1983) Selective flotation of scheelite from calcium minerals with sodium oleate as a collector and phosphates as modifiers. II. The mechanism of the interaction between phosphate modifiers and minerals. Int J Miner Process 10(3):219–235
53. Yu Y, Sun C, Lu S (2013) Study on floatibility of scheelite and calcium-bearing minerals and crystal chemical analysis. J China Univ Min Technol 42(02):278–283+313
54. Ai G, Liu C, Zhang W (2018) Utilization of sodium humate as selective depressants for calcite on the flotation of scheelite. Sep Sci Technol 53(13):2136–2143

55. Chen C, Hu Y, Zhu H et al (2019) Inhibition performance and adsorption of polycarboxylic acid in calcite flotation. Miner Eng 133:60–68
56. Chen Wei FQ, Zhang G, Yang Q et al (2017) The effect of sodium alginate on the flotation separation of scheelite from calcite and fluorite. Miner Eng 113:1–7
57. Dong L, Jiao F, Qin W et al (2019) Selective flotation of scheelite from calcite using xanthan gum as depressant. Miner Eng 138:14–23
58. Wang J, Bai J, Yin W et al (2018) Flotation separation of scheelite from calcite using carboxyl methyl cellulose as depressant. Miner Eng 127:329–333
59. Chen Wei FQ, Zhang G, Liu D et al (2018) Selective flotation of scheelite from calcite using calcium lignosulphonate as depressant. Miner Eng 119:73–75
60. Chen W, Feng Q, Zhang G et al (2017) The flotation separation of scheelite from calcite and fluorite using dextran sulfate sodium as depressant. Int J Miner Process 169:53–59
61. Zhang C, Wei S, Hu Y et al (2018) Selective adsorption of tannic acid on calcite and implications for separation of fluorite minerals. J Colloid Interface Sci 512:55–63
62. Chen W, Feng Q, Zhang G et al (2018) Investigations on flotation separation of scheelite from calcite by using a novel depressant: sodium phytate. Miner Eng 126:116–122
63. Hu Y, Sun W, Jiang Y et al (1998) Study on Inhibitory effect and mechanism of citric acid in flotation separation of scheelite fluorite. Met Ore Dressing Abroad 05:27–29
64. Chen C, Hu Y, Zhu H et al (2019) Inhibition performance and adsorption of polycarboxylic acids in calcite flotation. Miner Eng 133:60–68
65. Gao Z, Deng J, Sun W et al (2022) Selective flotation of scheelite from calcite using a novel reagent scheme. Miner Process Extr Metall Rev 43(2):137–149
66. Song S (2014) Study on flotation inhibition and mechanism of natural gum on three typical calcium-containing minerals
67. Zheng X, Smith RW (1997) Dolomite depressants in the flotation of apatite and collophane from dolomite. Miner Eng 10(5):537–545
68. Anziferova CA, Li C, Cui H (2007) Effects of Humic acid agents on fluorite, calcite and quartz under flotation conditions. Miner Process Met Ores Abroad 08:38–41
69. Xiong L (2013) Experimental study on gangue inhibitors containing calcium in scheelite flotation. Jiangxi University of Science and Technology
70. Guo W (2018) Effect and mechanism of polymer inhibitor on separation of scheelite from calcite. Jiangxi University of Science and Technology
71. Zhu H, Qin W, Chen C et al (2018) Flotation separation of fluorite from calcite using polyaspartate as depressant. Miner Eng 120:80–86
72. Foucaud Y, Filippova IV, Filippov LO (2019) Investigation of the depressants involved in the selective flotation of scheelite from apatite, fluorite, and calcium silicates: focus on the sodium silicate/sodium carbonate system. Powder Technol 352:501–512
73. Hu W (2013) Study on mechanism and application of combined collector flotation for fine wolframite. Jiangxi University of Science and Technology
74. Miller JD, Wadsworth ME, Mistra M et al (1984) Flotation chemistry of the fluorite/oleate system. Principles of mineral flotation: the Wark symposium, pp 31–42
75. Fuerstenau MCMJD, Pray RE et al (1965) Metal ion activation in xanthate flotation of quartz. Trans AIME 232:359–364
76. Kang J, Hu Y, Sun W et al (2019) Utilization of sodium hexametaphosphate for separating scheelite from calcite and fluorite using an anionic-nonionic collector. Minerals 9(11)
77. Xu Longhua TJ, Wu H, Lu Z et al (2017) Effect of Pb²⁺ ions on ilmenite flotation and adsorption of benzohydroxamic acid as a collector. Appl Surf Sci 425:796–802
78. Jiang Q, Dai Z (2012) Effect of mixed fatty acids on separation of schetungsten from fluorite and calcite. Min Metall Eng 32(02):42–44+8
79. Jianping. H (2013) Role of hydroxamic acid collector in the flotation of scheelite and wolframite
80. Gao Zhiyong BD, Sun W, Cao X et al (2015) Selective flotation of scheelite from calcite and fluorite using a collector mixture. Miner Eng 72:23–26

81. Dong L, Qin W, Jiao F et al (2018) Separation of scheelite and calcite by flotation with cationic-anionic collector. Min Metall Eng 38(04):61–64
82. Sreenivas T, Padmanabhan NPH (2002) Surface chemistry and flotation of cassiterite with alkyl hydroxamates. Colloids Surf A Physicochemical Eng Aspects 205(1–2):47–59
83. Wang P, Qin W, Ren L et al (2013) Solution chemistry and utilization of alkyl hydroxamic acid in flotation of fine cassiterite. Trans Nonferrous Met Soc China 23(6):1789–1796
84. Zhu H, Qin W, Chen C et al (2016) Study on the low temperature harvesting performance of scheelite and its application. Chin J Non-Ferrous Met 26(10):2188–2196
85. Wang L, Cao X, Song X et al (2012) Study on synergistic effect of sulfonates and nonionic surfactants. Oilfield Chem 29(03):312–316+30
86. Xu L, Tian J, Wu H et al (2017) The flotation and adsorption of mixed collectors on oxide and silicate minerals. Adv Colloid Interface Sci 250:1–14
87. Jiang H, Gao Y, Khoso SA et al (2019) A new approach for characterization of hydrophobization mechanisms of surfactants on muscovite surface. Sep Purif Technol 209:936–945
88. Wang L (2012) Study on flotation separation and adsorption mechanism of mica minerals and quartz. Central South University
89. Wang L, Sun W, Hu Y et al (2014) Adsorption mechanism of mixed anionic/cationic collectors in Muscovite—quartz flotation system. Miner Eng 64:44–50
90. Wang L, Hu Y, Liu J et al (2015) Flotation and adsorption of muscovite using mixed cationic-nonionic surfactants as collector. Powder Technol 276:26–33
91. Wang L, Hu Y, Liu R et al (2017) Synergistic adsorption of DDA/alcohol mixtures at the air/water interface: a molecular dynamics simulation. J Mol Liq 243:1–8
92. Wang Y, Bai Y, Hou B et al (2018) Mechanism of cationic/nonionic composite surfactants changing the surface wettability of oil-wet sandstone. J China Univ Petrol (Ed Nat Sci) 42(02):165–171
93. Yang Q, Jiang H, Ji W et al (2018) Flotation mechanism of mica with quaternary ammonium salt and octanol collector. Chin J Nonferrous Met 28(09):1900–1907
94. Nguyen KT, Nguyen TD, Nguyen AV (2014) Strong cooperative effect of oppositely charged surfactant mixtures on their adsorption and packing at the air-water interface and interfacial water structure. Langmuir 30(24):7047–7051
95. Fuerstenau MC, Rice DA, Somasundaran P et al (1965) Metal ion hydrolysis and surface charge in beryl flotation. Inst Min Metall Trans 74:381–391
96. Liu B, Wang X, Du H et al (2016) The surface features of lead activation in amyl xanthate flotation of quartz. Int J Miner Process 151:33–39
97. James RO, Healy TW (1972) Adsorption of hydrolyzable metal ions at the oxide—water interface. I. Co(II) adsorption on SiO2 and TiO2 as model systems. J Colloid Interface Sci 40(1):42–52
98. Hu Y, Wang D (1987) Mechanism of adsorption and flotation activation of metal ions at oxide mineral/water interface. J Cent S Univ Min Metall 05:501–508+90–91
99. Zhao W, Wei S, Haisheng H et al (2021) Improving the flotation efficiency of Pb-BHA complexes using an electron-donating group. Chem Eng Sci 234
100. Tian M, Zhang C, Han H et al (2018) Effects of the preassembly of benzohydroxamic acid with Fe (III) ions on its adsorption on cassiterite surface. Miner Eng 127:32–41
101. Free ML, Miller JD (1996) The significance of collector colloid adsorption phenomena in the fluorite/oleate flotation system as revealed by FTIR/IRS and solution chemistry analysis. Int J Miner Process 48(3):197–216
102. Fa KQ, Tao JA, Nalaskowski J et al (2003) Interaction forces between a calcium dioleate sphere and calcite/fluorite surfaces and their significance in flotation. Langmuir 19(25):10523–10530
103. Fa KQ, Nguyen AV, Miller JD (2005) Hydrophobic attraction as revealed by AFM force measurements and molecular dynamics simulation. J Phys Chem B 109(27):13112–13118
104. Fa K, Nguyen AV, Miller JD (2006) Interaction of calcium dioleate collector colloids with calcite and fluorite surfaces as revealed by AFM force measurements and molecular dynamics simulation. Int J Miner Process 81(3):166–177

105. Sun W, Han H, Sun W et al (2022) Novel insights into the role of colloidal calcium dioleate in the flotation of calcium minerals. Miner Eng 175
106. Wei Z, Han H, Hu Y et al (2017) Study on the flotation of Pb-BHA coordination collector mixed with scheelite and wolframite at room temperature. Nonferrous Met Eng 7(06):70–75
107. Han Haisheng HY, Sun W et al (2017) Fatty acid flotation versus BHA flotation of tungsten minerals and their performance in flotation practice. Int J Miner Process 159:22–29
108. He J, Han H, Zhang C et al (2018) New insights into the configurations of Lead(II)-benzohydroxamic acid coordination compounds in aqueous solution: a combined experimental and computational study. Minerals 8(9)
109. Tian Mengjie GZ, Sun W et al (2018) Activation role of lead ions in benzohydroxamic acid flotation of oxide minerals: new perspective and new practice. J Colloid Interface Sci 529:150–160
110. Tian Mengjie ZC, Han H et al (2018) Novel insights into adsorption mechanism of benzohydroxamic acid on lead (II)-activated cassiterite surface: an integrated experimental and computational study. Miner Eng 122:327–338
111. Hu W, Tian M, Cao J et al (2020) Probing the interaction mechanism between benzohydroxamic acid and mineral surface in the presence of Pb²⁺ ions by AFM force measurements and first-principles calculations. Langmuir 36(28):8199–8208
112. Hu Y, Han H, Tian M et al (2018) Action mechanism and application of lead benzohydroxamic acid metal-organic complex in flotation of oxidized ore. Conserv Utilization Miner Resour 01:42–47+53
113. Wei Z, Sun W, Wang P et al (2021) A novel metal–organic complex surfactant for high-efficiency mineral flotation. Chem Eng J 426:130853
114. He J, Han H, Zhang C et al (2018) Novel insights into the surface microstructures of lead(II) benzohydroxamic on oxide mineral. Appl Surf Sci 458:405–412
115. Zhong X, Yu S, Sun Y et al (2018) Research progress of organometallic framework compounds. J Chem Eng 39(05):53–58
116. Chen X, Cai J (2007) Principle and practice of single crystal structure analysis, 2nd edn
117. Wei Z, Sun W, Wang P et al (2023) The structure analysis of metal-organic complex collector: from single crystal, liquid phase, to solid/liquid interface. J Mol Liq 382
118. Tian M, Gao Z, Khoso SA et al (2019) Understanding the activation mechanism of Pb2+ ion in benzohydroxamic acid flotation of spodumene: experimental findings and DFT simulations. Miner Eng 143
119. Han H, Hu Y, Sun W et al (2018) Novel catalysis catalysis mechanisms of benzohydroxamic acid adsorption by lead ions and changes in the-surface of scheelite particles. Miner Eng 119:11–22
120. Han H, Xiao Y, Hu Y et al (2020) Replacing Petrov's process with atmospheric flotation using Pb-BHA complexes for separating scheelite from fluorite. Miner Eng 145:1–10
121. Cao M, Gao Y, Bu H et al (2019) Study on the mechanism and application of rutile flotation with benzohydroxamic acid. Miner Eng 134:275–280
122. Fang S, Xu L, Wu H et al (2019) Comparative studies of flotation and adsorption of Pb(II)/benzohydroxamic acid collector complexes on ilmenite and titanaugite. Powder Technol 345:35–42
123. Tian M, Hu Y, Sun W et al (2017) Study on the mechanism and application of a novel collector-complexes in cassiterite flotation. Colloids Surf A Physicochemical Eng Aspects 522:635–641
124. Yue T, Han H, Hu Y et al (2018) Beneficiation and purification of tungsten and cassiterite minerals using Pb-BHA complexes flotation and centrifugal separation. Minerals 8(12)
125. Zhao G, Zhong H, Qiu X et al (2013) The DFT study of cyclohexyl hydroxamic acid as a collector in scheelite flotation. Miner Eng 49:54–60
126. Jiang Y, Yin Z, Yi Y et al (2010) Synthesis and collecting properties of novel carboxyl hydroxamic acids for diaspore and aluminosilicate minerals. Miner Eng 23(10):830–832
127. Jiang Y, Zhao B, Zhou X et al (2010) Flotation of diaspore and aluminosilicate minerals applying novel carboxyl hydroxamic acids as collector. Hydrometallurgy 104(1):112–118

128. Jiang Y, Hu Y, Cao X (2001) Synthesis and structure-activity relationships of carboxyl hydroxidoxime in bauxite flotation. 11(04):706–709
129. Wei Z, Sun W, Han H et al (2020) Enhanced electronic effect improves the collecting efficiency of benzohydroxamic acid for scheelite flotation. Miner Eng 152
130. Ji Z, Li T, Yaghi OM (2020) Sequencing of metals in multivariate metal-organic frameworks. Science 369(6504):674–680
131. Gong X, Gnanasekaran K, Chen Z et al (2020) Insights into the structure and dynamics of metal-organic frameworks via transmission electron microscopy. J Am Chem Soc 142(41):17224–17235
132. Deng H, Doonan CJ, Furukawa H et al (2010) Multiple functional groups of varying ratios in metal-organic frameworks. Science 327(5967):846–850
133. Wang D, Lin Q, Jiang Y (1996) Mineral processing and molecular design of metallurgical reagents. Central South University of Technology Press
134. Zeng J (2012) First-principles study on flotation separation and process of Cu-Mo sulfide ore. Central South University
135. Tan X (2017) Targeting molecular design and mechanism of tungsten-tin mineral chelating collector. Northeastern University
136. Liu C, Zhu Y, Wu G et al (2018) First-principles study on electronic structure of cassitite surface and activation mechanism of lead. Conserv Utilization Miner Resour 03:17–21
137. Dai J, Huang P, Huang G et al (2023) Study on adsorption behavior of Gemini quaternary ammonium salt at the interface between fluorapatite and quartz. Compr Utilization Miner Resour 02:87–93
138. Wei Z, Sun W, Han H et al (2024) Molecular design of multiple ligand metal-organic framework (ML-MOF) collectors for efficient flotation separation of minerals. Sep Purif Technol 328
139. Wei Z, Sun W, Han H et al (2021) Probing a colloidal lead-group multiple ligand collector and its adsorption on a mineral surface. Miner Eng 160
140. Oliveira JF, Sampaio JA (1988) Development studies for the recovery of Brazilian scheelite fines by froth flotation//Plumpton AJ. Production and processing of fine particles. Amsterdam, Pergamon, pp 209–217
141. Deng R, Yang X, Hu Y et al (2018) Effect of Fe(II) as assistant depressant on flotation separation of scheelite from calcite. Miner Eng 118:133–140
142. Feng B, Guo W, Xu H et al (2016) The combined effect of lead ion and water glass in the flotation separation of scheelite from calcite. Sep Sci Technol 52(3):567–573
143. Hu Y, Cai D (1991) Effect of salinated water glass on the separation of gangue minerals from micrograined rhodochromite and illite. China Manganese Ind 05:30–35
144. Wang D, Hu YH (1988) Solution chemistry of flotation
145. Chen C (2011) Effect and mechanism of Inorganic anions on flotation behavior of three typical calcium-containing salts
146. Dong L, Jiao F, Qin W et al (2021) Utilization of iron ions to improve the depressive efficiency of tartaric acid on the flotation separation of scheelite from calcite. Miner Eng 168
147. Wei Q, Dong L, Jiao F et al (2019) Use of citric acid and Fe(III) mixture as depressant in calcite flotation. Colloids Surf A 578:123579

Chapter 6
Technology and Practice of Tungsten Flotation

Haisheng Han, Zhao Wei, and Wei Sun

Abstract The development of tungsten flotation technology has been significantly influenced by the evolution of flotation reagents, transitioning through various stages such as fatty acid flotation, chelating agent flotation, and metal–organic complex flotation. This chapter summarizes the progress and application of tungsten flotation technology, with a focus on the three main flotation technologies: fatty acid flotation, chelating collector flotation, and metal–organic complex flotation. The fatty acid method, which includes both normal temperature flotation and "Petrov" heating flotation, is widely used due to its strong collecting ability and adaptability to ore properties. Chelating agent flotation, primarily using hydroxamic acid, cupferron chelating agents, has been extensively applied in the flotation of wolframite and mixed ores of scheelite and wolframite. The metal–organic complex flotation method, utilizing a highly selective collector, has emerged as a novel technology capable of achieving normal temperature separation of complex tungsten resources, offering a new solution for efficient flotation separation of complex tungsten ores. The chapter also discusses the application of these technologies in various tungsten mines, highlighting the advantages and challenges associated with each method. The fatty acid method, despite its adaptability and low reagent cost, has a complex process and high energy consumption. The chelating agent method can achieve high-grade concentrates but requires heating and has high reagent costs. The metal–organic complex method simplifies the process and improves tungsten recovery, offering a promising direction for the future of tungsten flotation technology.

Keywords Tungsten ore · Flotation process · Fatty acid · Chelating agent · Metal–organic complex

H. Han (✉) · Z. Wei · W. Sun
Central South University, Changsha, China

Engineering Research Center of Ministry of Education for Carbon Emission Reduction in Metal Resource Exploitation and Utilization, Central South University, Changsha, China

State Key Laboratory of Mineral Processing, Changsha, China

H. Han
Technology Research Center of Hunan Province for Comprehensive Utilization of Associated Fluorite and Fluorine Chemical Engineering, Chenzhou, China

W. Sun et al., *Flotation Chemistry of Tungsten Minerals and Its Application*,
https://doi.org/10.1007/978-981-96-5077-4_6

6.1 Introduction

The development of tungsten flotation technology is often based on the development of flotation reagents, which has gone through several typical stages, such as fatty acid flotation technology, chelating agent flotation technology, as well as metal–organic complex flotation technology. This chapter summarizes the progress and application practice of tungsten flotation technology, focusing on the fatty acid flotation technology, chelating collector flotation technology and metal–organic complex flotation technology of three types of main flotation technology. Among them, fatty acid method includes normal temperature flotation and "Petrov" heating flotation. Scheelite has strong collecting ability and good adaptability to ore properties. It is the most widely used in scheelite flotation. Chelating agent flotation process is a flotation process with hydroxamic acid, cupric and other chelating agents as the main collector, which has been widely used in the flotation of wolframite and mixed ores of scheelite and wolframite. In recent years, the rise of the complex method is a metal–organic complex as a highly selective collector process, with good selectivity, can realize the normal temperature flotation separation of complex tungsten resources. The novel technology has been widely used for the scheelite, wolframite and mixed ores of scheelite and wolframite, which provides a new solution for efficient flotation separation of complex tungsten ore. It has become one of the important development directions of tungsten flotation technology.

6.2 Tungsten Flotation Technology and Its Development Trend

At present, most scheelite deposits found in China are skarn type tungsten deposits, and flotation is a classic mineral processing method for processing such scheelite. The characteristics of this kind of ore deposit are as follows: WO_3 grade is low, fine disseminated grain size, fine mesh or disseminated structure; scheelite is associated with a variety of useful metals, gangue minerals are similar to scheelite floating calcium salt minerals (e.g., fluorite, calcite). According to the mineral composition of tungsten, scheelite can be divided into two categories: scheelite—quartz type (or silicate mineral), scheelite—calcite, fluorite (barite) type. Due to the relatively large difference in floatability between scheelite and silicate minerals, for scheelite quartz type ores with simple composition, the flotation separation of scheelite and gangue minerals can be realized by using fatty acid as collector and water glass as inhibitor [1]. But the latter, due to scheelite, calcite and fluorite and other similar surface active particles and surface mutual transformation, conventional process is difficult to scheelite and calcium gangue minerals flotation separation, therefore, scheelite and calcium gangue minerals efficient flotation separation is a worldwide problem, the efficient comprehensive utilization of this kind of resources facing great challenges. For the mixed ores of scheelite and wolframite, due to the relatively poor floatability of wolframite, the collector of fatty acids that can better flotation wolframite has a weak collecting ability of wolframite, while the collector of chelate that can better flotation wolframite has a weak collecting ability of wolframite, which makes the mixed ores of scheelite and wolframite flotation also face a big technical problem [2, 3]. The efficient flotation of tungsten mineral resources depends on the progress of new flotation theories, new reagents and new technologies. In recent years, tungsten flotation chemistry has been deeply developed. Research on crystal chemistry and solution chemistry has promoted the development of tungsten flotation reagents and new processes towards the direction of high selectivity, clean and efficient. The application of new technologies in practice has significantly improved the level of tungsten beneficiation.

The core of the development of tungsten flotation technology is the continuous progress of flotation reagents. The first anionic collector (such as fatty acid) used in scheelite flotation has strong collecting ability and adaptability, but its selectivity is relatively poor. Chelate collectors have good selectivity, but their ability to collect is relatively weak, and metal ion activators are often needed to improve their flotation efficiency. Although cationic collector for scheelite and calcium bearing gangue minerals have good selectivity, but in practice to silicate and other gangue minerals also show a strong collection capacity, which makes cationic collector has not been widely used in production practice. Amphoteric collector has a strong ability to collect scheelite and pH adaptability, but the research of amphoteric collector is still in the laboratory stage. Metal–organic complex collector has strong collecting ability and good selectivity, and has been applied in practice, which will become one of the important development directions of tungsten flotation collector in the future. Based on the development of flotation agents, the flotation process of tungsten ore has roughly experienced several stages, such as fatty acid flotation process, chelating agent flotation process and metal–organic complex flotation process (Fig. 6.1). The fatty acid flotation process takes fatty acid as the main collector, including fatty acid roughing at room temperature—"Petrov" heating and concentration process and full room temperature flotation process. The former has strong adaptability to ore and is widely used in most scheelite at home and abroad. Chelating agent flotation process is based on hydroxamic acid, copper and iron reagent chelating agent as the main collector process, in wolframite and mixed flotation of scheelite and wolframite flotation has a wide range of applications. Metal–organic complex flotation process is a process in which metal–organic complex is a highly selective collector. It has good selectivity, strong adaptability, short flow and high concentration ratio, and can realize the normal temperature short flow flotation of complex tungsten resources. It is gradually applied in the industrial application of single scheelite and mixed ores of scheelite and wolframite.

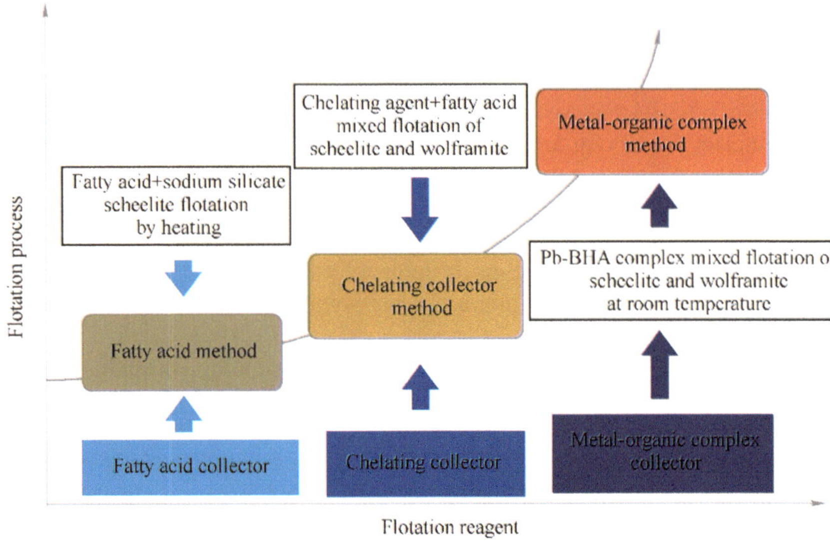

Fig. 6.1 Development of for tungsten flotation technology

6.3 Tungsten Flotation Technology Using Fatty Acid as Collector

The fatty acid method, which is the most common scheelite flotation method, is a flotation method with fatty acid collector as the core. Fatty acid method is often used as collector for fatty acids and their derivatives or mixtures, such as oleic acid, tall oil, naphthenic acid, oxidized paraffin soap, etc. Oleic acid (sodium oleate) and oxidized paraffin soap are widely used in flotation. Fatty acids have strong collecting ability of scheelite, but poor selectivity. Therefore, a large amount of water glass or modified water glass is often needed to be added in fatty acid flotation process as gangue mineral inhibitors. Fatty acid flotation process is usually divided into roughing operation and selection operation two stages, according to the difference of flotation temperature in the selection stage, the selection process is divided into two categories: heating flotation and normal temperature flotation process.

6.3.1 Fatty Acid Roughing—Heating and Selection Process

6.3.1.1 Basic Principle of "Petrov" Heating Flotation Process

Heated flotation, that is, the "Petrov" high-temperature pulp method [4], was invented by the Soviet expert Petrov in the late 1940s and widely used in the production practice of scheelite flotation. Petrov method is characterized by thick pulp, high temperature and high concentration of water glass [5]. The classical "Petrov" heating flotation process includes concentration of roughing concentrate, pulping at high temperature, and normal temperature flotation after dilution and demedication. First, concentrate the pulp of the roughing concentrate to a high concentration of 50–70%, and then add a large amount of water glass (usually 20–150 kg/t for the roughing concentrate), and in the high temperature (usually 85–95 °C) conditions for a long time (usually 60 min) intense agitation, the fatty acid collector on the gangue mineral surface for desorption. After diluting and deoxidizing, the flotation at room temperature can obtain a high-quality tungsten concentrate of more than 65% WO_3. The process is stable and the separation index is good. It has good adaptability to the flotation of scheelite—quartz type (or silicate mineral), scheelite—calcite, fluorite (barite) type scheelite.

The basic principle of the heating flotation method is to realize the separation of scheelite and gangue minerals according to the difference of the surface collector film resolution under special conditions. The inhibition mechanism of water glass on calcium-containing minerals is generally believed to be that hydrolyzed components $HSiO_3^-$ and SiO^{2-} of water glass react with mineral surface Ca^{2+} to generate calcium silicate precipitation, which inhibits the hydrophilicity of mineral surface. The solution chemical calculation of the condition of calcium silicate formation on mineral surface is shown in Fig. 6.2. At a pH of around 11.0, the product of calcium ions and the total concentration of silicate in the solutions of the three calcium-containing minerals is in order of magnitude: fluorite > calcite > scheelite. This indicates that the surface of fluorite is most prone to forming calcium silicate precipitates, while scheelite is the most difficult. Therefore, under the conditions of "Petrov method", the ability of water glass to desorption the collector from the surface of scheelite and calcium-bearing gangue minerals (calcite, fluorite, etc.) is different. Fluorite and calcite are inhibited due to a large amount of desorption of the surface collector, while scheelite still maintains good floatability, thus achieving efficient separation of scheelite from calcium-bearing gangue minerals. The high temperature conditions of "Petrov method" strengthen this process, so that gangue minerals are strongly inhibited, while scheelite is little affected. In addition, the flotation separation of mixed ores of scheelite and wolframite can be realized because wolframite is inhibited by a large amount of water glass in the thermal separation of scheelite and wolframite.

Subsequently, domestic and foreign researchers continue to improve the heating flotation process, which is mainly reflected in the optimization of pharmaceutical system, including: replacing single water glass inhibitor with combination adjustment inhibitor, strengthening the selective inhibition of gangue minerals and selective activation of scheelite; adding collector in advance to strengthen the selective capture of scheelite before demedicated, etc. In the pulp mixing process, sodium hydroxide and sodium sulfide are added to facilitate the analysis of the collector on the surface of calcite and fluorite by water glass [5, 6]. The combination of sodium hydroxide and water glass can form a high pH and strong alkaline pulp environment. The main components of sodium sulfide in the pulp are HS^- and S^{2-}. In the heating separation, the hydrolyzed components of sodium sulfide can repulse the collector originally adsorbed on the surface of sulfide ores such as pyrite and pyrrhotite, thus reducing

Fig. 6.2 Solution chemical calculation of calcium silicate formation condition on mineral surface [6]

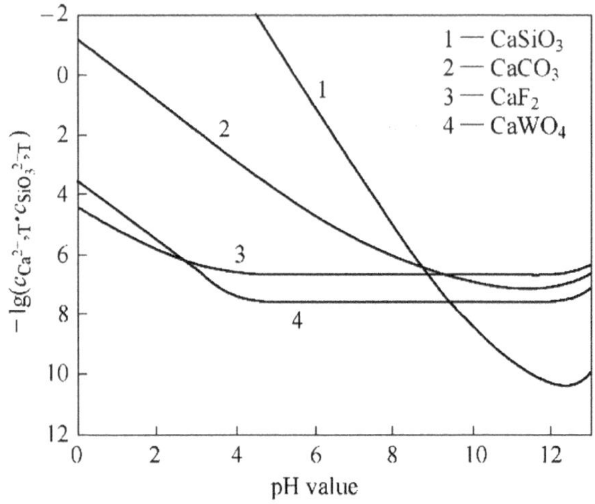

the hydrophobics of minerals and inhibiting their floating; sodium sulfide can also play a role in eliminating pulp activated ions, sodium sulfide in the hydrolysis of OH^- can form hydroxide precipitation with Ca^{2+}, reduce the impact of inevitable ions, so as to improve the separation efficiency. Therefore, the comprehensive use of sodium hydroxide, sodium sulfide and other agents combined with water glass is better than the use of water glass alone. In addition, it has been reported in the literature that lime, sodium fluosilicate and other agents can strengthen the inhibition of gangue minerals [7, 8]. In terms of process flow, mineral processing engineers try to flotation the pulp directly after heating treatment. The pulp is not diluted or deoxidized, which greatly simplifies the heating and cleaning operation and avoids tungsten loss in the process of multiple dilution, thus further improving the tungsten recovery [9]. However, the adaptation ability of this process is insufficient, and it is only used in some mines.

6.3.1.2 Fatty Acid Roughing—Heating and Concentration Process

In the process flow of scheelite heating flotation, the flotation of scheelite usually includes roughing and heating and cleaning processes. In roughing operation, fatty acid flotation process at room temperature is used to obtain roughing concentrate. In the heating and cleaning operation, the roughing concentrate first enters the heating and pulping operation, and then enters the flotation to get tungsten concentrate. The heating flotation operation is generally carried out in an independent loop, and the refined tailings are discharged directly as tailings. Heating flotation is widely used in scheelite flotation practice. Figure 6.3 shows the flowchart of typical heating flotation process of scheelite. After the crude concentrate is concentrated to a concentration of 50–70%, it enters the heated flotation process, adding sodium hydroxide, a large amount of water glass and fatty acids, and down-pulping for 60 min under the steam heating condition of 85–95 °C. After pulping, it enters the select stage to obtain high-grade tungsten concentrate.

Heating flotation has strong adaptability to the ore properties, and the heating process can be adapted to the flotation of different types of scheelite by adjusting the reagent system. Jin et al. [10] adopted the full flotation closed-circuit test process of "pre-desulfurization and decarbonization—heated flotation" for a graphite and high-calcium scheelite in Yunnan. In the heated flotation operation, sodium hydroxide, water glass, lime and fatty acid collector XP were used for heating at 60 °C for 1 h, and tungsten concentrate containing WO_3 57.45% with a recovery of 72.85% was finally obtained. Li et al. [6] carried out a heating flotation separation test of scheelite and wolframite on the mixed ores of scheelite and wolframite concentrate obtained by Pb-BHA collector in Shizhuyuan Polymetallic Mine. Under the condition of optimum dosage of GYR, sodium sodium sulfide, soda, etc., they obtained 65.99% scheelite concentrate and 30.71% wolframite concentrate by closed-circuit test. Wen et al. [11] used 731 as collector for roughing of a low-grade scheelite ore in Jiangxi province. The roughing concentrate was strengthened with water glass at 90 °C and then went through the heating flotation process of one roughing, five refining and one scavenging, and finally obtained the scheelite concentrate with a grade of 50.23% WO_3 and a recovery of 70.32%. Xu et al. [8] aimed at the problem that the grade and recovery of the flotation concentrate of a skarn scheelite were not ideal, they added lime to improve the grade of concentrate in the process of heating and decharging, optimized the amount of water glass and holding time, and finally increased the grade of tungsten concentrate by 5 percentage points, and the recovery increased to 97%. Yang et al. [12] used 731 as collector for Yaogangxian high-calcium scheelite to carry out heating flotation test. Under the condition that the grade of crude concentrate WO_3 was 4.56%, they used sodium sulfide, sodium hydroxide, 731 collector and water glass to hold heat for 40 min and adjust the pulp, and then obtained a grade of 65.48% through one roughing, three refined and one scavenging closed-circuit test. Tungsten concentrate with a recovery of 95.68%. Yang et al. [13] used the heating flotation process to recover low-grade scheelite from molybdenum tailings. Under the condition of raw ore grade WO_3 of 0.21%, sodium carbonate was used as pH adjustment agent, water glass as inhibitor and ZL as collector to obtain scheelite roughing concentrate through 1 roughing, 3 scavenging and 2 roughing separation. Then, scheelite roughing concentrate was obtained under the condition of heating and ZL as collector. With water glass inhibitor as the main agent and NS inhibitor as the auxiliary agent, tungsten concentrates with WO_3 content of 69.09% and recovery of 84.19% was finally obtained.

The advantages of heating flotation are strong adaptability to different types of scheelite, stable flotation index and good quality of scheelite concentrate. However, the disadvantages of this method are also very obvious: thick pulp, high temperature, high energy consumption, high cost, large dosage and high strength conditions on the corrosion of the equipment is more serious, the process is too long, complex, difficult to operate. Under the development trend of increasingly strict environmental protection requirements and increasing production efficiency, the heating flotation process needs to be optimized and improved or replaced.

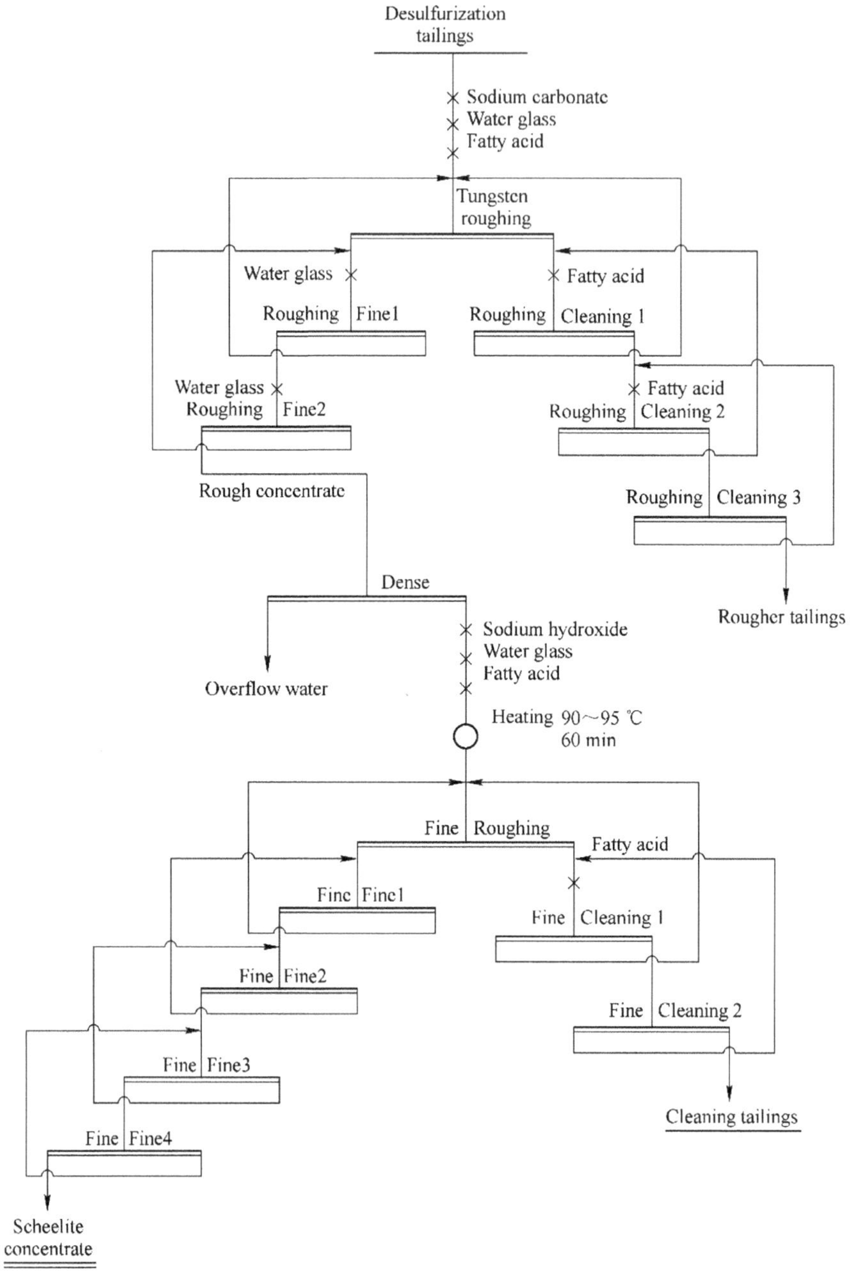

Fig. 6.3 Flowchart of typical scheelite heating flotation process

6.3.1.3 Application Practice of Roughage of Tungsten Fatty Acids—Heating and Concentration Process

The flotation process of fatty acid roughing and heating selective tungsten ore is widely used in large scheelite mines at home and abroad. Typical mines in China are introduced in the following examples, such as Luan Chuan Molybdenum Group in Luoyang, Xiaoliugou tungsten mine in Gansu, Yaogang xian tungsten mine, and Envision tungsten industry, etc.

1. **Scheelite of Luanchuan Molybdenum Industry Group, Luoyang**

Luanchuan County, Henan Province has huge reserves of low-grade scheelite resources. Luoyang Molybdenum Group is the largest and most representative mineral development enterprise in Luanchuan, with proven ore reserves of 2.1 billion tons, including 2.06 Mt of molybdenum metal with an average grade of 0.123%, ranking first in the world. Associated scheelite 620,000 t, with an average grade of 0.124%, belongs to the super-large scheelite deposit, and is accompanied by rich rhenium, sulfur, iron, gold, silver and other valuable minerals [14]. The first Company and the second Company of Luoyang Molybdenum Industry Group are mainly engaged in comprehensive recovery of low-grade scheelite resources.

(1) Ore properties

Table 6.1 shows the multi-element analysis results of ore chemistry, and Table 6.2 shows the chemical phase analysis results of tungsten in ore. The WO_3 grade of ore is 0.114%, the phase analysis of tungsten minerals shows that scheelite accounts for 95.2%, and a small amount of wolframite and tungsten-huite are also contained. The ore is mainly divided into molybdenum and tungsten, the main mineral of molybdenum is molybdenite, there are a small amount of oxide molybdenite, iron molybdenite, or to exist in scheelite like. Tungsten is mineralized to different degrees in various types of ores, with high content in skarn type and low content in hornrock type. The main metallic minerals in the ore are pyrite, pyrrhotite, molybdenite, scheelite, magnetite, chalcopyrite and so on. The main useful minerals are molybdenite and scheelite. Gangue minerals mainly include calcium-iron garnet, calcium-aluminum garnet, calcium-iron pyroxene, diopside, wollastonite and so on. The ore structure has flake, bundle, radial structure, euhedral—semi-euhedral granular structure, Mosaic structure, etc. The former three structures are the main ones. The ore structure has sparse disseminated structure, veinlike structure, brecciform structure.

Scheelite particle size is mainly fine, of which 8.37% are 0.1–0.2 mm particles, 17.25% are 0.074–0.1 mm particles, 43.09% are 0.019–0.074 mm particles, and

Table 6.1 Analysis of mineral chemical elements

Element	WO_3	S	Mo	Fe	SiO_2	Al_2O_3
Content/%	0.114	0.85	0.128	9.95	41.62	4.27
Element	CaO	MgO	CaF_2	Na_2O	Au	Ag
Content/%	23.39	1.82	6.85	0.19	0.5 g/t	2.20 g/t

Table 6.2 Tungsten phase analysis of ore

Tungsten phase	Content/%	Distribution rate/%
Tungsten in scheelite	0.107	94.53
Tungsten in wolframite	0.001	0.24
Tungsten central tungsten	0.006	5.23
Total tungsten	0.114	100.00

31.29% are < 0.019 mm particles. Under the condition that the content of < 0.074 mm is 75%, the dissociation degree of scheelite is 31.29%. The scheelite and garnet congeners accounted for 4.85% and scheelite and metalite congeners accounted for 8.21% in the congeners. Scheelite and other gangue mineral synbiotes accounted for 4.55%, the above synbiotes are mostly contiguous type, a few scheelite coated molybdenite, calcite and other minerals, the wrapped particle size is generally less than 0.03 mm. The molybdenite in scheelite is evenly distributed, with the presence of homogenous morphology, its content change of 5.3–0.36%, the average is 2.9%, with mechanical grinding method is difficult to make this part of tungsten molybdenum dissociation.

(2) Beneficiation process and technical indicators

The process flow of scheelite recovery workshop adopts full flotation process. Molybdenum tailings are floated for scheelite roughing and roughing concentrate is heated and dedusted to obtain scheelite concentrate, as shown in Fig. 6.4. The design and production process are as follows: the floating molybdenum tailings flow to the scheelite separation plant for classification, grinding, desulfurization, and then entering scheelite roughing; after roughing scheelite, roughing concentrate is obtained, which is concentrated to a concentration of about 70% and enters the process of heating and dedusting for selection. Scheelite concentrate is obtained by one roughing, two refined and two swept operation. After the concentrate is filtered and dried, it is the final scheelite concentrate. Flotation columns are used in most roughing and sorting operations.

The collector adopts a new flotation reagent FX-6 for scheelite, which has the characteristics of easy preparation, relatively low price and less dosage. All the properties are better than saponification flotation agent. In the heating and selection section, lime was originally used as detoxifying agent, which was large and polluted the environment. After continuous test and exploration, lime was cancelled, and only sodium hydroxide and water glass were used as detoxifying agent, and the grade of concentrate was significantly improved.

Mineral processing technical indicators: in 2017, the process in the raw ore grade of 0.11%, tungsten concentrate grade of 25–30%, comprehensive recovery reached 80%, including the recovery of 97% above.

In order to further optimize the heating flotation process of scheelite in CMOC Group Limited, researchers studied the optimization of reagent system in the normal temperature flotation section. Kang et al. [15] investigated the inhibition effect of single use of water glass, acidified water glass and ATM, as well as the synergistic

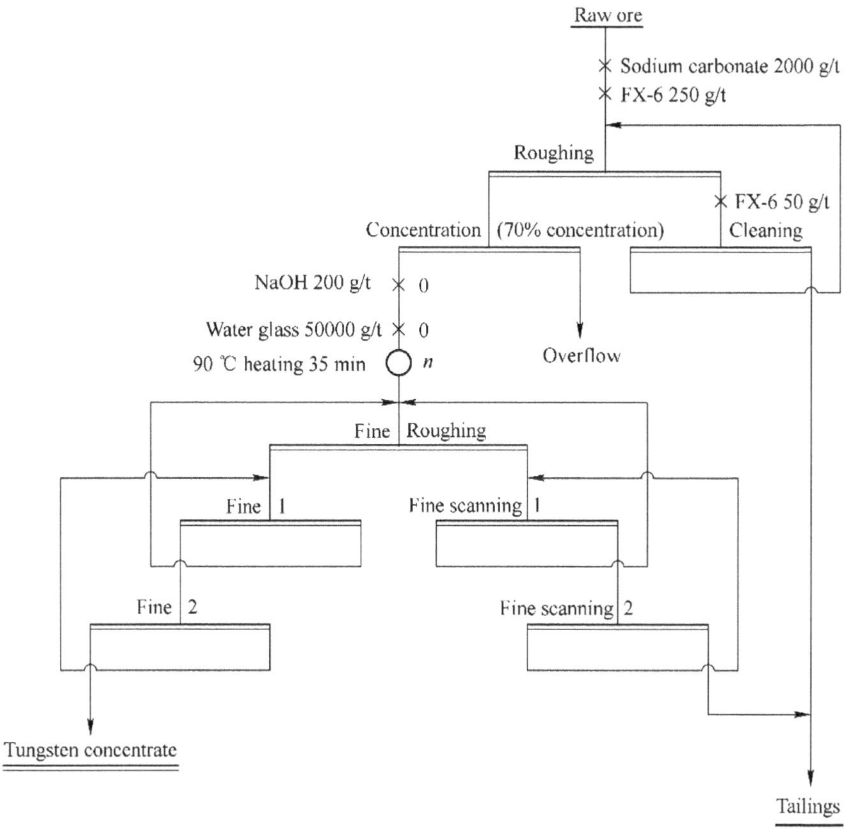

Fig. 6.4 Flotation process flow chart of scheelite heating in tungsten industry of CMOC Group Limited [15]

effect of acidified water glass and ATM. The results showed that the inhibitory effect of the three inhibitors was in the order of ATM > acidized water glass > ordinary water glass, and the simultaneous use of ATM and acidized water glass could play a synergistic effect. When the dosage of ATM+ acidified water glass was 20 g/t + 300 g/t, WO_3 grade of scheelite crude concentrate was significantly increased from 0.62 to 1.42%, and the recovery was increased from 84.74 to 87.28%, compared with no use of inhibitors. Wang et al. [16] studied the selective inhibition effect of the new inhibitor CC and water glass and its modified compounds, and increased the concentration operation to improve the roughing grade of scheelite. The results showed that the inhibition effect of the new inhibitor CC was good and the dosage was small. The scheelite roughing concentrate with grade of 3.12% and recovery of 82.87% could be obtained$_3$ through the closed-circuit test of single roughing, single cleaning and single scavenging. The industrial test results show that the roughage grade of scheelite is increased from 1.03 to 1.96% by using the new inhibitor CC and increasing the cleaning operation, and the recovery remains unchanged basically.

2. **Xiaoliugou Scheelite, Gansu Province**

Xiaoliugou tungsten mine, located in Sunan County, Gansu Province, belongs to Gansu Xinzhou Mining Co., LTD. After it was completed and put into operation in 2006, the designed production capacity is 2000 t/d, which is mainly used to treat scheelite ore of No. 1 and No. 4 orebodies in Xiaoliugou mining area.

(1) Ore properties

The multi-element analysis results of ore chemistry are shown in Table 6.3, and the chemical phase analysis results of tungsten in ore are shown in Table 6.4. The ore belongs to skarn type massive scheelite, and the valuable elements in the ore are mainly tungsten, sulfur, copper, molybdenum, bismuth, etc. The main tungsten ore is scheelite, accounting for 96.57%, and a small amount of tungstite and wolframite. The main useful minerals are scheelite, chalcopyrite, pyrite and pyrrhotite, followed by bismuthite, molybdenite and sphalerite. Gangue minerals mainly include tremolite, chlorite, quartz, diopside, sericite and feldspar. The ore structure has disseminated structure, patchy structure, veinlike structure, etc., the ore structure has eukaryotic—other omorphic granular structure, eukaryotic semi-eukaryotic granular structure, inclusion structure, metasomatic dissolution structure phase metasomatic residual structure. The scheelite in the ore has uneven distribution of thickness and fine grain, mostly disseminated, and some scheelite contains the same type of molybdenum. After grinding ore, the dissociation of scheelite monomer of 0.053–0.10 mm size can reach 95.32%. Part of scheelite contains homomorphic molybdenum in the ore, and the highest content of molybdenum reaches 0.44%. This part of molybdenum is difficult to be separated by dressing method.

(2) Beneficiation process and technical indicators

The technological process of Xiaoliugou Tungsten mine 2000 t/d concentrating plant is shown in Fig. 6.5. The desulfurized tailings are roughed by the roughing process of one roughing, two refined and three swept to obtain the roughing concentrate, and then the roughing concentrate is concentrated into the heating and cleaning

Table 6.3 Analysis of chemical elements in ore

Elements	WO_3	S	Zn	Bi	CaF_2	MgO	SiO_2	Al_2O_3
Content/%	0.73	0.82	0.039	0.016	3.94	4.6	45.32	8.39
Elements	TiO_2	Mn	Fe	F	P	Mo	Au	Ag
Content/%	0.48	0.19	2.89	0.54	0.067	< 0.01	0.029 g/t	3.26 g/t

Table 6.4 Tungsten phase analysis of ore

Phase	Scheelite	Wolframite	Tungstite	Whole tungsten
WO_3 grade/%	0.704	0.002	0.023	0.729
Percentage/%	96.571	0.274	3.155	100.00

section, and scheelite concentrate is obtained by one roughing, four refined and three swept. The beneficiation process mainly has the following characteristics: (1) scheelite roughing uses GYW new collector to replace 731 oxidized paraffin soap collectors with poor selectivity and large dosage, and uses sodium carbonate and water glass combination regulator to achieve efficient selective flotation separation of scheelite from gangue minerals containing calcium such as fluorite, calcite and diopside. In the separation of scheelite roughing concentrate, SN and GYW collector are added at the same time, which not only strengthens the selective inhibition effect of a large amount of water glass on calcium gangue minerals, but also strengthens the selective collection effect of scheelite, and realizes the direct selection at room temperature in an independent loop. (2) In the selection operation, the process of flotation after one or more dilution and depharming is cancelled, but directly flotation is carried out, which not only simplifies the process and reduces the loss of valuable metals, but also makes the heating flotation process more stable and the index is significantly improved. For the heated tailings, the shaker is used to further recover the lost tungsten in the tailings, and the process flow of flotation and gravity separation (shaker) is combined, so that the recovery of scheelite selective section can reach more than 97%.

Beneficiation technical indicators: under the condition of 0.65% raw ore, roughing concentrate grade of normal temperature roughing section is 5.74%, recovery is 86.13%; The concentrate grade of the heated concentrate section is 65.42%, the recovery is 94.90%, and the recovery of the whole process is 80.10%. The gravity concentrate grade of heated flotation tailings was 43.10% with a recovery of 3.12%, and the total recovery of tungsten in the whole process was 83.32%.

6.3.2 Fatty Acid Roughing—Normal Temperature Concentration Process

6.3.2.1 Principle and Flow of Normal Temperature Flotation Process

Normal temperature flotation started in the early 1970s and was pioneered in Gannan region of China. Different from the warm concentrate process, the normal temperature concentrate process does not need to be warmed up, and the roughing concentrate is cleaned directly under normal temperature conditions. The flow chart of typical scheelite flotation process at normal temperature is shown in Fig. 6.6. The normal temperature flotation process mainly uses collectors with relatively good selectivity such as 731 oxidized paraffin soap to carry out the flotation of tungsten ore under normal temperature conditions, and emphasizes the synergistic effect of sodium carbonate and water glass. By controlling pH value, $HSiO_3^-$ in pulp can be kept in the optimal concentration range and the roughing concentration ratio can be improved. In the selection process, a large amount of water glass is added for a long time (more than 30 min) to stir strongly, and the selectivity of inhibitors is improved

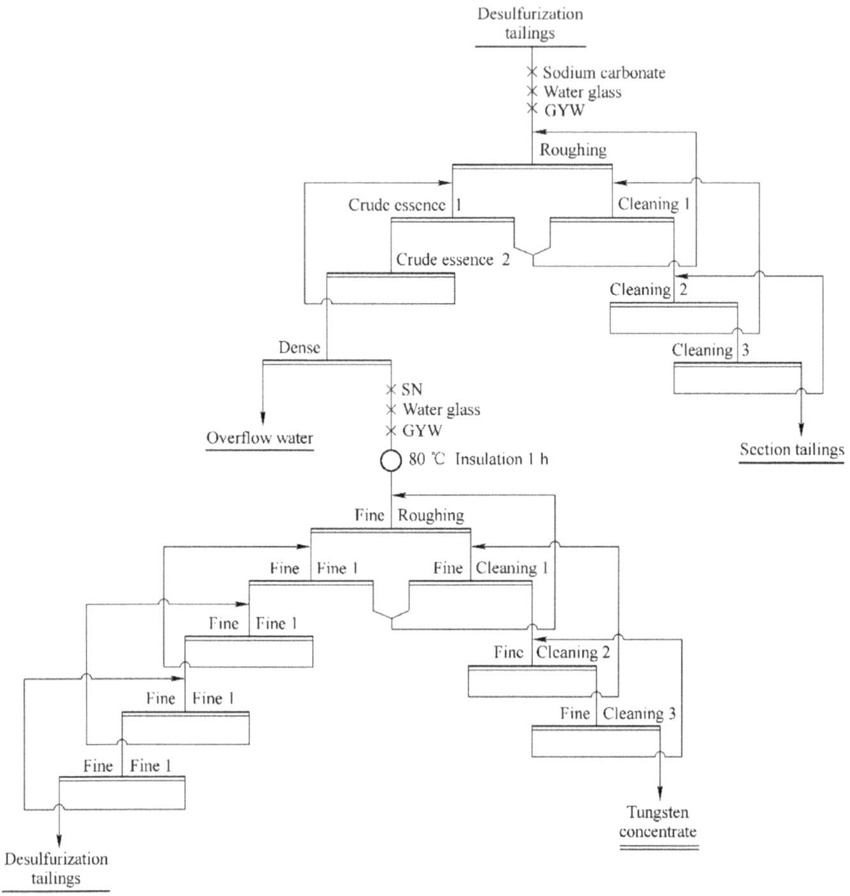

Fig. 6.5 Process flow of 2000 t/d tungsten concentrator in Xiaoliugou mine

by using the different resolution speed of collector film adsorbed on the surface of minerals. Then, the normal temperature selection is carried out to produce the final scheelite concentrate. The tailings are generally returned to the scheelite roughing section, and can also be discarded in the scheelite roughing section. According to the different properties of ore, the cleaning operation sometimes needs to concentrate the roughing concentrate to reach a certain concentration after adding water glass pulp, after pulp dilution for cleaning.

The normal temperature flotation process has been applied in some scheelite mines, and the process flow and reagent system are constantly being optimized. Ye 17] carried out experimental research on scheelite flotation of scheelite quartz type and scheelite fluorite calcite type respectively at normal temperature, and believed that normal temperature selection was feasible, but the concentrate quality was relatively poor and the recovery was low. Deng and Zhou [18] roughed out the calcite—fluorite

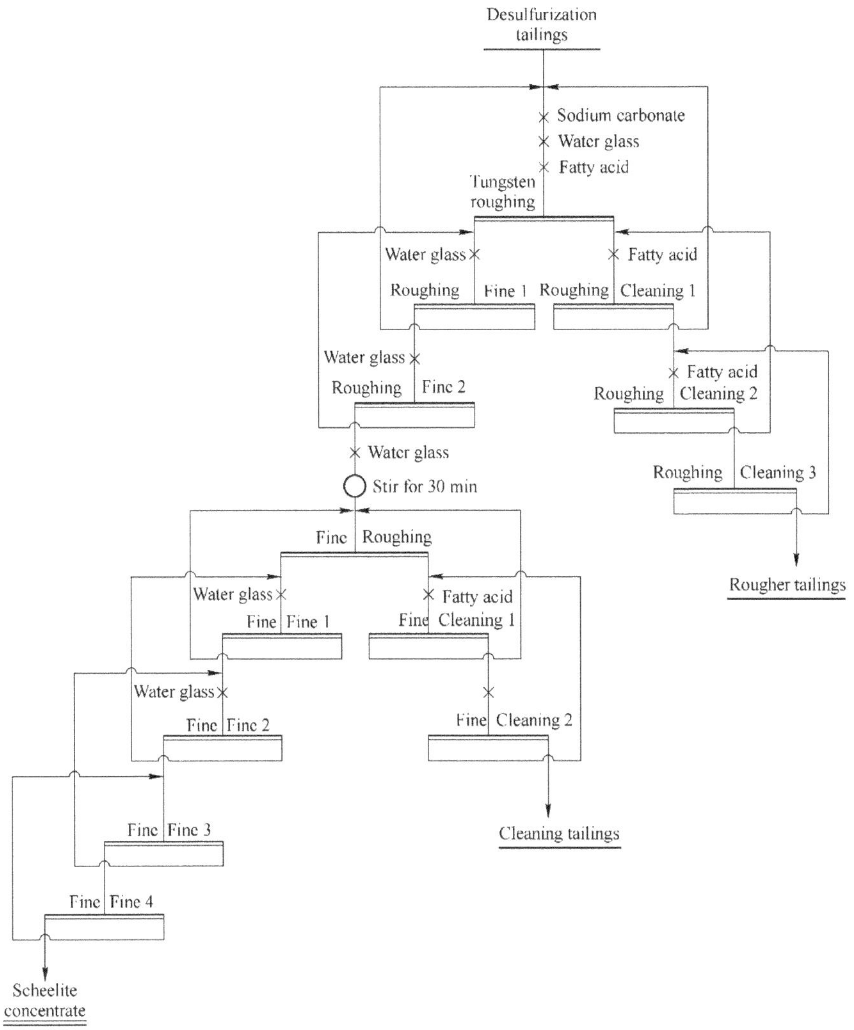

Fig. 6.6 Flow chart of typical scheelite flotation process at room temperature

scheelite containing 0.29% WO_3 by using sodium carbonate as modifier, water glass as gangue mineral inhibitor and TA-3 as scheelite collector. The scheelite roughing concentrate was concentrated to 40–60%, and TC combination inhibitor and a small amount of TA-3 agent were added. The scheelite concentrate with WO_3 65.17% was obtained by direct flotation after intense stirring for 40 min, and the recovery was 70.16%. Using fatty acid as collector and Y88 as inhibitor, Wang [19] in carried out normal temperature selection test on scheelite roughing concentrate with grade of 5.56%, realized the separation of scheelite from calcite and fluorite at normal temperature, and obtained scheelite concentrate with grade of 52.34%$_3$ and recovery of 86.05%. The flotation index could be comparable to the field heating flotation process index. Ding Dong carried out normal temperature flotation tests on a skarn-type scheelite ore with a grade of 7.81% from Ding [20]. The results showed that FX-6 was used as collector for normal temperature flotation, and roughing concentrate was obtained by one roughing and two scavenging. After the roughing concentrate was concentrated to 55–60%, water glass, sodium sulfide and FX-6 were added and stirred for 90 min. Then, scheelite concentrate with WO_3 grade of 64.13% and recovery of 70.02% was obtained by flotation process of one roughing, four refining and two scavenging. In order to improve the recovery of scheelite concentrate produced in a low-grade scheelite concentrator, Qi et al. [21] carried out an experimental study on the flotation process at room temperature. Roughing operation using sodium carbonate, water glass and CK-2 collector, after a roughing two fine three scavenging to get roughing concentrate, and then roughing concentrate added 5 kg/t water glass stirring at room temperature 45 min after a roughing four fine two scavenging selection operation, in raw ore WO_3 grade 0.3% condition, The separation index of scheelite concentrate with WO_3 grade of 63.33% and recovery of 86.30% was obtained. Zeng Yin-yin et al. carried out a normal temperature flotation test on a scheelite ore in Jiangxi, and determined the best process parameters and process flow. The roughing concentrate was obtained by using sodium carbonate, water glass and ZL collector in the roughing section through the process of "one roughing, three scavenging, two cleaning, and middle ore return in sequence" [22]. The roughing section added water glass and ZL and adopted the process of "one roughing, two scavenging, four cleaning, and concentrated return to the roughing and refining tailings". Under the condition of 0.77% WO_3 grade of raw ore, the flotation index of tungsten concentrate with WO_3 grade of 55.76% and recovery of 82.04% was obtained.

Compared with the heating method, scheelite flotation at room temperature has the characteristics of simple operation and low cost, but the adaptability of the heating flotation method is not as good as that of the ore. It is mainly applied to the scheelite mountain with relatively simple composition of scheelite and quartz skarn type, while the scheelite of the scheelite-fluorite (barite) type cannot obtain good separation index.

6.3.2.2 Application Practice of Roughing Separation of Fatty Acids from Scheelite at Room Temperature

Fatty acid flotation process at room temperature has good adaptability to quartz type and calcite type scheelite, and has been applied in Xianglushan tungsten mine in Jiangxi Province, Baoshan tungsten mine in Jiangxi Province, Glenn Separation Plant in Montana, Xintianling tungsten mine in Hunan Province, and other locations.

1. **Xianglushan tungsten mine in Jiangxi Province**

Xianglushan tungsten mine is located in Xiushui County, Jiangxi Province, with a designed mining and beneficiation capacity of 2050 t/d and four concentrator plants. The first concentrator has a processing capacity of 550 t/d, the second concentrator has a processing capacity of 300 t/d, and the third concentrator and the fourth concentrator have a processing capacity of 600 t/d.

(1) Ore properties

The multi-element analysis results of ore chemistry are shown in Table 6.5. The deposit is a skarn-type scheelite polymetallic deposit with three ore types, namely, diopside quartz keratinite scheelite (accounting for more than 95% of the total ore content), marbleized limestone scheelite and lead, zinc, sulfur scheelite. Metal ore is mainly scheelite, associated minerals are pyrrhotite, pyrite, sphalerite, chalcopyrite, wolframite, etc. Gangue minerals are quartz, potassium feldspar, plagioclase, tremolite, fluorite, muscovite, calcite and barite, etc. The main tungsten minerals are scheelite, accounting for about 95%, wolframite only accounts for about 4%, and the rest are tungstites. Scheelite is mostly irregular shape, particle size is generally between 0.1 and 0.15 mm, a few can reach more than 1 mm. Chalcopyrite is irregularly granular or vermicular, with a general rescue diameter of 0.023–0.046 mm, some of it is about 0.1 mm, and it is often wrapped by scheelite. Pyrrhotite is irregularly heteromorphic granular or flake with roughing grains and a few are coated with scheelite. The particle size of pyrite is usually 0.1–1 mm, and scheelite closely symbiosis, and often coated with scheelite.

(2) Process flow and technical indicators

Xianglushan tungsten mine has four beneficiation plants, and the ores all come from one ore body, so the technological process and pharmaceutical system are basically the same. Figure 6.7 shows the flotation process flow chart of scheelite in Xianglushan tungsten mine. After crushing, screening, grinding and grading of raw ore, the ore fineness is < 0.074 mm with 70% content. The pulp goes into flotation operations such as copper sulfur and copper sulfur separation operations for sulfide ore flotation operations. Desulphurized tailings into scheelite roughing operation, after one roughing, three scavenging and two cleaning to get scheelite roughing concentrate. About 65% of the scheelite roughing concentrate is produced through the operation of one roughing, three cleaning and five refining, and the tailings from the scheelite roughing operation are the final tailings. Roughing operation uses sodium carbonate, water glass and ZL collector, select operation adds water glass pulp, select tungsten concentrate under normal temperature conditions.

Beneficiation technical index: in the condition of raw ore grade of 0.566%, the concentrate grade of 58.29% was obtained, the recovery was 78.25%; at the condition of raw ore grade 0.631%, the concentrate grade was 53.45%, the recovery was

Table 6.5 Analysis of chemical elements in raw tungsten ore of Xianglushan

Element	WO_3	Cu	Zn	Bi	Sn	Pb	Mo	S	F
Content/%	1.35	0.13	0.17	0.06	0.004	0.044	0.006	6.62	3.0
Element	FeO	CaO	MgO	SiO_2	Al_2O_3	Fe_2O_3	TiO_2	Au	Ag
Content/%	6.30	12.02	5.80	40.76	7.71	9.71	0.16	0.06 g/t	21.9 g/t

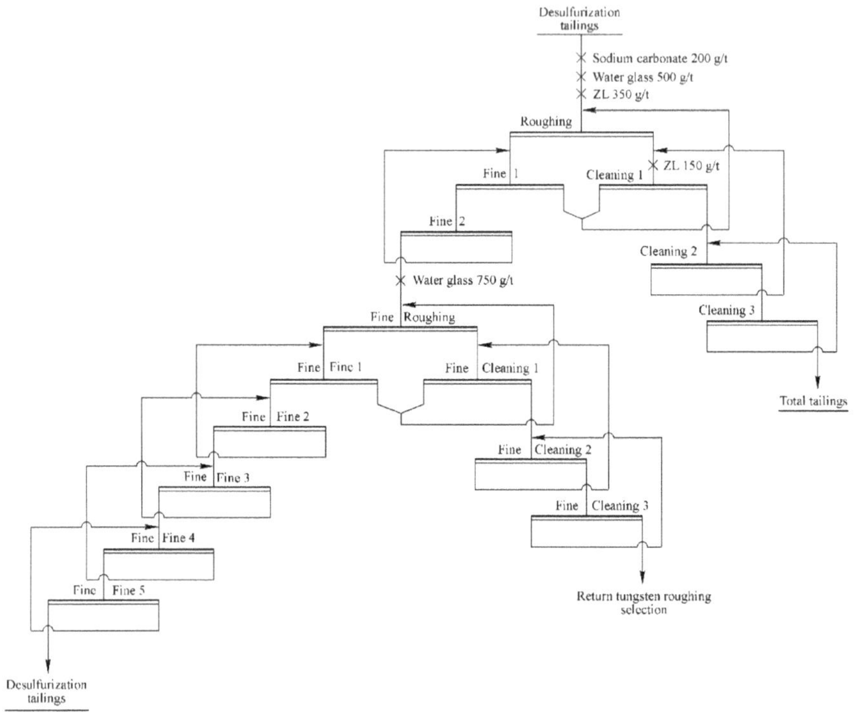

Fig. 6.7 Flotation process flow chart of scheelite in Xianglushan tungsten mine

78.08%; at the condition of raw ore grade 0.737%, the concentrate grade was 65.75% and the recovery was 77.29%. At the condition of raw ore grade 0.514%, the concentrate grade of 68.45% was obtained from the fourth concentrator with a recovery of 79.15%.

2. **Baoshan tungsten ore in Dangping, Jiangxi Province**

Dangping Baoshan scheelite mine is located in Jiangxi Province. The flotation plant was built in 1966. In the original flotation process, the final scheelite concentrate was selected by "Petrov method" for scheelite roughing concentrate. In 1970, the collector oleic acid of scheelite flotation was changed to 731 oxidized paraffin soap, and the roughing concentrate was successfully selected at room temperature.

(1) Ore properties

The multi-element analysis results of ore chemistry are shown in Table 6.6. Baoshan scheelite deposit is a contact metasomatic skarn type lead, zinc and scheelite deposit with complex orebody morphology. The surrounding rock is granite, limestone or marble with dense and hard structure. The orebody is composed of scheelite skarn ore and sulfide ore. Scheelite with a grade of 0.47% WO_3 is distributed in skarn with particles ranging from 0.05 to 0.1 mm, showing a disseminated structure. The sulfide ore is composed of pyrrhotite, pyrite, galena, sphalerite and chalcopyrite grades. It is produced in dense block or roughing block with uniform mineralization. Gangue minerals are mainly diopside, quartz, garnet, calcite and fluorite.

(2) Process flow and technical indicators

Figure 6.8 shows the normal temperature flotation process of scheelite in Baoshan Concentrator. In scheelite flotation process, after flotation of sulfide ore, sodium carbonate was added to adjust pulp pH to 9, water glass was added as inhibitor, and 731 or 733 oxidized paraffin soap was used as collector. Crude concentrate with grade of 15–20% was obtained through one roughing and two scavengings. In the case of non-concentration, water glass was added to 5–6 kg/t and stirred for 40 min at room temperature. Then carry on 5–6 times selection, get more than 65% scheelite concentrate grade.

Beneficiation technical index: Under the condition of raw ore grade of 0.354% WO_3, Baoshan concentrator can obtain tungsten concentrate with grade of 70.63% WO_3 and recovery of 79.65% through normal temperature flotation process.

3. **Glenn Concentrator, Montana, USA**

Glenn Separation Plant belongs to Montana Mineral Engineering Company, which is one of the most typical scheelite production plants in the world. It is characterized by the production of medium grade scheelite concentrate with high recovery through beneficiation, and then all scheelite concentrate is treated by chemical method to separate tungsten and molybdenum.

Glenn separation plant processing of raw ore WO_3 grade 1%, ore and molybdenite, chalcopyrite, pyrite, gangue minerals are quartz, calcite, biotite, amphibole and fluorite. Figure 6.9 shows the normal temperature flotation process of scheelite in Glenn concentrator. In this plant, a combined floating-heavy process is adopted to treat scheelite ore, and low-grade tailings are discarded by flotation roughing and scavenging, and the final scheelite concentrate is obtained by flotation concentrate and flotation concentrate. The pulp is blended with water glass, Dresinatet X, Enersol 211 and pine oil. Dresinatet X is a mixture of sodium oleate and sodium linoleate, and Emerson 221 is a low-freezing-point oleic acid that is used primarily to control foaming and recover scheelite in roughing grain grades. Scheelite roughing has a flotation concentration of 35%, and the selected concentrate is then separated by a shaker. Before the shaker separation, water glass is added to the roughing concentrate, so that the fine-grained material is completely dispersed, but also in order to further inhibit gangue minerals in the sorting operation. Five stages are required to

Table 6.6 Analysis of chemical elements in Baoshan raw ore

Elements	WO_3	Mo	Cu	Bi	Sn	Pb	Zn	S
Content/%	0.47	0.011	0.11	0.075	0.014	1.50	1.22	6.74
Elements	Fe	CaO	MgO	SiO_2	Al_2O_3	Au	Ag	As
Content/%	15.21	20.22	0.47	33.34	5.85	0.15 g/t	80 g/t	0.003

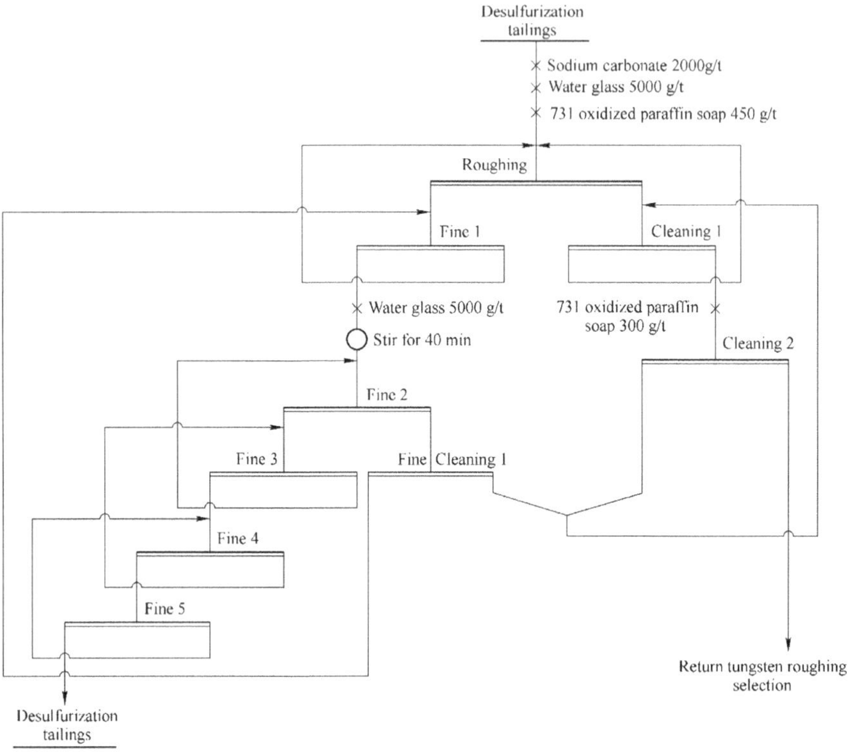

Fig. 6.8 Normal temperature flotation process of scheelite in Baoshan concentrator

obtain the final flotation concentrate, and tannic acid inhibition calcite is continuously added throughout the operation. After the shaker tailings are flotation by scheelite, the products in the tank are sorted by the shaker tailings. The shaker tailings are ground together with the tailings from the scheelite flotation roughing operation, and then the scheelite flotation scavenging is carried out after grinding. The mill and the hydraulic cyclone form a closed circuit, the grinding products are swept for scheelite flotation, the final tailings are swept and discarded, and the scavenging concentrate and roughing concentrate enter the shaker for separation together.

Flotation reagent dosage: Sodium carbonate 2.5 kg/t, water glass 2.5 kg/t, Dresinatet X 0.1 kg/t, Enersol 211 0.15 kg/t, pine oil 0.05 kg/t, tannic acid 0.05 kg/t.

Beneficiation technical indicators: the plant processing capacity of 277 t/d, after the separation of WO_3 grade 45% scheelite concentrate, tailings grade 0.08%, the total recovery of tungsten is 90–92%.

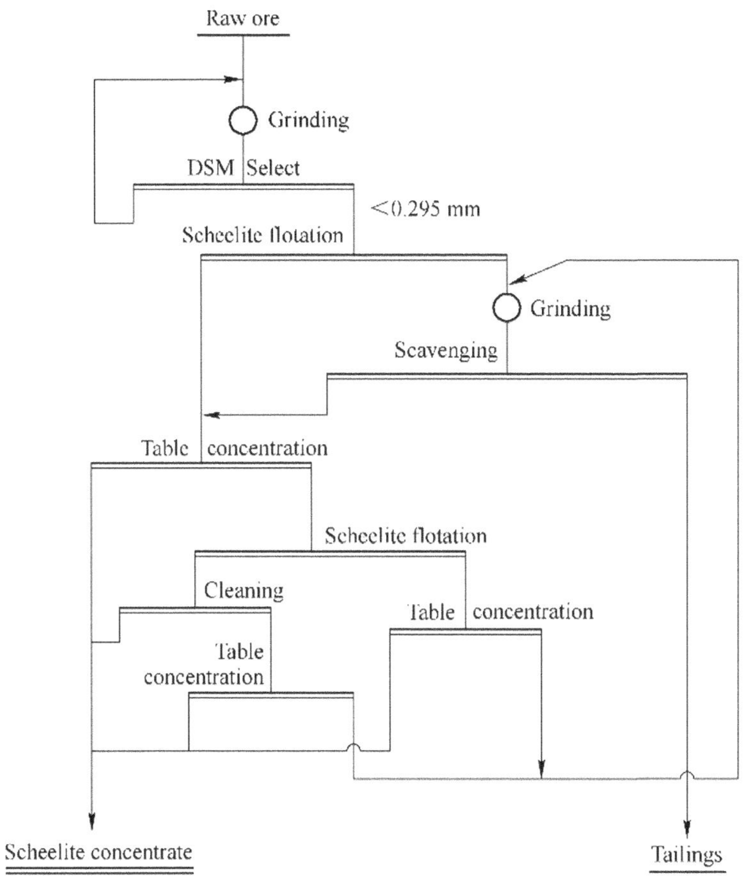

Fig. 6.9 Flotation process of scheelite at normal temperature in Glenn concentrator

6.3.3 *"Lime Method" Normal Temperature Flotation Process of Fatty Acids*

6.3.3.1 Principle and Flow of "Lime Method" Normal Temperature Flotation Process

Vazguez et al., Union Carbide Company, studied a lime process to achieve highly selective flotation separation of scheelite and fluorite. A good flotation index can be obtained by adding lime to the rough-sorted ore, and then adding sodium carbonate and water glass to adjust the pulp. This lime process, with good selectivity and high concentrate grade, is especially suitable for skarn type ore, and is currently adopted by some tungsten concentrators in the world.

The study of the action mechanism of lime process shows that: adding lime into the system, calcium ions adsorb on the surface of fluorite, calcite and quartz, causing the surface charge to change from negative to positive, while scheelite still maintains negative charge. Then, when sodium carbonate was added and pulp was stirred, calcium carbonate precipitated on the surface of quartz, fluorite and calcite, while scheelite retained negative charge without precipitation. After adding water glass, the inhibition of calcite is enhanced, so as to improve the floatability of scheelite and gangue minerals, and realize its flotation separation [23, 24]. Yang [25] studied the process parameters of flotation scheelite by lime method. The results showed that the amount of lime and the mixing time of refined water glass had great influence on the flotation of scheelite, so the amount of lime and the mixing time of water glass should be strictly controlled. Liu and Xu [26] obtained scheelite concentrate with WO_3 grade of 42.12% and recovery of 46.26% by using lime, sodium carbonate, water glass and 733 as reagents, aiming at low grade scheelite with WO_3 grade of 0.39% from Shizhuyuan Polymetallic Mine. Li et al. [27] used lime flotation for refractory high-calcium schetungsten from a concentrator in Yunnan province. The dosage of roughing reagent was 440 g/t lime, 3 kg/t sodium carbonate, 2.3 kg/t water glass and 300 g/t fatty acid collector. Through the closed-circuit roughing process of one roughing, three cleaning and one finishing at room temperature, WO_3 grade of 6.30% was obtained. With a recovery of 79.76%, the grade of tungsten crude concentrate was increased by 2.5 percentage points and the recovery was almost unchanged compared with that of sodium carbonate pulping method alone.

However, the lime flotation method also has certain limitations. Zhang et al. [28] compared the effect of lime, caustic soda and soda ash in the flotation of scheelite-quartz type ore, and the results showed that lime had the worst effect, which confirmed that lime flotation method was not suitable for the flotation of scheelite-quartz type ore. This method has remarkable effect on scheelite-calcite—fluorite type ores. In addition, lime produces a large amount of Ca^{2+} in the pulp, which will increase the consumption of fatty acid collectors. At present, the "lime process" is relatively rarely used in industrial production.

6.3.3.2 Application Practice of "Lime Method" Flotation Process at Room Temperature

Tempiute tungsten separation plant is located in Nevada, USA. The size of the separation plant is 1000 t/d, the grade of raw ore WO_3 is 0.4–0.5%, and the proven ore reserves are 44,000 kt. The main rock formation in the Tempiute area is the limestone deposits of the Paleozoic, and the contact rock ore bodies currently mined can be divided into four main types: (1) Garnet contact rock, which is the most abundant in the Moody zone, consists of red to reddish brown garnet, accompanied by pyroxene, containing a small amount of sulfide, and scheelite with fine grain size. (2) Sulfide contact rocks, mainly composed of garnet and pyroxene, contain a large amount of pyrite and pyrrhotite, and occur oxidation along the fault surface. (3) Epidotite and chlorite contact rocks, mainly calcite, epidotite and fluorite, and rich scheelite, calcite crystal recrystallization makes this contact rock similar to granite, or similar to limestone. (4) Pyroxene or amphibole contact rock, mainly containing pyroxene, amphibole and a small amount of garnet, dark green color, local contains trace molybdenite. To sum up, the main minerals in the deposit are scheelite, pyrrhotite, pyrite, sphalerite, garnet, calcite, fluorite, etc.

The production principle flow of Tempiute tungsten concentrator is shown in Fig. 6.10. The overflow of hydraulic cyclone passes magnetic flocculation, and a large amount of magnetite and pyrrhotite are first discharged by wet cylinder magnetic separator. The tailings of magnetic separation enter flotation, and collector Z-6 and foaming agent pine oil are added to the pulp mixing barrel to flotation pyrite. The roughing concentrate is obtained by three times of selection. The tailings of pyrite flotation are scheelite flotation. In four tandem pulping drums, lime, sodium carbonate, water glass and collector (oleic acid and naphthenic acid 1:1) are added. Dosage of agents: lime 900 g/t, sodium carbonate 1810 g/t, water glass 3628 g/t, oleic acid, naphthenic acid 272 g/t. Scheelite roughing is carried out in Aketel No. 48 flotation machine, a total of 14 slots. An appropriate amount of collector is added to the third and seventh tanks, and the foam products scraped out of the first three tanks are no longer sent to the concentrator for concentration. After that, the foam products scraped out are collected twice in the flotation machine of Aketel No. 24 to get tungsten concentrate, which is sent to the ammonium paratungstate plant near Bishop, California for treatment.

The process is characterized by magnetic flocculation of the pulp before scheelite flotation. A large amount of magnetite and pyrrhotite are excluded by magnetic separation, and then scheelite is flotation by lime method to obtain low-grade concentrate and then sent to chemical dressing treatment to ensure a high recovery.

Beneficiation technical indicators: raw ore grade 0.4–0.5%, concentrate grade WO_3 is greater than 15%, recovery is greater than 85%.

In general, the fatty acid method has relatively strong adaptability to ore, especially the heating flotation process has good adaptability to the ore type and ore properties, and has the advantages of low reagent cost and stable flotation index, so it occupies a dominant position in scheelite mines at home and abroad. However, its process is long and complex, and the heating flotation process has high energy consumption and cost. Most importantly, the poor selectivity of fatty acids makes the system often need to add a lot of water glass to inhibit gangue minerals, and the "fatty acid—water glass" circulation system restricts the improvement of scheelite recovery.

6.4 Flotation Process of Tungsten Ore with Chelating Agent as Collector

In order to solve the defects of fatty acid flotation process, chelate was employed as collector and developed tungsten ore flotation process based on chelate collector. Chelate collector mainly includes hydroxamic acid, ararsenic acid and copper iron,

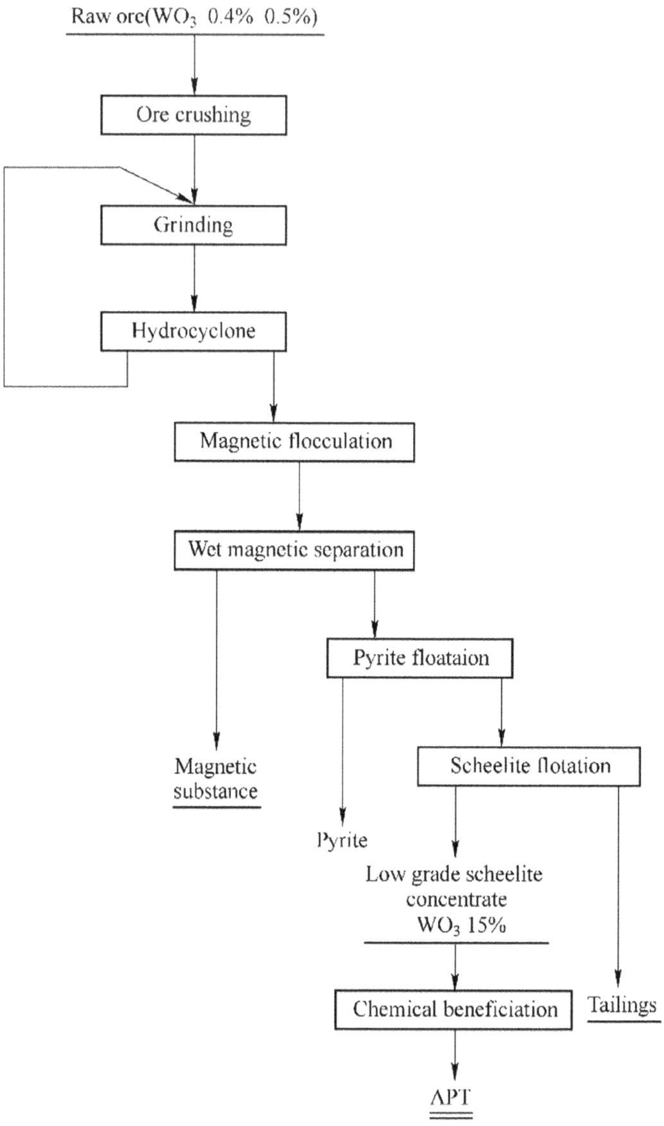

Fig. 6.10 Production principle flow of Tempiute tungsten concentrator

etc., can form stable complexes with calcium, iron, manganese and other metal particles on the mineral surface so as to adsorb on the mineral surface. Chelating collectors show good selectivity in the flotation of mixed ores of scheelite and wolframite, and can be used in the flotation of scheelite and scheelite and wolframite. Hydroxamic acid is the most commonly used chelate collector, its molecular structure can be expressed as R-CONHOH, common hydroxamic acids such as benzohydroxamic acid, salicylic hydroxamic acid, octylhydroxamic acid, etc. Arsine acids commonly include benzyl arsenate and toluene arsenate, etc. Due to the strong toxicity of arsenate chelating agents and their great impact on the ecological environment, the chelating agents of arsenate have been rarely used at present. Cupferron is also a common chelating collector and has been used in the flotation of tungsten ore. Compared with fatty acid collectors, chelate collectors have good selectivity. Chelate collectors represented by hydroxamic acid and cupric ferriol reagent play an important role in the development of tungsten flotation. In the 1990s, Shizhuyuan Nonferrous Metals Co., Ltd, Guangzhou Nonferrous Metal Research Institute, Beijing General Research Institute of Mining and Metallurgy, Changsha Nonferrous Metallurgy Design and Research Institute, etc., jointly tackled the key problems, and successfully developed a new comprehensive beneficiation technology of molybdenum-bismuth flotation, scheelite and wolframite synchronous flotation, scheelite heating flotation, wolframite fine mud flotation and fluorite flotation based on the main full flotation process with chelator capture as the core—Shizhuyuan method [29]. The Shizhuyuan method has solved the two major problems recognized in the world for many years, that is, wolframite and scheelite must be recovered step by step by different beneficiation methods, and scheelite and calcium-containing minerals are difficult to be separated by flotation. It has made a major technical breakthrough in the field of tungsten beneficiation in the world, and vigorously promoted the development of tungsten, tin, molybdenum and bismuth polymetallic ore beneficiation technology. The Shizhuyuan method, which takes chelating collector as the core, makes China's tungsten flotation technology reach the world's leading position.

6.4.1 GY Method Flotation Process

6.4.1.1 Principle and Flow of Flotation Process by GY Method

The most typical flotation process of chelating collector is tungsten selection process of "GY method". GY flotation process is based on GY series reagents (hydroxamic acid and fatty acid) independently developed by Guangzhou Institute of Nonferrous Metals as the flotation collector process, which was successfully tested in Shizhuyuan Polymetallic Mine 380 concentrator in 1997. GY method is in neutral or weak alkaline medium conditions, with modified water glass selectively inhibit fluorite, calcite and other calcium-containing minerals, with lead salt as activator activated tungsten minerals, with GYB benzohydroxamic acid as the main collector, modified fatty acids as assistant collector, scheelite and wolframite mixed flotation. Figure 6.11 shows the principle flow of the monochrome tungsten mixed flotation process by GY method. In the flotation process of GY method, roughing concentrate is obtained through the chemical system of GY method, and then scheelite concentrate is obtained by heating flotation process. Tailings are obtained by gravity separation method such as shaking machine. Or the roughing concentrate is separated by strong magnetic, the scheelite roughing concentrate is heated to get scheelite concentrate, and the wolframite roughing concentrate is obtained by gravity separation or flotation method.

GY flotation process can realize the mixed flotation of wolframite, for both wolframite and mixed ores of scheelite and wolframite can achieve good results. The roughing concentrate obtained from tungsten flotation by GY method has high grade and low yield, which can greatly reduce the ore amount of heating and select section, reduce energy consumption and production cost, and simplify the production process. Tungsten roughing tailing pulp by GY method is neutral, good clarification, and can be directly discharged by adding a small amount of lime to help sedimentation, which is conducive to environmental protection. Compared with 733 method, tailings treatment costs can be greatly saved. In addition, GY floatation tungsten in neutral medium, about 90% of fluorite into tungsten roughing tailings, for the subsequent comprehensive recovery of fluorite created good conditions.

The tungsten ore flotation process by GY method is widely used in the mixed flotation of scheelite and wolframite. Zhou and Lin [30] used the activation agent ZP, the chelating collector GY and the modified water glass to design the "New flotation process of scheelite and wolframite by GY method". Aiming at a complex tungsten ore containing 0.59% WO_3 in Hunan, they obtained scheelite concentrate grade WO_3 73.26% with a recovery of 73.20%, wolframite concentrate grade 66.25% with a recovery of 13.55%. The total recovery of tungsten reached 86.73%. Guo [31] replaced the original heating flotation process with the GY method of constant-temperature flotation and centrifuge beneficiation, which improved the recovery of 43.91%, and solved the problem of tungsten fine mud recovery. Zhao [32] used water glass and aluminum sulfate as inhibitors, lead nitrate as activators, GYB and GYR as combined collectors to conduct separation tests on scheelite ore with WO_3 grade of 0.46% from Shizhuyuan Polymetallic Mine. The final scheelite ore concentrate with grade of 28.32% and recovery of 78.04% was achieved.

6.4.1.2 Application Examples of Tungsten Flotation by GY Method

1. **Shizhuyuan Polymetallic Mine in Hunan Province**

The superlarge tungsten polymetallic deposit in Shizhuyuan Polymetallic Mine is the largest bismuth, tungsten and fluorite deposit in China. The deposit is mainly caused by the superposition of multi-stage granitic magma and hydrothermal contact metasomatism of multi-grade marble. The main industrial minerals are scheelite, wolframite, citroite, molybdenite, bismuthite, fluorite. The ratio of scheelite to wolframite is about 2:1, in addition, it is also associated with a large amount of molybdenum, lead, plating, zinc, tin, copper, sulfur and a small amount of gold and silver. It is a rare super-large tungsten and molybdenum polymetallic deposit in the world, known as the "world nonferrous metal Mineral Museum".

(1) Ore properties

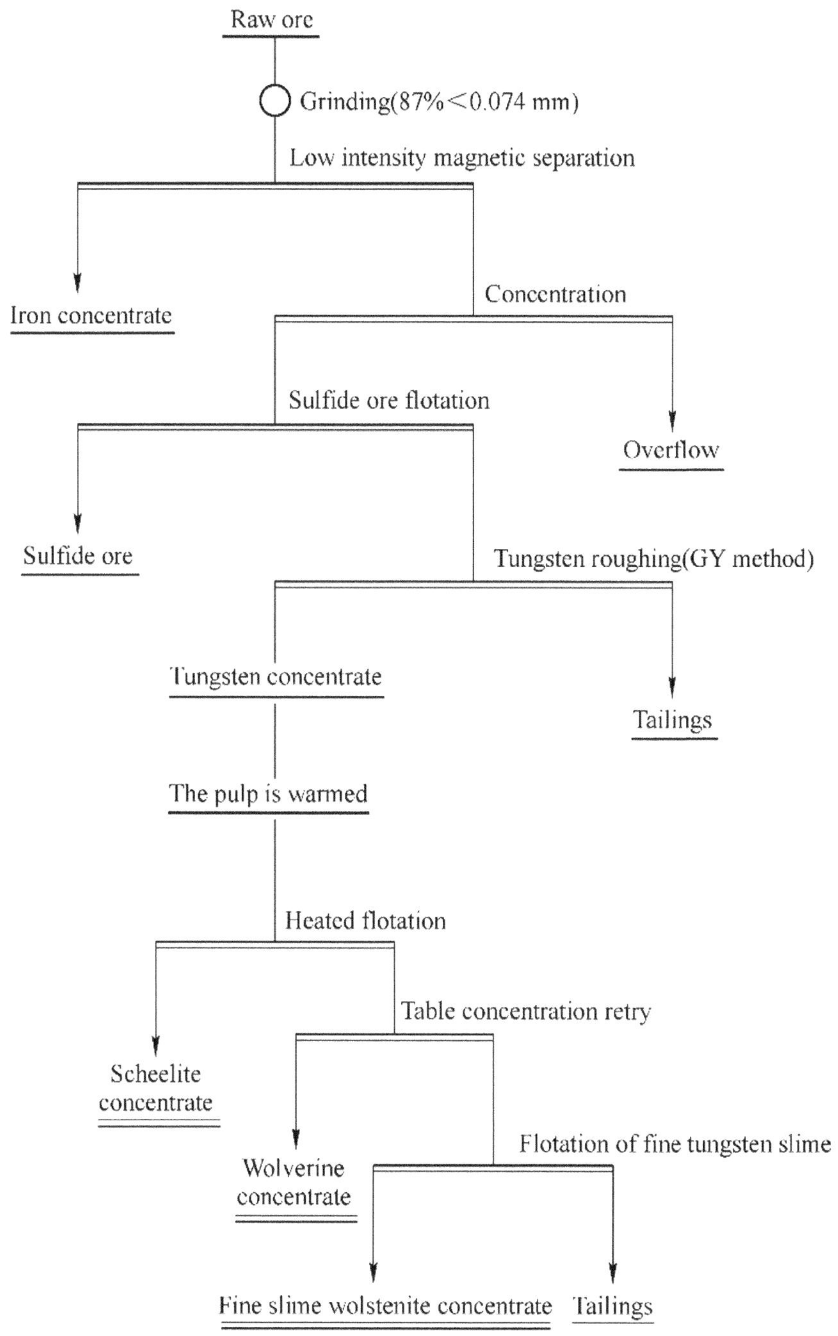

Fig. 6.11 Principle flow chart of mixed flotation process of scheelite and wolframite by GY method

The technological mineralogy characteristics of Shizhuyuan Polymetallic Mine are as follows:

(1) wolframite and scheelite are the main tungsten-containing minerals, with a ratio of about 1:2.
(2) Complex mineral types, 143 kinds of minerals have been identified, with more than 20 kinds of minerals of industrial value.
(3) Mineral distribution granularity is fine and uneven, about 80% of < 200 mesh dissociation, some minerals even less than 0.01 mm is not completely dissociated.
(4) The occurrence state and symbiotic relationship of minerals are complex. Scheelite often dissolves replacement wolframite, and is closely associated with molybdenite, bismuthite, fluorite, etc. Cassiterite has a very fine distribution size, and only accounts for 32% of the total tin in ore zone III. Most tin is dispersed in other minerals in the form of homogenous substances. The chemical composition analysis of raw ore is shown in Table 6.7, and mineral composition is shown in Table 6.8.

(2) Process flow and technical indicators

The primary reason for the difficult separation of tungsten minerals in Shizhuyuan Polymetallic Mine is the symbiosis of wolframite and scheelite, followed by the fine dissemination size of wolframite and wolframite, brittle and easy to grind, and it is easy to over pulverize and slime in the grinding process, especially wolframite. The floatability of scheelite and wolframite differs greatly, and the fatty acid collector is difficult to achieve the simultaneous flotation enrichment of wolframite and wolframite. To realize the efficient and comprehensive recovery of tungsten resources in Shizhuyuan Polymetallic Mineis a challenge in the field of tungsten beneficiation.

Table 6.7 Chemical composition analysis of Shizhuyuan polymetallic mine

Components	WO_3	Bi	Mo	Sn	Cu	Pb	Zn	Fe	Be
Content/%	0.35	0.11	0.056	0.11	0.024	0.019	0.034	7.17	0.0095
Components	K_2O	Rb	$CaCO_3$	CaF_2	SiO_2	Al_2O_3	TiO_2	MgO	MnO
Content/%	1.82	0.087	2.79	20.82	41.65	7.31	0.15	0.89	1.74

Table 6.8 Mineral composition of Shizhuyuan polymetallic mine

Minerals	Scheelite	Wolframite	Molybdenite	Bismuthite	Cassiterite	Garnet	Pyrite/pyrrhotite
Content/%	0.351	0.182	0.130	0.158	0.070	14.058	2.380
Minerals	Chalcopyrite	Magnetite	Feldspar/kaolinite	Fluorite	Calcite	Quartz	Sericite/biotite
Content/%	0.111	3.500	18.673	21.600	3.821	23.661	6.695

In the 1970s, the Hunan Institute of Metallurgy, Beijing General Research Institute of Mining and Metallurgy, Changsha Institute of Mining and Metallurgy, Central South College of Mining and Metallurgy and other research units designed five programs, seven times of the whole process of industrial tests, failed to achieve a qualitative breakthrough in mineral processing indicators. The State included the improvement of Shizhuyuan Polymetallic Mine beneficiation technology in the "8th Five-Year Plan" and "9th Five-Year Plan" National Science and Technology Projects. From 1990 to 1995, many research institutes and colleges and universities in China proposed the main process flow such as "flotation", "gravity separation-flotation", "flotation-gravity separation- flotation" and "flotation-magnetic separation-flotation" after a lot of experimental research, and the recovery of tungsten, molybdenum and bismuth in Shizhuyuan Polymetallic Mine has been greatly improved. Zhang et al. [33] Guangzhou Institute of Nonferrous Metals, studied the simultaneous mixed flotation of wolframite and scheelite by using benzohydroxamic acid as collector, a small amount of fatty acids as collector, lead nitrate as activator, water glass and aluminum sulfate as combination inhibitor, and the tungsten concentration ratio of crude concentrate was greatly improved. The CF process developed by Academician Sun Chuanyao of Beijing General Research Institute of Mining and Metallurgy and the CKY process proposed by Yu Jun have obtained good flotation indexes. Through a series of experiments, the tungsten beneficiation process of Shizhuyuan Polymetallic Mine has made great progress, and the comprehensive recovery of tungsten has been gradually improved. The 1000 t/d Shizhuyuan Polymetallic Mine concentrator designed according to the whole floating production process was completed in October 1998, and put into operation smoothly. In 2002, the "Shizhuyuan method" of full flotation of scheelite and wolframite won the second prize of National Science and Technology Progress. In 2008, the 2000 t/d concentrator designed by Changsha Nonferrous Metallurgy Design and Research Institute in accordance with the whole floating main process and flotation column scheme was built and reached the production standard. In the whole floating stem process, the tungsten selection operation adopts the process of "scheelite heating flotation—wolframite shaker—desliming—wolframite flotation". Because scheelite heating flotation, adding a large number of water glass, wolframite is strongly inhibited, wolframite floatability is greatly reduced, the recovery of wolframite in flotation operation is very low. At the same time, wolframite must be deslimed before flotation, but the pulp contains a lot of water glass, wolframite is difficult to settle, resulting in wolframite easy loss to the desliming process. In March 2013, the production transformation of the polymetal concentrator was carried out. In the tungsten roughing section, strong magnetic separation was used to separate wolframite and scheelite for separate systematic flotation. The comprehensive recovery of tungsten was greatly improved, thus forming the flotation process as shown in Fig. 6.12. In 2014, the production transformation was carried out for the gravity separation section of heated tailings. Suspension vibration gravity separation equipment was used to recover part of the tungsten and tin lost in the heated tailings. Compared with the traditional gravity separation equipment, the separation effect was remarkable, but it was still difficult to make breakthrough in the recovery of micro-fine tungsten and tin. For the mixed tungsten ore in the rich section of Shizhuyuan Polymetallic Mine III ore belt, the "new comprehensive beneficiation technology of tungsten, molybdenum and bismuth polymetallic ore—Shizhuyuan method" has been successfully completed after the national scientific and technological research in the Eighth and Ninth Five-Year Plan Period, making the processing technology of the mixed ores of scheelite and wolframite reach the international leading level. After several scientific researches, the "Shizhuyuan method" was improved and optimized, and the comprehensive recovery of tungsten in the rich section of III ore belt was increased to about 70%.

2. **Xingluokeng Tungsten Mine, Ninghua, Fujian Province**

Xingluokeng Tungsten Mine in Ninghua is one of the four tungsten mines in China, and its tungsten fine slime is a symbiosis resource of scheelite and wolframite. Xingluokeng Tungsten Mine in Ninghua is a molybdenum bearing wolframite and scheelite deposit with large granitic veins and fine veins. The total reserves of WO_3 reach 295,500 t, accounting for 5.3% of the national reserves. The daily processing capacity of Xingluokeng Tungsten Mine tungsten concentrator is about 5000 t. After

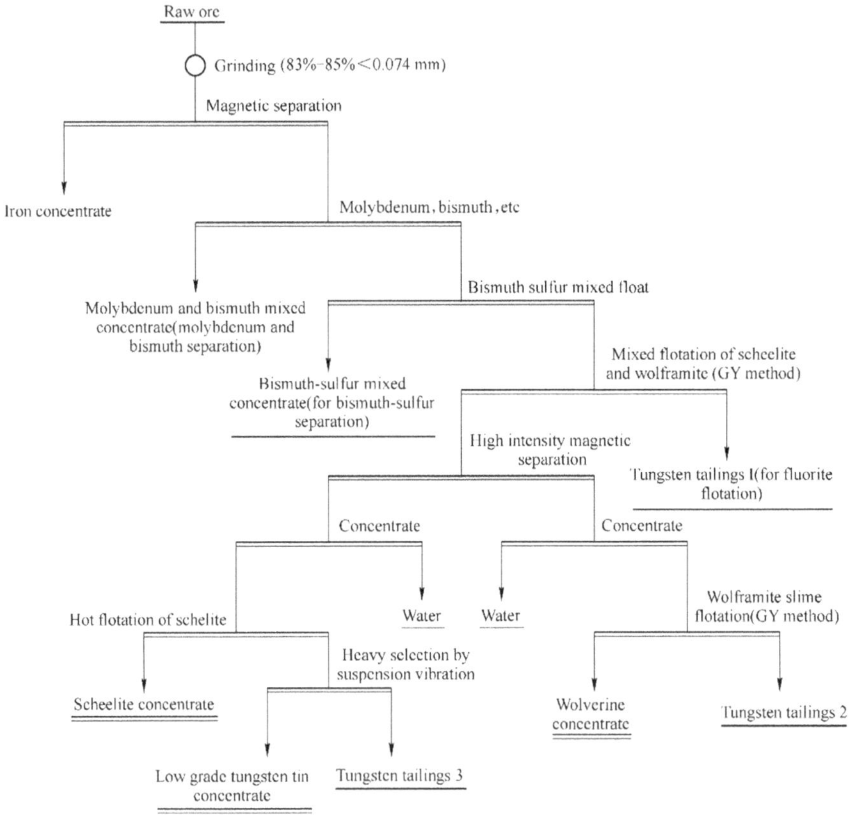

Fig. 6.12 Main flotation process of tungsten ore in Shizhuyuan polymetallic mine by GY method

the raw ore is broken and graded, the roughing grade is recovered by gravity separation method of shaker + spiral chute, and the fine grade is recovered by flotation method.

(1) Ore properties

The chemical composition analysis results of tungsten ore are shown in Table 6.9. It can be seen that the main chemical components in raw ore are SiO_2, Al_2O_3 and K_2O, while the secondary chemical components are Fe_2O_3, Na_2O, MgO, CaF_2 and WO_3. The main valuable component is WO_3. The main mineral compositions of raw ore and their relative contents are shown in Table 6.10, indicating that the main metallic minerals in the tungsten slime are wolframite, scheelite, pyrite, followed by molybdenite, chalcopyrite, arsenopyrite, limonite, magnetite, hematite, siderite, a small amount of bismuth, galena, sphalerite, natural bismuth, and the main useful minerals are wolframite and wolframite. The main ganglion minerals are quartz, feldspar, Muscovite, biotite, sericite, followed by chlorite, fluorite, apatite, dolomite, and a small amount of epidotite, kaolinite, zircon, etc. The results show that tungsten mainly exists in the form of wolframite and scheelite minerals in the primary ore. The proportion of wolframite and scheelite tungsten is approximately 4:6, and the share of tungsten in the primary ore is only 1.15%. Wolframite is usually associated with biotite, followed by quartz. Scheelite is granular, several grains or single grains or euhedral grains scattered in quartz, feldspar, sometimes connected with wolframite, disseminated particle size is very uneven, the general particle size is 0.02–1.28 mm The process mineralogy study of raw ore shows that the mineral composition and mineral characteristics of the ore, the distribution relationship is far more complex than quartz vein type tungsten ore, with low grade of useful metals, a wide variety of minerals, tungsten mineral distribution granularity is not the same, and fine characteristics.

(2) Process flow and technical indicators

After the selection plant was completed and put into operation in August 2007, according to the design, the principle process of three-stage and one-closed circuit crushing—one-stage rod grinding—classification gravity separation—fine mud flotation was adopted. Primary and secondary fine tungsten slime (< 38 μm size) accounts for about 25% of the raw ore in the main process, and the average grade of WO_3 in the fine slime is 0.17%. The mineral composition of the fine tungsten slime is complex, and the ratio of primary ore to weathered ore varies greatly, so it is difficult to recover. The tungsten fine mud originally used the process of "desulfurization—mixed flotation of scheelite and wolframite—heating cleaning selection—weak magnetic separation—strong magnetic separation—shaking table re-election" to recover the scheelite and wolframite minerals. There are problems such

Table 6.9 Analysis results of chemical composition of raw ore

Components	WO_3	SiO_2	Al_2O_3	Fe_2O_3	CaO	MgO	Na_2O	K_2O	CaF_2	P_2O_5
Content/%	0.16	74.18	12.20	2.65	2.13	1.88	1.10	4.41	0.76	0.05

Table 6.10 Main mineral compositions of raw ore and their relative contents

Minerals	Content/%	Minerals	Content/%
Wolframite	0.150	Muscovite/sericite	10.654
Scheelite	0.146	Quartz	52.860
Molybdenite	0.027	Feldspar	25.336
Bismuthite	0.009	Kaolinite	0.089
Pyrite	0.351	Zircon	0.006
Plain sand	0.031	Chlorite	1.478
Chalcopyrite	0.036	Ferrodolomite/siderite	1.467
Apatite	0.257	Limonite	0.073
Epidote	0.002	Biotite	6.241
Fluorite	0.787	Total	100.00

as complex process flow, low tungsten beneficiation index, tungsten recovery of less than 30%, high beneficiation costs, tailing water settlement difficulties, and serious production losses. In response to the problems of the original production process. In 2015, Xingluokeng tungsten mine cooperated with relevant domestic research units, and through experimental research and production practice, optimized the process flow and technical transformation, and adopted the process of "tungsten fine mud two-stage pre-treatment desliming—normal temperature flotation column—flotation machine flotation—centrifuge classification selection", and optimized the separation chemical The process flow and chemical system are shown in Fig. 6.13. The SLon-1600 type centrifugal separator is used for the concentrate of tungsten fine mud. When the feed WO_3 content is 4–8%, the concentrate of the centrifugal separator with grade of 18–22% WO_3 can be obtained, and the working recovery is above 85%. By adding a two-stage cyclone desliming operation, the desliming effect is significantly improved, and the tungsten metal loss rate of desliming overflows is reduced by 30%, which lays a foundation for the improvement of tungsten recovery of fine tungsten slime recovery system. The new process satisfies the demand of the production capacity of the fine tungsten slime separation system, and optimizes the process flow to realize the approximate water–glass tungsten float. The production practice shows that the tungsten recovery of the new process is up to 76.52%, the production cost is greatly reduced, and good economic benefits are obtained.

Beneficiation technical index: raw ore grade 0.16%, roughing concentrate grade 4–8% WO_3, centrifugal concentrate grade 18–22% WO_3, operational recovery of more than 85%, fine mud separation process tungsten recovery of 76.52%.

6.4.2 Flotation Process by CF Method

The CF flotation process for mixed ores of scheelite and wolframite is a new flotation process developed by Beijing General Research Institute of Mining and Metallurgy

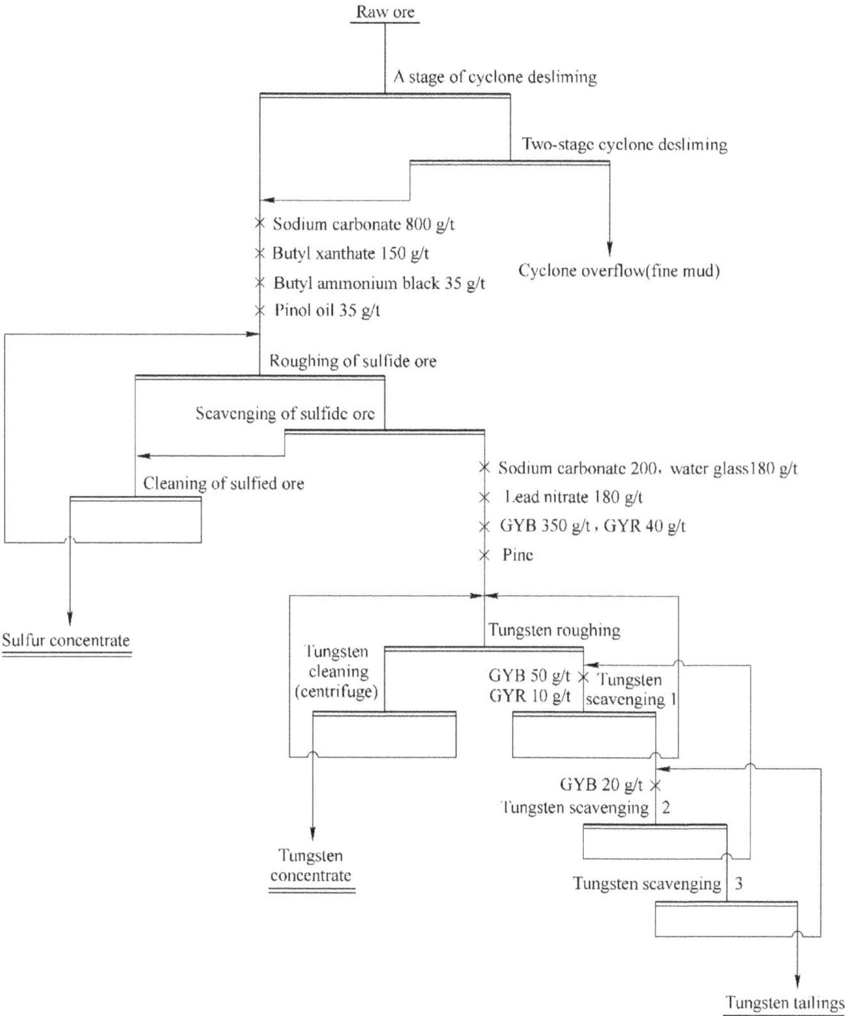

Fig. 6.13 Process flow chart of fine mud flotation by GY method in Xingluokeng tungsten mine [34]

with the clasp collector CF. The process can realize the mixed flotation of scheelite and wolframite under the condition of natural pH value, which changes the condition that mixed ores of scheelite and wolframite must be recovered under alkaline condition in the past. The process uses a small amount of water glass as the regulator, lead nitrate as the activator of tungsten ore, CF as the collector of tungsten ore, emulsified oleic acid or oleic acid as the collector and foaming agent, tungsten ore flotation under the condition of pH = 7–9. With $Pb(NO_3)_2$ as activator, CF as collector, and water glass and CMC as inhibitors, Cheng Xinchao's CF method realized the flotation separation of tungsten minerals from fluorite and calcite at pH = 8.2 [35]. Xiao et al. [36] have carried out comparative experiments of lime method, caustic soda method and CF method respectively for Shizhuyuan Polymetallic Mine, and the test results show that the CF method has the best technological index. In addition, the CF method can also adapt to low temperature flotation conditions. Figure 6.14 shows the main flow of CF flotation process for Shizhuyuan Polymetallic Mine. In the tungsten flotation section, a small amount of water glass is used as adjusting agent (100–400 g/t), lead nitrate is used as activator (1000 g/t), CF is used as collector (700 g/t), and a very small amount of emulsified oleic acid is used as foaming agent. The pH value of flotation pulp is generally 7–9. The semi-industrial test results of CF method showed that under the condition of feeding grade of 0.653%, crude concentrate with grade of 12.94% and recovery of 89.17% could be obtained. In the end, scheelite concentrate with grade of 68.63% and recovery of 65.76% was obtained, wolverine concentrate with grade of 67.79% and recovery of 17.86%, tungsten concentrates with total grade of 68.41% and recovery of 83.62%.

The chelator flotation process can realize the mixed flotation of wolframite and scheelite. Compared with fatty acid method, the chelator-based flotation process can obtain high grade of crude concentrate and low yield, which can greatly reduce the ore amount in the heating and sorting section and simplify the production process. Moreover, the loss of fluorite in the tungsten flotation section is low, which creates conditions for the comprehensive recovery of fluorite in the future. But scheelite must be heated flotation to obtain high grade concentrate, resulting in a long process, the use of chelating agent also makes the flotation reagent cost is high. At the same time,

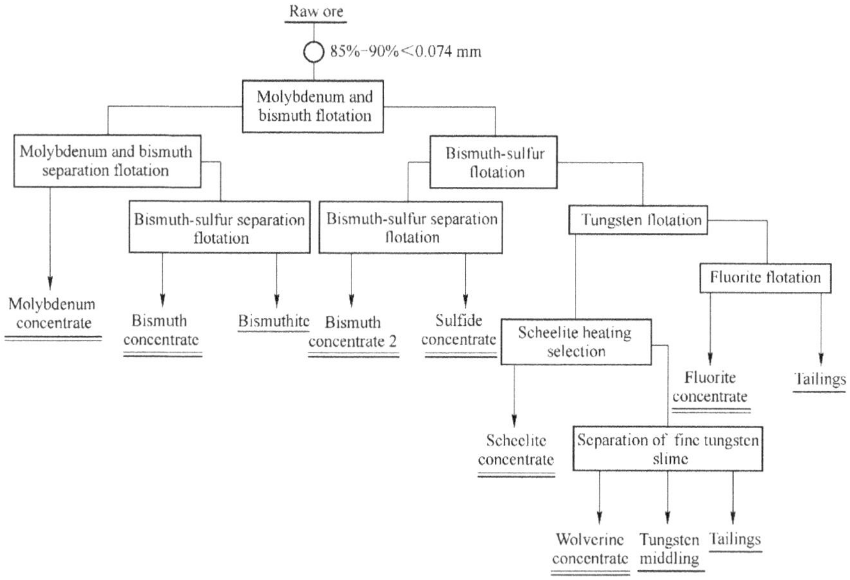

Fig. 6.14 Main flow of CF flotation process for Shizhuyuan polymetallic mine [36]

the large amount of water glass has caused problems such as the difficulty of tailing water settlement reuse and the difficulty of tailing fluorite recovery by flotation.

6.5 Flotation Process of Tungsten Ore Using Metal–Organic Complex as Collector

With the depletion of easily selected high-quality tungsten resources, the main body of tungsten ore development gradually evolved into complex and difficult to deal with low-grade ores. The occurrence characteristics of this type of ore are more complex, the content of calcite gangue minerals such as fluorite and calcite are increasing, and the distribution size is finer and the symbiotic relationship is more complex, such as polymetallic ore bodies such as Shizhuyuan Polymetallic Mine and Huangshaping. Taking Shizhuyuan Polymetallic Mine as an example, the difficulty of ore body sorting increases with the passing of time, the comprehensive utilization of resources is more difficult, the traditional beneficiation process is more and more difficult to adapt to the change of ore properties, and the production index continues to deteriorate, which puts forward higher requirements for tungsten beneficiation process [37]. The development of a new collector with high selectivity has become an inevitable choice to solve the difficult problem of efficient development and utilization of complex resources.

6.5.1 Metal–Organic Complex Collector

Focusing on the efficient development and utilization of complex scheelite and wolframite resources, the mineral processing team of Central South University has developed a new metal-organic complex collector (lead ion-benzohydroxamic acid Pb-BHA) with strong selectivity. Pb-BHA complex collector has strong selective harvesting ability of scheelite and wolframite, while fluorite does not float (Fig. 6.15), indicating that the complex has good selectivity, which is conducive to the efficient flotation separation of tungsten/fluorite associated resources. Han et al. [38] carried out a detailed study on complex collectors, and the results showed that the ratio of Pb to BHA in the complex was in the concentration range of 1:2–2:1, scheelite and wolframite had stronger collecting ability, while in the range of $pH = 8–10$, scheelite and wolframite had better floating ability. By adjusting the ratio of Pb/BHA and pH of pulp reasonably, the floatability of the complex to wolframite and gangue minerals can be changed, so as to realize the efficient separation of tungsten minerals and calcium gangue minerals.

In order to overcome the problem of insufficient selectivity of water glass in tungsten flotation and inhibition of tungsten ore, Wei et al. [39, 40] designed and

Fig. 6.15 Pb-BHA shows the floatability of the three calcium-bearing minerals in the collector system, reprinted from Ref. [39], copyright 2018, with permission from Minerals Engineering

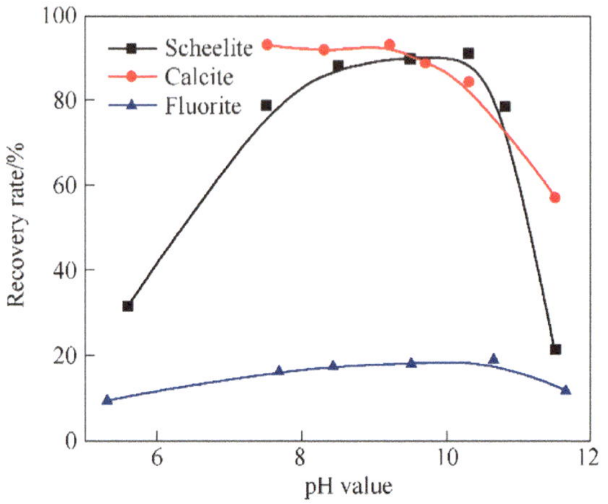

developed Al-Na_2SiO_3 polymer inhibitor based on the mechanism of metal ion regulation of silicic acid colloidal self-assembly, and applied it to the flotation separation of scheelite and calcite. The results showed that when Pb-BHA complex was used as collector, Al-Na_2SiO_3 polymer could achieve a good flotation separation of scheelite from calcite and fluorite (Fig. 6.16). Therefore, when Pb-BHA organometallic complex is used as collector and Al-Na_2SiO_3 polymer is used as inhibitor for scheelite flotation, the selectivity advantages of the two will be mutually enhanced, showing synergistic selectivity, thus realizing efficient flotation separation of scheelite and calcite.

Fig. 6.16 Effect of pH on the flotation of scheelite and calcite by Al-Na_2SiO_3, reprinted with permission from Ref. [40]

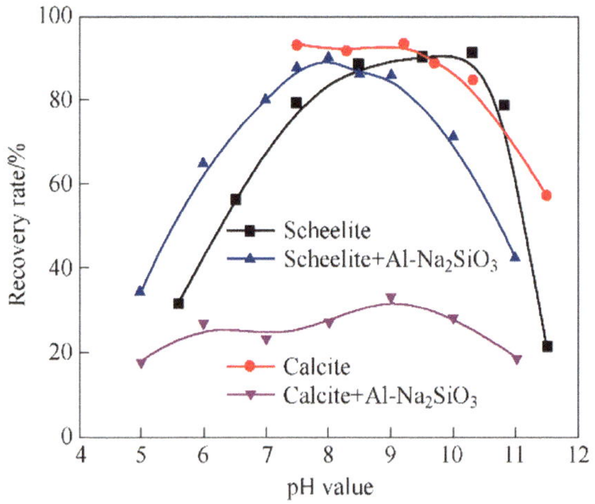

6.5.2 *Flotation Process by Metal–Organic Complex Method*

The flotation process of tungsten ore at room temperature with highly selective Pb-BHA metal–organic complex as collector is shown in Fig. 6.17. After the desulfurized tailings are pulped with sodium carbonate, the complex is used as the only collector, and a small amount of Al-Na_2SiO_3 is used as the inhibitor. The final high-grade tungsten concentrate is obtained through multiple selection and scavenging under normal temperature conditions. The complex flotation process uses the flotation column for roughing and sorting, and the flotation machine for scavenging. Tungsten concentrates with grade of more than 40% can be obtained through one roughing and two refining. The process is short and the efficiency is high. At present, complex flotation process has been applied industrially in Shizhuyuan Polymetallic Mine, Huangshaping, Xingluokeng Tungsten Mine and other large mines, and the production index is good.

The complex flotation process realizes the high efficiency flotation of tungsten ore in normal temperature and short flow, which can replace the "Petrov" heating flotation method to a certain extent, simplifies the working flow and improves the working efficiency. Due to the cancellation of the addition of a large amount of water glass in the traditional process, the comprehensive recovery of tungsten is significantly

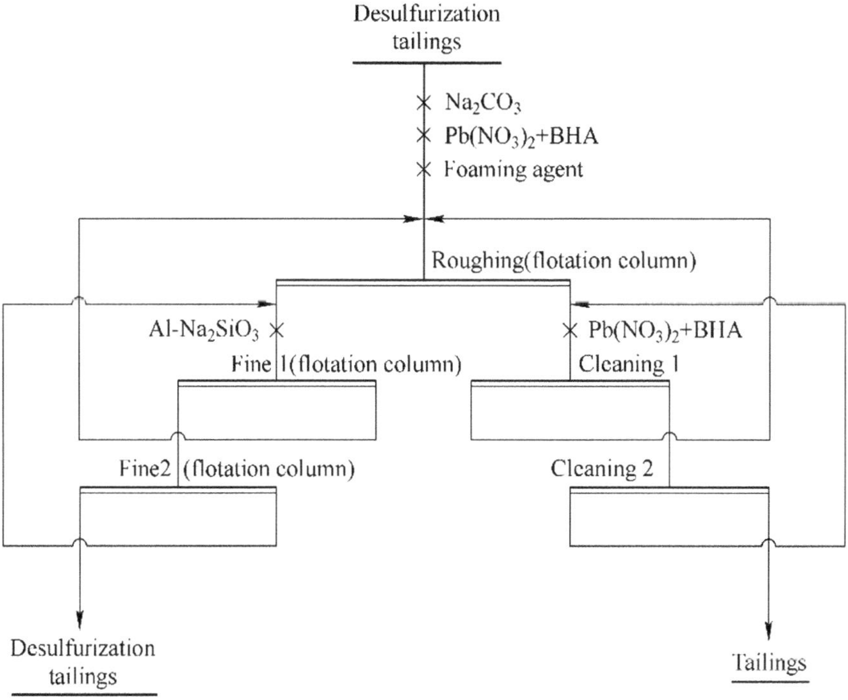

Fig. 6.17 Typical complex flotation process flow chart

improved, and at the same time, it is conducive to the recovery of associated minerals in tailings, such as fluorite, feldspar, etc. However, complex flotation process also has some defects, the dosage of reagent is large, the cost is relatively high, how to reduce cost and increase efficiency through the optimal design of reagent is the direction of future development.

6.5.2.1 Shizhuyuan Polymetallic Mine in Hunan Province

As a resource associated with polymetallic scheelite and wolframite, Shizhuyuan Polymetallic Mine is the most representative mine in the field of tungsten beneficiation. Its "Shizhuyuan method" beneficiation process makes the processing technology of mixed scheelite and wolframite ore reach the international leading level. However, after years of development, the resource utilization of the rich ore section of the original III ore belt has come to an end, and the low-grade, complex and difficult ore, which accounts for the majority of resource reserves, has become the main body of future development. The ore in this section has more complex occurrence characteristics: the grade decreases gradually, the content of calcite gangue minerals such as fluorite and calcite are increasing, the symbiotic relationship is complex, and the distribution size is finer. As a result, it is more difficult to distinguish the ore body and make comprehensive utilization of resources. The "Shizhuyuan method" is more and more difficult to adapt to the changes of ore properties. The comprehensive recovery of tungsten gradually decreases to 63–65%, and the production index continues to deteriorate. In the flotation section of the "Shizhuyuan method", chelating collector and fatty acid are used as mixed collectors, and a large amount of water glass is used as inhibitors, and scheelite heating flotation process is adopted. Tungsten mines represented by Shizhuyuan Polymetallic Mine generally face the following difficult problems: Although the heating and refining of scheelite can further improve the concentrate grade, the tungsten loss caused by heating and refining seriously restricts the improvement of tungsten recovery; A large amount of water glass can inhibit tungsten and fluorite to a certain extent, which fundamentally affects their comprehensive recovery. It is difficult to reuse tail water caused by a large amount of water glass, which causes prominent environmental problems. With the difficulty of ore body sorting further increasing, the original process is more and more difficult to adapt to the change of raw ore properties, and the recovery gradually decreases.

From 2015 to 2016, three polymetal separation plants in Shizhuyuan Polymetallic Mine carried out complex flotation process commissioning. The new complex process industrial commissioning was carried out in Shizhuyuan Polymetallic Mine1000-ton polymetal processing plant. After about one month of production commissioning, the production index of roughing processing reached the standard and tended to be stable. The grade of crude WO_3 concentrate was about 15–25%, and the recovery was 78–80%. The tin recovery was increased by more than 10%. In August 2015, the 2000-ton metal sorting plant in Shizhuyuan switched to the new process. In July 2016, the 3000-ton multi-metal sorting plant in Dongbo switched to the new process, and the comprehensive recovery of tungsten was significantly improved. The beneficiation process of Shizhuyuan Polymetallic Mine Dongbo 3000 t/d plant is two-stage continuous grinding—iron processing—full flotation of sulfide ore, molybdenum—bismuth-sulfur separation, bismuth-sulfur separation—mixed ores of scheelite and wolframite, mixed ores of scheelite and wolframite at room temperature—fluorite flotation. Figure 6.18 shows the process flow chart of the mixed flotation of scheelite and wolframite. The separation process of scheelite and wolframite mixed flotation operation is one roughing, two cleaning, two cleaning, one cleaning, and the cleaning foam is returned to the rough selection, and the rest of the middle ore is returned in sequence. The flotation column is used for the roughing and cleaning separation of the scheelite and wolframite mixed flotation, and the flotation machine is used for the scavenging separation. The foam is selected as the scheelite and wolframite mixed concentrate in the second column, and the flotation tailings are sent to the fluorite flotation system. In this process, desulfurized tailings are adjusted by sodium carbonate pulp pH value is about 9.8, adding complex collector (lead nitrate 600 g/t, benzohydroxamic acid 400 g/t), No. 2 oil 10 g/t for pulping, and then roughing concentrate is obtained through flotation column once roughing, select section adding a small amount of salinated water glass (water glass 200 g/t, 70 g/t aluminum sulfate) for two times to obtain tungsten concentrate grade of more than 40%. The process does not need to add water glass in the tungsten roughing process, and realizes the simultaneous flotation enrichment of monochrome tungsten ore through the extremely high selectivity of Pb-BHA complex collector. The concentrate grade and recovery are significantly improved, and the enrichment of fluorite in tungsten roughing concentrate is avoided, and fluorite is not obviously suppressed, which creates favorable conditions for the subsequent flotation recovery of fluorite.

Beneficiation technical indicators: raw ore grade WO_3 0.36%, Sn grade 0.12%, scheelite and wolframite mixed concentrate grade WO_3 40.37%, Sn 2.11%, recovery WO_3 72.88%, Sn 11.08%.

In the traditional fatty acid flotation process, a large amount of water glass and fatty acids are added in rough sorting operation, resulting in low grade of crude concentrate, high content of calcite and fluorite gangue minerals, and greater pressure of subsequent heating and sorting. And the recovery cannot be further improved due to the influence of a large amount of water glass. The new process cancels the addition of water glass and fatty acids in the crude selection operation. With the help of the strong selective collection ability of scheelite and calcite gangue minerals in Pb-BHA complex collector, and the strong selective inhibition ability of scheelite, calcite and silicate minerals in Al-Na_2SiO_3 polymer inhibitor, the mixed floatation of scheelite and wolframite at room temperature was realized, the grade of crude concentrate and final concentrate was greatly improved, and the recovery was also significantly improved. There is almost no enrichment of fluorite in tungsten concentrate, which creates a good condition for the subsequent tungsten selection operation and the recovery of fluorite, which is conducive to the comprehensive recovery and utilization of resources. The flow comparison between Shizhuyuan GY method and the new tungsten flotation process based on Pb-BHA complex collector is shown in

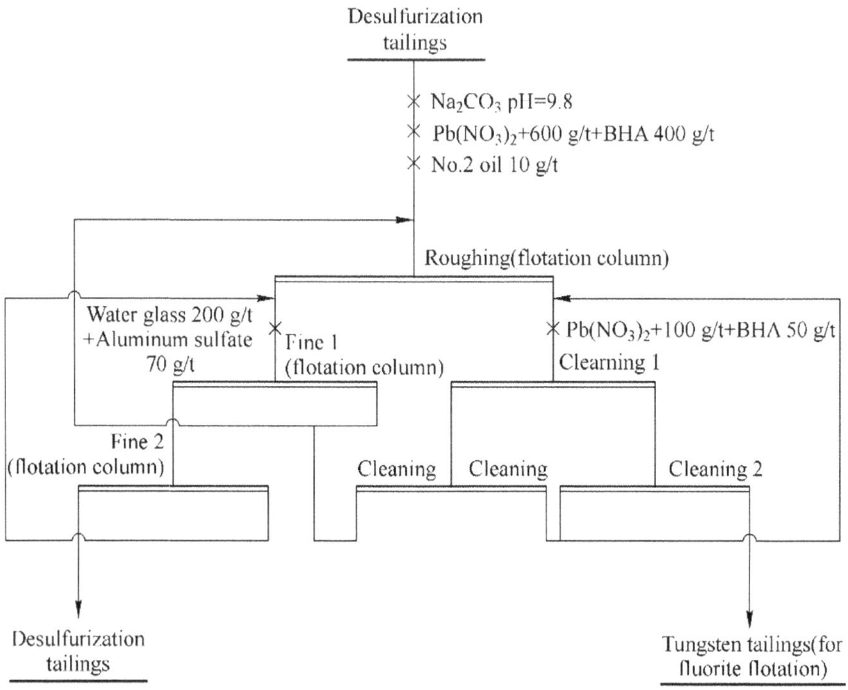

Fig. 6.18 Process flow chart of complex in Shizhuyuan polymetallic mine 3000 t/d separation plant

Fig. 6.19. The advantages of complex collector are as follows: first, it has good selectivity and strong collecting ability for wolframite and scheelite, but has no collecting ability or weak collecting ability for fluorite and silicate minerals. It is especially suitable for the separation of tungsten minerals and calcium gangue minerals. Second, normal temperature flotation, instead of the "Petrov" heating flotation process, greatly simplified the flotation process, shorten the operation process; Third, the addition of a large amount of water glass in the traditional process was canceled, which greatly improved the comprehensive recovery of tungsten. The application in Shizhuyuan polymetal processing plant made the recovery of tungsten increased from 63 to 65% of the original process to more than 70%, and was conducive to the efficient flotation enrichment of fluorite in tailings. Fourth, the process is short, the energy consumption is small, and the beneficiation water is easy to reuse. Compared with the traditional fatty acid method and chelating collector method, metal–organic complex flotation process has significant advantages, and has become one of the important development directions of tungsten flotation process.

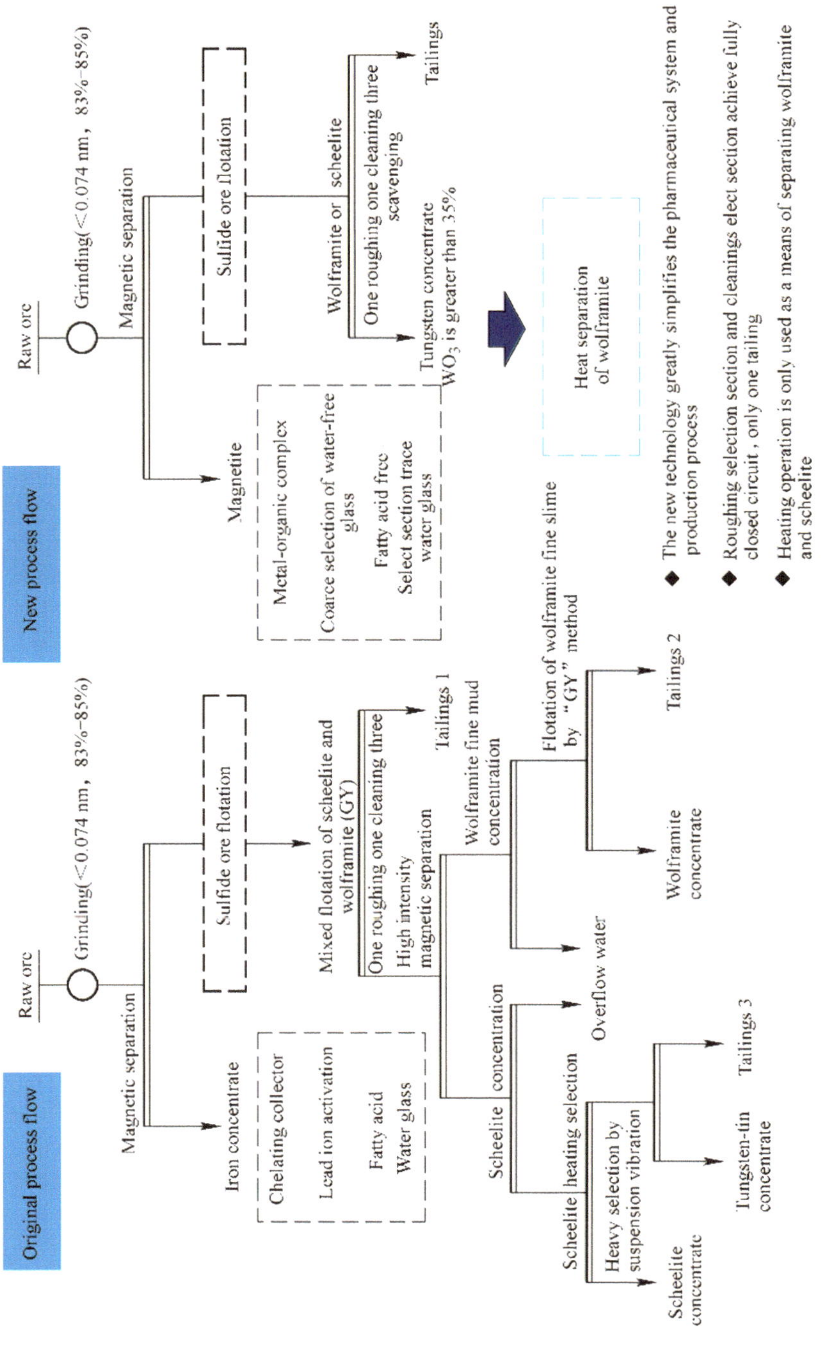

Fig. 6.19 Tungsten flotation process by GY method (left) and complex tungsten flotation process (right)

6.5.2.2 Xingluokeng Tungsten Mine, Ninghua, Fujian

The original flotation process of Xingluokeng Tungsten Mine fine mud is desliming—desulphurization—normal temperature flotation—centrifugal separator selection. The problem is that wolframite and scheelite have fine dissemination particle size, $flotation_3$ concentrate grade is only 3–5%, further concentrate grade improvement is not obvious, can only be further enriched by centrifuge gravity separation. Through the analysis of roughing concentrate mineral facies, the main gangue in the roughing concentrate with WO_3 grade of 4.35% is calcite (about 20%) and fluorite (about 30%). Therefore, the main contradiction of tungsten concentrate grade improvement is focused on the selectivity of collector to calcite and fluorite. According to the properties and characteristics of raw ore, the new flotation process of tungsten fine slime in Xingluokeng Tungsten Mine adopts the process flow of "roughing by GY method—centrifuge cleaning—complex re-cleaning". The flotation process of "GY method" is used in the roughing section of the new process to ensure that the tungsten roughing obtains a higher recovery; Then the roughing concentrate enters the centrifuge for preliminary enrichment of tungsten concentrate, and the fine calcite, fluorite and other calcite gangue minerals are removed; The centrifuge concentrate then enters the flotation process of the complex, and the concentrate grade is further improved by virtue of the high selectivity of the complex collector, and high grade tungsten concentrate is obtained. In this scheme, in order to maximize the recovery of tungsten concentrate, the concentrate grade of roughing separation section by "GY method" and centrifuge preconcentration section in the original process was reduced by adjusting the reagent system and process. In the separation section by complex method, high selectivity of Pb-BHA complex collector was used to achieve efficient enrichment of tungsten ore.

In March 2020, the tungsten slime flotation workshop carried out the industrial production commissioning of the new flotation process of tungsten slime by metal–organic complex method, as shown in Fig. 6.20. Roughing adopts the "GY method" reagent system with lead ion as activator and benzohydroxamic acid TW705 and fatty acid as collector. Roughing concentrate is firstly sorted by SLon-1600 centrifugal concentrator, roughing tailings are swept three times to get final tailings, and centrifuge sorted tailings and scavenging concentrate are returned in sequence. Among them, roughing, scavenging one and scavenging two use flotation column, and scavenging three use flotation machine. Centrifuge concentrate WO_3 grade is about 12% and enters the select operation section. In complex sorting operation, the complex of lead nitrate and TW705 was used as collector, and CU was used as foam regulator. Tungsten concentrate was obtained by two roughing and two scavenging. The results show that the new process has good adaptability to ore feeding, good index and stable production.

Beneficiation technical index: raw ore grade is 0.16%, concentrate grade is 45.25%, select recovery reaches 95.51%, comprehensive recovery of the whole process reaches 81.30%.

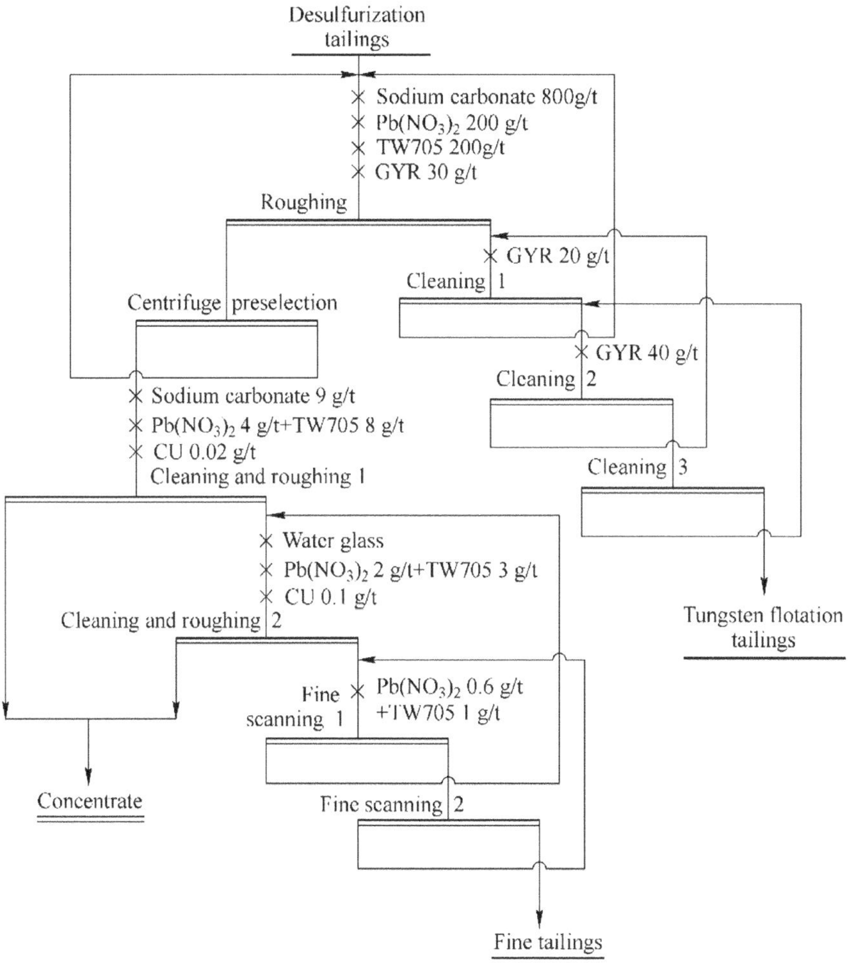

Fig. 6.20 Process flow chart of Xingluokeng tungsten mine complex flotation [41]

6.5.2.3 Huangshaping Tungsten-Molybdenum Polymetallic Mine in Hunan Province

Hunan Nonferrous Polymetallic MineHuangshaping Tungsten-Molybdenum Polymetallic Mine is one of the representative mines in China. The mining area is an important lead and zinc producing area in China. After nearly 60 lead–zinc mining, lead and zinc resources are nearly depleted. The proved reserves of ore resources are about 80 Mt, which is an important and continued resource.

1. Ore properties

The ore type is mainly skarn type tungsten-molybdenum ore, accounting for 75% of the total. Tungsten minerals are mainly scheelite, which are generally distributed in diopside, fluorite and garnet skarn in the form of medium roughing-grained disseminated and veined. Scheelite generally contains molybdenum, molybdenum and tungsten have an obvious reciprocal relationship, molybdenum exists in the scheelite lattice in the form of homomorphism. Particle size analysis shows that the average particle size of scheelite is 0.12 mm, and it belongs to medium fine-grained embedment. The chemical composition analysis results of Huangshaping Tungsten-Molybdenum Polymetallic Mine are shown in Table 6.11. The main elements in raw ore are O, Ca, Si, F, Fe, useful elements are W, F and WO_3 grade is 0.25%. The mineral composition and relative content of Huangshaping Tungsten-Molybdenum Polymetallic Mine are shown in Table 6.12, indicating that the target minerals in the raw ore are mainly scheelite and fluorite, as well as molybdenite, sphalerite, pyrite and magnetite, etc. Gangue minerals are mainly calcium ferric garnet, calcite, feldspar and quartz. The scheelite has uniform particle size, ranging from 0.04 to 0.32 mm. It belongs to the fine to micro-fine grained uniform distribution type.

2. Process flow and technical indicators

Huangshaping Tungsten-Molybdenum Polymetallic Mine was originally heated flotation process, but production was suspended in 2015 due to unsatisfactory technical indicators. In 2020, Huangshaping Tungsten-Molybdenum Polymetallic MinePolymetallic Mine carried out complex flotation process commissioning, and achieved good production indicators. The current production process of the concentrator is shown in Fig. 6.21. The raw ore is first desorred by magnetic separation, which removes iron minerals and magnetic impurities in the pulp to obtain iron

Table 6.11 Chemical composition analysis results of Huangshaping tungsten-molybdenum polymetallic mine

Elements	WO_3	SiO_2	CaF_2	$CaCO_3$	Fe	Zn	Mo	Bi	Al_2O_3	S
Content/%	0.25	35.95	18.59	3.46	8.87	0.18	0.021	0.008	7.90	0.62

Table 6.12 Mineral composition and relative content of Huangshaping tungsten-molybdenum polymetallic mine

Minerals	Content/%	Minerals	Content/%	Minerals	Content/%	Minerals	Content/%
Scheelite	0.224	Chalcopyrite	0.003	Mica	0.933	Apatite	0.037
Wolframite	0.006	Galena	0.004	Chlorite	1.622	Quartz	3.550
Molybdenite	0.236	Topaz	0.014	Sphene	0.042	Feldspar	7.522
Sphalerite	0.032	Fluorite	12.91	Siderite	0.262	Zircon	0.006
Cassiterite	0.01	Calcite	6.46	Ilmenite	0.011	Thoriite	0.004
Toxic sand	0.186	Diopside	8.869	Magnetite	1.879	Others	0.124
Pyrite	0.392	Tremolite	1.310	Limonite	0.356	Total	100.000
Pyrrhotite	0.022	Calcium iron garnet	51.954	Epidote	1.020		

concentrate. Then it enters the flotation process of sulfide ore. Molybdenum and bismuth mixed flotation adopts the process of one rough, three fine and three scavenging to get molybdenum and bismuth mixed concentrate. Copper sulfate is used as zinc activator for sulfur roughing, xanthate and kerosene are used as collector, and BK205 is used as foaming agent. Molybdenum, bismuth and sulfur mixed concentrate was separated by the process of one roughing, four refined and two swept. The desulfurized tailings go into the tungsten flotation system. Tungsten flotation adopts the process flow of one roughing, five fine, three scavenging, and middling ore return in sequence. Tungsten roughing Al-SBL1000 g/t, Pb-BHA 800 g/t, tungsten scavenging Pb-BHA 200 g/t; Tungsten selection for a total of five times, the dosage of Al-SBL is 200 g/t, 100 g/t, 50 g/t. The final tungsten concentrate product is obtained, and the tailings enter the fluorite flotation system.

Beneficiation technical index: raw ore grade WO_3 is 0.28%, final tungsten concentrate grade WO_3 is 31.76%, recovery is 72.36%.

6.6 Summary

The process of tungsten ore flotation has experienced several main stages, such as fatty acid flotation process, chelating agent flotation process and metal–organic complex flotation process, which promoted the continuous improvement of the flotation technology level of tungsten ore. The adaptability of fatty acid method to ore is relatively strong, especially the heating flotation process has a good adaptability to the ore properties, at the same time has the advantages of low agent cost, stable flotation index and so on, so it occupies a dominant position in scheelite mines at home and abroad. But its process is long and complex, heating flotation process energy consumption and cost is high, "fatty acid—water glass" circulation system also restricts the improvement of scheelite recovery; Chelator-based flotation process can realize the mixed flotation of scheelite and wolframite ore. Compared with the fatty acid method, the chelator-based flotation process can obtain high grade of crude concentrate, which can greatly reduce the ore amount in the heating and select section, simplify the production process, and the loss of fluorite in the tungsten flotation section is low. But scheelite must be heated flotation to obtain high grade concentrate, resulting in a long process, the use of chelating agent also makes the flotation reagent cost is high. The complex flotation process realizes the efficient flotation of tungsten ore in normal temperature and short flow, which can replace the heated flotation process to a certain extent, simplifies the working flow, and provides a new technical solution for the efficient development and utilization of complex tungsten resources.

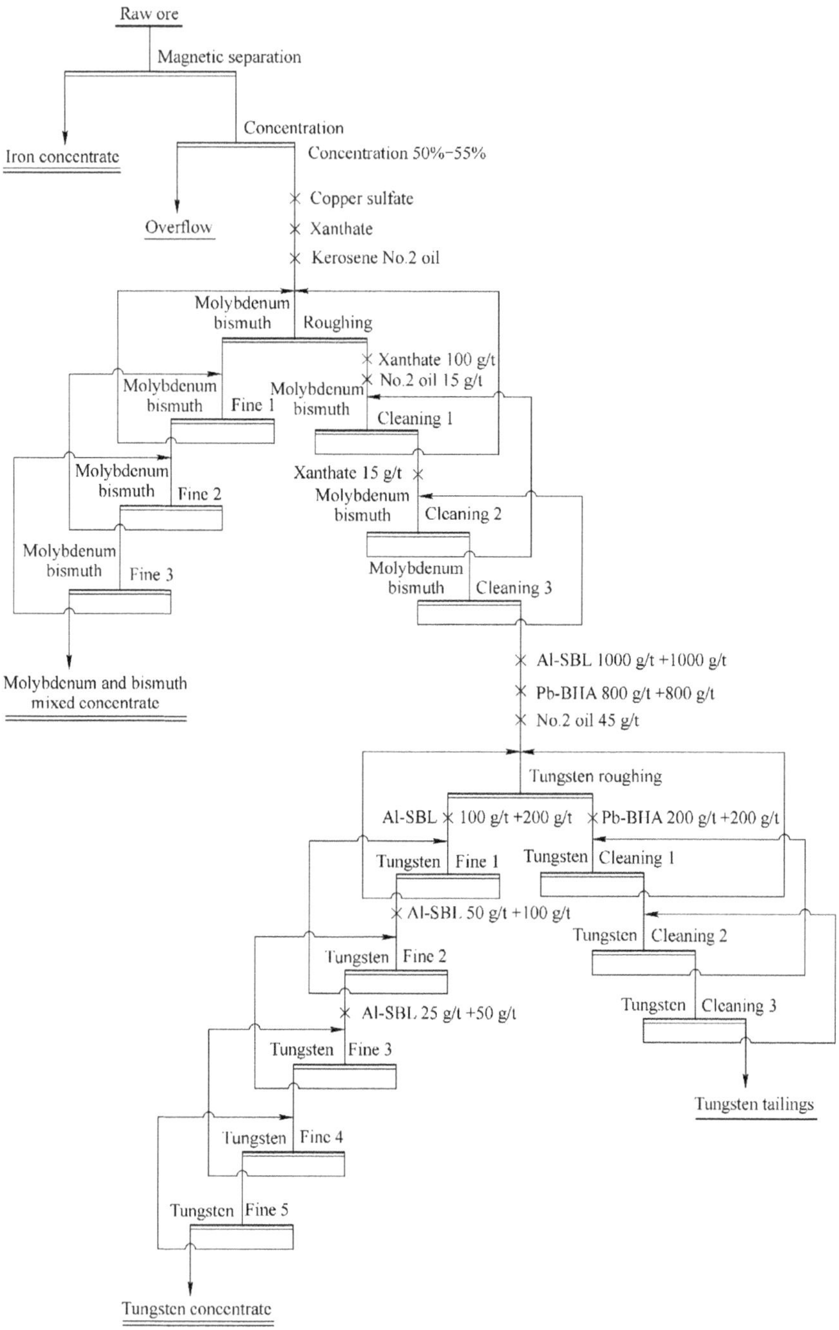

Fig. 6.21 Flotation process flow of Huangshaping tungsten-molybdenum polymetallic mine [42]

Acknowledgements This work was supported by National key research center and development Program of the 14th Five-Year Plan (No. 2022YFC2905105 and No. 2022YFC2905104), National Natural Science Foundation of China (NSFC) (No. 52122406 and No. 52374291), Hunan High-tech Industry Technology Innovation Leading plan (No. 2022GK4056), Hunan Innovative Province construction special Project (No. 2020RC3001).

References

1. Zhang Y (2012) Study on performance and mechanism of flotation separation inhibitor for scheelite and gangue minerals containing calcium. Central South University
2. Yang X (2018) Beneficiation studies of tungsten ores—a review. Miner Eng 125:111–119
3. Sun W, Hu Y, Qin W et al (2000) Research progress of flotation reagents for tungsten ore. Prot Utilization Miner Resour 3:42–46
4. Liu H (2010) Research on new flotation technology of refractory scheelite at room temperature. Mineral Processing Engineering of Central South University
5. Kang J, Liu Y, Ahmed Khoso S et al (2018) Significant improvement in the scheelite heating flotation with sodium sulfide. Minerals (Basel) 8(12):587
6. Li P, Sun W, Cao X et al (2016) Experimental study on temperature separation process of mixed flotation of with scheelite and wolframite. China Tungsten Ind 31(04):32–36
7. Gao Y, Liu Q, Dong J et al (2021) Experimental Study on Flotation of a scheelite from Yunnan. China Tungsten Ind 36(03):31–35
8. Xu G, Wang P, Zhao T (2011) Optimization of pre-heating dedusting of scheelite. China Molybdenum Ind 35(01):23–25
9. Qiu X, Dong T (2012) Modern tungsten ore dressing
10. Jin Q, Zhang X, Yang M (2019) Experimental study on heating flotation of refractory scheelite. Min Metall 28(02):32–36
11. Wen S, Wang L, Guo L et al (2018) Flotation test of a low-grade tungsten ore in Jiangxi. Met Mine 11:91–94
12. Yang C, Tan X, Guo J et al. Application of 731 collector in high calcium scheelite thermal flotation. China Tungsten Ind 20,35(03):38–41
13. Yang J, Che W, Wang H et al (2021) Experimental study on flotation of scheelite from molybdenum tailings. Mod Min 37(03):105–109
14. Guo M, Wang Y, Cheng C (2016) Research on separation process of a scheelite ore in Henan Province by joint columns-machine process optimization. China Tungsten Ind 31(03):50–54
15. Kang J, Sun W, Chen C et al (2016) Experiment on improving scheitrotungsten roughing concentrate grade of a tungsten-molybdenum ore in Henan Province. Met Mine 03:91–94
16. Wang Y, Kang J, Sun W et al (2017) Application of novel inhibitors in flotation of a low-grade schetungsten ore in Luanchuan. Conserv Utilization Miner Resour 04:44–47
17. Ye X-J (1999) Research on flotation process of scheelite at room temperature. China Tungsten Ind Z1:3–5
18. Deng L, Zhou X (2008) Research on scheelite flotation process at room Temperature. China Tungsten Ind 05:20–22
19. Wang Q, Zhou J, Liu Z et al (2003) Study on preparation of Y88 scheelite at room temperature. Hunan Nonferrous Met 05:11–12 (in Chinese)
20. Ding D (2021) Experiment on tungsten recovery by flotation from scheelite ore. Mod Min 37(06):241–243
21. Qi Z, Wang L, Sun D et al (2021) Research on flotation of a low-grade scheelite with a new collector CK-2. Min Metall Eng 41(02):66–69
22. Zeng Y, Cao Y, Huang G et al. Experimental study on flotation process of scheelite in Jiangxi Province at normal temperature. Min Metall Eng 20,40(3):47–49, 53

23. Huang W (1989) Study on flotation of scheelite by lime method. Jiangxi Metall 01:16–19
24. Yang Z, Xie X, Tong X et al (2015) The role of lime in flotation process. Compr Utilization Miner Resour 02:17–21
25. Yang S (1982) Flotation of scheelite by lime process. Jiangxi Metall 02:39–41
26. Liu H, Xu Z (2013) Study on flotation process of low-grade scheelite by lime process at constant temperature. Compr Utilization Miner Resour 02:33–35
27. Li M, Gao Y, Yang Y (2020) Experimental study on flotation of scheelite by lime method. China Met Bull 10:49–50
28. Zhang A, Li G, Wang R (2009) Research on flotation Process of a scheelite. Mod Min 25(4):34–35
29. Sun C-Y, Cheng X, Li C-G (2004) New comprehensive beneficiation technology of complex polymetallic ore with tungsten-bismuth-molybdenum-fluorite—Shizhuyuan method. China Tungsten Ind 19(5):8–14
30. Zhou X, Lin R (2000) Research on new process of GY method of mixed floatation of scheelite and wolframite. Compr Utilization Miner Resour 02:1–4
31. Guo J (2011) Modification of tungsten slime separation process in Xingluokeng tungsten mine. Met Mine 06:97–100
32. Zhao J (2014) Study on new technology and mechanism of separation of low-grade scheelite slime sand. Central South University
33. Zhang Z, Zhang X, Ye Z et al (1999) Research on new tungsten flotation technology of Shizhuyuan polymetallic ore by GY method. Min Metall Eng 1999(4):22–25
34. Li A, Yang M (2017) Optimization and production practice of separation process of complex refractory tungsten slime. Nonferrous Met (Miner Process Sect) 2:46–51
35. Cheng X (2000) Study on a new method for separation of tungsten minerals and calcium-bearing minerals and the mechanism of chemical action I. Study on a new method for separation of tungsten minerals and calcium-bearing gangue minerals by flotation -CF method. Miner Process Met Ores Abroad 06:21–25
36. Xiao Q, Li C, Kang G (1996) Research on beneficiation process of tungsten stem flotation by CF method in Shizhuyuan polymetallic ore. Min Metall 03:26–32
37. Deng H (2011) Study on the new process and technical and economic evaluation model of separation and smelting combined with low grade complex refractory tungsten ore. Central South University
38. Han H, Hu Y, Sun W et al (2017) Fatty acid flotation versus BHA flotation of tungsten minerals and their performance in flotation practice. Int J Miner Process 159:22–29
39. Wei Z, Hu Y, Han H et al (2018) Selective flotation of scheelite from calcite using Al-Na2SiO3 polymer as depressant and Pb-BHA complexes as collector. Miner Eng 120:29–34
40. Wei Z, Hu Y, Han H et al (2019) Selective separation of scheelite from calcite by self-assembly of H2SiO3 polymer using Al3+ in Pb-BHA flotation. Minerals 9(1):43
41. Li A, Wei, Han H et al (2021) Production practice of new flotation process for complex collector of Xingluokeng mixed flotation of scheelite and wolframite. Metal Mine 06:73–79
42. Hu Z, Huang S, Zhou H (2021) Experimental study on improving the recovery of a tungsten polymetallic ore. Min Metall Eng 41(05):75–78

Chapter 7
Development and Utilization of Fluorite Resources Associated with Tungsten Tailings

Haisheng Han, Zhao Wei, Jianhua Kang, and Wei Sun

Abstract Fluorite, a key strategic mineral, is indispensable in high-tech industries such as aerospace, semiconductors, and environmental protection. Global fluorite reserves, led by China, Mexico, and South Africa, underscore its strategic value, especially as high-tech demands rise. This chapter focuses on the development and utilization of fluorite resources associated with tungsten tailings, which are often overlooked due to their complex mineralogy and challenging separation processes. The various flotation reagents, including anionic collectors like fatty acids and sulfonic acids, cationic collectors, amphoteric collectors, and combined collectors were reviewed, which have shown promise in improving the selectivity and recovery of fluorite. The use of inhibitors such as water glass, sodium hexametaphosphate, and organic inhibitors like tannic acid and starch is also discussed, emphasizing their role in achieving effective separation of fluorite from gangue minerals. The chapter further explores the challenges in the development and utilization of associated fluorite resources, particularly the need for new technologies to address the comprehensive recovery of complex and low-grade ores. The case studies of Shizhuyuan Polymetallic Mine in Hunan Province and Huangshaping Tungsten-Molybdenum Polymetallic Mine in Hunan demonstrate the successful application of these technologies, resulting in significant improvements in fluorite recoveries and economic benefits. The efficient development and utilization of fluorite resources from tungsten tailings are vital for ensuring the sustainable supply of fluorine chemicals, which are essential for the growth of strategic emerging industries.

Keywords Fluorite · Tungsten tailings · Flotation reagents · Flotation technology

H. Han (✉) · Z. Wei · J. Kang · W. Sun
Central South University, Changsha, China

Engineering Research Center of Ministry of Education for Carbon Emission Reduction in Metal Resource Exploitation and Utilization, Central South University, Changsha, China

State Key Laboratory of Mineral Processing, Changsha, China

H. Han
Technology Research Center of Hunan Province for Comprehensive Utilization of Associated Fluorite and Fluorine Chemical Engineering, Chenzhou, China

W. Sun et al., *Flotation Chemistry of Tungsten Minerals and Its Application*,
https://doi.org/10.1007/978-981-96-5077-4_7

7.1 Introduction

Fluoride, known as "industrial monosodium glutamate", is often widely used as an additive in aerospace, semiconductor, communications, environmental protection technology, clean energy and other strategic emerging industries. It is also an important raw material for traditional chemical industry, metallurgy, building materials, optics and other industries, and has an irreplaceable strategic position [1]. Clark value of fluorine element is 0.046%, only fluorite (fluorine content 48.9%), cryolite (fluorine content 54.3%), fluorite (fluorine content 61%) and other minerals in the earth fluorine element has the value of industrial exploitation [2]. The chemical formula of fluorite is CaF_2. As one of the highest fluoride minerals in the world, fluorite can be formed in magmatic, sedimentary, hydrothermal and other geological conditions. It has the characteristics of wide distribution and large scale [3, 4]. By the end of 2020, the global reserves of fluorite have reached 320 Mt, and fluorite has become the most important mineral raw material source of fluorine resources in the world [5]. The strategic emerging industry led by fluorine-chemical industry has been highly valued by all countries [6–9]. Fluorite, as the main raw material, is known as the "second rare earth" and has been included in the list of strategic mineral resources by China, the United States, the European Union, Japan and other countries or economies.

7.2 Fluorite Resources and Their Strategic Significance

7.2.1 Uses of Fluorite and Its Strategic Significance

Fluorite is commonly referred to as fluorite. It is one of the typical non-metallic minerals, isometric system, composed of two elements, calcium (Ca) and fluorine (F) to form calcium fluoride (CaF_2), with a relative molecular mass of 78.07. Pure fluorite is colorless, and the change in color is caused by different impurities. Calcium is easily replaced by yttrium and cerium. In addition, iron, sodium and barium are also common impurities. Therefore, the appearance of fluorite is often green, purple, white, blue, yellow, rose and other colors, and some can emit fluorescence, hence the name. Table 7.1 shows the basic properties of fluorite.

Fluorite is an important raw material of fluorine chemical industry. It has been widely used in metallurgy, chemical industry, medicine, agriculture, aerospace, ceramics and precision instruments. With the development of deep processing of materials, the use of fluorite has been extended to fine, involving atomic energy, rockets, aircraft and other advanced scientific frontiers.

Table 7.1 Physical and chemical properties of fluorite

Chemical formula	CaF_2	Chinese name	Calcium fluoride
Often containing impurities	Yttrium, cerium, uranium, thorium, etc.	Solubility	Slightly soluble in water
Crystal system	Equiaxed crystal system	Space group	$Fm\overline{3}m$
Number of cell atoms	4	Relative illuminance value	1.433–1.448
Toughness	Crunchy and brittle	Transparency	Complete transparency
Cleavage	Complete cleavage, the {111} plane is the normal cleavage plane	Cleavage	Cracks are rough and fuzzy under the {011} crystal plane family
Mohs hardness	4	Melting point	1360 °C
Color	Colorless, purple, green, blue, brown, etc.	Specific gravity	3.00–3.25

7.2.1.1 Fluorine Chemical Industry

Fluorite is consumed most in the fluorine chemical industry, with a consumption ratio of 52% [8]. The most important use of fluorite in the field of fluorine-chemical industry is the production of basic raw material—hydrofluoric acid, the consumption of fluorite accounts for about 50–60% of the total production of fluorite in the world. In the process of producing hydrofluoric acid, the quality of raw materials from the source is required to be high, and the impurity content must be strictly controlled. The purity of CaF_2 in China is generally required to be 93–98%. This kind of fluorite is called chemical grade fluorite or acid grade fluorite.

The most mature production method of hydrofluoric acid is mainly fluorine-sulfuric acid method, through acid grade fluorite (fluorite concentrate) and sulfuric acid in the heating furnace or reaction tank generation. The process is divided into two steps. At present, the process of Buss in Switzerland is the most common abroad. Hydrofluoric acid as a colorless liquid, volatile, has a strong irritating odor and strong corrosion. As it is the most important product in fluoride, the domestic development rate is very fast. In the field of inorganic fluorine chemical materials, hydrofluoric acid is used to produce aluminum fluoride, artificial cryolite, fluorine refrigerants and secondary fluoride such as fluorine plastics, fluorine rubber and so on. In terms of organic, hydrofluoric acid can be used with chloroform and carbon tetrachloride. Different combination conditions can be selected according to the different application of the products produced. Common fluorine compounds with low toxicity can be used as refrigerants, aerosol accelerators, and some solvent polymerization intermediates are used for fluorocarbon resin [10].

As shown in Fig. 7.1, based on the production of hydrofluoric acid, the fluorine chemical industry can be divided into inorganic fluorine chemical industry and organic fluorine chemical industry according to its end product application field.

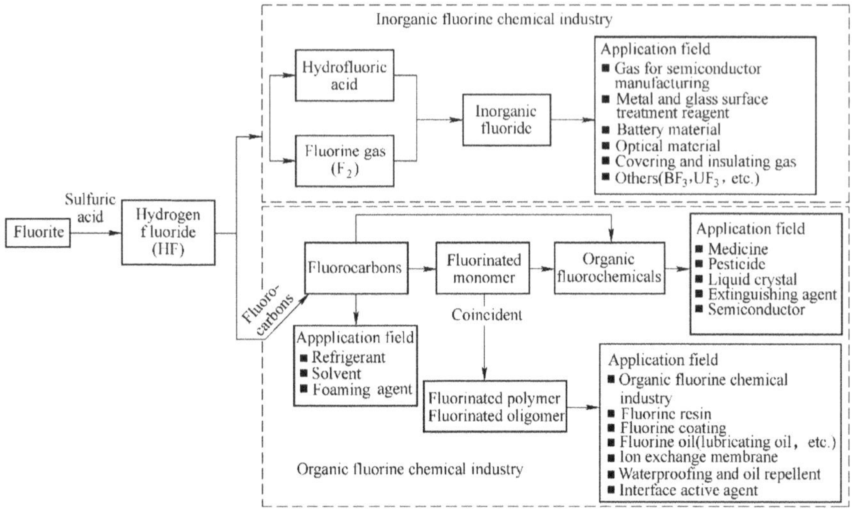

Fig. 7.1 Flow chart of fluorine chemical industry [8]

Inorganic fluorine chemical industry is mainly used in fluorine-containing electronic chemicals, fluorine-containing special gases, other inorganic fluoride, etc. The products are widely used in semiconductor manufacturing, battery materials, optical materials, insulating gases and other fields [11, 12]. Organic fluorine chemical industry is mainly used in the production of fluorocarbon products, fluorinated polymer, organic fluorinated chemicals, products are used in refrigerants, foaming agents, fluorine oil, medicine, pesticides, liquid crystal, ion exchange membrane, semiconductor manufacturing and other fields [13]. With the booming development of the fluorine chemical industry, the demand for fluorite in the fluorine chemical industry may continue to increase in the future.

7.2.1.2 Metallurgical Industry

The metallurgical industry has a huge demand for massive fluorite. In steelmaking (roughing smelting), flux is usually added according to the slag melting situation. Fluorite can help other slag-making raw materials to melt and reduce viscosity, so as to solve the problem of liquid metal and slag spatter caused by poor slag melting performance in the process of steelmaking. It plays an irreplaceable role in ensuring the safe operation of smelting, metallurgical quality and energy saving [14]. In the converter, electric furnace steelmaking, fluorite has other mineral raw materials can not match the effect of slag melting. In China, fluorite as a flux used in converter and electric furnace steelmaking is very common, the general usage is 1.5–8 kg/t, and special steel, high quality steel demand for fluorite is greater. Therefore, with the adjustment of the product structure of the steel industry in the future, the demand for fluorite will show a steady and rising pattern.

7.2.1.3 New Material Industry

As "industrial monosodium glutamate", fluorine-containing materials are becoming increasingly important in strategic emerging industries such as new energy, new energy vehicles, emerging information, new medicine, energy conservation and environmental protection, aviation and aerospace. For example, in the new energy and new energy automobile industry, fluorine-containing materials are used in photovoltaic power generation, secondary lithium-ion batteries, proton exchange membrane batteries, wind power coatings and other manufacturing because of their chemical corrosion resistance, heat resistance, aging resistance, insulation and low refractive index [15]. In the emerging information industry, fluorine chemical products are widely used in electronic products photolithography, etching, fine film preparation, cleaning, impurity removal and other processes [12]. In the new pharmaceutical industry, fluorine-containing compounds are widely used in the preparation of new pharmaceutical products such as anticancer agents, anesthetics, fungicides and animal reagents due to their characteristics of being soluble in lipids, safe, healthy and with few side effects.

In the field of aviation industry, PTFE as an indispensable sealing material in the aerospace industry, has many excellent properties, such as good corrosion resistance, weather resistance, non-toxic, no pollution and high and low temperature resistance [16]. Due to the continuous development of rubber products containing fluorine materials, at present, low energy consumption, high performance, long life of multi-functional fluorine rubber is also widely used in the field of aerospace.

In addition, anhydrous hydrogen fluoride is the main chemical used in the preparation of fluoride used in the missile program, and the propellant of fluorine is able to control the combustion rate of solid propellant particles. Therefore, hydrogen fluoride as a necessary raw material for the production of jet liquid propellant, missile jet fuel propellant, directly determines the total cost of the rocket, but also controls the energy of the whole propellant system [14].

Fluorite resources are related to the stable development of various industries, especially new energy and new material industries, and are important strategic minerals. In order to ensure the rational, orderly and economic development and utilization of fluorite resources, a series of policies, measures and regulations concerning the development and utilization of fluorite resources and its import and export trade have been introduced in China since the twentieth century. In 2010, The General Office of the State Council of China issued the Notice on Taking Comprehensive Measures to Control the Exploitation and production of refractory clay fluorite, which defined fluorite as a "precious resource that can be used up and is non-renewable". According to the 12th Five-Year Development Plan of China's Fluorine-chemical Industry, fluorite "is a world-class scarce resource similar to rare earth". In the National Mineral

Resources Plan (2016–2020), fluorite was listed in China's "strategic mineral catalog". In 2018, the US Department of the Interior included fluorite in the list of 35 key minerals vital to the US economy and national security, and many other countries and regions, including the European Union and Japan, also listed it as a strategic mineral or key mineral.

7.2.2 *Distribution and Development of Fluorite Resources*

7.2.2.1 Fluorite Resources in the World

Reserves of the world's fluorite resources continue a slow upward trend, with total world fluorite reserves at 320 Mt at the end of 2020, according to the newly released Mineral Commodity Summaries 2021 by the U.S. Geological Survey [5]. It is mainly distributed in Mexico, China, South Africa, Mongolia, etc., while the United States, the European Union, Japan, the Republic of Korea and India have almost no reserves of fluorite resources, forming structural scarcity.

As shown in Table 7.2, by the end of 2020, Mexico has 68 Mt of reserves, accounting for 21.25% of the world's fluorite reserves, ranking first in the world, mainly concentrated in Coahuila, SAN Luis Tosi and Guanajuato, with an average grade of about 60% fluorite. China's reserves of 42 Mt account for 13.13% of the world's total, ranking second in the world; South Africa has 41 Mt of fluorite reserves, accounting for 12.81% of the global reserves, ranking the third in the world. The fluorite deposits are mainly located in Delantuva Province and Northwest Province, and the fluorite deposits are shallow. Mongolia's reserves of 22 Mt account for 6.88% of the world, ranking the fourth in the world, and its reserves are mainly concentrated in the central and eastern provinces of Kent, Central Gobi and East Gobi. The remaining known reserves are mainly in Spain, Vietnam, the United States, The United Kingdom, Iran, Thailand, Kazakhstan, France, Namibia and other countries. According to the evaluation of metallogenic geological conditions, the Pacific Rim metallogenic belt accounts for more than half of the global fluorite reserves, which is the main distribution area of fluorite resources in the world.

7.2.2.2 Fluorite Resources in China

Fluorite ore is the dominant mineral species in China, and fluorite deposits are widely distributed in China [3]. According to the China Mineral Resources Report 2021, China's fluorite resources are distributed in 19 provinces (autonomous regions) by the end of 2020. Zhejiang, Jiangxi, Hunan, Fujian, Inner Mongolia and Anhui provinces all have relatively high fluorite reserves [17]. Compared with the global fluorite resources, China's fluorite resources have higher development value due to lower impurity content, especially arsenic, sulfur, phosphorus and other contents, and better mining conditions, so it occupies a pivotal position in the global fluorite resources

Table 7.2 World fluorite production and reserves statistics from 2019 to 2020 [5]

Countries	Fluorite production/kt		Reserves/kt
	2019	2020	
China	4300	4300	42,000
United States	NA	NA	4000
Germany	50	50	NA
Iran	55	55	3400
Mexico	1230	1200	68,000
Mongolia	718	720	22,000
Morocco	88	88	210
Pakistan	100	100	NA
South Africa	210	320	41,000
Spain	139	140	10,000
Vietnam	238	240	5000
Canada	80	100	NA
Myanmar	53	53	NA
Other	107	110	120,000
Total	7460	7600	320,000

[18]. Nanling metallogenic Belt is one of the most important fluorite accumulation areas in China. The proven reserves of fluorite mineral are 27.93 Mt, accounting for more than half of the reserves in China [17, 19].

According to the metallogenic series theory of Academician Chen Yuchuan and considering the genetic types and industrial types of fluorite deposits, the fluorite deposits in China can be divided into three types: sedimentary modification type, hydrothermal filling type and associated type [4, 20]. According to the same or similar combination of second-order ore-forming elements, 11 deposit subtypes—deposit types are further divided, as shown in Table 7.3.

There are about 230 major fluorite deposits in China, of which large and medium-sized fluorite deposits are mainly concentrated in the eastern coastal areas, central China and the Bayan Obo and Erenjiete area of Inner Mongolia. 83% of fluorite resources are distributed in Hunan, Zhejiang, Jiangxi, Inner Mongolia, Fujian and Yunnan provinces [21]. A single fluorite deposit is currently under development and utilization in China, accounting for 83% of the total deposits and 57% of the total reserves in China, mainly distributed in Inner Mongolia, Zhejiang, Jiangxi, Fujian and other provinces [22]. There are about 40 associated fluorite deposits, accounting for less than 20% of the total deposits and 43% of the total reserves. They are mainly distributed in Hunan, Inner Mongolia, Yunnan and Jiangxi provinces. China has large reserves of associated fluorite resources, but it is difficult to recover and utilize them comprehensively due to the constraints of the technical conditions of beneficiation, smelting and processing as well as the cost of beneficiation [23].

Table 7.3 Main types of fluorite ore deposits in China

Mineral deposits	Mineral deposit type	Essential elements of mineralization	Typical ore deposit
Depositional modified type	Sumochagan aobao sedimentary modified fluorite ore	Rifted basin + limestone + submarine volcanic eruption + fold (fracture) + magmatic activity	Sumo Chagan aobao, Inner Mongolia
			Bei Aobao Tu, Inner Mongolia
	Qinglong sedimentary modified fluorite ore	Sedimentary Basin + limestone + submarine volcanic eruption + fold (fault)	Dachang, Qinglong, Guizhou Province
			Fuyuan old factory in Yunnan Province
Hydrothermal filling fluorite ore	Qiba spring type hydrothermal filled fluorite ore	Intrusive rock + fracture	Qibaquan, Gansu Province, Qiyi Mountain, Inner Mongolia, Honganhua River, Hubei, Jiangle Changkou, Fujian, Chen Lou, Song County, Henan, and HeyuanZhiji, Guangzhou
	Wuyi type hydrothermal filling fluorite ore	Volcanic rock + fault	Yangjia, Wuyi, Zhejiang; Haojialou, Pingquan, Hebei; Ningguozhuang Village, Anhui; Sanbaotun, Yixian, Liaoning
	Bayanshan-type hydrothermal filling fluorite ore	Limestone + fracture + intrusive rock	Bayan Mountain, Changshan, Zhejiang Province, and Hongxi Plate, De'an, Jiangxi Province
	Hushan type hydrothermal filling fluorite ore	Volcanic rock + intrusive rock (subvolcanic rock) + fracture	Hushan, Suichang, Zhejiang Province
	Shuangjiangkou hydrothermal filling fluorite ore	Intrusive rock + fracture + limestone (xenolith)	Shuangjiang Estuary, South Hengnan, Hunan Province
Associated type fluorite ore	Bayan obo type iron niobium rare earth associated fluorite ore	–	Bayan obo type iron niobium rare earth associated fluorite ore, Inner Mongolia
	Shizhuyuan type tungsten, tin, molybdenum and bismuth associated fluorite ore	–	Shizhuyuan type tungsten, tin, molybdenum and bismuth associated fluorite ore in Hunan Province

(continued)

Table 7.3 (continued)

Mineral deposits	Mineral deposit type	Essential elements of mineralization	Typical ore deposit
	Taolin type lead–zinc associated fluorite ore	–	Taolin type lead–zinc associated fluorite ore, Hunan
	Bitter lawn barite is associated with fluorite ore	–	Chongqing Kuoxao type barite associated with fluorite ore

The fluorite ore in China is characterized by more lean ore, less rich ore, more refractory ore and less easy ore. The reserves of reserved resources that can be directly regarded as metallurgical grade rich ore (CaF_2 grade greater than 65%) only account for 11.6% of the total reserves of reserved resources, 70% of which are distributed in Zhejiang, Hubei, Inner Mongolia, Jiangxi and other provinces. The reserves of retained identified resources with CaF_2 grade between 30 and 65% account for 45.3% [3].

In summary, the characteristics of fluorite resources in China can be summarized as follows:

(1) Although China has great potential for fluorite resources, its geological work is not high at present due to the wide distribution of the resources.
(2) The proved deposits of fluorite are mostly distributed in Zhejiang, Jiangxi, Hunan, Inner Mongolia and Fujian provinces. The total reserves of fluorite in these five provinces accounted for about 90% of the total reserves of our country.
(3) There are a large number of single type fluorite ore deposits in China, but the distribution is extremely dispersed and the reserves are small, which makes it difficult to form a scale of product production. Associated (or symbiotic) fluorite deposits are concentrated and have extremely rich reserves, accounting for a large proportion of fluorite reserves, but it is difficult to recover and utilize them.
(4) With the continuous development of fluorite resources, high-grade rich ore gradually disappeared, more and more poor ore exploration, fluorite resources guarantee degree is seriously insufficient.

Although China is the largest producer and exporter of fluorite in the world, the long-term mining of fluorite has caused the domestic reserves of fluorite resources to decline gradually. The comprehensive utilization of associated fluorite resources has become the only way to relieve the shortage of fluorite resources in China. Since the 1970s, the recovery technology of associated fluorite has been carried out one after another in China. However, due to the complex mineral assemblage of associated fluorite ore deposit, fluorite is often disseminated in the form of fine veins, low grade and great difficulty in sorting, so it is difficult to realize the comprehensive recovery of associated fluorite [19, 24]. Until the twenty-first century, with the breakthrough and development of mineral processing agents, technology and equipment, the associated fluorite resources with huge reserves gradually began to be comprehensively exploited and utilized. According to the data of Land and Resources Bureau of the National Bureau of Statistics in 2018, the identified resources are 221 Mt, the basic reserves are 49.79 Mt, and the single fluorite resources that can be exploited and utilized are about 22 Mt, indicating that the degree of resource security is "seriously insufficient". The reserves of fluorite resources in Hunan Province are more than 100 Mt, but most of them are "symbolic associated ores that are difficult to use" under the current economic and technological conditions. The typical representative mine is Shizhuyuan Polymetallic Mine in Chenzhou, with more than 75 Mt of fluorite resources.

7.2.3 *Bottleneck of Development and Utilization of Associated Fluorite Resources*

Different from a single fluorite ore, associated fluorite resources often contain a variety of valuable metals, complicated separation process, high calcium carbonate content, difficult flotation separation, for a long time it is difficult to form high-quality fluorite concentrate ($CaF_2 > 95\%$), difficult to meet the needs of traditional fluorine chemical industry, greatly limiting the comprehensive utilization level of associated fluorite resources and the development of fluorine chemical industry. For example, the development and utilization of associated fluorite resources in Nanling metallogenic belt are mainly difficult as follows: (1) The traditional fluorite chemical industry requires fluorite concentrate grade to be at least 95%, and the production of fluorite chemical products with lower grade fluorite concentrate is a huge challenge. (2) The fluorite concentrate contains excessive sulfur and silicon, which seriously affects the downstream fluorite chemical industry, and the ultrafine fluorite concentrate dust is serious and the conversion rate of hydrofluoric acid is low. (3) The fluorite resources in Nanling metallogenic belt contain a large number of impurities, such as calcium carbonate and mica, which are the most difficult to separate in fluorite ore dressing. (4) The beneficiation process of main metal at the front end has a great influence on the flotation of fluorite at the back end, limiting the recovery of fluorite resources. (5) It is difficult to treat the fluorine-containing wastewater produced in the process of mineral processing and fluorine-chemical industry, and the environment is seriously polluted. Generally speaking, the mineral processing technology of associated fluorite resources in China is relatively disconnected from the technology of producing hydrofluoric acid from refined fluorite powder, and the comprehensive utilization rate of associated fluorite resources is not high. Therefore, it is urgent to develop a new technology for efficient and comprehensive utilization of complex and low grade associated fluorite resources, so as to improve the comprehensive utilization rate of fluorite resources and provide a strong guarantee for the fluorine chemical industry in China.

7.3 Fluorite Flotation Agents and Their Mechanism of Action

Fluorite is a typical calcium salt mineral; flotation technology is the main technology for the development and utilization of this kind of resources. How to solve the problem of flotation separation of fluorite and impurity minerals, to meet the needs of back-end fluorine chemical industry, has become the key to revitalization of associated fluorite resources, flotation agents (especially collectors and inhibitors) is the core of this technology.

7.3.1 Fluorite Flotation Collector and Its Mechanism of Action

7.3.1.1 Anionic Collector

Anionic collectors commonly used in fluorite flotation mainly include fatty acids, alkyl sulfuric acid or sulfonic acid, phosphoric acid and organic arsenate.

1. Fatty acids

Fatty acid collectors are the most widely used in the flotation process of fluorite minerals, including various fatty acids and their soaps, such as oleic acid, linoleic acid, tall oil, oxidized paraffin soap, etc. Oleic acid, the scientific name of octadecenoic acid, is a natural unsaturated acid, widely exist in animal and plant oils, rich in content, because of the strong collecting ability and low price, become the most typical collector for fluorite flotation.

Chen et al. [25] used oleic acid as collector, sodium carbonate and acid water glass as adjusting agent to flotation separation of fluorite from tungsten tailings of Xianglushan, Jiangxi Province, and obtained fluorite concentrate with CaF_2 grade of 91.88% and recovery of 50.26%. They not only created good comprehensive economic benefits, but also improved the comprehensive utilization rate of mine resources, which was conducive to reducing the stock of tailings. And extended the service life of the tailings pond. Li et al. [26] carried out flotation recovery test of fluorite from tungsten tailings of Shizhuyuan Polymetallic Mine. In view of the inhibition of the fluorite in tailings by high-alkali tungsten, they activated fluorite with sulfuric acid to neutralize the high-alkali pulp, took 733 as collector and water glass as inhibitor, and successfully realized the recovery of fluorite through the process of one rough, two cleaning and five concentrate. Zhou et al. [27] conducted a fluorite recovery test on low-grade tungsten-molybdate bismo-bismo-molybdenum flotation tailings from Huangshaping, using 733 as the fluorite collector, sulfuric acid + water glass +SF-1 as the inhibitor of fluorite roughing, and SF-2 and LP as the inhibitor of fluorite concentration. Through the process of first roughing and second scavenging, regrinding of roughing concentrate, reseparation of fine first and fine second middle ore, return of other middle ore in sequence and magnetic separation of concentrate, concentrate product with fluorite grade of 97.36% and recovery of 57.23% was obtained.

The deficiency of oleic acid collector lies in low solubility, poor activity and slow dispersion at low temperature, so many experts and scholars have made a lot of positive improvements on it, including pulp heating, strong stirring, saponification, sulfonation, emulsification and so on. Zhang et al. [28] modified the fatty acid collector. By saponifying the mixed fatty acid and then mixing it with surfactant SP and a small amount of additive F, a new collector containing carboxyl, hydroxyl and other active groups was obtained, which was used for flotation of a refractory fluorite ore in India. After the flotation process of one roughing, ten refined and two scavenging, the fluorite concentrates with CaF_2 grade of 91.12% and recovery of 77.26% was obtained. Zhou Weizhi successfully replaced oleic acid with sodium oleate in the flotation of fluorite from Taolin lead–zinc ore [29], which not only made the flotation concentrate grade up to 98.06%, but also reduced the content of silica in concentrate by about 30% compared with the same period last year. An et al. [30] used the high fraction of nucleolus oil from the natural oil family, and mixed it with the lower part of rapeseed oil and the lower part of bran oil to form a mixed collector mainly composed of binary carboxylic acid. The recovery of fluorite increased from 48.46% to 84.88–90.97%. The fluorite products could reach the super grade or the first grade, and the sorting effect was remarkable.

Fatty acids belong to carboxylic acid compounds, and the hydrocarbon group in its structure is a kind of oleophilic group, which can promote the hydrophobic surface of minerals [31]. It is believed that the mechanism of flotation of salt minerals by oleic acid and other fatty acid collectors lies in the formation of pharmaceutical adsorption layer or fatty acid-metal ion precipitation through coordination between carboxyl groups and metal active sites on mineral surfaces [32, 33]. Kellar et al. [34] analyzed the interaction between oleic acid and fluorite. The study showed that oleic acid was chemically adsorbed with fluorite at low solubility, forming a single layer film on the surface of fluorite. With the increase of sodium oleate concentration, oleic acid reacted with the Ca atom on the surface of fluorite to form calcium dioleate, which covered the surface of fluorite to form a multilayer film. As shown in Fig. 7.2, due to the surface similarity of fluorite calcite, the flotation behavior of the two minerals is basically the same in the absence of inhibitor action, which is also the difficulty of flotation separation of fluorite calcite. Quartz is basically not floatable in the absence of metal ion activation. In the case of inhibitor intervention, the addition sequence of collector and inhibitor has an obvious influence on the flotation behavior of fluorite and calcite: when the collector first interacts with minerals, the inhibition effect of inhibitor on gangue minerals decreases significantly, that is to say, after the collector first interacts with minerals, the inhibition of calcite will become more difficult.

As can be seen from the results of infrared spectrum analysis in Fig. 7.3, sodium oleate is adsorbed on the surface of fluorite and calcite by chemisorption, and this adsorption is quite firm. There was competitive adsorption between sodium oleate and water glass on the surface of fluorite and calcite, which hindered the adsorption of fatty acids to a certain extent.

When the pH in solution is higher than the isoelectric point of fluorite, the surface of fluorite is negatively charged, and the flotation of fluorite by anion such as sodium oleate also proves the existence of chemical adsorption between the anionic collector and the surface of fluorite [36]. By means of density functional theory, Jiang et al. [37] studied the influence of impurity Ce, Th, U and Y in fluorite crystal on the flotation performance of fluorite. Through the analysis of impurity defect frontier

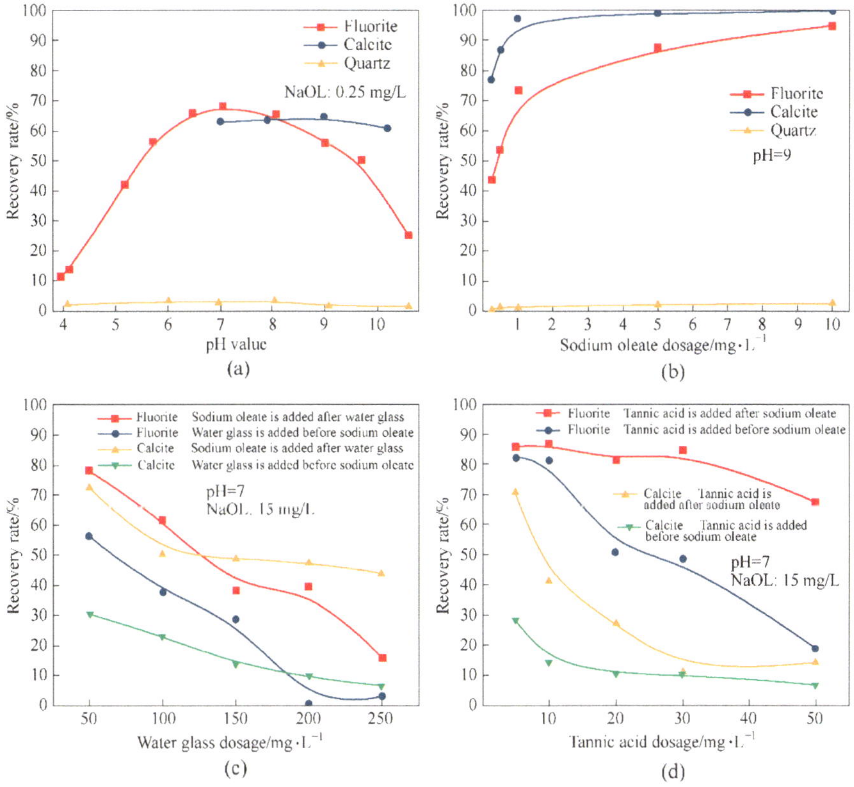

Fig. 7.2 Flotation behavior of fluorite and main gangue minerals under the action of sodium oleate and inhibitors: **a** effect of pH on flotation behavior of fluorite and gangue minerals; **b** effect of sodium oleate dosage on flotation behavior of fluorite and gangue minerals; **c** influence of water glass on flotation behavior of fluorite and gangue minerals; **d** effect of tannic acid on flotation behavior of fluorite and gangue minerals [35]

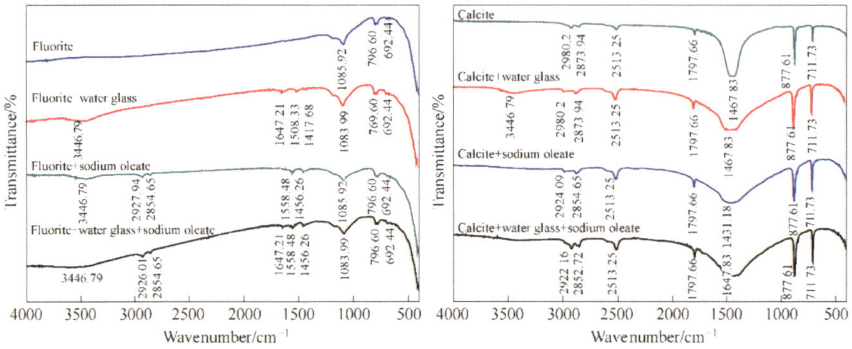

Fig. 7.3 Infrared spectrogram of the interaction of fluorite, calcite and reagent

orbit, state density, Fermi level, etc., it was shown that when there were Ce, Th, U and Y defects in fluorite crystal, the interaction between fluorite and sodium oleate could be enhanced, which was beneficial to the flotation of fluorite.

2. Sulfonic acid class

Sulfonic acid collector refers to a class of anionic surfactants with sulfonic acid group, which is rarely used in industry. Compared with fatty acid collectors, sulfonic acid collectors have better selectivity, rich foam, good water solubility, low temperature resistance, hard water resistance and other properties, but the collection ability is weaker than fatty acid. The most common sulfonate collector is sodium petroleum sulfonate, which can be divided into two categories: aromatic and aliphatic. Petroleum sodium sulfonate usually takes petroleum roughing oil, top roughing oil, middle distillate oil and the residue produced in the production of synthetic detergent process as raw material, with gaseous SO_3, liquid SO_3 and fuming sulfuric acid as sulfonating agent, after sulfonation, neutralization, separation, purification and a series of processes. The main component of sodium petroleum sulfonate (active substance) is sulfonate, which has a high hydrophilic sulfur base connected with alkyl, resulting in the structural formula of RSO_3Na, in which R is alkyl. The surface activity of sodium petroleum sulfonate is closely related to the length and structure of the alkyl part, the number and location of the sulfonic acid base group, etc. When the number of alkyl or alkyl upper branch chains increased, the surface activity decreased obviously. The hydrophobicity of sodium petroleum sulfonate decreases obviously when it contains double sulfonic groups. The longer the hydrocarbon chain part of sodium petroleum sulfonate, the larger its molecular weight, the lower the solubility in water.

Zhu et al. [38] obtained fluorite flotation concentrate with grade of 97.45% and CaF_2 recovery of 89.54% by using PSK-27 of sodium petroleum sulfonate as collector through "one roughing and four cleaning" open-circuit test of the whole process. Using sodium petroleum sulfonate as collector and sodium carbonate, sulfuric acid and water glass as adjusting agents, Wang et al. [39] studied the flotation of a quartz type fluorite ore in Jiangxi province, and obtained fluorite concentrate with CaF_2 grade of 97.45% and recovery of 89.54% through the flotation process of one roughing and four concentrate. The test proved that there were significant differences in the collection effect of sodium petroleum sulfonate obtained by different process procedures. Li et al. [40] used an anionic collector of unsaturated fatty acids mixed with sulfonated unsaturated fatty acids to flotation a single fluorite ore. The grinding finity was 70% < 0.074 mm, sodium carbonate was used as the regulator, water glass was used as the inhibitor, and the flotation process was one roughing, six cleaning and two scavenging. Finally, fluorite concentrate with CaF_2 grade of 95.37% and recovery of 85.82% was obtained.

Zhu [38] found that the residual ethanol in the purification process of sodium petroleum sulfonate would affect the chemical properties and the content of active matter. The higher the active matter content of sodium petroleum sulfonate, the higher the flotation recovery of pure fluorite minerals, the better the flotation performance of the reagent. With the increase of sulfonation degree of sodium petroleum sulfonate, the steric hindrance effect of non-polar end of sodium petroleum sulfonate molecules is gradually weakened, and the water molecules in the flotation system can have a greater probability of passing through the gap between collector molecules and the fluorite surface, reducing the hydrophobicity of mineral particles surface, resulting in the reduction of mineral flotation recovery. When the active matter content and sulfonation degree of sodium petroleum sulfonate change at the same time, the influence of the active matter content change on the flotation effect of collector is dominant. Zheng et al. [41] demonstrated the effect of hydrophobic end of alkyl base on the binding ability of sulfonate base collector to fluorite and related flotation reactions. By using alkyl sulfonates with different hydrophobic ends, it was found that the binding ability of alkyl sulfonates to fluorite increased with the increase of alkyl chain length from C10 to C16. The surface tension of sulfonates can be reduced to improve the reactivity of the reagent to fluorite when phenyl group is present in the hydrophobic tail of sulfonates. The alkyl chain of sulfonate is too long, which will lead to its low solubility in the pulp, limiting its interaction with fluorite ore.

3. Phosphoric acids

Phosphonic acid collector refers to a class of anionic surfactants with phosphonic acid groups, mainly monophosphonic acid and bisphosphonic acid and their derivatives, including alkyl phosphate esters, phosphonous lipids, alkyl phosphonic acid, alkylarylphosphonic acid and so on. Phosphonic acid collector is weaker than other types of collectors, but this kind of collector is more selective, can form a relatively stable complex with its own phosphoric acid group and metal ions, chelating use, has been successfully used in the selection of citrine and fluorite [42, 43]. For example, when the collector Flolotol-7,9 (1-hydroxy-C7-9 alkyl -1, 1-hydroxy-bisphosphonic acid) is used to float fluorite ore with CaF_2 grade greater than 34% and $CaCO_3$ grade of 8–14% in raw ore at room temperature, without water glass floating as inhibitor, Fluorite concentrate with CaF_2 grade of more than 93% and $CaCO_3$ grade of 1–1.5% can be obtained, and the recovery is above 78% [42]. Hu et al. [44] used benzbenzylphosphonic acid as collector to float calcite fluorite and control the pH of the pulp to separate fluorite from calcite.

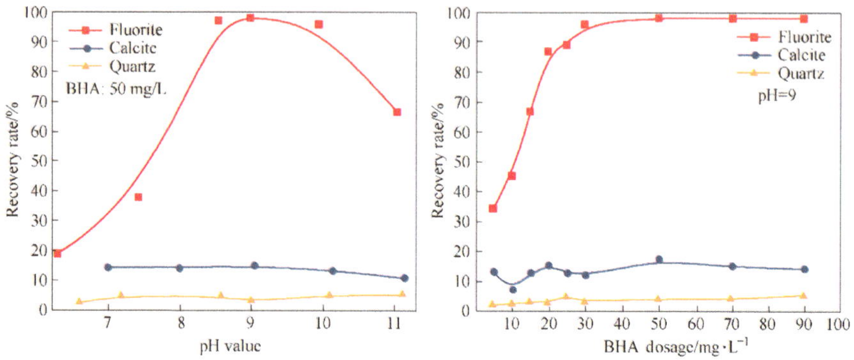

Fig. 7.4 Effect of pH and dosage of benzohydroxamic acid as collector on flotability of fluorite, calcite and quartz, reprinted with permission from Ref. [46]

Deng et al. [45] synthesized an aminophosphoric acid collector, N-(4-methylphenyl)-α-aminophenyl phosphate, using benzaldehyde, p-toluidine and phosphite as raw materials by Mannich method, and used it to study the flotation performance of actual fluorite ore. Single mineral flotation test results show that the recovery of roughing fluorite is also 97.66%. Compared with the traditional collector oleic acid, the recovery is increased by 1.38%, and the grade is increased by 0.83%.

4. Chelating collector

Jiang et al. [46] studied the ability of chelating hydroxamic acid collectors to collect fluorite. When benzohydroxamic acid was used as collector, the influence of pH and dosage on the floatability of fluorite, calcite and quartz was shown in Fig. 7.4. Benzohydroxamic acid is selective to the flotation separation of fluorite and calcite, and its ability to collect fluorite is much better than that of calcite. It is possible to achieve the separation of fluorite and calcite without the use of any inhibition.

Salicylhydroxamic acid and octylhydroxamic acid were used as collectors. The influence of pH on the buoyancy of fluorite and calcite was shown in Fig. 7.5. It can be seen from the figure that within the range of experimental pH value, when salicylhydroxamic acid and benzohydroxamic acid were used as collectors, the buoyancy of calcite was very poor, and the recovery was less than 20%. However, for fluorite, both salicylhydroxamic acid and benzohydroxamic acid showed strong collecting ability, and the recovery of fluorite reached 95% when the pH was in the range of 9–10, indicating that these two hydroxamic acids could be used as selective collectors for the flotation separation of fluorite and calcite. Although octylhydroxamic acid also showed strong collecting ability for fluorite, it also showed good collecting ability for calcite. When pH was 9–11, the recovery of calcite was about 70%, indicating that octylhydroxamic acid was not suitable for the selective collector for flotation separation of fluorite and calcite.

When benzohydroxamic acid was used as collector and $Pb(NO_3)_2$ was 50 mg/L, the effect of pH on the floatability of fluorite, calcite and quartz was shown in Fig. 7.6. As can be seen from the figure, the recovery of quartz gradually increases

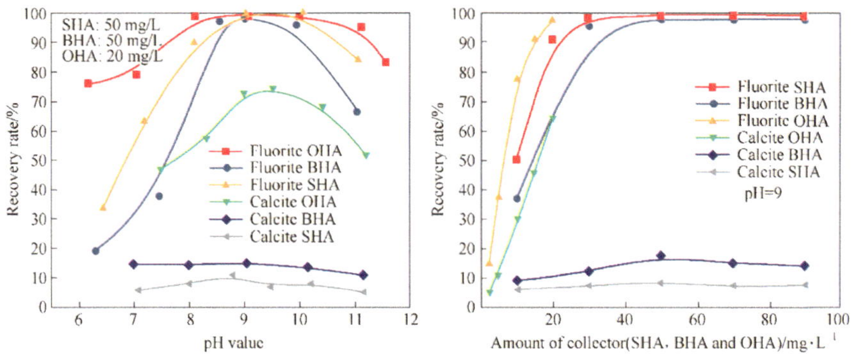

Fig. 7.5 Effect of pH and dosage of three hydroxamic acids (salicylhydroxamic acid, benzohydroxamic acid, octyl hydroxamic acid) on flotation effect of fluorite and calcite as collector [47]

with the increase of pH. When pH exceeds 9, the recovery of quartz exceeds 90%. With the presence of lead ions, the buoyancy of calcite also becomes better. At pH = 7–10, the recovery of calcite exceeds 90%, and when pH exceeds 10, the recovery of calcite rapidly declines. The recovery of fluorite decreases in the presence of lead nitrate and is about 70% at pH = 9–10.5. Therefore, Pb^{2+} activated the flotation of gangue minerals calcite and quartz, but had a certain inhibitory effect on fluorite.

Jiang et al. [47] studied the mechanism of flotation separation of fluorite and calcite by using benzohydroxamic acid as collector through crystal structure chemistry. The results showed that the density and percentage content of fluorite Ca in unit cell were higher than that of calcite Ca, indicating that fluorite had more adsorption sites for benzohydroxamic acid collector compared with the surface of calcite. Since both fluorite and calcite have a suspended bond (unsaturated broken bond) in unit

Fig. 7.6 Effect of pH on floatability of fluorite, calcite and quartz when $Pb(NO_3)_2$ was 50 mg/L as collector with benzohydroxamic acid

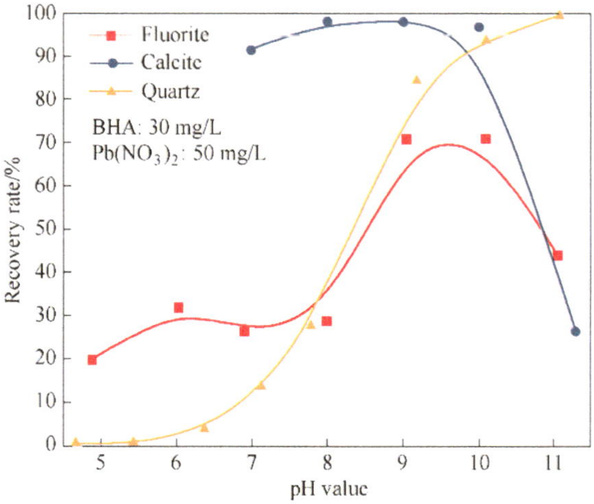

cell, and the suspended bond energy of fluorite is greater than that of calcite, the interaction strength of benzohydroxamic acid with fluorite is stronger than that of calcite. Filippova et al. [48] indicated that the adsorption of chelating agents also depended on the arrangement of cationic sites on the surface of Ca-containing minerals. The more orderly the cationic sites are arranged on the surface of Ca-containing minerals, the more anionic reagents are adsorbed on the surface of Ca-containing minerals, and the more Ca-containing minerals are recovered by flotation. Fluorite is a cubic crystal, in which Ca^{2+} is located in the center of the face-centered cubic lattice, and F^- is located on the tetrahedral site, the Ca^{2+} on the surface of fluorite is very well arranged, and after the anion collector interaction may form a neatly arranged anionic reagent monolayer on the surface. Therefore, fluorite can achieve a better flotation effect with less collected dose conditions.

7.3.1.2 Cationic Collector

Cationic collector mainly refers to amine collector, whose functional group is positively charged amine $R\text{-}NH_3^+$, such as fatty amine, aromatic amine, etc., common dodecylamine, dodecyltrimethyl ammonium chloride and trioctyl methyl ammonium chloride, etc. Cationic collector is applied to fluorite mineral flotation in the way of reverse flotation. The adsorption type is mainly electrostatic adsorption, which is greatly affected by slurry pH and collector concentration [49]. The action time is short, the flotation effect is good, and there is no need to select again. Fatty amines with carbon atomic number of more than 12 are solid at room temperature and difficult to dissolve in water. The ether group is introduced into the alkyl group of amine to form liquid ether amine, and the dissolution point can be reduced, and the flotation effect is remarkable [50, 51].

The collectors dodecylamine, dodecyl dimethyl benzyl ammonium chloride and dodecyl trimethyl ammonium chloride have large group electronegativity and strong collecting ability for calcium-containing minerals. Under the condition of pH = 6–10, the flotation separation effects of the three collectors are as follows: dodecylamine, dodecyldimethyl benzyl ammonium chloride, dodecyltrimethyl ammonium chloride, and the maximum recovery of dodecylamine is 90%. The results show that the maximum recovery of fluorite and scheelite by flotation with dodecylamine as collector is 50% different under neutral or weak alkaline conditions, which provides a potential scheme for the flotation separation of scheelite and fluorite [52]. But the amount of this kind of collector should not be too large, when the slime, the collector will be adsorbed on the slime, the formation of a large number of viscous foam, not only reduce the quality of concentrate, but also increase the consumption of agents, so before use to do a good job of desliming work [53].

7.3.1.3 Amphoteric Collector

Amphoteric collector molecules also have amphophilic groups, which react with alkali to form soap, react with acid to form amine salt [54], so in acid and alkali medium solubility is good and low temperature resistance, in addition, this kind of collector has low sensitivity to hard water and seawater, not only can occur on the surface of the mineral electrostatic adsorption, chemical adsorption, but also can chelate with some metal ions [55]. The characteristics of amphoteric collector are large water solubility, good selectivity and wide range of pH. N-octyl-β-aminopropionate (SF_8): $CH_3(CH_2)_7NHCH_2CH_2COOCH_{33}$ and n-decyl-β-methyl propionate (SF_{10}): $CH_3(CH_2)_9NHCHCHCOOCH$, applied to quartz type fluorite ore and garnet type fluorite ore (containing Al, Mg and Fe) flotation effect is remarkable. The reagent does not dissociate in the pulp during flotation, and it is easy to form complex valency bond with the mineral surface, and has good selectivity [51]. The results show that the acid grade fluorite concentrate with CaF_2 more than 97% and SiO_2 less than 1% can be obtained by using water glass as inhibitor. By contrast, SF_8 is suitable for quartz fluorite ore, while SF_{10} is more suitable for garnet fluorite ore [56]. The research and development of amphoteric collector mainly has the problems of high cost and complex synthesis process, so it is seldom used in fluorite flotation industry at present [57].

7.3.1.4 Combined Collector

Taking advantage of the synergistic effect among various agents, collector combination is used to improve flotation index, and good results have been achieved in recent years [58]. The combination of sodium oleate with sodium octanoate, sodium laurate and sodium stearate can improve the selectivity of collector, and the difference of floatability between fluorite, scheelite and calcite is significantly enhanced [59]. The results showed that the combination of reagents had higher separation efficiency and lower dosage of reagents [60].

The collector DW-1, which was mixed by modified sodium fatty acid and solubilizing dispersing agent, strengthened the chelating collector group through the complex system, and the flotation effect was greatly improved. The flotation test of a quartz type fluorite ore (CaF_2 17.32%) showed that the flotation process of 1 roughing, 1 scavenging and 6 finishing was adopted at 6 °C, and the fluorite concentrate with CaF_2 98.37% and recovery 80.12% was obtained. The flotation index was better than that of single sodium oleate, and the dosage of reagent was greatly reduced [61].

Cao et al. [62] mixed sodium oleate and sodium dodecyl sulfate as collector (sodium oleate and sodium dodecyl sulfate 1:10), flotation of a carbonate fluorite ore, sodium oleate, sodium dodecyl humate and citric acid combined inhibition, obtained fluorite concentrate CaF_2 93.89%, recovery 75.79% [63].

KY-108 collector prepared by combination of oleic acid and hydroxamic acid has high selectivity for flotation of a middle and low grade single silicate fluorite ore.

The results show that fluorite concentrate with CaF_2 97.59% and recovery 97.03% can be obtained by using the closed circuit process of regrinding and reconcentration of roughing concentrate [64].

A large number of practices have shown that mixed reagents are conducive to improving beneficiation indexes, saving flotation reagents, and improving the adaptability of reagents to various beneficiation conditions, which is an important research direction of fluorite flotation in the future.

7.3.1.5 Adsorption Mechanism of Anionic Collector on Fluorite Surface

Although there are many kinds of fluorite flotation collectors, the most widely used in the actual production is still fatty acid collectors or the combination of mainly fatty acid collectors. Sun et al. [65, 66] studied the solution chemical behavior of fluorite flotation by oleic acid, and found that calcium ions dissolved by fluorite could react with oleic acid anion to form calcium oleate colloid, which had a significant impact on flotation behavior. As shown in Fig. 7.7, the presence conditions of calcium oleate are jointly determined by oleic acid concentration, pH and calcium ion concentration. Under alkaline conditions, when the concentration of calcium ion is higher than 10^{-6} mol/L, calcium oleate exists as the dominant component in the flotation solution system. The colloid size of Ca-OL is between 0.100 and 1 μm, and compared with sodium oleate, Ca-OL has the characteristics of lower surface tension, positive Zeta potential and poor foaming.

As shown in Fig. 7.8, calcium oleate colloid collector was used for fluorite flotation, and it was found that calcium oleate colloid collector has a strong ability to collect fluorite. When pH = 10.0 and collector concentration was 3×10^{-4} mol/L, the recovery of flotation of fluorite by calcium oleate was 16% higher than that of sodium oleate, and the recovery of flotation of calcite was 11% lower than that of sodium oleate, indicating that the selectivity of calcium oleate colloid was obviously better than sodium oleate.

Antti [68] found that calcium oleate formed monolayers at low concentrations mainly through chemisorption. With the increase of ion concentration and reagent concentration, calcium oleate precipitates on the surface of fluorite to form a multilayer adsorption layer gradually. Miller et al. [69] further studied the formation conditions and adsorption mechanism of calcium oleate and found that the existence conditions of calcium oleate were determined by calcium ion concentration and pH value in the solution. When the concentration of calcium oleate is between 4×10^{-6} mol/L and 4×10^{-5} mol/L, calcium oleate is generated in the solution and then transported to the mineral surface, rather than forming and agglomerating directly on the surface. The formation and adsorption of calcium oleate have an important impact on the flotation effect. Fa et al. [70–73] used AFM colloidal probe to study the interaction force between calcium oleate, calcite and fluorite surface, and compared it with the classical DLVO force. Combined with molecular dynamics simulation, they revealed the interfacial water structure of the fluorite surface, and found that the

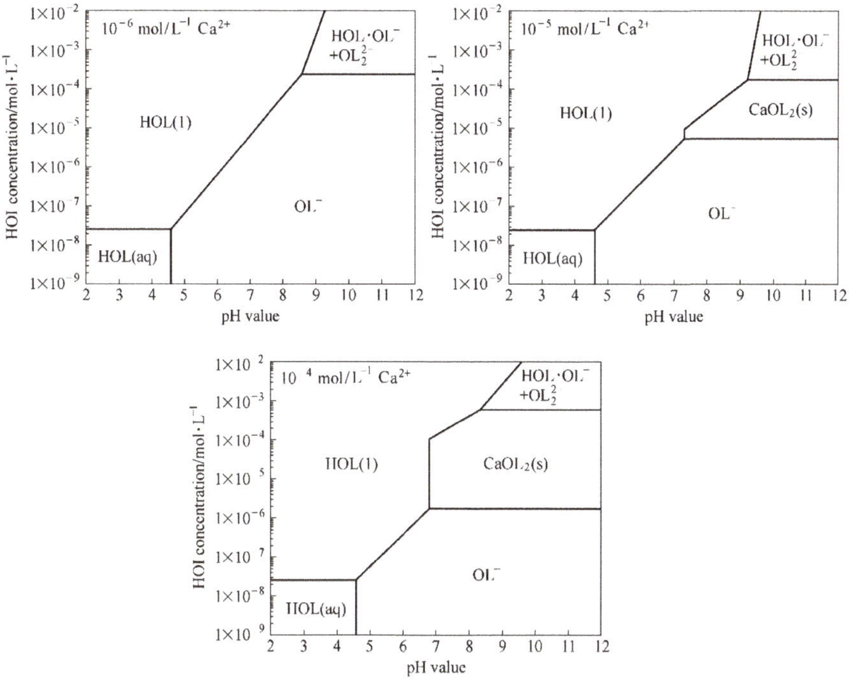

Fig. 7.7 Dominant components of oleic acid solution in the presence of calcium ions at different concentrations, reprinted from Ref. [66], copyright 2022, with permission from Minerals Engineering

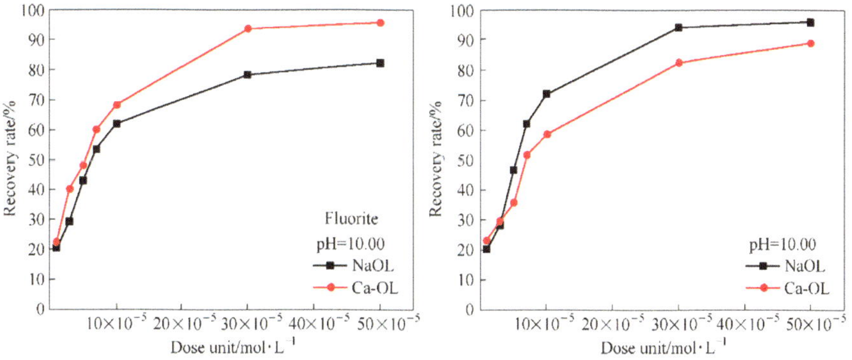

Fig. 7.8 Flotation recoverys of fluorite and calcite under the action of calcium oleate collector, reprinted from Ref. [67], copyright 2022, with permission from Minerals Engineering

colloid collector of calcium oleate had a stronger adsorption effect on the fluorite surface (Fig. 7.9).

Sun et al. [74] found that the adsorption of calcium oleate colloid had a significant effect on the micro-nano structure of mineral surface. As shown in Fig. 7.10, after calcium oleate colloid action, the mineral surface adsorption layer height increases, the surface adsorption height of scheelite and fluorite minerals after calcium oleate colloid action is significantly higher than that of calcite surface, surface roughness is significantly increased, and hydrophobicity is enhanced, which is beneficial to the flotation of scheelite and fluorite.

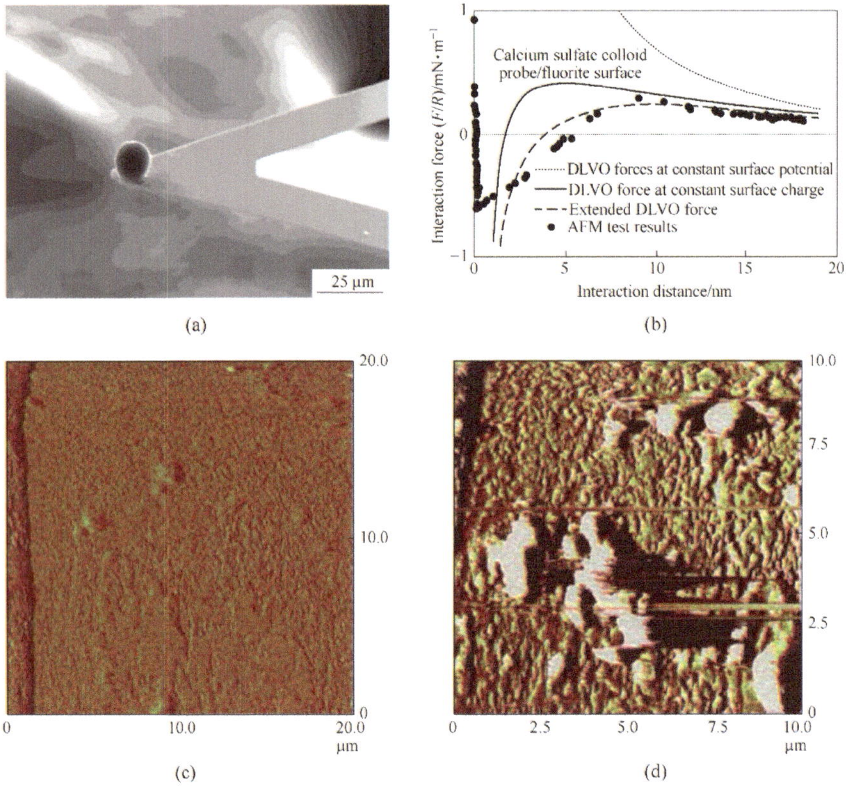

Fig. 7.9 Calcium oleate colloidal particles, the force on the fluorite (111) surface, and the AFM adsorption morphology of Ca-OL on the fluorite surface, reprinted from Ref. [70, 71], copyright 1996 and 2006, with permission from International Journal of Mineral Processing

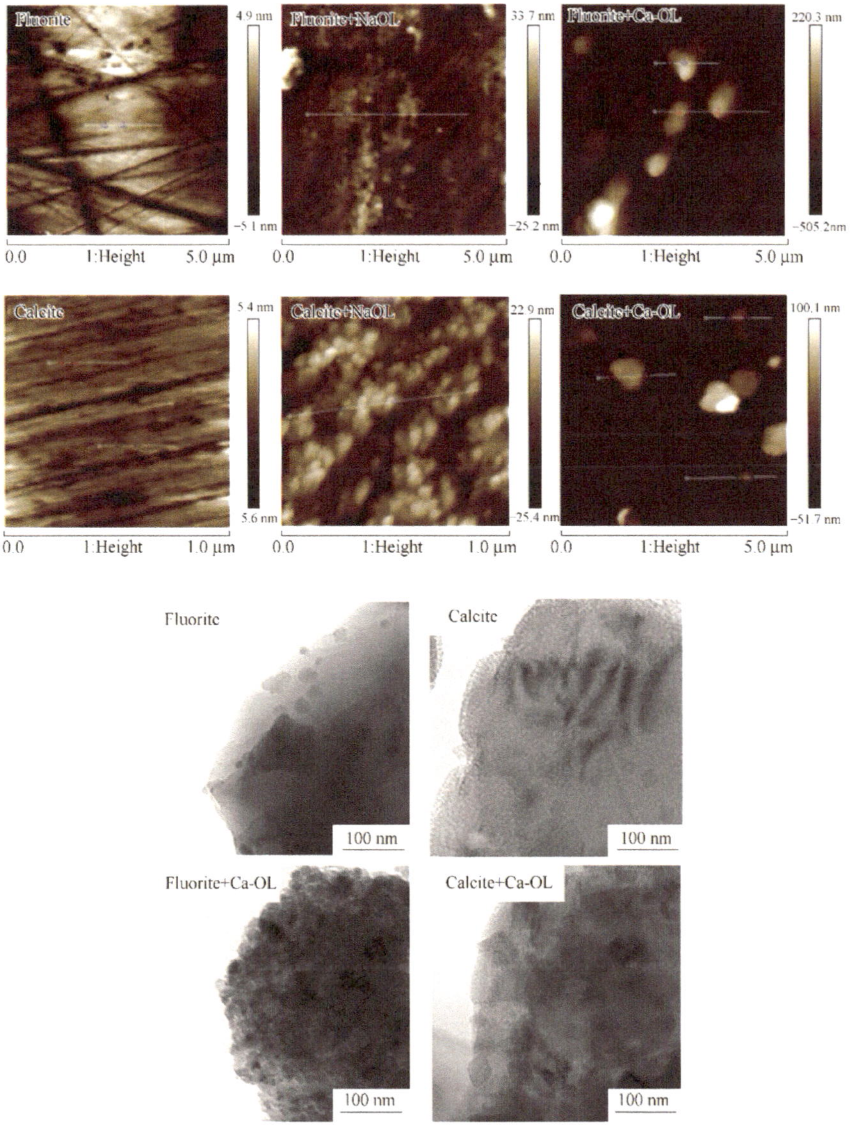

Fig. 7.10 Effect of calcium oleate colloid on micro/nano structure of mineral surface, reprinted from Ref. [74], copyright 2022, with permission from Minerals Engineering

7.3.2 *Fluorite Flotation Inhibitors and Their Mechanism of Action*

Fluorite is often associated with gangue minerals such as calcite, dolomite, barite, etc., with similar floatability, suitable inhibitors must be selected to achieve efficient separation. The commonly used inhibitors in fluorite flotation include inorganic inhibitors: water glass and modified water glass, sodium hexametaphosphate, etc. Organic inhibitors: tannin extract, tannin, lignosulfonate, organic phosphonic acid, starch, etc. In addition, the use of combined inhibitors makes the separation efficiency of gangue minerals such as fluorite and calcite greatly improved.

7.3.2.1 Inorganic Inhibitors

At present, the most common inorganic inhibitor in production is water glass, which is an aqueous solution of water glass. The chemical formula is $Na_2O{\cdot}nSiO_2$, where n is the ratio of mole number of silicon dioxide to sodium oxide, which is called the ratio of sodium silicon or modulus. The different modulus of water glass, its properties will have a large difference, the general n value is 1.5–3.5. In the flotation separation of fluorite, water glass is the most widely used inhibitor, it can effectively inhibit the majority of silicate minerals, but also to some calcium gangue minerals play an inhibitory effect, a large number of scientific research workers have studied water glass on the surface of silicate minerals adsorption mode, adsorption strength and other factors [75, 76]. Nowadays, the mechanism of silicate adsorption by water glass is widely mentioned:

(1) Water glass in the dissolution process produces a large number of colloids in the form of SiO_2, such colloids are easy to adsorb on the gangue mineral surface and strong hydrophilic, which will form a hydrophilic film on the mineral surface, so as to achieve the effect of inhibiting gangue minerals.
(2) According to the analysis of flotation solution chemistry, water glass not only exists in the form of colloid during the dissolution process, but also has the dissociation of $HSiO_3^-$, or SiO_3^{2-}. Through the interaction with dissolved Mg^{2+} and Ca of gangue minerals such as calcite, hydrophilic $MgSiO_3$ and $CaSiO_3$ sediments are adsorbed on the mineral surface.

Sodium silicate is widely used in mineral processing, and there are plenty of literature reports in fluorite flotation field. Zhang et al. [77] conducted experimental research on a fluorite ore in Inner Mongolia, and found that when NMG was used as collector, water glass as inhibitor, and No. 2 oil as foaming agent, and through the process of one rough, one scavenging and two cleaning, the fluorite ore with a raw ore grade of 43.36% was enriched to 97.35% concentrate grade, and the recovery was as high as 98.13%. Ge [76] studied the flotation separation mechanism of fluorite and hematite by water glass, and the results showed that the solubility and ion composition of water glass in solution were related to the solution concentration, pH value and the

Table 7.4 Flotation effect of salinized water glass on fluorite and calcite

Polyvalent metallic salts	Fluorite recovery/%	Calcite recovery/%
$Al_2(SO)_4)_3 \cdot 18H_2O$	65.4	47.4
$MgSO_4 \cdot 7H_2O$	23.7	66.3
$CuSO_4 \cdot 5H_2O$	86.4	20.9
$FeSO_4 \cdot 7H_2O$	87.9	18.3
$ZnSO_4 \cdot 7H_2O$	20.8	85.4

ratio of silicon to sodium. Under the condition of pH = 6–9, the adsorption activity of hematite and fluorite to water glass ions is different, and the separation of minerals is realized.

Based on the extensive application of water glass, many successful applications of modified water glass have been widely developed. By using acidified water glass as an inhibitor, Zhou et al. [78] proved that acidified water glass had better inhibition effect than CMC, which not only improved the grade and recovery of concentrate, but also solved the problems of slow settlement of tailings and turbidity of mineral return water. At the same time, the inhibition mechanism of acidified water glass on quartz was discussed. The main reason for the inhibition effect of acid water glass on quartz was that H_2SiO_3 colloidal particles formed under acid aqueous solution could selectively adsorb on the surface of quartz, thus producing inhibition effect on quartz. In addition, the selectivity of colloidal silicate ions is significantly higher than that of single water glass ions, which can have strong interaction with the Ca^{2+} energy on the surface of calcite. Feng Bo et al. successfully achieved the inhibition of calcite by preparing acid water glass with a 3:1 ratio of water glass and oxalic acid [79].

The experimental study of pure minerals showed that single water glass had a certain inhibition effect on calcium-containing minerals including fluorite [80]. In order to enhance the inhibition selectivity of water glass, water glass was mixed with polyvalent metal salts in a certain ratio to form salinated water glass. The flotation effects of various salinated water glass on fluorite and calcite were shown in Table 7.4. Among them, the selectivity of $CuSO_4 \cdot 5H_2O$ and $FeSO_4 \cdot 7H_2O$ salinized water glass is greatly improved [78]

7.3.2.2 Organic Inhibitors

The commonly used organic inhibitors in fluorite flotation include starch, dextrin, CMC (sodium carboxymethyl cellulose), extract, sodium lignosulfonate, tannin and other macromolecular inhibitors as well as small molecular inhibitors with hydrophilic groups such as –OH, $-NH_2$, –COOH and –CSS. Macromolecular organic inhibitors generally have a large number of polar solidophilic groups and hydrophilic groups that can be fixated on gangue minerals, which rely on solidophilic groups adsorbed on the surface of minerals, and then use its long chain strong hydrophilicity to inhibit minerals, so as to achieve selective separation effect. Strong chemical interaction can occur between small molecule inhibitors and gangue, which can be selectively adsorbed on the mineral surface to make the collector desorption or prevent the collector from adsorbing on the mineral surface and make the mineral inhibited.

1. Tannic acid

Tannic acid, also known as tannin, is an effective inhibitor of calcite. It is a polymeric phenolic compound that can be extracted from the seeds, fruits, leaves and bark of plants. Tannin extract is also a tannin inhibitor. Because this natural biomass contains multiple adjacent hydroxyl groups and has a special affinity for metal ions, it can be used as a highly effective adsorbent for metal ions and as an inhibitor of metal minerals. There are two main viewpoints on the inhibition mechanism of tannin on calcite. One is that the carboxyl group of tannin acts on the surface of calcite, while the hydroxyl group is arraying outwards with water molecules to form a water film, resulting in the hydrophilicity of calcite. The other is that phenolic groups in tannic acid molecules interact with calcite phase through physical or chemical interaction, hindering the action between collector molecules and calcite, resulting in the inhibition of calcite.

Zhang et al. [81] systematically studied the inhibition effect and mechanism of tannic acid on calcite. As shown in Fig. 7.11, the recovery of fluorite decreased from 93.09 to 77.83% when tannic acid was in the range of 0–10 mg/L. However, the recovery of calcite decreased rapidly from 96.04 to 9.18%, which is a very large decrease in recovery. When tannic acid was in the range of 10–50 mg/L, calcite almost did not rise. The recovery of fluorite is higher than 75% at pH = 6–8. When pH = 6, the recovery of fluorite can reach about 78%, and then with the increase of pH, the recovery of fluorite slowly decreases. When pH is higher than 10, the recovery of fluorite will show an obvious downward trend. Tannic acid has little inhibition effect on fluorite under weak acidic and neutral conditions. Tannic acid has a strong selective inhibition on calcite, which is not affected by the pH value in solution.

Zhang et al. [81] studied the selective adsorption of tannic acid on calcite surface by surface adsorption experiment, ζ potential measurement, UV–vis analysis and X-ray photoelectron spectroscopy (XPS). The adsorption capacity test showed that tannic acid could block the interaction between collector and calcite, and Zeta potential test showed that the negative charge shift of the interaction between calcite and tannic acid was greater than that of sodium oleate. Ultraviolet visible spectroscopy, XPS and solution chemistry analysis showed that the Ca^{2+} and $Ca(OH)^{+}$ components on the mineral surface were the key factors for the selective adsorption of tannic acid. The tannic polyhydroxyphenol complexes were selectively adsorbed on the calcite surface through the chemical interaction with the $Ca(OH)^{+}$ on the calcite surface [81].

2. Starches

As early as in the 1930s, people found the selective inhibition of starch, mainly used in the reverse flotation of iron ore. Wu et al. [80] verified the effect of starch on

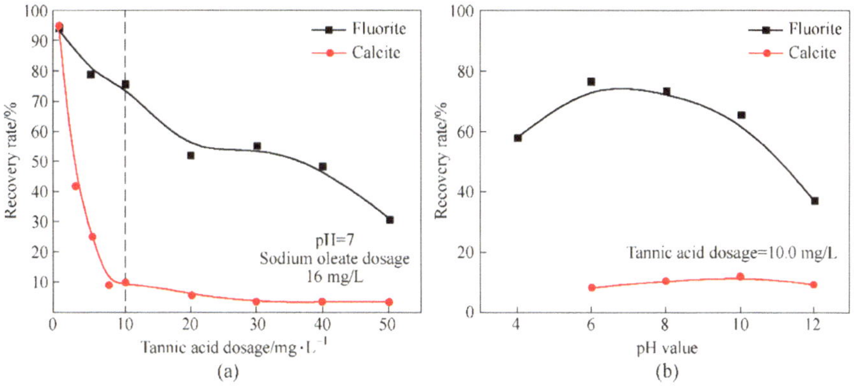

Fig. 7.11 Effect of tannic acid dosage and pH on the recovery of fluorite and calcite, reprinted from Ref. [81], copyright 2018, with permission from Journal of Colloid and Interface Science

fluorite, calcite and barite when oleic acid was used for flotation in alkaline medium, and the inhibition order was calcite > barite > fluorite. The study of Li et al. [81] showed that the inner electron binding energy of Ca atoms on the surface of fluorite was higher than that of calcite, so it was more difficult for starch polysaccharides to form chemical bonds on the surface of fluorite than on the surface of calcite.

Dextrin is the product of starch hydrolysis. When common starch is boiled in water and hydrolyzed, it will produce dextrin with small molecular weight, neutral and good water solubility. Liu et al. [82] studied the adsorption mechanism of dextrin on the surface of non-metallic minerals, and the results showed that dextrin did not directly interact with the metal active site on the surface of minerals, but was chemically adsorbed on the surface of minerals through the hydroxyl compound formed by metal ions on the surface of minerals in aqueous solution.

3. Polyacrylic acid

Zhang [83] systematically studied the inhibition effect and mechanism of polyacrylic acid on calcite. As shown in Fig. 7.12, polyacrylic acid has a significant effect on the buoyancy of fluorite and calcite minerals. When the dosage of polyacrylic acid is 0–10 mg/L, the recovery of fluorite decreases from 92.91 to 72.71%, with a small decreasing range. However, the recovery of calcite dropped sharply from 95.87 to 12.34%, indicating a very large decrease. The results showed that the inhibition ability of polyacrylic acid to calcite was stronger than that of fluorite at this dosage. When the polyacrylic acid is in the range of 10–40 mg/L, the recovery of fluorite decreases from 72.71% to about 10%. In this range, the recovery of calcite is basically lower than 10%, and calcite is completely inhibited by polyacrylic acid. It can be seen that the separation of fluorite and calcite is possible only when the appropriate dosage of polyacrylic acid (7.5 mg/L).

When the dosage of fixed polyacrylic acid was 7.5 mg/L, the recovery of fluorite increased from 61.32 to 79.98% when pH increased from 4 to 8. When pH = 8, the recovery of fluorite can still reach about 80%, and then with the increase of pH, the

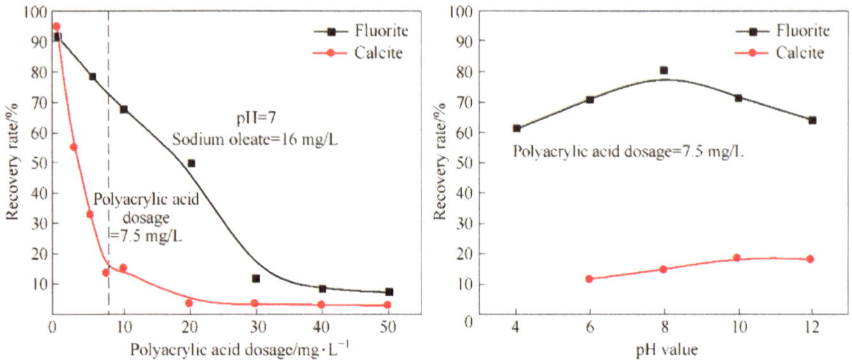

Fig. 7.12 Effect of pH value and dosage on the inhibition performance of polyacrylic acid, reprinted from Ref. [83], copyright 2018, with permission from Physicochemical Problems of Mineral Processing

recovery of fluorite slowly decreased. It shows that the inhibition effect of polyacrylic acid on fluorite is small, and the reduction of recovery of fluorite is mainly due to the influence of pH value. In addition, in the range of pH = 6–12, the recovery of calcite changes little, the inhibition effect of polyacrylic acid on calcite is strong, not affected by the change of pH value in solution. It can be seen that the separation of fluorite and calcite is possible under the appropriate dosage of polyacrylic acid (7.5 mg/L) and pH value.

The results of wetting Angle and adsorption capacity analysis show that polyacrylic acid can be uniformly adsorbed on the surface of fluorite, which increases its hydrophilicity, but has limited effect on the adsorption of sodium oleate on the surface of fluorite. In addition, polyacrylic acid can be a large number of uneven adsorption on the surface of calcite, so that its surface is strongly hydrophilic, and may hinder the subsequent single-layer uniform adsorption of sodium oleate. Potentiodynamic, infrared and XPS results show that polyacrylic acid can be replaced by sodium oleate on the surface of fluorite, but does not affect the adsorption of sodium oleate on the surface of calcite. Polyacrylic acid has physical adsorption and chemisorption on the surface of fluorite and calcite, but the chemisorption strength is different, leading to the difference in the competitive adsorption of sodium oleate on the surface of the two minerals. The presence of bound water was clearly found in the XPS diagram of O1s, which proved the reaction of the carboxyl group of polyacrylic acid with the hydrolyzed product $Ca(OH)^+$ of calcite [83].

7.3.2.3 Combinatorial Inhibitors

The combination of organic and inorganic inhibitors can often play a synergistic role to enhance the inhibition effect on gangue such as calcite and improve the grade and recovery of fluorite. Yu et al. [84] used water glass, aluminum sulfate and extract as inhibitors of barite and sodium oleate as fluorite collector to achieve effective separation of fluorite and barite from lead and zinc tailings, and finally obtained fluorite concentrate with CaF_2 grade of 95.06% and CaF recovery of 96.58%. Zhang et al. [81] used polyacrylic acid and tannic acid in combination to selectively inhibit calcite flotation in Jibailing fluorite flotation, and the inhibition effect was better than acidized water glass. Zhou et al. [85] used conventional sodium oleate as collector, T29 and acidified water glass for a certain proportion of the mixed solution as inhibitors, flotation Jinta County a high calcium fluorite ore, fluorite concentrate CaF_2 grade reached 98.02%, carbonate and silicate impurities are very low, meet the fluorite concentrate grade 1 product requirements. Niu et al. [86] flotation Qinglong carbonate fluorite ore through the combined action of salinated water glass and sodium metaphosphate inhibitors. The combined inhibition effect is remarkable, and the CaF_2 grade of fluorite concentrate is up to 98.1%, and the recovery is 83.68%.

Are water glass and lead nitrate mixed inhibitors? As shown in Fig. 7.13, Jiang et al. [47] found that when sodium oleate was used as collector and pH was 9, the salinated water glass prepared with the ratio of water glass and lead nitrate in the range of 5:1 to 20:1 had strong inhibition on calcite, but had no obvious effect on the flotation of fluorite. When the proportion of water glass and lead nitrate was 10:1, the inhibition effect on calcite was the best, and the recovery was only about 10%, while the recovery of fluorite was about 80%. The difference between the two recoveries was 70%, which proved that the salinated water glass was used as the inhibitor, and the separation effect of fluorite and calcite was the best.

When sodium oleate was used as collector, the ratio of water glass to lead nitrate was 10:1, and pH was 9, the influence of water glass dosage on the floatability of fluorite and calcite was shown in Fig. 7.14. With the addition of salinized water glass, the recovery of calcite decreased sharply, while that of fluorite decreased slightly. When the dosage of salinized water glass is 0.5 mg/L, the recovery of fluorite is about 90%, and the recovery of calcite is about 10%, indicating that the dosage of

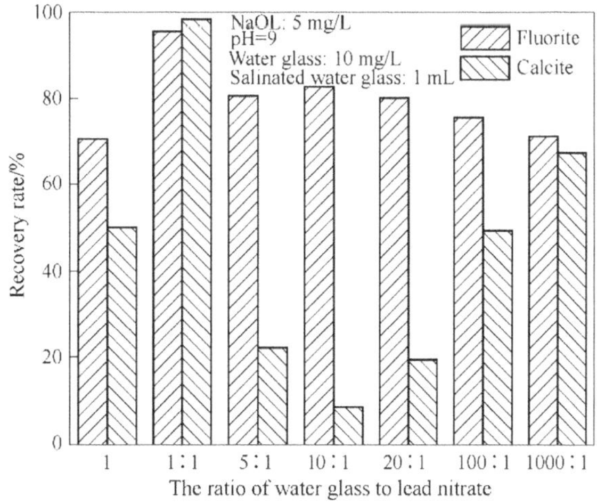

Fig. 7.13 Effect of sodium oleate as collector, pH 9, on the floatability of fluorite and calcite, reprinted from Ref. [87], copyright 2021, with permission from Colloids and Surfaces A: Physicochemical and Engineering Aspects

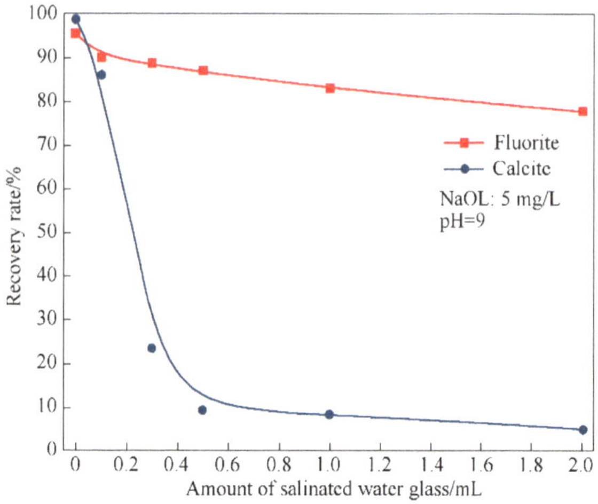

Fig. 7.14 Effect of sodium oleate on flotability of fluorite and calcite when the ratio of water glass to lead nitrate was 10:1 and pH was 9 as collector, reprinted from Ref. [87], copyright 2021, with permission from Colloids and Surfaces A: Physicochemical and Engineering Aspects

salinized water glass and lead nitrate 10:1 can achieve effective separation of fluorite and cubic solution at 0.5 mL.

7.4 Flotation Technology of Fluorite Associated with Tungsten Ore

In the traditional flotation of tungsten/fluorite co-associated resources, the main process flow of sulfide ore flotation, tungsten flotation and fluorite flotation are often adopted. Therefore, the flotation process of tungsten ore inevitably affects the subsequent flotation of fluorite. The current flotation processes of tungsten ore mainly include fatty acid method and metal complex method. For tungsten flotation process of fatty acid method, on the one hand, part of fluorite with good floatability floats on the tungsten section and loses in tungsten concentrate, on the other hand, tungsten flotation by adding a large number of inhibitors (such as water glass) to inhibit gangue minerals, fluorite is also strongly inhibited, poor floatability, resulting in the back-end fluorite flotation process is difficult to obtain the ideal index. For the tungsten flotation process of metal complex method, although the metal complex collector has good selectivity, it does not have the ability to collect fluorite, and the fluorite is not inhibited in the flotation process of tungsten ore, but the residual metal complex flotation reagent also has a certain impact on the subsequent flotation of fluorite, such as the adsorption of collector, foam structure and viscosity. Therefore, how to solve the negative effects of tungsten flotation on fluorite flotation is the key to realize the efficient utilization of associated fluorite.

7.4.1 Activated Flotation Technology of Fluorite from Tungsten Flotation Tailings by Fatty Acid Method

Shizhuyuan Polymetallic Mine in Hunan Province began to try to recycle fluorite by flotation in the 1970s. Until 2014, the recovery of fluorite was less than 40% before the new flotation process for tungsten ore. Tungsten ores such as Huangshaping, Yaogangxian, Xintianling and Xianglushan adopted the process of fatty acid separation, so fluorite could not be recovered. The essential reason lies in the contradiction between flotation of tungsten ore and flotation of fluorite. The exploitation and utilization of tungsten and fluorite resources are described in detail, taking the high calcium tungsten/fluorite deposit in Shizhuyuan Polymetallic Mine, Hunan Province as an example. WO_3 0.3%, CaF_2 18–22% and $CaCO_3$ 20–30% of Chaishan ore belong to high calcium refractory tungsten and fluorite resources.

7.4.1.1 Influence of Flotation of Tungsten Ore on Flotation of Fluorite

As shown in Fig. 7.15, the tungsten ore adopts the classic flotation process of high-alkali fatty acid method, which can restrain gangue minerals (including fluorite) through high alkalinity and a large amount of water glass. In the roughing concentrate, WO_3 grade of 6.43%, CaF_2 grade of 58.96 and 7.02% fluorite are lost in the tungsten roughing concentrate, which is finally heated and selected, and difficult to recover (Table 7.5). In fact, this part of fluorite often has high monomer dissociation and good floatability, so it is easy to recover fluorite. However, due to the limitation of flotation process of tungsten ore, recovery is difficult.

In order to investigate the influence of tungsten flotation operation on fluorite flotation, fluorite flotation tests were carried out on tungsten tailings and tungsten included raw ore, and the results are shown in Table 7.6. According to the closed-circuit process shown in Fig. 7.16, tungsten ore is selected for fluorite, and a good concentrate grade of 91.04% with a recovery of 65.3% can be obtained. In contrast, for fluorite separation from tungsten tailings, the concentrate grade is only 86.64% with a recovery of only about 45%. The above experimental results indicate that the addition of "fatty acid + water glass" in the front-end tungsten flotation process has a great influence on the back-end fluorite flotation: the floatability of fluorite minerals inhibited by a large amount of water glass in the tungsten flotation section is greatly reduced, and the back-end fluorite flotation effect is poor.

7.4.1.2 Hydrochloric Acid Stripping Activation with Enhanced Adsorption of Fatty Acid Calcium

Solution chemistry calculation shows that there must be a large number of Ca-OL colloid in the saturated solution system containing calcium minerals, and these colloid components play an important role in the flotation process. Sodium oleate and Ca-OL

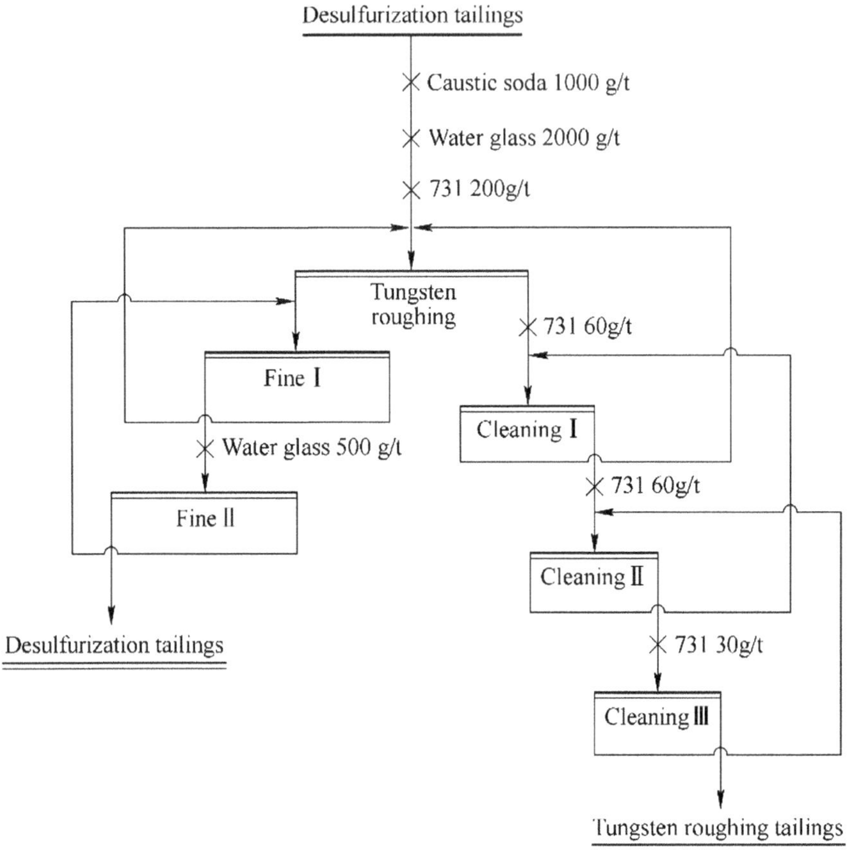

Fig. 7.15 Closed circuit test flow of tungsten roughing

Table 7.5 Closed circuit test results of tungsten roughing

Product name	Yield/%	Grade/%		Recovery/%	
		WO_3	CaF_2	WO_3	CaF_2
Roughing tungsten concentrate	2.51	6.43	58.96	80.54	7.02
Tungsten roughing tailings	97.49	0.04	20.10	19.46	92.98
Feed mine	100	0.20	21.08	100	100.00

colloid were used as collectors to flotation scheelite, fluorite and calcite, respectively. The results of flotation recovery at different pH values are shown in Fig. 7.17. With the increase of pH, the recovery of scheelite increased with Ca-OL colloid flotation. When pH = 11, the recovery of scheelite reached 73.38%, which was 18.69% higher than that of sodium oleate. When pH < 10, the recovery of Ca-OL colloid flotation fluorite is always higher than sodium oleate, and when pH = 8, the recovery of Ca-OL

Table 7.6 Closed circuit test indicators of fluorite (tungsten tailings and tungsten raw ore as feedstock, respectively)

Selected raw ore	Product	Yield/%	Grade/%	Recovery/%
Tungsten tailings	Concentrate	9.45	86.34	45.30
	Tailings	90.55	10.88	54.70
	Feed mine	100.00	18.01	100.00
Tungsten raw ore	Concentrate	13.93	91.04	65.30
	Tailings	86.07	7.83	34.70
	Ore feeding	100.00	19.42	100.00

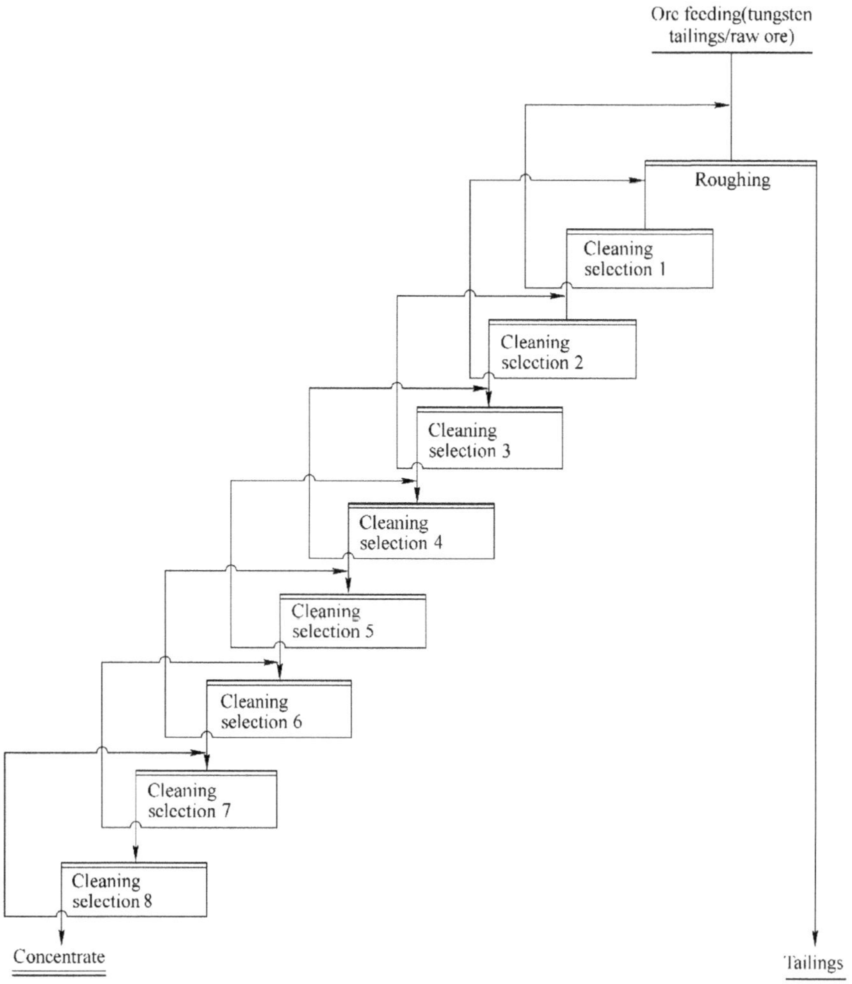

Fig. 7.16 Flow chart of closed-circuit comparison test for fluorite

colloid flotation fluorite reaches 95.45%. When pH > 10.00, the flotation recovery of fluorite decreased significantly, because the fluorite surface F^- was replaced by OH^- under the strong alkaline condition, which affected the adsorption of collector on the surface. For calcite, the recovery of colloid flotation with Ca-OL is always lower than that with sodium oleate. When pH = 11, the recovery of calcite by colloid flotation with Ca-OL is 54.17%, which is 15.99% lower than that with sodium oleate.

Under the action of Ca-OL colloid, the flotation recovery of scheelite, fluorite and calcite changes with pH similar to that of sodium oleate, but the flotation selectivity of Ca-OL colloid is stronger than sodium oleate. The flotation effect of Ca-OL colloid on scheelite and fluorite is better than sodium oleate.

Therefore, adding a large amount of acid (hydrochloric acid or sulfuric acid) to tungsten tailings can, on the one hand, release calcium ions and promote the formation of a large amount of Ca-OL colloid; on the other hand, peel off the surface adsorption agents (such as collectors and inhibitors, etc.) of calcium-containing minerals, which is conducive to the flotation of fluorite, as shown in Fig. 7.18. At the same time, studies have shown that the adsorption products of water glass on the surface of fluorite are easy to desorption under acidic conditions, but difficult to desorption under alkaline conditions [88]. In addition, under acidic conditions, the surface of

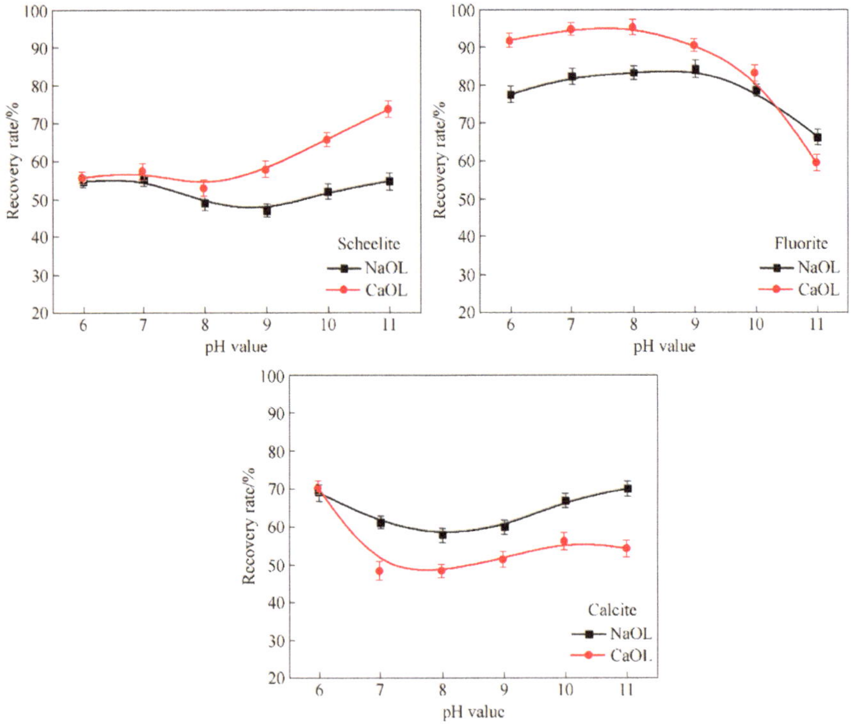

Fig. 7.17 Ca-OL colloid and sodium oleate as collectors, the relationship between the flotation recovery of three calcium-containing minerals and pH ($c_{Ca\text{-}OL} = c_{NaOL} = 1 \times 10^{-3}$mol/L)

calcite will slightly dissolve and release new surface, which can provide a new adsorption site for water glass, tannic acid and other inhibitors, thus strengthening the inhibition of calcite.

Based on the above studies, hydrochloric acid or sulfuric acid was used as the pre-treatment agents for flotation of fluorite, and the amount of acid was about 10 kg/t. The results are shown in Table 7.7. The results show that the recovery of fluorite directly flotation without pretreatment is only about 47%, and the enrichment ratio is not high. After pretreatment with sulfuric acid or hydrochloric acid, the floatability of fluorite was enhanced and the recovery was significantly improved, especially hydrochloric acid. Therefore, for traditional fatty acid tungsten tailings, surface activation of fluorite with a large amount of acid is helpful to improve the flotation efficiency of fluorite.

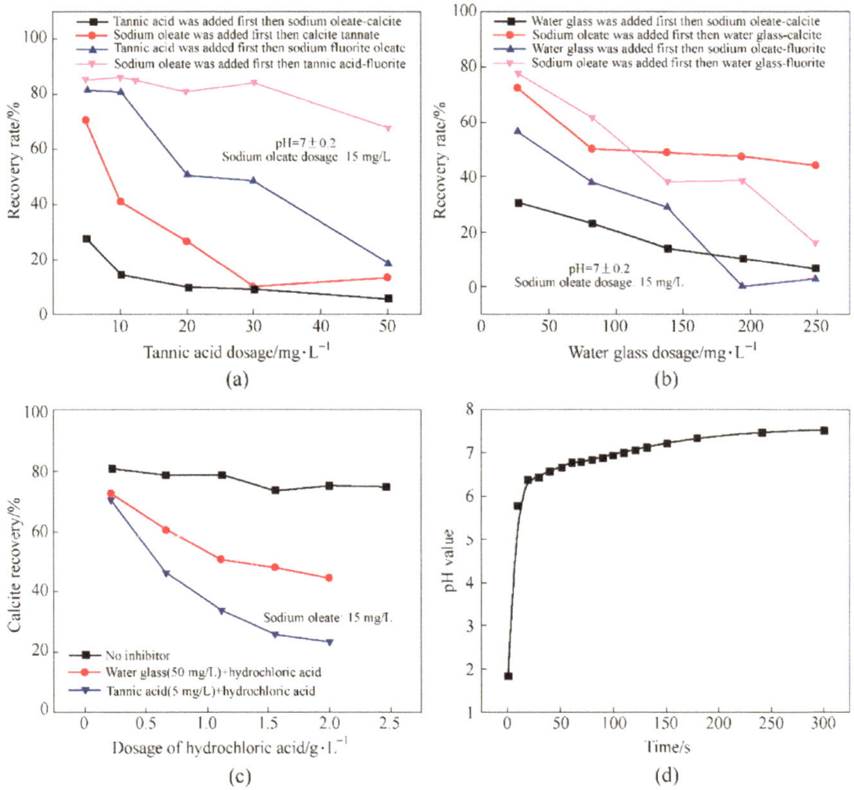

Fig. 7.18 Effect of reagent addition sequence and acid dosage on flotation behavior of fluorite and calcite, reprinted from Ref. [89], copyright 2018, with permission from Minerals Engineering

Table 7.7 Effects of acid activation treatment on fluorite flotation of tungsten tailings

Activator	Products	Yield/%	CaF_2 grade/%	CaF_2 recovery/%
Hydrochloric acid	Roughing concentrate	43.27	40.54	77.71
	Tailings	56.73	8.87	22.29
	Feed mine	100.00	22.57	100.00
Sulfuric acid	Roughing concentrate	32.57	43.38	64.49
	Tailings	67.43	11.54	35.51
	Feed mine	100.00	21.91	100.00
Unacidified	Roughing concentrate	22.56	46.54	47.26
	Tailings	77.44	15.13	52.74
	Feed	100.00	22.22	100.00

7.4.1.3 Flotation Practice of Fluorite from Tungsten Tailings of High Calcium Carbonate Type

The key to the flotation of fluorite resources associated with tungsten tailings lies in how to realize efficient flotation separation of fluorite, calcite and silicate minerals in a ternary system. However, as calcium-containing minerals, fluorite and calcite have similar surface-active sites, and the convergent surface transformation of minerals leads to similar surface properties, which makes it difficult to separate. As calcium ions dissolved from semi-soluble salts, fluorite and calcite have great influence on the flotation separation of mineral bases. To solve this problem, scientific research institutions such as Central South University and Changsha Research Institute of Mining and Metallurgy put forward the technical concepts of the flotation process system of fluorite, namely, roughing separation and activation of hydrochloric acid, inhibition of calc-bearing gangue in acidic zone and inhibition of silica-bearing gangue in alkaline zone, as shown in Fig. 7.19. The three-stage closed-circuit flotation process of "Enhancement of roughing of fluorite by hydrochloric acid extraction and activation combined with low-temperature resistant collector—enhancement of silicon by alkaline extraction and inhibition of calcium by hydrochloric acid system" was designed and developed, which solved the problem of interaction of "multiple system" of fluorite/calcium-bearing gangue/silicate gangue in fluorite flotation [90]. The technology has been applied in industrial applications in Hunan Shizhuyuan Chaishan Polymetal concentrator and Huangniao Polymetal concentrator, and the comprehensive recovery of fluorite from tungsten tailings has been greatly improved. For example, the CaF_2 grade of fluorite concentrate from Shizhuyuan Chaishan is 18–22% and $CaCO_3$ is 20–30%. After flotation operation, the CaF grade of fluorite concentrate is higher than 88% and the recovery is about 50%.

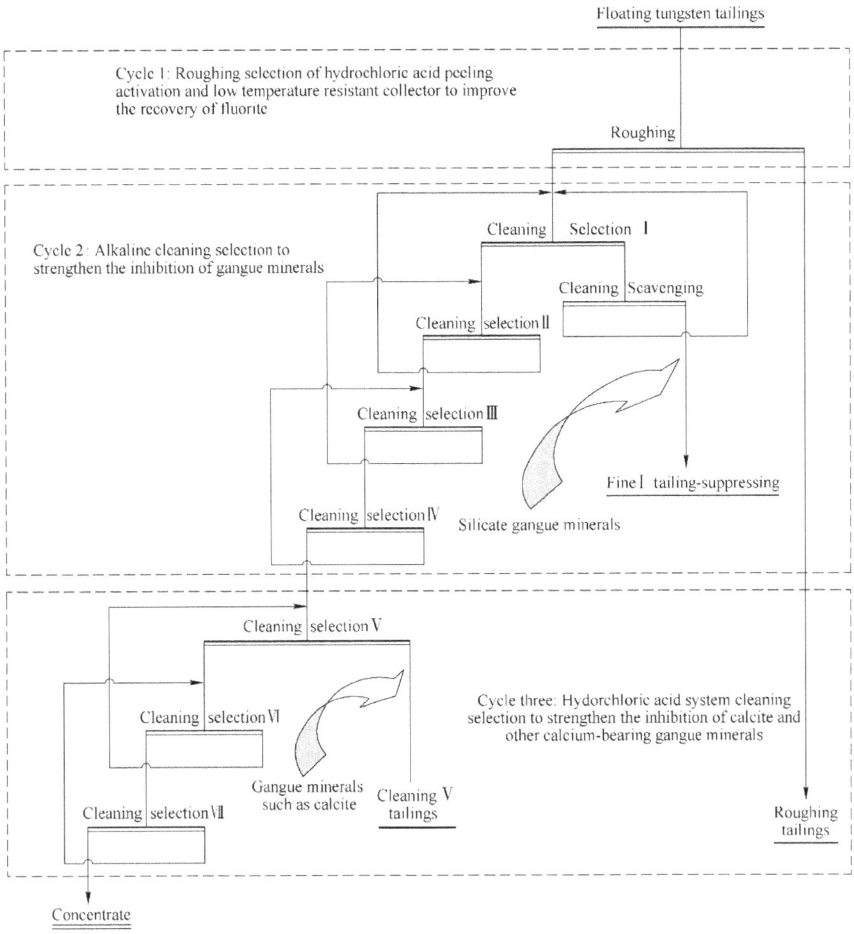

Fig. 7.19 Flowchart of the new flotation process of fluorite with three-stage closed cycle

7.4.2 Mixed Flotation of Scheelite and Fluorite—Selective Separation Technology

Before 2014, Shizhuyuan polymetallic Separation plant adopted the main process of flotation of molybdenum, bismuth and sulfide ore—tungsten flotation—fluorite flotation ("GY" method). Due to the addition of fatty acids in the tungsten flotation process, part of fluorite with good floatability would float and eventually lose in tungsten concentrate. At the same time, the addition of a large amount of water glass strongly inhibited fluorite, making fluorite in the subsequent flotation poor floatability, difficult to obtain the ideal mineral processing index. Scheelite and fluorite are typical calcium-containing minerals with similar surface properties, so it is difficult to separate them by flotation. Therefore, the idea of using the similarity of surface properties of calcium-bearing minerals to achieve mixed flotation technology is put forward. Under the support of the "Twelfth Five-Year" National Science and Technology Support Program, Shizhuyuan Nonferrous Metals Co., Ltd. has carried out the exploration of mixed flotation process of tungsten-fluorite together with domestic universities and research institutes.

7.4.2.1 Mixed Flotation of Scheelite and Fluorite

The test ore sample is the non-magnetic product of sulfide ore flotation tailings from 380 concentrator of Hunan Shizhuyuan Nonferrous Metals Co., Ltd. after high intensity magnetic separation. The test ore sample contains WO_3 0.54%, CaF_2 25.24% and $CaCO_3$ 11.76%. The main useful minerals are scheelite and fluorite, and the main gangue minerals are quartz and calcite, etc. The chemical composition analysis results of the ore are shown in Table 7.8, and the main minerals and their relative contents are shown in Table 7.9. The ore fineness is 72% of < 0.074 mm (58% of < 0.037 mm), and the content of WO_3 and CaF_2 in < 0.037 mm is significantly higher than that in > 0.037 mm, and most scheelite and fluorite are distributed in < 0.037 mm. There is no significant difference in the distribution of calcite in roughing and fine grain size.

After experimental exploration, the process plan of roughing flotation of scheelite fluorite by mixed flotation—reverse flotation of mixed concentrate to remove impurity—inhibition of fluorite flotation by reverse flotation—fluorite flotation by swinging tailings of scheelite—scheelite flotation to recover fluorite—scheelite tungsten heating and concentration is finally determined. The test principle process is shown in Fig. 7.20. The test ore samples were firstly mixed flotation of scheelite

Table 7.8 Results of ore chemical multi-element analysis

Element	WO_3	CaF_2	$CaCO_3$	SiO_2	Al_2O_3	MgO	Na_2O	K_2O	S
Contents/%	0.54	25.24	11.76	43.12	6.83	0.96	0.91	2.21	0.084

Table 7.9 Mineral composition and relative content of ores

Minerals	Relative content/%	Minerals	Relative content/%
Sulfide ore	0.2	Garnet	4.4
Scheelite	0.65	Yunmu	2.5
Firefly stone	25.5	Feldspar	2.5
Quartz	35.6	Amphibole	2.6
Calcite	11.5	Fushan stone	0.8
Dolomite	3.5	Clay minerals	6.0
Chlorite	1.2	Other	3.05

and fluorite to obtain scheelite and fluorite mixed concentrate, and a large number of gangue minerals such as quartz, calcite and garnet were abandoned. After the mixed concentrate after reverse flotation to remove impurities, further remove calcite and other impurities, and re-grinding, grinding fineness selection < 38 μm accounting for 76%, so that scheelite, fluorite can be more fully dissociated monomer, then scheelite flotation, flotation concentrate and shaker tailings scheelite heating selection, flotation tailings using shaker gravity separation, shaker concentrate for high grade fluorite concentrate, After the ore in the shaker is combined with the heated tailings, the fluorite is flotation, and a low-grade fluorite concentrate is obtained.

After test verification, the conditions of mixed flotation of scheelite fluorite were determined: flotation pH = 7.5, water glass as inhibitor, dosage of 600 g/t, K3 as collector, dosage of 200 g/t. Then the closed-circuit test of the mixed flotation process was carried out. The closed-circuit test process is shown in Fig. 7.21, and the test results are shown in Table 7.10.

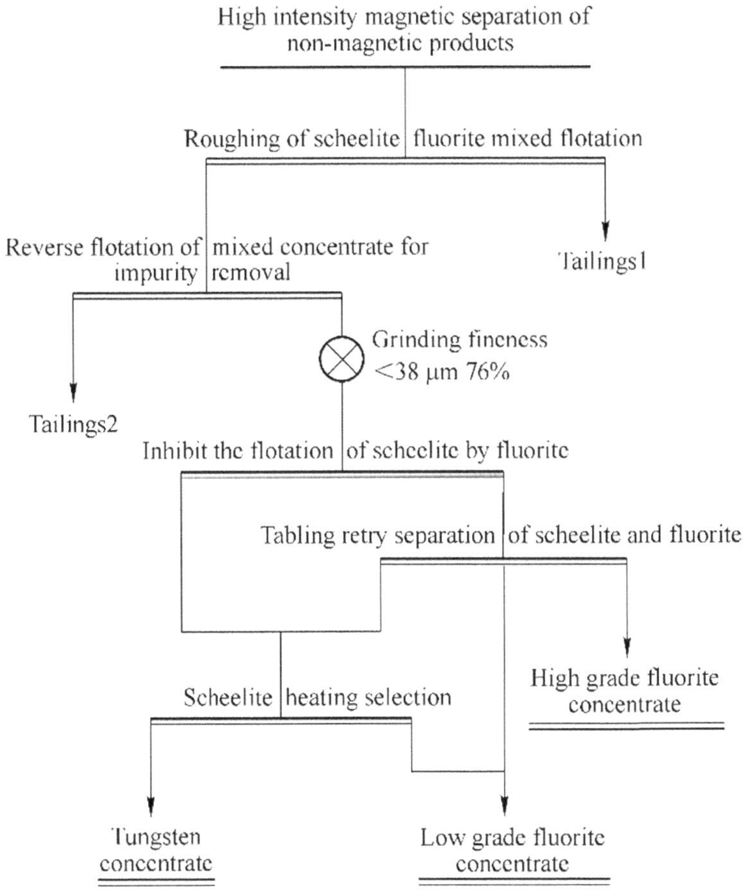

Fig. 7.20 Principle flow of mixed flotation reseparation of scheelite fluorite

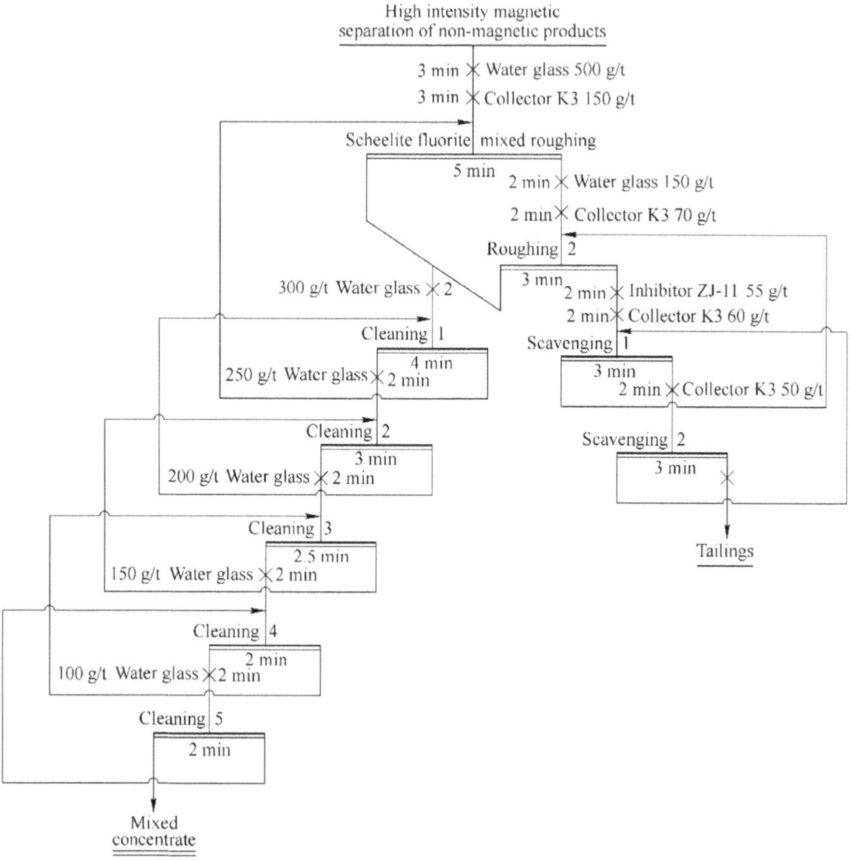

Fig. 7.21 Closed circuit test flow chart of roughing selection section of scheelite fluorite mixed flotation

Table 7.10 Closed-circuit flotation test results of mixed roughing section (%)

Product name	Yield	Grade			Recovery		
		WO_3	CaF_2	$CaCO_3$	WO_3	CaF_2	$CaCO_3$
Mixed concentrate	34.54	1.37	68.38	21.85	86.79	93.40	63.74
Tailing mine	65.46	0.11	2.55	6.56	13.21	6.60	36.26
Combined count	100.00	0.54	25.29	11.84	100.00	100.00	100.00

According to the test results, for the test ore sample containing 0.54% WO_3, 25.24% CaF_2 and 11.76% $CaCO_3$, the mixed concentrate of 1.37% WO_3, 68.38% CaF_2 and 21.85% $CaCO_3$ can be obtained through the closed-circuit process of two roughing, two scavenging and five selections. The recoveries of WO_3 and CaF_2 in the mixed concentrate were 86.79% and 93.40%, respectively. Tailings contained 0.11%

WO_3, 2.55% CaF_2 and 6.56% $CaCO_3$, and the loss rates of WO_3 and CaF_2 in tailings were 13.21% and 6.60%, respectively. The process of tungsten and fluorite mixed flotation—separation of tungsten and fluorite can significantly reduce the loss of fluorite and improve the recovery of fluorite flotation. However, the scheelite fluorite mixed concentrate needs scheelite and fluorite separation and flotation to finally get scheelite and fluorite products. Subsequent separation is difficult, and it is difficult to obtain high grade fluorite concentrate and tungsten concentrate at the same time.

7.4.2.2 Experimental Study on Impurity Removal by Reverse Flotation of Mixed Concentrate

1. Open circuit test on impurity removal by reverse flotation of mixed concentrate

After mixed roughing flotation, the calcite content of scheelite and fluorite mixed concentrate obtained is relatively high, reaching 21.85%, indicating that calcite has been enriched to a certain extent in the mixed flotation operation, which is unfavorable to the subsequent separation of scheelite and fluorite. After many programs' exploration, it was finally determined that mixed concentrate should be processed by impurity removal before separation operation. The results of comparative test show that the effect of reverse flotation is better than that of direct flotation. The open circuit test flow of reverse flotation impurity removal for scheelite fluorite mixed concentrate is shown in Fig. 7.22, and the test results are shown in Table 7.11.

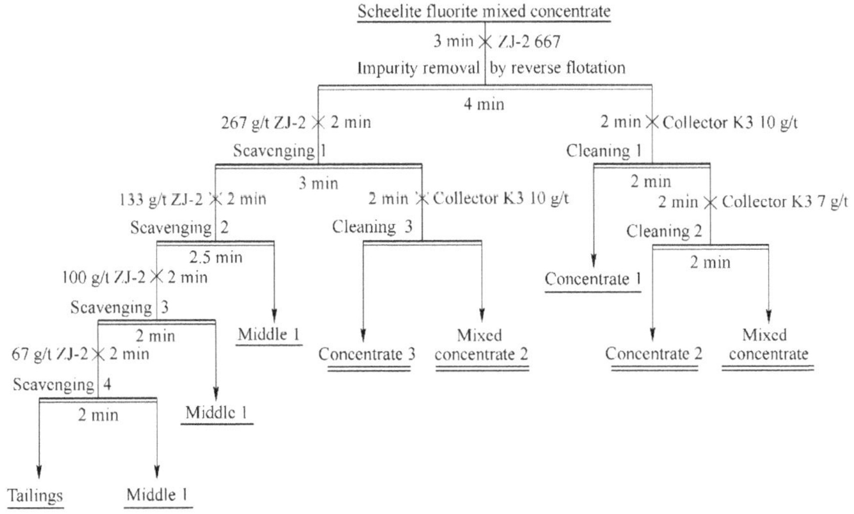

Fig. 7.22 Flow chart of reverse flotation impurity removal open-circuit test for scheelite fluorite mixed concentrate

Table 7.11 Reverse flotation impurity removal open circuit test results of scheelite fluorite mixed concentrate (%)

Product name	Yield	Quality position			Job recovery		
		WO_3	CaF_2	$CaCO_3$	WO_3	CaF_2	$CaCO_3$
Mixed concentrate 1	37.81	2.20	81.76	6.64	61.62	45.57	11.60
Mixed concentrate 2	14.96	2.35	90.15	4.24	26.04	19.88	2.93
Cleaning 3 concentrate	2.27	1.24	67.54	18.59	2.09	2.26	1.95
Cleaning 2 concentrate	3.04	0.75	62.48	27.83	1.69	2.80	3.91
Cleaning 1 concentrate	2.11	0.57	64.30	23.18	0.89	2.00	2.26
Medium mine 1	6.20	0.68	81.41	6.39	3.12	7.44	1.83
Medium mine 2	4.34	0.47	78.47	9.67	1.51	5.02	1.94
Medium mine 3	3.74	0.36	76.00	13.66	1.00	4.19	2.36
Tailing mine	25.53	0.11	28.80	60.37	2.04	10.84	71.22
Combined count	100.00	1.35	67.84	21.64	100.00	100.00	100.00

From the test results in Table 7.11, it can be seen that the mixed concentrate can be obtained by the anti-flotation removal heterogeneous open circuit test of scheelite-tungsten fluorite, which can obtain scheelite-fluorite mixed concentrate 1 containing WO_3 2.20%, CaF_2 82.95%, $CaCO_3$ 6.64%, and scheelite-fluorite mixed concentrate 2 containing WO_3 2.35%, CaF_2 90.50%, $CaCO_3$ 4.24%, and tailings with yield of 25.53%, tailings containing WO_3 0.10%, CaF_2 30.26% and $CaCO_3$ 60.37%, the loss rates of WO_3 and CaF_2 in tailings were 1.89% and 11.20%, respectively. The results of the anti-flotation removal and open circuit test results of scheelite-fluorite mixed concentrate show that scheelite-fluorite anti-flotation removal can be used as a process scheme to improve the grade of scheelite-fluorite mixed concentrate, and the concentrate products need to be separated from scheelite.

2. Closed circuit test for impurity removal by reverse flotation of mixed concentrate

On the basis of open circuit test, closed-circuit test of reverse flotation impurity removal of mixed concentrate was carried out. The closed-circuit test process was shown in Fig. 7.23, and the test results were shown in Table 7.12. It can be seen that the mixed concentrate with WO_3 2.04%, CaF_2 81.04%, $CaCO_3$ 7.55% and WO_3 1.59%, CaF_2 87.92%, $CaCO_3$ 4.87% can be obtained through the closed-circuit test for impurity removal by reverse flotation of scheelite fluorite, and the mixed concentrate with WO_3 1.59%, CaF_2 87.92%, and $CaCO_3$ 4.87%. Mixed concentrate 1 + 2 contained 1.86% WO_3, 83.72% CaF_2 and 6.51% $CaCO_3$ on average. The recoveries of WO_3 and CaF_2 were 97.45% and 86.29%, respectively. Gangue minerals, especially calcite, were effectively removed from the mixed concentrate, which provided good conditions for the separation of scheelite and fluorite.

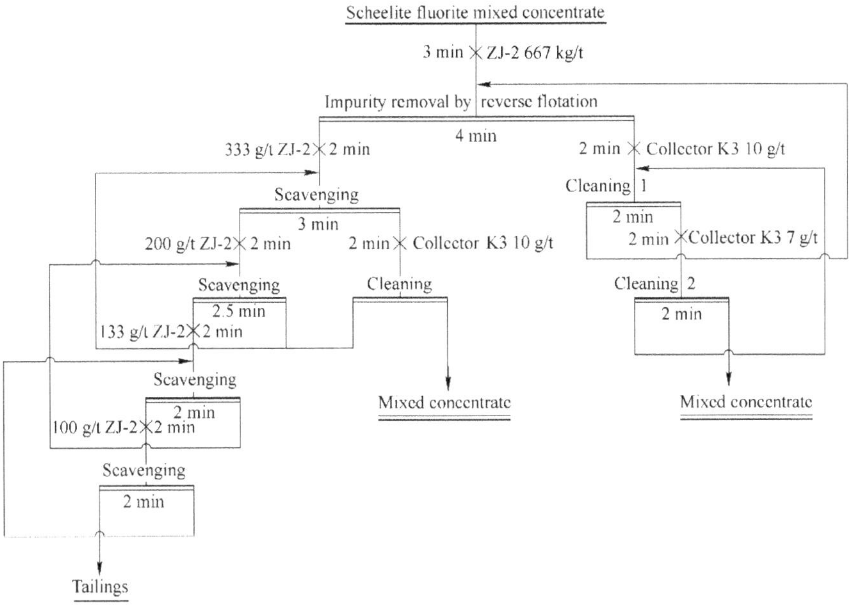

Fig. 7.23 Flow chart of closed-circuit test for impurity removal by reverse flotation of scheelite fluorite mixed concentrate

Table 7.12 Closed circuit test results of reverse flotation for impurity removal of scheelite fluorite mixed concentrate (%)

Product name	Yield	Quality position			Job recovery		
		WO_3	CaF_2	$CaCO_3$	WO_3	CaF_2	$CaCO_3$
Mixed concentrate 1	43.78	2.05	80.96	7.71	67.00	52.19	15.73
Mixed concentrate 2	27.33	1.51	87.79	4.77	30.78	35.33	6.07
Tail mine	28.89	0.10	29.34	58.09	2.22	12.48	78.20
Combined total	100.00	1.34	67.91	21.46	100.00	100.00	100.00

7.4.2.3 Separation Test of Scheelite and Fluorite in Mixed Concentrate

1. Open-circuit separation test of scheelite and fluorite in mixed concentrate

The experiment explored two scheme processes of flotation of scheelite by inhibiting scheelite and flotation of scheelite by inhibiting fluorite. After comparing the closed-circuit test results, the scheme process of flotation of scheelite by inhibiting fluorite was finally selected. The open-circuit test flow of the scheme is shown in Fig. 7.24, and the test results of the whole open-circuit process are shown in Table 7.13.

It can be seen from the test results in Table 7.13 that tungsten roughing concentrate containing WO_3 12.90%, CaF_2 54.49% and $CaCO_3$ 31.96% and tailings containing WO_3 0.32%, CaF_2 88.22% and $CaCO_3$ 2.24% can be obtained from

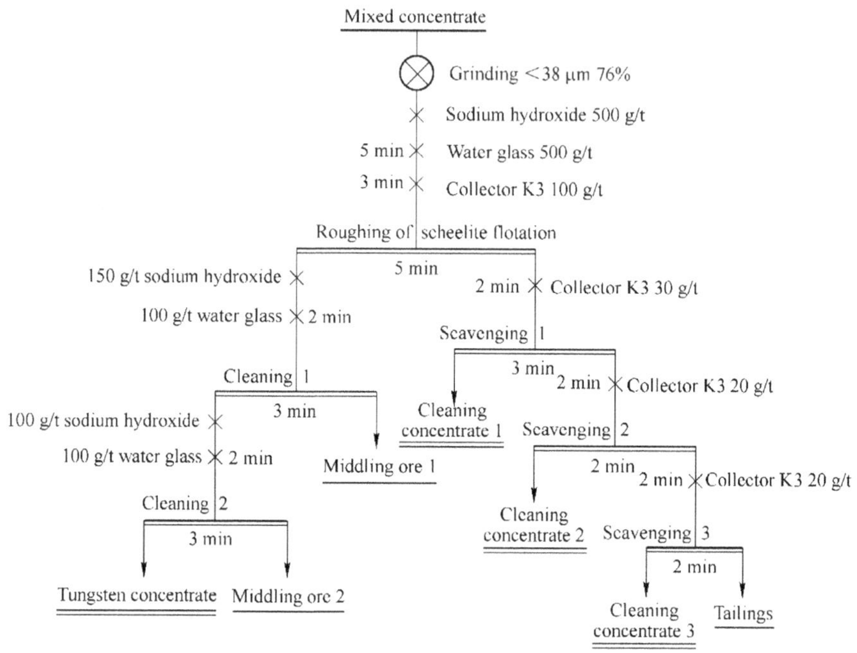

Fig. 7.24 Flow chart of open-circuit flotation test of scheelite

Table 7.13 Open-circuit flotation test results of scheelite (%)

Product name	Yield	Quality position			Job recovery		
		WO_3	CaF_2	$CaCO_3$	WO_3	CaF_2	$CaCO_3$
Roughing tungsten concentrate	8.32	12.88	54.47	31.97	58.25	5.42	41.24
Medium mine 2	2.76	7.62	67.55	19.47	11.43	2.23	8.33
Medium mine 1	4.36	3.14	81.69	12.10	7.43	4.26	8.18
Scavenging 1 concentrate	5.59	2.33	78.23	11.40	7.07	5.23	9.88
Scavenging 2 concentrate	3.50	1.26	80.50	8.59	2.39	3.37	4.66
Scavenging 3 concentrate	2.33	0.58	83.25	6.37	0.73	2.32	2.30
Tailing mine	73.14	0.32	88.22	2.24	12.70	77.17	25.41
Combined total	100.00	1.84	83.61	6.45	100.00	100.00	100.00

mixed concentrate 1 + 2 through the open-circuit test of restraining fluorite flotation of scheelite.

2. Closed circuit test for separation of scheelite and fluorite in mixed concentrate

The closed-circuit test flow of the scheme is shown in Fig. 7.25. As can be seen from the test results in Table 7.14, mixed concentrate 1 + 2 can obtain a tungsten roughing concentrate containing 10.86% WO_3, 57.62% CaF_2 and 30.03% $CaCO_3$ after a closed-circuit test to inhibit scheelite flotation by fluorite, and the WO_3 recovery of tungsten roughing concentrate is 83.06%. The flotation tailings were fluorite roughing concentrate containing WO_3 0.36%, CaF_2 87.71% and $CaCO_3$ 2.62%. The WO_3 and CaF_2 recoveries of fluorite roughing concentrate were 16.94% and 90.35%, respectively.

Although scheelite and fluorite have similar flotation behavior, they differ greatly in enrichment efficiency. Synchronous roughing and enrichment are easy, but subsequent separation is difficult, difficult to obtain high quality concentrate products, and the process is too complicated, difficult to realize engineering application.

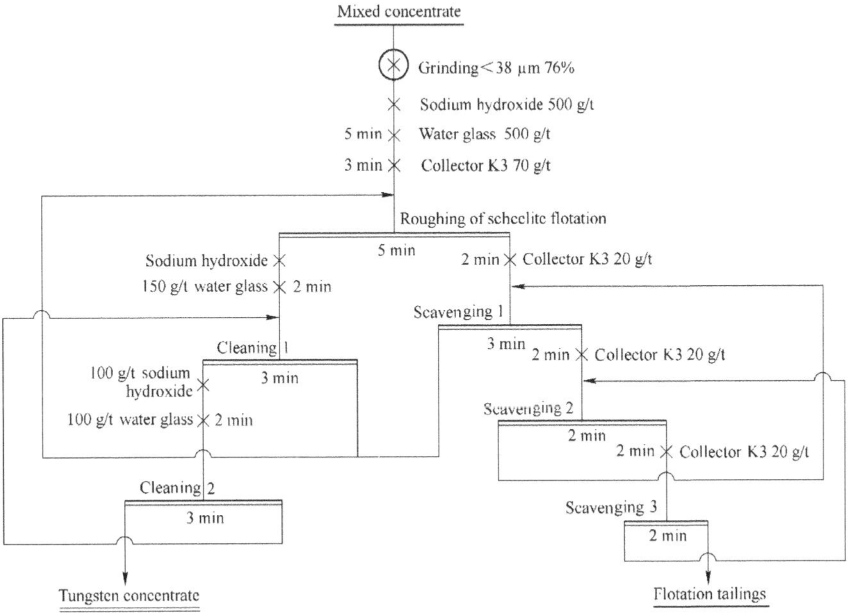

Fig. 7.25 Flow chart of closed-circuit flotation test of scheelite

Table 7.14 Closed circuit test results of scheelite flotation/%

Product name	Yield	Quality position			Operating rate		
		WO_3	CaF_2	$CaCO_3$	WO_3	CaF_2	$CaCO_3$
Roughing tungsten concentrate	14.12	10.86	57.46	30.21	83.79	9.73	66.13
Flotation tailings	85.88	0.35	87.65	2.54	16.21	90.27	33.87
Combined total	100.00	1.83	83.39	6.45	100.00	100.00	100.00

7.4.3 Enhanced Flotation Technology of Fluorite from Tungsten Flotation Tailings by Metal Complex Method

The new mixed flotation process of scheelite and wolframite in normal temperature short process based on highly selective Pb-BHA complex collector has made a major breakthrough in Hunan Shizhuyuan Non-ferrous Metals Co., LTD., which has been popularized and applied in Nanling area, and the recovery of tungsten and fluorite has been greatly improved. The new technology realizes the simultaneous flotation enrichment of wolframite, scheelite and cassiterite through the new metal–organic complex collector, avoids the enrichment of fluorite in tungsten roughing concentrate, and greatly improves the recovery of tungsten and tin. The fluorite in the tungsten tail is not inhibited by water glass in the process of tungsten selection, and has high surface activity, so it is easy to concentrate by flotation. The new technology breaks through the traditional mineral processing route, focuses on the comprehensive recovery of tungsten and fluorite, and greatly improves the comprehensive utilization rate of fluorite while taking into account the efficient extraction of tungsten.

In the main process flow of sulfide ore flotation, tungsten flotation and fluorite flotation, the recovery of tungsten separation stage is improved and the loss of fluorite in tungsten products is reduced by using collector with strong selectivity for tungsten ore. At the same time, the amount of inhibitor can be reduced or cancelled, and the influence on subsequent flotation of fluorite can be reduced. For example, compared with the traditional fatty acid and water glass heating system, water glass is eliminated in the roughing selection section of the new flotation process of Shizhuyuan scheelite and wolframite by using lead benzohydroxamic acid (Pb-BHA) as collector. The surface of fluorite in the mixed floating tailings is not inhibited, and the fluorite maintains good floatability. With HS as collector and water glass and acidified water glass as inhibitors, the flowsheet of tungsten tailings flotation test is shown in Fig. 7.26, and the results are shown in Table 7.15. Under the open-circuit condition, the grade of fluorite concentrate was 93.83% with a recovery of 56.27%. Compared with the fatty acid tungsten separation process (about 45% recovery), the recovery of fluorite was significantly improved. However, compared with direct flotation of

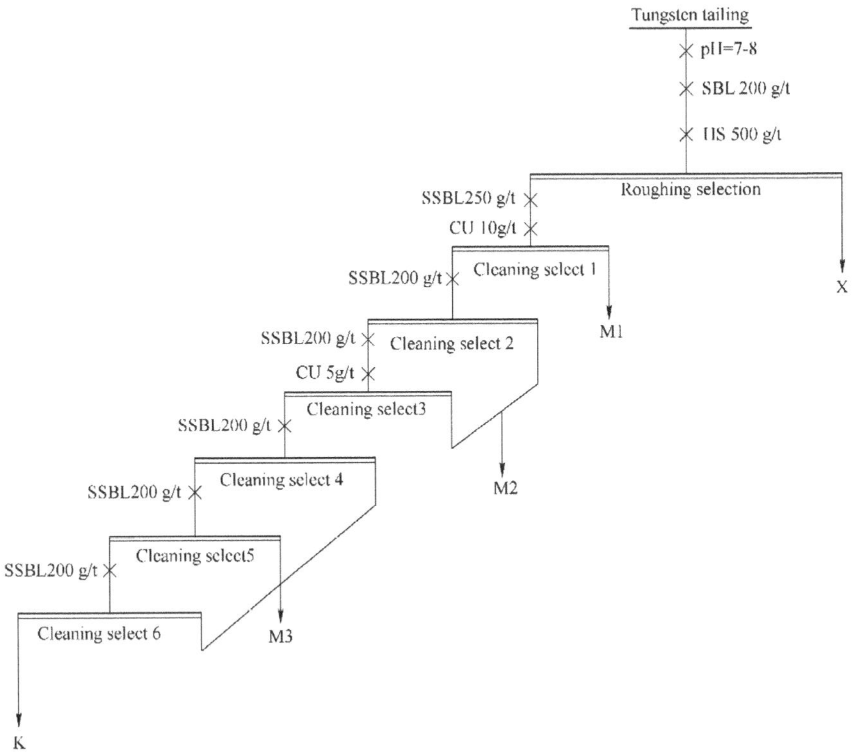

Fig. 7.26 Flotation recovery process of tailing fluorite by the new process of monochrome tungsten mixing flotation

fluorite from tungsten raw ore, the recovery of fluorite is slightly lower. This is because, although a large amount of water glass is eliminated in the tungsten separation section of tungsten flotation process to avoid the strong inhibition of fluorite, lead ions used in the process of tungsten separation still have a significant impact on the flotation of fluorite. In order to explore the effects of residual chemicals on the flotation of fluorite from tailings, it is necessary to systematically study the effects of benzohydroxamic acid, lead nitrate and their complexes on the flotation behavior of minerals such as fluorite, calcite and quartz when sodium oleate is used as collector.

7.4.3.1 Effect of Residual Chemicals in Tungsten Tailings on Fluorite Flotation

The benzohydroxamic acid-lead metal–organic complex collector has high selective collecting ability for tungsten ore, cancelates the addition of water glass and fatty acid in the process of tungsten separation, avoids the inhibition of fluorite, and creates good conditions for the efficient recovery of fluorite. In order to explore the effects

Table 7.15 Test results of flotation recovery of tailing fluorite by the new process of scheelite and wolframite mixed flotation and comparison with other processes (%)

Tungsten selection process	Products	Yield	CaF_2		$CaCO_3$	
			Grade	Recovery	Grade	Recovery
Tungsten selection process by complex method	K	12.38	93.83	56.54	0.9	1.90
	M_1	8.42	31.35	12.85	15.02	21.51
	M_2	2.98	40.98	5.94	21.78	11.04
	M_3	1.78	41.42	3.59	2.35	0.71
	X	74.44	5.82	21.09	5.12	64.84
	Feed	100	20.55	100.00	5.88	100.00
Fatty acid process for tungsten selection	K	9.96	90.22	46.75	1.55	2.77
	M_1	9.88	32.35	16.63	14.33	25.41
	M_2	3.21	41.33	6.90	18.97	10.93
	M_3	1.85	45.42	4.37	1.57	0.52
	X	75.1	6.49	25.35	4.48	60.37
	Feed mine	100	19.22	100.00	5.57	100.00
Direct flotation of fluorite	K	13.33	94.32	60.98	0.76	1.69
	M_1	9.44	28.99	13.27	17.42	27.51
	M_2	2.45	38.98	4.63	22.58	9.26
	M_3	1.49	40.33	2.91	1.89	0.47
	X	73.29	5.12	18.20	4.98	61.07
	Feed mine	100	20.62	100.00	5.98	100.00

of residual chemicals on the flotation of fluorite in tailings, the effects of benzohydroxamic acid, lead nitrate and their complexes on the flotation of fluorite, calcite, quartz and other minerals were systematically studied when sodium oleate was used as collector.

1. The influence of lead nitrate on the flotation behavior of fluorite, calcite and quartz

When sodium oleate is used as collector and the dosage of lead nitrate is 50 mg/L, the effect of pH value on the floatability of fluorite, calcite and quartz is shown in Fig. 7.27. It can be seen that the floatability of quartz is greatly improved after the addition of lead ion, which indicates that lead ion can live fossil quartz when sodium oleate is used as collector. In the range of pH = 7–10, the floatability of calcite is better than fluorite. In the presence of lead ion, the optimal flotation pH of fluorite changes from 7 to about 5.5.

Figure 7.28 shows the effect of $Pb(NO_3)_2$ dosage on the recovery of fluorite, calcite and quartz at pH 9 as collector. With the increase of the dosage of lead nitrate, the recovery of fluorite decreased gradually, indicating that the presence of lead ion inhibited the flotation of fluorite. Quartz changed from unfloatable to floatable with the addition of lead nitrate, but too much lead nitrate was not conducive to flotation

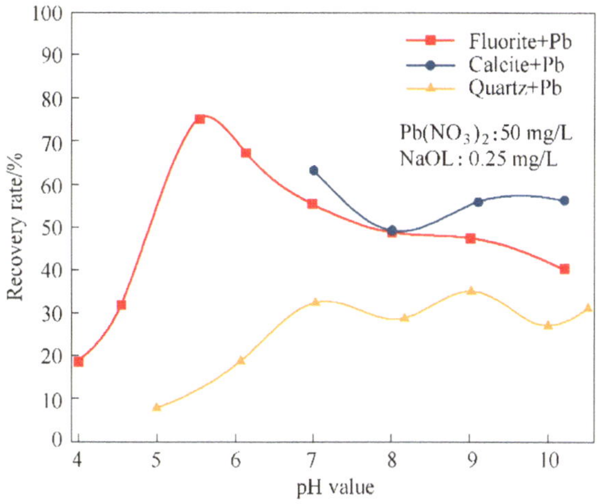

Fig. 7.27 Effect of pH value on floatability of fluorite, calcite and quartz as collector with sodium oleate ($Pb(NO_3)_2$ 50 mg/L)

of quartz. The recovery of calcite was almost unchanged, indicating that lead ion had little effect on the flotation of calcite. When the dosage of lead nitrate is 200 mg/L, the recovery of calcite is about 55%, and that of fluorite and quartz is about 20%.

Figure 7.29 shows the effect of Pb $(NO_3)_2$ dosage on the floatability of fluorite, calcite and quartz at pH = 9 when the dosage of lead nitrate is 200 mg/L. The recovery of the three minerals increases with the increase of sodium oleate dosage, and the buoyancy of calcite is better than that of fluorite and quartz. When the dosage of sodium oleate was 8 mg/L, the recovery of all three minerals was more than 90%. This indicated that sodium oleate as a collector, in the presence of lead ions, fluorite, calcite, quartz three minerals difficult to separate effectively. Residual lead ions are inevitable in the tungsten tailings, which will affect the subsequent flotation recovery of fluorite.

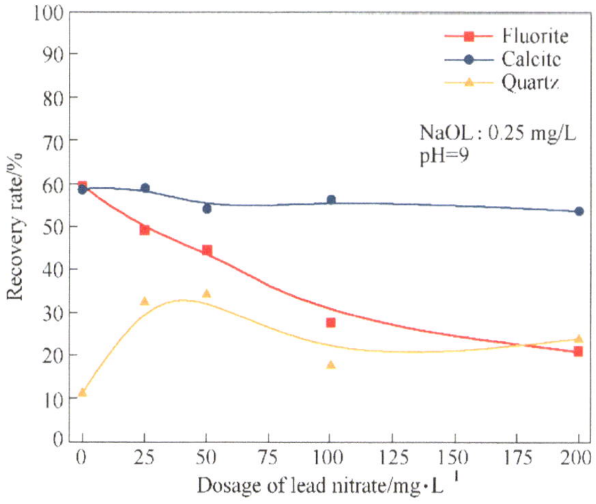

Fig. 7.28 Effect of $Pb(NO_3)_2$ dosage on floatability of fluorite, calcite and quartz when sodium oleate was used as collector and pH was 9

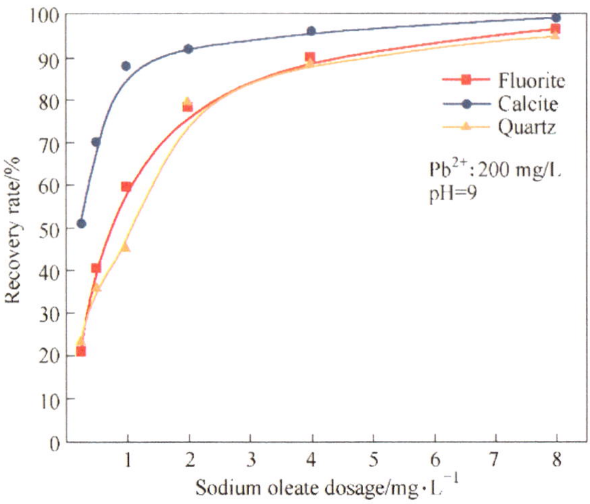

Fig. 7.29 Influence of sodium oleate dosage on the floatability of fluorite, calcite and quartz when the dosage of lead nitrate is 200 mg/L and pH = 9

2. Effect of BHA on flotation behavior of fluorite, calcite and quartz

Figure 7.30 shows the effect of pH on floatability of fluorite, calcite and quartz when sodium oleate is used as collector and the dosage of benzohydroxamic acid is 50 mg/L. With the increase of pH, the recovery of fluorite firstly increases and then decreases, and reaches the maximum when pH is 9 (about 95%). The recovery of calcite remained at about 60%, which was similar to that of calcite when sodium oleate was used as collector alone, indicating that the existence of benzohydroxamic acid had little effect on the flotation of calcite by sodium oleate. In the whole pH range, quartz is basically not floating. This indicates that the residual agent benzohydroxamic acid has little effect on the flotation separation of fluorite, calcite and quartz in tungsten tailings.

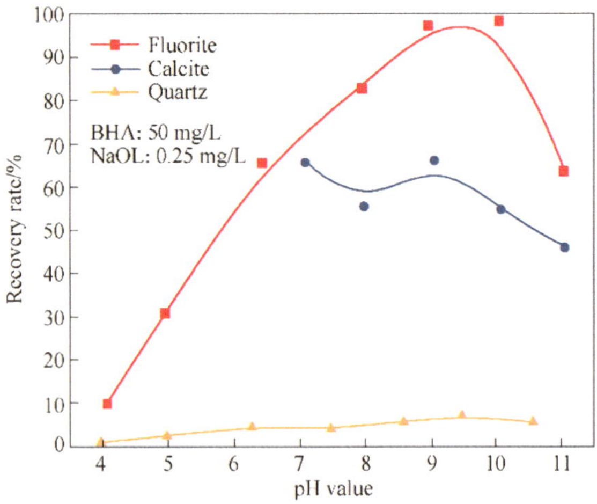

Fig. 7.30 Effect of pH on floatability of fluorite, calcite and quartz with sodium oleate as collector (benzohydroxamic acid 50 mg/L)

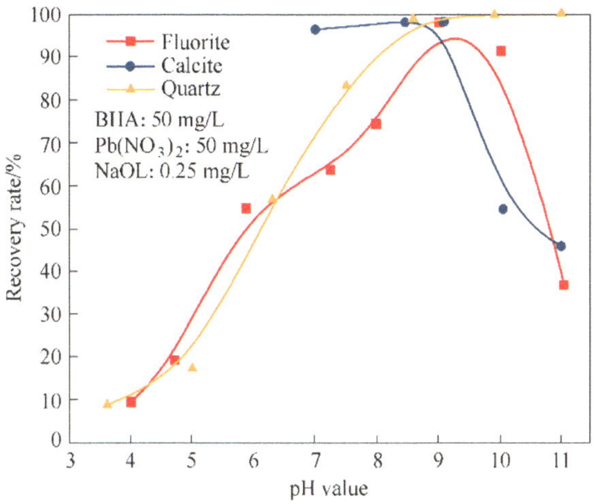

Fig. 7.31 Effect of pH on floatability of fluorite, calcite and quartz as collector with sodium oleate (benzohydroxamic acid 50 mg/L, lead nitrate 50 mg/L)

3. Effects of BHA and $Pb(NO_3)_2$ complexes on flotation behavior of fluorite, calcite and quartz

Figure 7.31 shows the effect of pH on the floatability of fluorite, calcite and quartz when sodium oleate is used as collector and the dosage of benzohydroxamic acid is 50 mg/L and the dosage of lead nitrate is 50 mg/L (benzohydroxamic acid and lead nitrate are mixed and then added). In the presence of benzohydroxamic acid and lead nitrate, the recovery of quartz gradually increases with the increase of pH, and reaches more than 98% when pH exceeds 9. The recovery of calcite is more than 95% when pH is 7–9, and decreases rapidly when pH is greater than 9. The recovery of fluorite increases gradually in the pH range of 4–9 and decreases rapidly when pH exceeds 10. At this point, flotation separation of fluorite from calcite and quartz is difficult to achieve.

The above studies showed that the residual lead ions and benzohydroxamic acid in tungsten ore flotation tailings had a great impact on the subsequent flotation recovery of fluorite. Among them, lead ion and the complex formed by lead ion and benzohydroxamic acid are the main interfering factors. Lead ion can live flotation of quartz and inhibit the flotation of fluorite. Pb-BHA complexes can activate the flotation of calcite.

4. Effect of Pb-BHA complex structure on flotation behavior of fluorite, calcite and quartz

The structure of Pb-BHA complex determines its adsorption behavior on mineral surface, which is closely related to the reaction ratio of lead ion, benzohydroxamic acid and reaction pH. Figure 7.32 shows the effect of the dosage of benzohydroxamic acid on the floatability of fluorite, calcite and quartz at pH 9. As can be seen from Fig. 7.32, when there is no lead ion, the flotation recovery of fluorite is 45% when the dosage of benzohydroxamic acid is 10 mg/L (approximately equivalent to the

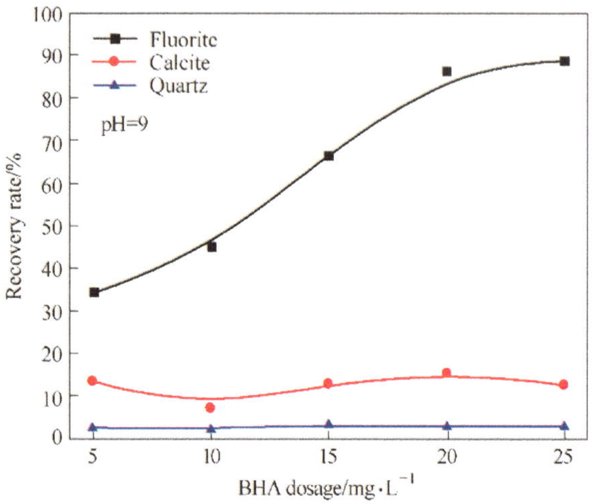

Fig. 7.32 Effect of benzohydroxamic acid dosage on floatability of fluorite, calcite and quartz when pH = 9

concentration of residual collector in tungsten tailings flotation), while calcite and quartz do not float at all.

Figure 7.33 shows the effect of the amount of lead nitrate (added after the mixture of benzohydroxamic acid and lead nitrate) on the buoyancy of three minerals at pH = 9 and 10 mg/L benzohydroxamic acid. As can be seen from the figure, when the dosage of lead nitrate is 15 mg/L (approximately equivalent to the concentration of residual lead ion in tungsten ore flotation tailings) and the dosage of benzohydroxamic acid is 10 mg/L, the flotation recovery of calcite is increased to 60%, that of quartz is increased to 20%, and that of fluorite is decreased to 25%. This further indicates that the residual lead ions and the complex formed by lead ions and benzohydroxamic acid (Pb-BHA) in the flotation process of tungsten tailings will have an impact on the subsequent flotation recovery of fluorite, and the influence of different structures of the complex is different.

The experiment further explored the effect of sodium oleate concentration on the flotation behavior of fluorite, calcite and quartz at the concentration of residual chemicals in tailings (lead ion concentration 15 mg/L, benzohydroxamic acid concentration 10 mg/L, benzohydroxamic acid and lead nitrate were mixed and then added). The experimental results were shown in Fig. 7.34. As can be seen from the figure, with the increase of sodium oleate dosage in tailings with the concentration of residual chemicals, the recovery of the three minerals increases. When the concentration of sodium oleate is more than 2 mg/L, the recovery of fluorite and calcite is more than 95%, with good floatability. With the increase of sodium oleate dosage, the recovery of quartz was increased from about 50% to about 90%. This indicates that the existence of benzohydroxamic acid and lead nitrate in the tungsten tailings has a strong activation effect on the flotation of quartz and calcite by sodium oleate, which is not conducive to the flotation separation of fluorite from other minerals.

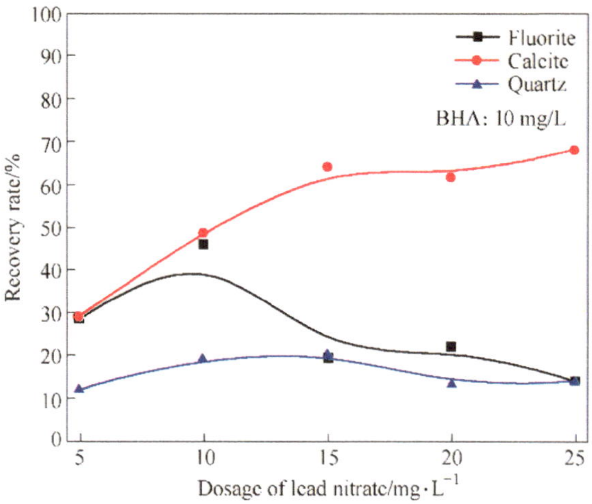

Fig. 7.33 Effect of Pb nitrate dosage on floatability of fluorite, calcite and quartz when pH = 9 and benzohydroxamic acid is 10 mg/L

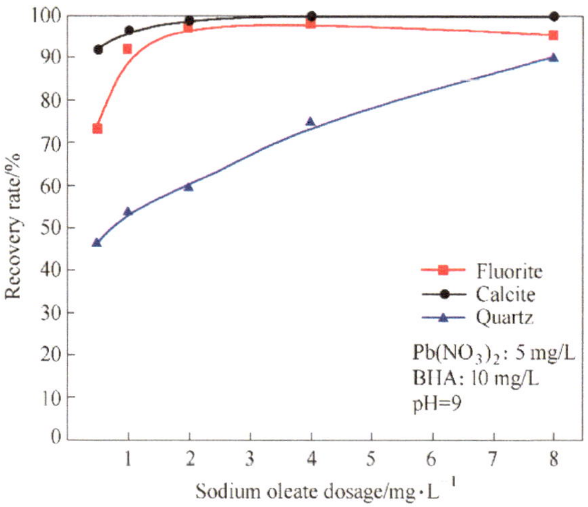

Fig. 7.34 Effect of sodium oleate dosage on floatability of fluorite, calcite and quartz at pH = 9 (benzohydroxamic acid 10 mg/L, lead nitrate 15 mg/L)

7.4.3.2 Influence of Demedicated Treatment on Flotation Behavior of Fluorite, Calcite and Quartz

The above studies showed that benzohydroxamic acid and lead ions will inevitably remain in tungsten ore flotation tailings. The flotation activation of quartz and calcite makes it difficult to separate them from fluorite, which brings great difficulties to the subsequent recovery and utilization of fluorite. In order to eliminate the influence of the residual agents, the mineral demedicated after the action of benzohydroxamic acid and lead was studied (in the demedicated experiment, benzohydroxamic acid and lead nitrate were mixed and then added).

1. Demedicating method

Table 7.16 Effects of demedicated methods on quartz buoyancy

Demedicating method	Quartz recovery/%
Concentrated demedicated	9.87
EDTA-2Na dosage 50 mg/L	6.28
Na_2S dosage 400 mg/L	57.89
HCl	77.64

Concentrated dedosing, EDTA-2Na dedosing, Na_2S dedosing and HCl dedosing were compared in the experiment. According to the effects of the residual agents on the floatability of the three minerals, the effect of dedusting was judged by comparing the recovery of quartz after dedusting.

Concentrated dedusting refers to the pure mineral after BHA and Pb^{2+} treatment through static precipitation, dump the supernuant, and then add deionized water for flotation. EDTA-2Na depharmacizing means that the chelation of EDTA-2Na and Pb^{2+} is used to add EDTA-2Na to the pulp to react with lead ions to generate stable complex precipitation, reduce the concentration of Pb^{2+} in the solution, and reduce the activation of Pb^{2+} on quartz. Na_2S depharmacization refers to the use of OH^- ions and S^{2-} ions produced during its hydrolysis to form hydroxide precipitate or sulfide precipitate with Pb^{2+} in the pulp, so as to reduce the influence of Pb^{2+} on the flotation process of pure minerals. HCl dedusting refers to the use of HCl acid dissolved mineral surface may be covered by Pb-BHA complex substances, to achieve the purpose of eliminating the effect of chemicals.

Table 7.16 shows the effects of various dedusting methods on quartz buoyancy. It can be seen from the table that the dedosing effect of concentrated dedosing and EDTA-2Na among the four dedosing methods is better, which can effectively eliminate the influence of residual benzohydroxamic acid and lead nitrate on quartz. Therefore, the floatability changes of the three minerals after demedicating were further explored in the following two demedicating methods, concentrated demedicating and EDTA-2Na demedicating.

2. Effect of EDTA-2Na dereagent on flotation behavior of fluorite, calcite and quartz

The core of fluorite flotation is the efficient flotation separation of fluorite and calcite. The main inhibitors used are water glass, sodium hacetaphosphite, tannin, baking glue, starch, acid water glass, etc. The residual agents will certainly have a greater impact on the inhibition effect of inhibitors. In the experiment, the most commonly used water glass, sodium hexametaphosphate, acid water glass three inhibitors were selected to explore their influence on the flotation behavior of three kinds of minerals after different demedicating treatment.

Figure 7.35 shows the effect of the dosage of benzohydroxamic acid (10 mg/L), lead nitrate (15 mg/L) and sodium oleate (5 mg/L) on the floatability of fluorite, calcite and quartz when pH is 9. It can be seen from the figure that with the increase of the dosage of EDTA-2Na, the recovery of quartz decreases continuously. When EDTA-2Na dosage is 30 mg/L, quartz is basically not floating. However, the recoveries of fluorite and calcite did not change significantly. The recoveries of fluorite and calcite

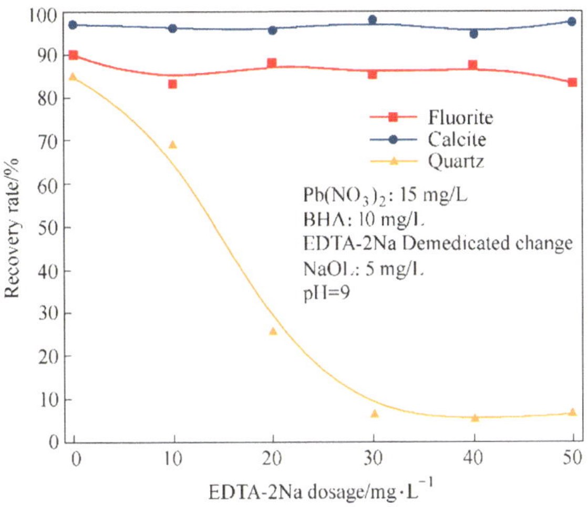

Fig. 7.35 Effect of EDTA-2Na dosage on floatability of fluorite, calcite and quartz at pH = 9 (benzohydroxamic acid 10 mg/L, lead nitrate 15 mg/L, sodium oleate 5 mg/L)

were maintained at about 95% and 90% respectively. From the perspective of quartz flotation, EDTA-2Na can eliminate the activation of Pb^{2+} on quartz, but activate the flotation of calcite. Therefore, the flotation separation of fluorite and calcite can not be realized by using EDTA-2Na demedicating.

Figure 7.36 shows the effect of dosage of benzohydroxamic acid (10 mg/L), lead nitrate (15 mg/L), EDTA-2Na (30 mg/L), sodium oleate (5 mg/L). At pH = 9, the effect of dosage of sodium hexametaphosphite on the floatability of fluorite, calcite and quartz. As can be seen from the figure, in the whole dosage range, quartz basically does not float; The recoveries of fluorite and calcite decreased with the increase of sodium hexametaphosphate dosage, but they decreased to the same extent. When sodium hexametaphosphate dosage was 8 mg/L, fluorite, calcite and quartz almost did not float. Therefore, the flotation separation of fluorite and calcite cannot be realized by using sodium hexametaphosphate as inhibitor after demedicated EDTA-2Na.

Figure 7.37 shows the effect of water glass dosage on the floatability of fluorite, calcite and quartz at pH = 9 when the dosage of benzohydroxamic acid is 10 mg/L, the dosage of lead nitrate is 15 mg/L, the dosage of EDTA-2Na is 30 mg/L and the dosage of sodium oleate is 5 mg/L. As can be seen from the figure, quartz is almost impossible to float in the whole dosage range; The recovery of fluorite and calcite decreased with the increase of water glass dosage. The degree of decline of fluorite is greater than that of calcite, calcite has better floatability than fluorite. Therefore, after demedicated EDTA-2Na, the separation of fluorite and calcite cannot be realized by using glass as an inhibitor.

Figure 7.38 shows the effects of dosage of benzohydroxamic acid (10 mg/L), lead nitrate (15 mg/L), EDTA-2Na (30 mg/L) and sodium oleate (5 mg/L) on the floatability of fluorite, calcite and quartz when pH is 7. As can be seen from the figure, quartz almost does not float in the whole range of dosage; With the increase of the amount of acid water glass, the recovery of fluorite and calcite decreases sharply at first, and then slowly. When the dosage of water glass was 0–0.2 mg/L, the decline

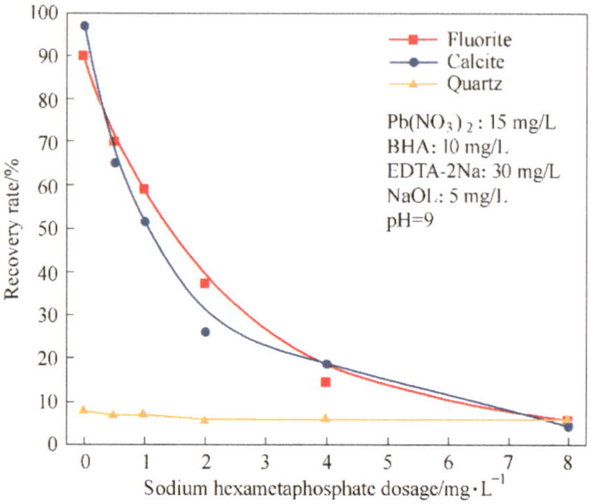

Fig. 7.36 Effect of sodium hexametaphosphate dosage on floatability of fluorite, calcite and quartz at pH = 9 (benzohydroxamic acid 10 mg/L, lead nitrate 15 mg/L, EDTA-2Na 30 mg/L, sodium oleate 5 mg/L)

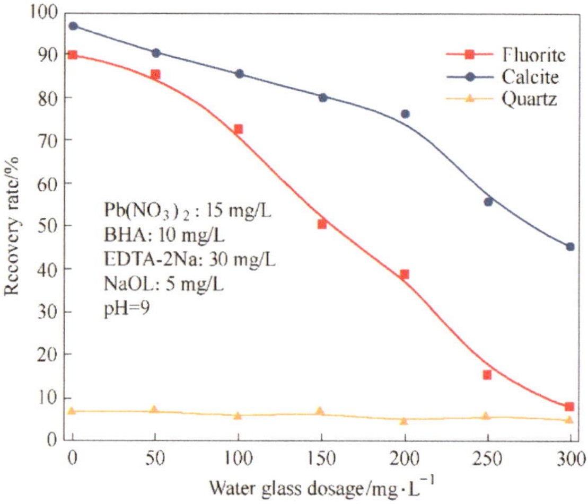

Fig. 7.37 Effect of water glass dosage on the floatability of fluorite, calcite and quartz at pH = 9 (benzohydroxamic acid 10 mg/L, lead nitrate 15 mg/L, EDTA-2Na 30 mg/L, sodium oleate 5 mg/L)

degree of fluorite and quartz was basically the same. When the dosage of water glass was 0.2 mg/L, the decline rate of fluorite was significantly lower than that of calcite. However, the recovery of fluorite was low, which dropped to below 40%, and the difference between them was only about 20%. Therefore, after demedicated EDTA-2Na, acid water glass has a good inhibition ability to calcite, but the floatability of fluorite is greatly affected.

The above experimental results show that although EDTA-2Na can effectively remove residual agents, it is not conducive to the selective inhibition of sodium hacetaphosphate, water glass and acid water glass on calcite, and it is difficult to realize the efficient flotation separation of fluorite from calcite and quartz.

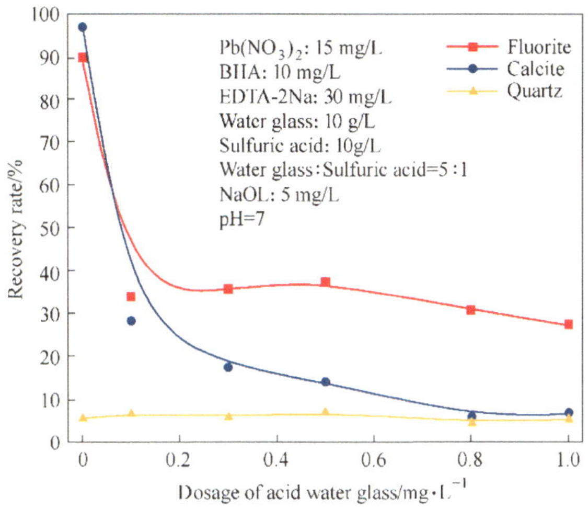

Fig. 7.38 Influence of dosage of acid water glass on floatability of fluorite, calcite and quartz at pH = 7 (benzohydroxamic acid 10 mg/L, lead nitrate 15 mg/L, EDTA-2Na 30 mg/L, sodium oleate 5 mg/L)

3. The effect of concentration and demedicating removal on the flotation behavior of fluorite, calcite and quartz

Figure 7.39 shows the effects of dosage of benzohydroxamic acid (10 mg/L) and lead nitrate (15 mg/L) on the floatability of fluorite, calcite and quartz after concentration and demedication, sodium oleate (5 mg/L) and pH (9). As can be seen from Fig. 7.39, after concentration and demedicating, quartz becomes unfloatable; with the increase of sodium hempetaphosphate dosage, the recoveries of fluorite and calcite gradually decrease, and their recoveries are almost the same. When sodium hexametaphosphate dosage is 2 mg/L, the recovery of fluorite, calcite and quartz is about 15%, 25% and 10% respectively. Therefore, after concentration and demedicated, sodium hexametaphosphate as inhibitor can not achieve flotation separation of fluorite and calcite.

Figure 7.40 shows the effects of dosage of benzohydroxamic acid (10 mg/L) and lead nitrate (15 mg/L) on the floatability of fluorite, calcite and quartz after concentrated and demedicated sodium oleate (5 mg/L). At pH = 9, water glass dosage affects the floatability of fluorite, calcite and quartz. As can be seen from the figure, the recoverys of fluorite and calcite decrease gradually with the increase of water glass dosage, but the decline trend of fluorite is much more moderate than that of calcite, water glass has more significant inhibition effect on calcite, and quartz becomes unfloatable. When the dosage of water glass is 300 mg/L, the recovery of fluorite is about 65%, the recovery of calcite is about 25% and the recovery of quartz is about 10%. The results showed that the flotation separation of demedicated fluorite from calcite and quartz could be achieved to a certain extent by using water glass as an inhibitor.

Figure 7.41 shows the effect of dosage of benzohydroxamic acid (10 mg/L), lead nitrate (15 mg/L), sodium oleate (5 mg/L) after concentration and demedicated, and pH 7, the effect of dosage of acid water glass on the floatability of fluorite, calcite

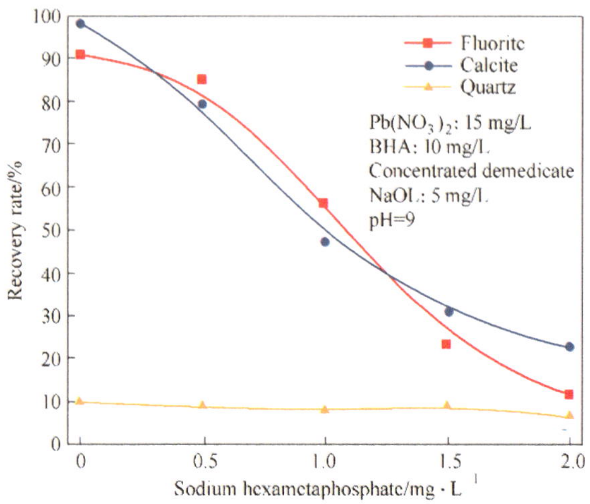

Fig. 7.39 Effect of sodium hexametaphosphate dosage on floatability of fluorite, calcite and quartz at pH = 9 (benzohydroxamic acid 10 mg/L, lead nitrate 15 mg/L, concentrated demedicated, sodium oleate 5 mg/L)

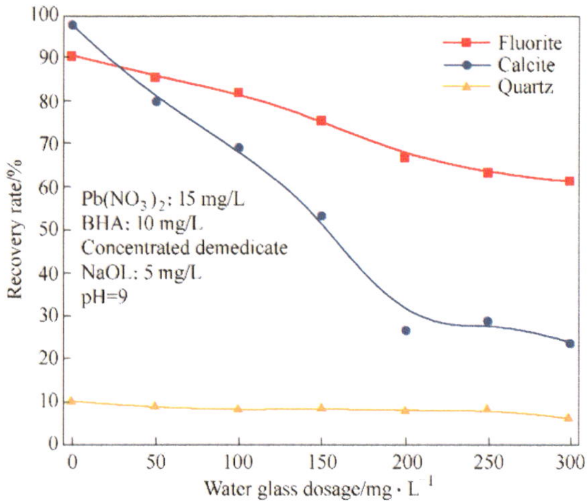

Fig. 7.40 Effect of water glass dosage on floatability of fluorite, calcite and quartz at pH = 9 (benzohydroxamic acid 10 mg/L, lead nitrate 15 mg/L, concentrated demedicated, sodium oleate 5 mg/L)

and quartz. As can be seen from the figure, quartz hardly floats after concentration and demedicating. With the increase of the dosage of acid water glass, the recovery of calcite decreases rapidly, while that of fluorite decreases slowly. When the dosage of water glass was 1.0 mL, the recovery of calcite dropped from 90% to about 10%, while the recovery of fluorite remained at about 80%. The results showed that the acid water glass had strong inhibition on calcite after concentration and demedicated, and the flotation separation of fluorite and calcite could be achieved well.

The above experimental results show that the concentration method can greatly remove the residual reagent and reduce the influence of tungsten flotation reagent on the subsequent flotation of fluorite. Using sodium oleate as collector and water glass and acid water glass as inhibitors, the flotation separation of fluorite from calcite

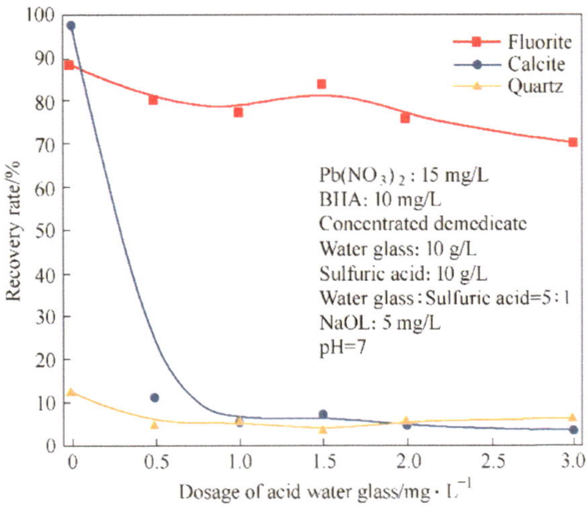

Fig. 7.41 Influence of the dosage of acid water glass on the floatability of fluorite, calcite and quartz when sodium oleate was 5 mg/L and pH = 7 (benzohydroxamic acid 10 mg/L, lead nitrate 15 mg/L, concentrated and demedicated)

and quartz can be realized efficiently, and the acid water glass is the best separation effect.

7.4.3.3 Flotation Practice of Associated Fluorite from Tungsten Flotation Tailings

1. Efficient flotation process and practice of associated fluorite from tungsten tailings in Shizhuyuan Polymetallic Mine, Hunan

The content of CaF_2, $CaCO_3$ and SiO_2 is 20.67, 5.33 and 43.42% in the tungsten tailings of Nonferrous Chenzhou fluorite separation plant in Hunan Province, which belong to low grade fluorite ore with high calcium carbonate. The residue of benzohydroxamic acid and lead nitrate will activate quartz and calcite and inhibit fluorite, so that the flotation recovery of fluorite is not good, the total recovery of fluorite is only about 40%. After the actual process investigation, the flotation process of the separation plant into the optimization design, before flotation added dedusting operation, using water glass, acid water glass as inhibitors for a month of industrial commissioning. The process flow of the site is concentrated dedusting, 1 rough selection, 1 scavenging selection and 8 times selected closed circuit. The reformed flow chart is shown in Fig. 7.42. After using the new technology, fluorite concentrate with CaF_2 grade of 92.05% and recovery of 60.04% can be obtained. The concentrate grade is increased by more than 3 percentage points and the recovery is increased by 20 percentage points.

Since 2015, Chenzhou Fluoride Chemical Co., LTD. Fluorite separation plant uses flotation demedicating process, the production index gradually improved, in 2017, the fluorite output increased from 130,000 to 150,000 t/a; 2021 through further technical transformation of fluorite concentrate breakthrough 180,000 t/a, metallurgical grade

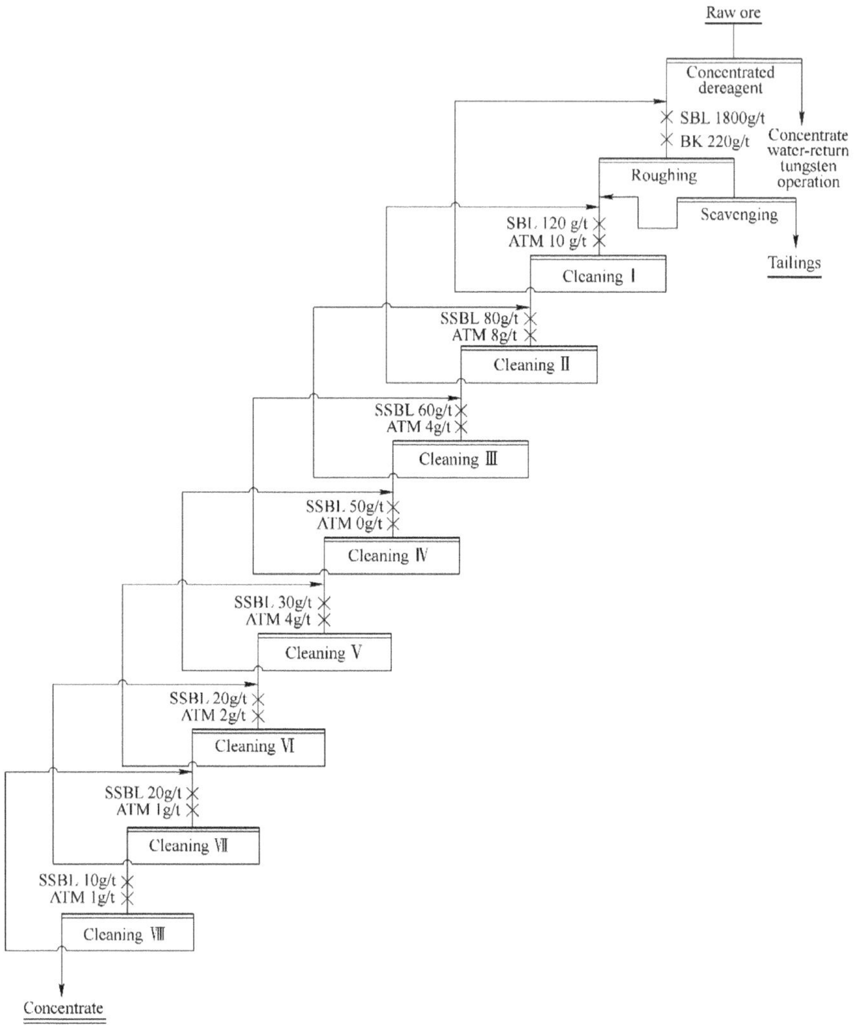

Fig. 7.42 Process flow chart of Chenzhou Fluoride Chemistry Co., Ltd. after modification of fluorite separation plant

fluorite recovery of about 15%, acid grade fluorite recovery of more than 60%, fluorite comprehensive recovery of more than 75%.

2. High efficiency flotation process and practice of associated fluorite from Hunan Huangshaping polymetallic ore

Huangshaping polymetallic mine area is one of the representative mines in China, and the mine area is an important lead and zinc producing area in China, known as the "Pearl of Nanling". After nearly 60 years of mining, the resources of lead and zinc are almost exhausted. However, the skarn type tungsten-stann-molybdenum-bismuth deposit in Huangshaping mining area is associated with a large number of valuable elements, among which WO_3 content is more than 0.18%, and associated fluorite content is 10–15%. The ore resources explored at present are about 80 Mt, which is an important subsequent resource. Since 2006, Pingmine has been conducting trial mining and beneficiation of tungsten and molybdenum ore. In 2011, a 1500 ton/day scale multi-metal beneficiation plant was built. With fatty acids and their derivatives as collectors and a large amount of water glass as inhibitors, the comprehensive recovery of tungsten is about 53–55% through the "Petrolov method" heating flotation process, and the economic benefit is poor, and the tungsten dressing tailing water has a serious impact on the production index and the environment; At the same time, due to the mass use of water glass, tungsten tail fluorite enrichment is difficult, economic benefits are difficult to meet the production requirements. Due to poor technology, low comprehensive recovery of tungsten and fluorite, and poor economic benefits, production was halted in 2015, and the Huangshaping Replacement Resource Project was forced to stop.

The polymetallic resources in Huangshaping Tungsten-Molybdenum Polymetallic Mine and Shizhuyuan Polymetallic Mine belong to Nanling metallogenic belt, and the resources are similar. The new technology provides a new hope for the efficient development and utilization of the polymetallic resources in Huangshaping. In 2020, Hunan Nonferrous Huangshaping Mining Co., Ltd. and Central South University jointly conducted a key test study on improving the recovery of tungsten and fluorite in Huangshaping polymetallic mine. The comprehensive recovery process is shown in Fig. 7.43, and the comprehensive recovery of tungsten and fluorite has been greatly improved.

Among them, the main body of fluorite flotation adopts the process of one roughing, five cleanings and one scavenging. The concentrate of two cleaning tail cleaning enters three cleaning, the tailings of two tail cleaning and one cleaning tailings are merged and discharged directly after one scavenging, and the rest of the middle ore returns in sequence. Fluorite roughing using flotation column, flotation pH control at 10.5, YZ-4 used for roughing control of calcite enrichment, the amount of 100 g/t, SBL dosage 800 g/t, HS dosage 250 g/t; Fluorite scavenging HS dosage 50 g/t. Acidized water glass and ATM were used to inhibit the floating of silicate gangue and calcite in the stage of fluorite selection, and more than 85% fluorite concentrate was obtained. The industrial test flow is shown in Fig. 7.44.

After the industrial commissioning from late September to the end of October 2020, the flotation system of Huangshaping polymetallic mine fluorite is stable and the index keeps improving, reaching the expected industrial commissioning target. Statistics of production indicators of 6 shifts in 3 working days from October 26 to 28 are shown in Table 7.17. In the production of the six shifts, a total of 3714 t of raw ore were treated, and the grade of raw ore fluorite was 16.22%. After the new process, the final fluorite concentrate grade is 87.62% and the recovery is 59.64%. Compared with the original process, the comprehensive recovery of fluorite was increased by 30 percentage points, and the economic benefit was remarkable.

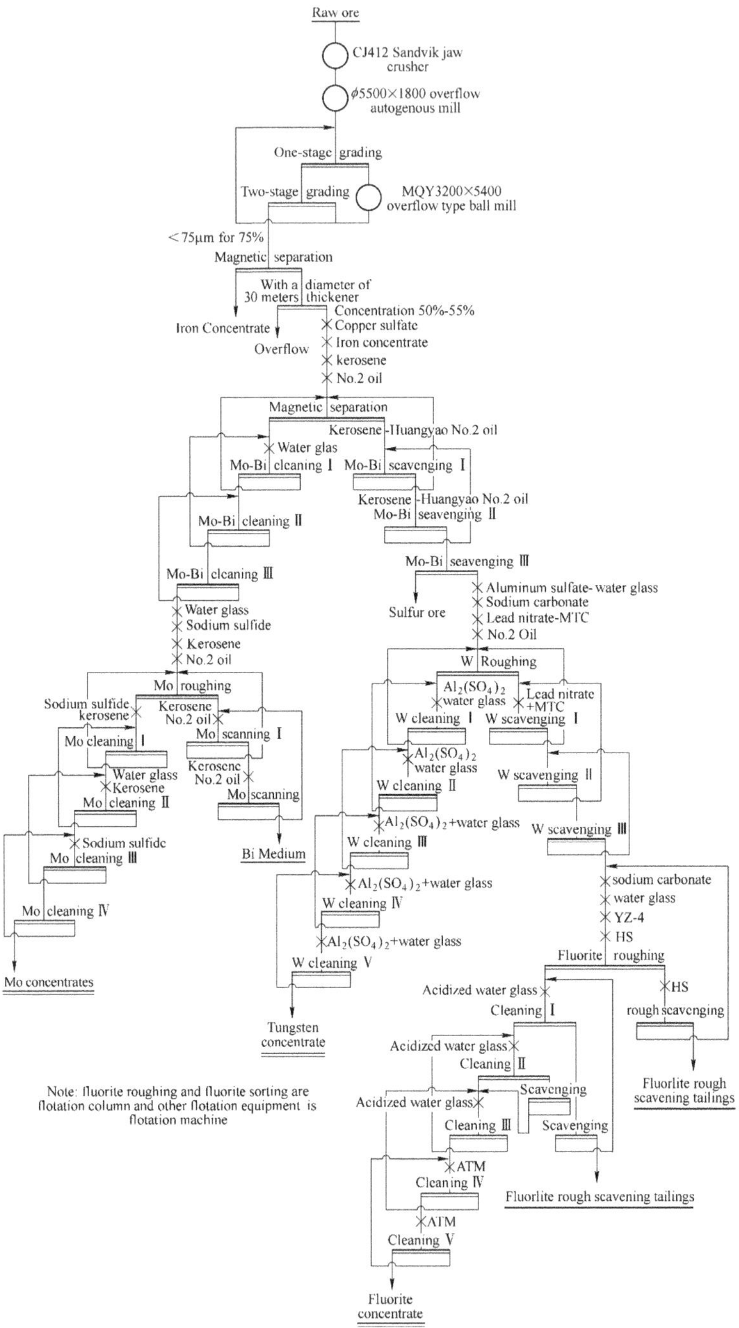

Fig. 7.43 Comprehensive recovery process flow chart of tungsten-molybdenum-fluorite in Huangshaping

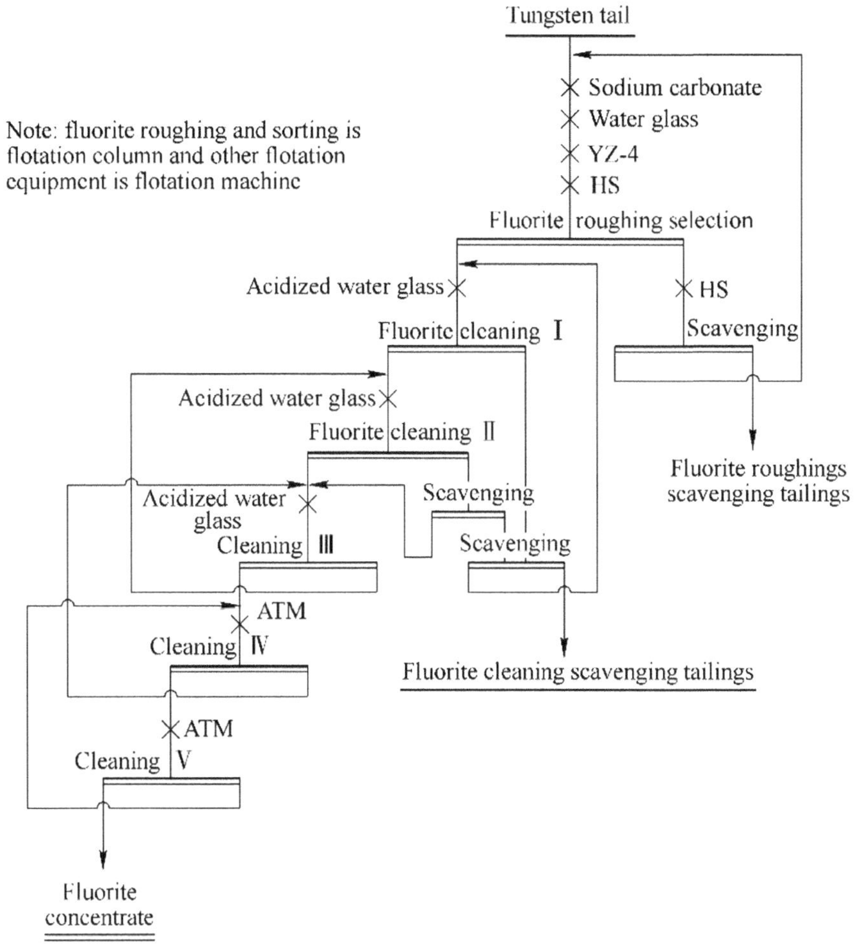

Fig. 7.44 Process flow chart of Huangshaping fluorite flotation

Table 7.17 Industrial commissioning indicators of fluorite in Huangshaping polymetallic mine (2020)

Shift	Handling capacity/t	Raw ore grade/%	Recovery/%	Concentrate grade/%
26th night shift	699	11.96	56.00	89.30
26th day shift	722	18.29	60.47	89.85
27th night shift	721	19.81	60.6	88.47
27th day shift	706	15.35	61.36	89.16
28th night shift	495	16.65	61.22	84.65
28 day shift	371	15.27	58.21	84.30
Total	3714	16.22	59.64	87.62

7.5 Impurity Control Technology in the Process of Producing Hydrogen Fluoride with Fluorite Concentrate Powder

Due to the poor endowment of associated fluorite resources, high impurities such as calcite, silicate and sulfide, and the fineness of fluorite concentrate powder far exceeds the national fine powder requirements, serious problems such as dust, agglomeration and low conversion rate will occur in the chemical application process of these fine powder, restricting the acid grade utilization of low-grade fluorite concentrate powder. In view of this bottleneck problem, Hunan Nonferrous Chenzhou Fluoride Chemical Co., Ltd. developed impurities control technology and equipment in the process of producing hydrogen fluoride from low grade fluorite concentrate powder (Fig. 7.45), which broke the technical bottleneck of direct acid production from low grade fluorite concentrate powder, realized the large-scale acid grade utilization of 90% grade fluorite concentrate powder, and reactivated nearly 100 Mt of high calcium and low grade associated fluorite resources.

The main core technologies are as follows:

(1) Through the enhancement of temperature field and pressure field, as well as the clever collocation of 105 acid and 98 acid, the problem of efficient decomposition and conversion of high-calcium fluorite concentrate powder was solved. The process of producing hydrofluoric acid from fluorite concentrate powder is a typical shrinkage nuclear reaction model, which is controlled by the efficiency of mass transfer process. However, the low order fluorite powder reacts with sulfuric acid to produce calcium sulfate, which easily covers the surface of fluorite particles and hinders the reaction. At the same time, a large amount of calcium carbonate in the fluorite powder reacts faster with sulfuric acid to produce a large amount of calcium sulfate, which further hinders the acid reaction. It was found that the purity of acid and the way of adding acid had a significant effect on the crystal type and porosity of calcium sulfate. 105 acid could help improve the crystallization property of calcium sulfate, greatly improve the porosity, and improve the reaction efficiency of sulfuric acid and fluorite particles. Therefore, through the clever combination of 105 acid and 98 acid, supplemented by temperature field enhancement, the high efficiency conversion of fine fluorite powder with high calcium was achieved, with the conversion rate as high as 97%.

(2) The closed negative pressure dust extraction—forced circulation reactor was designed and developed to solve the problems of dust, slag discharge and waste gas purification. Aiming at the problems of fine particle size of low grade fluorite (over 80% particle size less than 0.038 mm) and high content of calcium carbonate, the reactor end face sealing device was invented: By changing the sealing structure of the reactor end face, changing the compensator on the sealing end face into the pilot pipe and installing the compensator at the feed screw, a unique sealing technology of the reactor end face was developed and the sealing system of the end face was changed, so that the pressure was increased from the

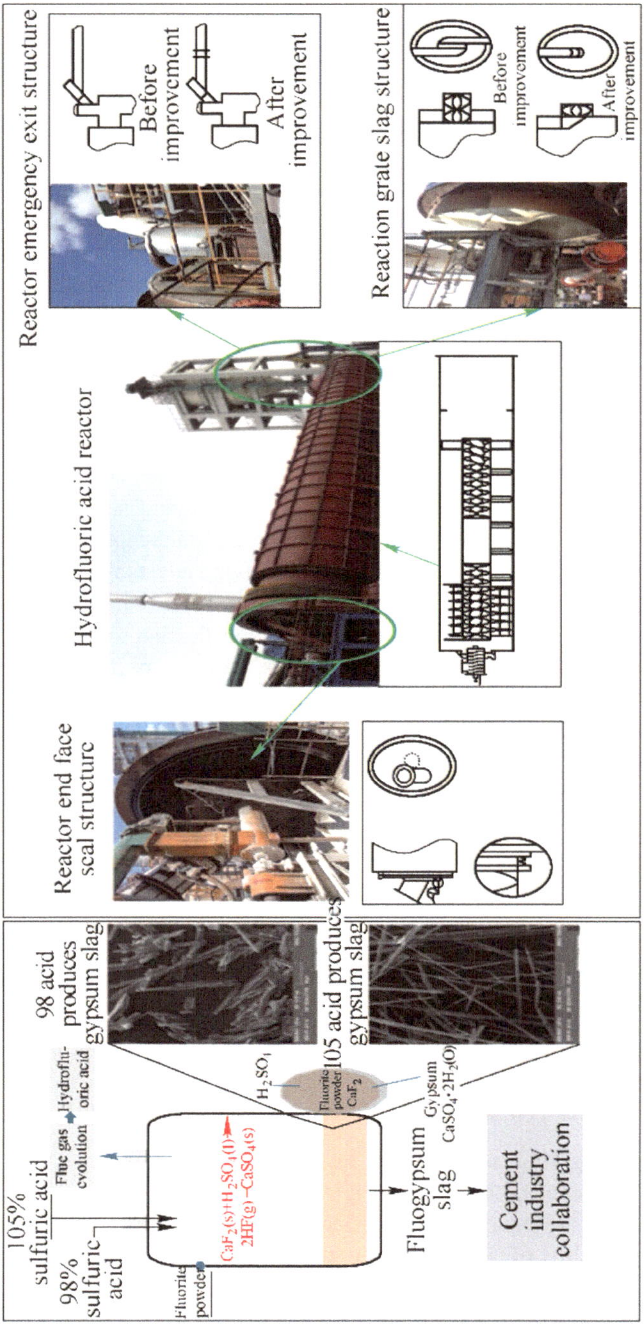

Fig. 7.45 Technology and equipment for the preparation of hydrofluoric acid by low grade fluorite concentrate powder

original 0.5–2.0 kPa to avoid the leakage of HF gas from the reactor. To achieve safe and environmentally friendly stable production. Aiming at the problems of high sulfuric acid content, uneven particle size, large viscosity and poor fluidity in by-product gypsum, a low grade fluorite reaction grate slag device has been developed and applied: According to the characteristics of low grade fluorite gypsum, the structure of the feed spoon is changed, the double scoop is changed into a single scoop and the slag discharge mechanism combined with the furnace tail cover, which solves the problem of poor discharge of gypsum, greatly increases the production capacity, and solves the problem that the use of low grade fluorite in the same diameter reactor can not reach the production. To solve the problem of safe production of fluorine chemical industry, a unique reactor emergency discharge and absorption system is developed: once the front-end production system fails and the pressure in the furnace exceeds 1.0 kPa, the emergency discharge system automatically opens immediately, and the HF gas in the furnace is discharged to the emergency absorption system, which adopts two-stage washing and one-stage alkali washing to absorb and discharge to the standard.

(3) Designed and developed the roughing cooler intensive desulphurization and rectification system, realized the rapid cooling and interception of elemental sulfur in the preparation of hydrofluoric acid. Taking advantage of the fact that the water temperature of Cuijiang is only about 16 °C, the pre-purification tower is separately designed into scrubber and purification tower. Increase the packing in the purification tower, and through the unique process design, using roughing hydrofluoric acid for gasification and cooling, so that most of the gas phase sulfur precipitation in the equipment, the unprecipitated sulfur gas and roughing HF gas into the roughing cooler, the roughing cooler using low temperature water cooling, and designed an efficient sulfur removal device, the remaining gas phase sulfur conversion removal, the purification tower and the roughing cooler uniform open a ready, To ensure the continuity of production, the power consumption per ton of products is reduced to 450 kW h, and the same industry acid grade fluorite ($CaF_2 > 97\%$) production enterprises. Increase the diameter of the roughing cooler line pipe, in order to ensure the effective intercession of elemental sulfur at the same time convenient cleaning, so that the use of the roughing cooler becomes both energy saving and sulfur removal, prolong the service cycle of the follow-up device. At the same time, a special distillation system is developed to further remove trace sulfur, so as to meet the quality requirements of low-grade fluorite acid production products.

Through the integration of the above technologies, the impurity control technology in the process of producing hydrogen fluoride with low grade fluorite fine powder was formed. Compared with the industry FC-97 fluorite acid production process, the applicability of this technology to the quality of fluorite has been greatly increased. It can realize the large-scale acid grade utilization of about 90% CaF_2 fluorite fine powder, and it is the only technology in China that uses low grade fluorite to produce anhydrous hydrogen fluoride. The technology greatly improves the comprehensive utilization rate of fluorite resources, and provides strong guarantee for the development of fluorine chemical industry and strategic new industries in China.

7.6 Summary

Fluorite, as an industrial source of F element (non-renewable non-metallic resource), is known as the "second rare earth". It has been included as a strategic mineral or key resource by China, the United States, the European Union and other countries or regions, and is an important strategic resource for the development of national economy and high-tech industry in the future. Associated fluorite resources have become the main body of fluorine resources development and utilization, but most of these fluorite resources are "symbolic associated fluorite ores that are difficult to use". On the one hand, the resource endowment of associated fluorite is poor (CaF_2 16–21%, $CaCO_3 > 10\%$), and the grade of fluorite concentrate (CaF_2 88–93%) is not up to the requirement of acid grade fluorite ($CaF_2 \geq 97\%$) due to the influence of upstream main metal beneficiation process, so the resource utilization faces great challenges. Around this problem, domestic research institutes broke through the technology of direct production of anhydrous hydrogen fluoride (AHF) from low quality fluorite concentrate, and developed a series of efficient flotation technologies for associated fluorite resources. By combining mineral processing with chemical engineering, the acid grade utilization of low grade associated fluorite resources was realized on a large scale, and more than 100 Mt of complex associated fluorite resources were reactivated in Nanling. The efficient utilization of this resource alleviates the shortage of fluorite resources in China, and establishes advantaged resource advantages for the development of fluorine-chemical industry, helping China's fluorine-chemical industry transition to high-end industry.

Acknowledgements This work was supported by National key research center and development Program of the 14th Five-Year Plan (No. 2022YFC2905105 and No. 2022YFC2905104), National Natural Science Foundation of China (NSFC) (No. 52122406and No. 52374291), Hunan High-tech Industry Technology Innovation Leading plan (No. 2022GK4056), Hunan Innovative Province construction special Project (No. 2020RC3001).

References

1. Hayes TS, Miller MM, Orris GJ et al (2017) Fluorine. Reston, VA, p 92
2. Dill HG (2010) The, "chessboard" classification scheme of mineral deposits: mineralogy and geology from aluminum to zirconium. Earth Sci Rev 100(1–4):1–420
3. Wang J, Shang P, Niu G (2010) Main concentration areas and resource potential of fluorite ore in China. Chem Miner Geol 32(2):9

4. Wang J, Shang P, Xiong X et al (2014) Classification of fluorite deposits in China. Geol China 41(02):315–325
5. Usgs (2021) Mineral commodity summaries 2021. Reston, VA
6. Wu Q, Xue S (2019) Analysis on supply security of key mineral resources under Sino-US trade change. J China Univ Geosci (Soc Sci Ed) 19(05):69–78
7. Chen Q, Zhang Y, Ni S et al (2017) Analysis on the strategy of strengthening Japan's mineral resources strategy. China Min 26(12):8–15
8. Li J, Zhang S, Shang P et al (2019) Analysis of fluorite resources status and strategic value. Conserv Utilization Miner Resour 39(6):7
9. Key minerals: international trends and thoughts. Miner Deposits Geol 38(04):689–698 (2019)
10. Tian H (2009) Resources and human settlements 5:3
11. Chen S, Zhang S (2013) Utilization status and development strategy of fluorite resources in fluorinated chemical industry in China. Resour Ind 15(02):79–83 (in Chinese)
12. Hou H (2013) Current situation and development trend of inorganic fluoride industry. China Petrol Chem Ind Econ Anal 8:4 (in Chinese)
13. Luo Y-M (2010) Current situation and development trend of fluorine-containing materials industry in China. New Chem Mater 38(11):31–34 (in Chinese)
14. Wang W, Bai Z (2014) Status quo of fluorite resources and industry development in China. Met Mine 3:9
15. Zhang G, Zhang Q, Zhan X et al (2012) Research progress of fluorine-containing materials in new energy field. Chem Ind Prod Technol 19(5):6
16. Xu B, Zhu G, Zhu M (2013) Research progress of expanded polytetrafluoroethylene sealing materials for aerospace. China Plast 8:5
17. Ministry of Land and Resources of the People's Republic of China (2021) China mineral resources report 2021. Beijing
18. Zhao P, Zheng H, Zhang X et al. Status quo and development suggestions of fluorite industry resources in China. Chem Miner Geol 20,42(02):178–183
19. Fang G, Wang D, Chen Y et al. Metallogenic regularity and genesis of Nanling fluorite deposit. Acta Geol Sin 20,94(01):161–178
20. Wang J, Shang P, Xiong X et al (2015) Metallogenic regularity of fluorite deposit in China. Geol China 42(01):18–32
21. Li J, Gao Y, Zhang H (2017) Status quo and sustainable development strategy of fluorite resources in China. China Min 26(10):7–14
22. Lin Z (2018) Geological characteristics of fluorite deposit in Fujian Province. Chem Miner Geol 40(3):4
23. Ministry of Land and Resources of the People's Republic of China (2018) China Mineral Resources Report 2018. Geological Publishing House, Beijing
24. Li H (2010) Research on development and utilization prospect of fluorite ore and resource potential evaluation of strategic constituency in China. China University of Geosciences, Beijing
25. Li C, Gao Z (2017) Effect of grinding media on the surface property and flotation behavior of scheelite particles. Powder Technol S778446952X
26. Li J (2012) Experimental study on comprehensive recovery of fluorite from Shizhuyuan scheelite flotation tailings. Nonferrous Met Miner Process Sect 1:3
27. Zhou J, Zhu Y (2012) Experimental study on recovery of fluorite by Flotation from Huangshaping low-grade molybdenum and bismuth tungsten flotation tailings. Min Metall Eng 32(001):29–31
28. Zhang X, Pu Y, Ai J et al (2016) Application of a new collector in flotation of refractory fluorite ore from India. Nonferrous Met Miner Process Sect 1:5
29. Zhou W (1985) Study on Improving flotation index of fluorite in Taolin lead-zinc mine. Nonferrous Min 12:24–29
30. An Shunchen (1987) Lho-collector and flotation of fluorite. Nonferrous Met (Miner Process Sect) 01:17–30

31. Zhang Q, Liu R, Cao X et al (2013) Study on the relationship between structure and properties of fatty acid scheelite collectors. Nonferrous Met Sci Eng 4(05):85–90
32. Hu Y, Wang D (1987) Adsorption and Flotation activation Mechanism of metal ions at oxide mineral/water interface. J Cent South Univ Min Metall 05:501–508
33. Gao X-H, Gao Z-Y, Sun W (2017) Research progress on effect of metal ions on mineral flotation behavior and mechanism. Chin J Nonferrous Met 27(04):859–868
34. Kellar JJ, Cross WM, Miller JD (1990) In-situ internal reflection spectroscopy for the study of surfactant adsorption reactions using reactive internal reflection elements. Sep Sci 25(13–15):2133–2155
35. Lu Q, Han H, Chen Z et al (2023) Action Mechanism and application of typical organic inhibitors in flotation separation of fluorite and calcite. Met Mine 559(01):216–222
36. Moulin P, Roques H (2003) Zeta potential measurement of calcium carbonate. J Colloid Interface Sci 261(1):115–126
37. Gao Z, Wei S, Gao J et al (2017) A density functional theory study on the effect of lattice impurities on the electronic structures and reactivity of fluorite. Minerals 7(9):160
38. Zhu X-Y (2019) Effect and mechanism of sodium petroleum sulfonate on fluorite flotation. Wuhan University of Technology
39. Wang Z, Ren Z, Gao H et al (2019) Application of sodium petroleum sulfonate in flotation of a quartz fluorite ore in Jiangxi Province. Non Metallic Ores 42(6):5
40. Ai G, Li J, Wu H et al (2016) Experimental study on recovery of fluorite from a scheelite and wolframite tailings. Non Metallic Min 39(3):4
41. Zheng R, Ren Z, Gao H et al (2018) Evaluation of sulfonate-based collectors with different hydrophobic tails for flotation of fluoritc. Mincrals 8(2)
42. Hu B (2012) Synthesis of fatty acid soap collector and its flotation performance for fluorite ore. Central South University
43. Lin H (1993) Research progress of fluorite flotation reagents. Prot Utilization Miner Resour 1:4
44. Hu Y, Xu Z (2003) Interactions of amphoteric amino phosphoric acids with calcium-containing minerals and selective flotation. Int J Miner Process 72(1–4):87–94
45. Deng X, Wang W, Guo D et al (2012) Synthesis and flotation performance of N-(4-methylphenyl) -α-amino-benzyl phosphate. Appl Chem Ind 41(10):4
46. Zeng L, Jiang W, Chen W et al (2021) Research and practice of tungsten flotation residual reagent on flotation of associated fluorite. China Tungsten Ind 36(06):32–40
47. Jiang W, Gao Z, Khoso SA et al (2018) Selective adsorption of benzhydroxamic acid on fluorite rendering selective separation of fluorite/calcite. Appl Surf Sci 435:752–758
48. Filippova IV, Filippov LO, Duverger A et al (2014) Synergetic effect of a mixture of anionic and nonionic reagents: Ca mineral contrast separation by flotation at neutral pH. Miner Eng 66–68:135–144
49. Abramoff AA, Li C, Cui H (2007) Theoretical basis and regularity of cationic collector action mechanism in mineral flotation. Met Ore Dressing Abroad 44(8):5
50. Zhang Y, Zhong H, Tan X et al (2011) Research progress of cationic collectors. Conserv Utilization Miner Resour 3:6
51. Zhang X, Wang Z (2007) Development, utilization and separation of fluorite. Min Express 23(7):3
52. Li S, Wang Y (2010) Study on flotation behavior of calcium bearing minerals by amine collectors. Min Metall Eng 30(5):5
53. Peng J, Zhong H, Wang S et al (2014) Research progress of unsulfide ore collector. Mod Chem Ind 41(001):39–42
54. Gong M (2007) Foam flotation
55. Tian J, Xiao G, Huang G et al (2012) Research progress on synthesis of amphoteric flotation collector. Hunan Nonferrous Met 28(1):5
56. Wang H, Zhong H (1999) Study on ester amphoteric collector. Nonferrous Met Miner Process Part 1:7 (in Chinese)

57. Jin H, Shangguan Z (1997) Research and production practice on the separation performance of T-69 fluorite flotation agent. Met Mine 12:3
58. Lu Y, Sun S (2007) Mining engineering 5(6):3
59. Qian Y, Cui T, Zhang J (2015) Research progress of fluorite flotation collector. Liaoning Chem Ind 2:5 (in Chinese)
60. Jiang Q-M, Dai Z-L (2012) Effect of mixed fatty acids on separation of schetungsten from fluorite and calcite. Min Metall Eng 32(2):4
61. Deng H, Ren H, Xu X et al (2012) Research on flotation process and application of low temperature collector for quartz type fluorite ore. Non Metallic Ores 35(5):3
62. Study on mixed collector in flotation separation of fluorite and barite (1995) Min Metall Eng 15(2):5
63. Cao X, Zhu Y, Lu J (2014) Experimental study on beneficiation of a refractory fluorite ore in Hunan Province. Non Metallic Ores 37(3):3
64. Cao Z, Song Y, Zhong H et al (2012) Experimental study on flotation of Hangkou fluorite ore in Suichang. Compr Utilization Miner Resour 3:4
65. Wei Z, Hu Y, Han H et al (2018) Selective flotation of scheelite from calcite using Al-Na2SiO3 polymer as depressant and Pb-BHA complexes as collector. Miner Eng 120:29–34
66. Sun W, Han H, Sun W et al (2022) Novel insights into the role of colloidal calcium dioleate in the flotation of calcium minerals. Miner Eng 175:107274
67. Sun W, Han H, Sun W et al (2022) New insights into the role of calcium dioleate in selectively separating fluorite from calcite during cleaning process. Colloids Surf A 648:129245
68. Antti BM, Forssberg E (1989) Pulp chemistry in industrial mineral flotation. Studies of surface complex on calcite and apatite surfaces using FTIR spectroscopy. Miner Eng 2(2):217–227
69. Free ML, Miller JD (1996) The significance of collector colloid adsorption phenomena in the fluorite/oleate flotation system as revealed by FTIR/IRS and solution chemistry analysis. Int J Miner Process 48(3)
70. Free ML, Miller JD (1996) The significance of collector colloid adsorption phenomena in the fluorite/oleate flotation system as revealed by FTIR/IRS and solution chemistry analysis. Int J Miner Process 48(3):197–216
71. Fa K, Nguyen AV, Miller JD (2006) Interaction of calcium dioleate collector colloids with calcite and fluorite surfaces as revealed by AFM force measurements and molecular dynamics simulation. Int J Miner Process 81(3):166–177
72. Fa KQ, Tao JA, Nalaskowski J et al (2003) Interaction forces between a calcium dioleate sphere and calcite/fluorite surfaces and their significance in flotation. Langmuir 19(25):10523–10530
73. Fa KQ, Nguyen AV, Miller JD (2005) Hydrophobic attraction as revealed by AFM force measurements and molecular dynamics simulation. J Phys Chem B 109(27):13112–13118
74. Sun W, Han H, Sun W et al (2022) Novel insights into the role of colloidal calcium dioleate in the flotation of calcium minerals. Miner Eng 175 (in Chinese)
75. Zhang Y, Hu Y, Wang Y et al (2014) Effect and mechanism of water glass on flotation behavior of calcium-containing minerals. Chin J Nonferrous Met 24(9):7
76. Sun W, Song S (2013) Water glass and its application and analysis in scheelite flotation. China Tungsten 28(4):4
77. Zhang G, Lu H, Ren D et al (2009) Flotation test method of a fluorite ore in inner Mongolia. Inner Mongolia Sci Technol Econ 21:2
78. Zhou W, Cheng J, Feng Q et al (213) Effect of acidified water glass on beneficiation of a high calcium fluorite ore from Mexico. Non Metallic Ores 36(03):31–32
79. Zhou W, Moreno J, Torres R et al (2013) Flotation of fluorite from ores by using acidized water glass as depressant. Miner Eng 142–145
80. Yao W (2020) Theoretical and experimental study on enhancement of flotation separation of calcium-bearing minerals based on grinding media and metal ion AIDS. Wuhan University of Science and Technology
81. Zhang C, Wei S, Hu Y et al (2018) Selective adsorption of tannic acid on calcite and implications for separation of fluorite minerals. J Colloid Interface Sci 512:55–63

82. Meilin S (2005) Practice of recovering fluorite from flotation tailings. China Mine Eng 34(6):33–34, 55
83. Zhang C, Gao Z, Hu Y et al (2018) The effect of polyacrylic acid on the surface properties of calcite and fluorite aiming at their selective flotation. Physicochemical Probl Miner Process 3(54):868–877
84. Yu F, Gao H, Shi W et al (2011) Beneficiation and recovery of fluorite from a lead-zinc tailing in Hunan Province. Met Mine 08:162–165
85. Zhou T, Shi W (2011) Experimental study on beneficiation of a high calcium fluorite ore in Jinta County. Met Mine 03:102–104
86. Niu Y, Huang M (2010) Beneficiation and production practice of Qinglong carbonate fluorite ore. Conserv Utilization Miner Resour 03:16–19
87. Sun R, Liu D, Liu Y et al (2021) Pb-water glass as a depressant in the flotation separation of fluorite from calcite. Colloids Surf A 629:127447
88. Shi J, Li M, Yang Z et al (2023) Effect of ultrasonic treatment on flotation behavior of calcite in water glass and sodium oleate system and its mechanism. Min Metall Eng 43(01):45–49
89. Gao J, Hu Y, Sun W et al (2019) Enhanced separation of fluorite from calcite in acidic condition. Miner Eng 133:103–105
90. Ji D, Kang J, Sun W et al (2021) Experimental study on high-value utilization of associated fluorite in Huangshaping polymetallic mine. Met Mine 06:80–85
91. Yongxiong GE (1990) Study on the chemistry of water glass Solution and flotation separation mechanism of fluorite and hematite. Min Metall Eng 02:24–27

Chapter 8
Development and Utilization of Associated Feldspar and Quartz Resources in Tungsten Ore

Li Wang, Haisheng Han, and Wei Sun

Abstract Quartz and feldspar are the main gangue minerals in the tungsten ore, which accounts for 60–90%. Feldspar and quartz are also valuable materials in the manufacture of glass, ceramic, and fillers. Their comprehensive recycles from the tungsten ore is conducive to the efficient utilization of mineral reksources and environmental protection. However, similar crystal structure and chemical composition make feldspar and quartz similar in physical and chemical properties, resulting in the separation of quartz and feldspar difficult. In recent years, flotation has proved to be the most effective method for the separation of feldspar and quartz, which mainly consist of hydrofluoric acid flotation, sulfuric acid flotation, neutral flotation, and alkaline flotation. But the clean separation of feldspar and quartz remains a challenge. This chapter introduces the flotation separation methods and the flotation reagent combinations for the separation of feldspar and quartz whereas their mechanisms are also analyzed. The results showed that the alkaline flotation with metal ions as activators and soaps as collectors has good potential whereas hydrofluoric acid flotation and sulfuric acid flotation have great environmental harm and strong corrosion of equipment. Moreover, the practical cases of the alkaline flotation of quartz from feldspar in the tungsten ore are introduced, and the practical results show that the alkaline flotation achieves the efficient separation of quartz from feldspar, indicating the alkaline flotation have great industrial application potential.

Keywords Tungsten ore · Quartz · Feldspar · Alkaline flotation · Industrial application

---

L. Wang (✉) · H. Han · W. Sun
Central South University, Changsha, China

Engineering Research Center of Ministry of Education for Carbon Emission Reduction in Metal Resource Exploitation and Utilization, Central South University, Changsha, China

State Key Laboratory of Mineral Processing, Changsha, China

H. Han
Technology Research Center of Hunan Province for Comprehensive Utilization of Associated Fluorite and Fluorine Chemical Engineering, Chenzhou, China

W. Sun et al., *Flotation Chemistry of Tungsten Minerals and Its Application*,
https://doi.org/10.1007/978-981-96-5077-4_8

Table 8.1 Contents of quartz and feldspar in typical tungsten tailings in China (%)

Sources of tungsten tailings	Quartz	Feldspar	Other	References
Tungsten tin tailings from Jiangxi	39.08	30.18	2.26	[1]
Tungsten tin tailings from Guangxi	37.00	44.00	3.00	[2]
A dolomite tungsten tailing from Guangdong	68.6	2.00	29.4	[3]
Tungsten tin tailings from southern Jiangxi	38.84	24.50	4.00–7.00	[4]
A quartz vein type tin tungsten tailing	63.30	15.78	20.92	[5]
Quartz vein tungsten tailings from Jiangxi	27.35	66.72	6.93	[6]
Mineral from Limu Jinzhuyuan deposit	38.00	50.00	12.00	[7]

8.1 Introduction

Feldspar and quartz are the minerals with the largest proportion and the widest distribution in the surface rocks. They are the main gangue minerals in the Yunying type tungsten ore and quartz vein tungsten ore, as well as the main constituent minerals of the vast majority of granite. The quartz, feldspar and other non-metallic mineral resources accumulated in tailings ponds in Chinese metal mines account for the vast majority of the total tailings resources. At the same time, the demand for feldspar and quartz is growing rapidly in China. Therefore, the comprehensive utilization of metal mine tailings is not only conducive to the efficient utilization of mineral resources and environmental protection, but also important to avoid the excessive mining of non-metallic mines. In terms of geochemical mineralization process, quartz and feldspar are the main gangue minerals of tungsten deposit. At present, the tailing dumping rate of associated feldspar and quartz gangue in tungsten ore is more than 95% in China, and the tailing amount is huge. Table 8.1 shows the content of feldspar and quartz in typical tungsten tailings in China. The total content of feldspar and quartz in Yunying type and quartz vein type tungsten tailings range from 60 to 90%, which has the potential of resource utilization and is conducive to solving the problem of tailings stacking.

8.2 Feldspar and Quartz Resource Properties

8.2.1 Basic Properties of Feldspar

Feldspar is the general name of feldspar group minerals, which is the most important rock-forming mineral constituting the surface rocks, accounting for about 60% of the total mineral content in the crust. It is a shelf-like aluminosilicate mineral containing calcium, sodium, potassium and barium, which is widely used in ceramics, glass and other industries. The following three types of feldspars are frequently encountered in tungsten tailings [8]. The first is soda feldspar: soda feldspar is a series of orthopedic solid solution of sodic minerals, widely exist in all kinds of rocks, generally glass crystals, usually gray, white, mainly used to make glass, ceramic glaze, soap, ceramic tiles, etc. The second is potassium feldspar: usually meat red, yellow, white, gray, etc., widely used in ceramic billet, electric porcelain, glass, potash fertilizer, grinding materials, ceramic glaze and other industries. The third is calcium feldspar, usually produced in the basic igneous rocks, glassy crystals, more brittle, mainly colorless, white or light gray, usually used in glass, ceramics, chemical, welding rod and other industries. The chemical composition and physical properties of feldspar minerals are shown in Table 8.2.

Due to the similar structure, in tungsten tailings, potassium feldspar and soda feldspar can mix with each other to form a eutectic. Common types of potassium feldspar are: glassy feldspar (K, Na) [$AlSi_3O_8$], containing up to 50% soda feldspar molecules; common feldspar (K, Na) [$AlSi_3O_8$], containing soda feldspar molecules can reach 30%; microcline feldspar (K, Na) [$AlSi_3O_8$], which can contain up to 20% soda feldspar molecules. Feldspar has good fusibility, easy grinding, malleability and other properties. Glass made of Potassium feldspar and soda feldspar has a high degree of chemical stability, in addition to hydrofluoric acid and sulfuric acid, not by other acids, alkali corrosion [10]. In addition, because potassium feldspar has the advantages of low melting point, wide melting interval and high viscosity of molten liquid, industrial use is more extensive [9] than other feldspar.

8.2.2 *Basic Properties of Quartz*

Quartz is the general name of silica crystalline mineral in nature. It is one of the most widely distributed minerals on the earth and also one of the main rocks forming minerals [11]. There are many minerals containing silica in nature (the abundance is about 60%), most of which form rocks [12] with silicate minerals, such as gangue (SiO_2 > 99%) in the form of minerals in magmatic rocks; Quartz sandstone (SiO_2 90–95%) in sedimentary rocks; and weathered quartz sand on the surface. The main chemical composition of quartz is SiO_2, with the highest purity of crystal, usually also contains a small amount of Fe_2O_3, MgO, Al_2O_3, TiO_2 and other impurities. The color of quartz in tungsten tailings is often colorless, milky white, gray, and transparent or translucent because of impurities. The Mohs hardness is about 7, the section has glass or fat luster, the specific gravity varies according to the crystal type, the density is 2.22–2.65 g/cm^3, and the melting point is 1750 °C [12]. In addition, quartz has strong acid resistance (except hydrofluoric acid), but it is not alkali resistant, it can react with alkaline substances to produce soluble silicate [13]. In nature, quartz has seven crystalline states and one glassy state under atmospheric pressure. See Table 8.3 [12, 14] for their names and morphology. Among them, low temperature quartz (i.e., α-quartz) is the most widely distributed quartz minerals in nature and the most common of a mineral species, so without special instructions, quartz is specifically referred to as "α-quartz" [15]. The polycrystalline transition between the different forms of quartz is shown in Fig. 8.1. Quartz crystal transformation has

Table 8.2 Chemical composition and physical properties of feldspar minerals [9]

Name		Potassium feldspar	Soda feldspar	Calcium feldspar	Barium feldspar
Chemical general form		$K_2O{\cdot}Al_2O_3{\cdot}6SiO_2$	$Na_2O{\cdot}Al_2O_3{\cdot}6SiO_2$	$CaO{\cdot}Al_2O_3{\cdot}2SiO_2$	$BaO{\cdot}Al_2O_3{\cdot}2SiO_2$
Crystal structure formula		$K[AlSi_3O_8]$	$Na[AlSi_3O_8]$	$Ca[Al_2Si_2O_8]$	$Ba[Al_2Si_2O_8]$
Theoretical chemical composition (%)	SiO_2	64.70	68.70	43.20	32.00
	Al_2O_3	18.40	19.50	36.70	27.12
	$RO(R_2O)$	KO_2: 16.90	Na_2O: 11.80	CaO: 20.10	BaO: 40.88
Crystal system		Monocline	Triple oblique	Triple oblique	Monocline
Density (g/cm^3)		2.54–2.57	2.61–2.64	2.74–2.76	2.72–2.79
Mohs hardness		6–6.5	6–6.5	6–6.5	6–6.5
Coefficient of thermal expansion $\alpha/°C^{-1}$		7.5×10^{-8}	7.4×10^{-8}	0.048×10^{-8}	0.023×10^{-8}
Melting temperature/°C		1150	1100	1550	1715
Melting interval/°C		1130–1530 Wide spacing	1120–1250 Narrow spacing	1250–1550 Narrow spacing	–
Solution viscosity		Big	Small	Small	–
Miscibility		Alkaline feldspar: $KAlSi_3O_8$—$NaAlSi_3O_8$; Plagioclase: $NaAlSi_3O_8$—$CaAl_2Si_2O$			

the following two characteristics: the first characteristic is the slow transformation of high temperature type (reconstruction type transformation), the transformation starts from the surface to the interior gradually, the transformation structure changes after the transformation, so the transformation process is slow, the volume changes, and the need for high temperature, the transformation process is irreversible; The second characteristic is the rapid transformation of low temperature type (displacement type transformation): the transformation speed is fast, after reaching the transformation temperature, the crystal surface instantaneous transformation, no special changes in the structure after transformation, small volume change, and the transformation reversible.

Quartz crystal transformation will cause a series of physical changes, such as volume, relative density, among which volume change is the most influential factor to the ceramic production industry. Generally, the slower the quartz crystal conversion rate, the larger the volume effect value, and the faster the speed, the smaller the volume effect value. For example: β-quartz converts quickly at low temperatures, so the volume effect value is small during the conversion process. Some quartz crystal transformation at low temperature is rapid, although the volume expansion is small, but in the formation of no liquid phase transformation, so it leads to strong destructive, harmful, it is easy to make ceramic products deformation and cracking consequences. See Table 8.4 [15] for the temperature and volume changes of quartz crystal transformation.

8.2.3 Uses and Quality Standards of Feldspar

Feldspar is mainly used in the glass and ceramic industry. The amount of feldspar in the glass industry accounts for 50–60%, the amount of feldspar in the ceramic industry accounts for about 30%, and the rest is used in the potash fertilizer, filler and other industries [16, 17].

(1) Glass solvent. Feldspar is one of the main ingredients in glass mixtures, usually potassium and soda feldspar, and is used mainly to increase the alumina content and alkali content of glass ingredients, to reduce the melting temperature of glass production and to reduce the amount of alkali [18]. In addition, the process of molten feldspar condensation into glass is slower, and the crystallization capacity of feldspar is small, the performance of these feldspar can adjust the viscosity of the glass, to avoid glass in the formation process of feldspar crystal and destroy the product.
(2) Raw materials of ceramic billet body. Potassium feldspar is usually used in the firing process of ceramic embryo body, and the amount of addition generally accounts for about 20% of the raw material. It plays the role of barren raw material before firing, reducing the drying shrinkage deformation of the blank body, improving the drying performance and shortening the drying time [14, 19]. When burning as a flux to reduce the burning temperature, to promote

Table 8.3 Crystal morphology of quartz under atmospheric pressure [12]

Names	α-quartz	β-quartz	α-squamous quartz	γ-squamous quartz	β-squamous quartz	α-quartzite	β-quartzite	Quartz glass
Form	Tripartite crystal system	Hexagonal system	Rhombic system	Hexagonal system	Hexagonal system	Tetragonal system	Equiaxed system	Amorphous state

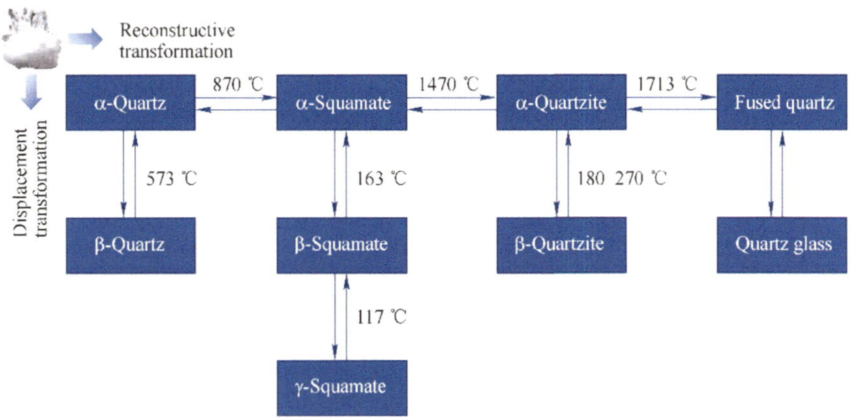

Fig. 8.1 Polycrystalline transition of quartz

Table 8.4 Quartz crystal transition temperature and volume changes [15]

Crystal transformation	Temperature/°C	Volume expansion/%
β-quartz → α-quartz	573	0.82
α-quartz → α-squamous quartz	870	16.0
α-squamous quartz → α-quartzite	1470	4.7
α- quartzite → fused quartz	1713	0.1
α-squamous quartz → β-squamous quartz	163	0.2
β-squamous quartz → γ-squamous quartz	117	0.2
β-quartzite → α-quartzite	150	2.8

the melting of quartz and kaolin, accelerate the formation of mullite, so that the compact body and reduce the gap, improve its mechanical strength and dielectric properties, improve the light transmittance of the body.

(3) Ceramic glaze. The raw materials of ceramic glaze are mainly composed of potassium feldspar, quartz and clay, among which the content of potassium feldspar in raw materials can reach 10–35%. Due to the solvent action of potassium feldspar, the glaze can be fully molten, and the glaze of feldspar is good, and the glaze is smooth and transparent [19, 20].

(4) Manufacture potash fertilizer. The chemical properties of potassium feldspar are extremely stable. In addition to hydrofluoric acid, it is almost not decomposed by acid and alkali at normal pressure and room temperature. And potassium feldspar contained in potassium oxide, can not be directly absorbed as potassium fertilizer for plants. In order to make potassium feldspar a useful resources of potassium both at home and abroad, the potassium oxide is made water soluble or structure-soluble, and potassium feldspar is made from potassium fertilizer [21].

Table 8.5 General quality requirements of Chinese feldspar products (%)

Grade	Fe_2O_3	Primary use
Premium	< 0.3	White glaze, etc.
Grade I	< 0.5	Glazes, white billets, plate glass, etc.
Grade II	< 0.8	Billet, electric porcelain, etc.
Grade III	< 1.0	Enamel glaze, etc.

(5) Others. Feldspar is also utilized as a crucial component in the production of white cement, paper, refractory materials, machinery manufacturing, coatings, and other essential raw materials. The quality standard of feldspar products mainly depends on its chemical composition and use. In addition, according to the different uses of feldspar, the quality standard of the particle size and uniformity of the products will also have requirements [22]. Feldspar mining distribution around the world, different regions of commercial feldspar quality standards are not the same. At present, the quality standards of Chinese feldspar products have not yet developed a unified standard, only mainly for the iron content of feldspar products have more strict requirements. The general quality requirements of Chinese feldspar products are shown in Table 8.5 [23]. The product quality requirements of feldspar in its main fields are shown in Tables 8.6, 8.7 and 8.8.

Table 8.6 Quality requirements for feldspar and feldspar powder in the glass industry (%)

Grade	Quality requirements			
	$K_2O + Na_2O$	SiO_2	Al_2O_3	Fe_2O_3
Premium	12.00 or higher	65.00 or less	18.00 or higher	0.10 or less
Grade I	11.00 or higher	70.00 or less	16.00 or higher	0.20 or less
Grade II	–	–	15.00 or higher	0.35 or less
Qualified	–	–	14.00 or higher	0.50 or less
Remarks	Moisture: dry processing ≤ 1, wet processing ≤ 5 Particle size: < 0.6 mm, superior product < 0.1 mm grain size < 15%; first grade product < 25%			

Table 8.7 Quality requirements for feldspar in the ceramic industry (%)

Grade	Quality requirements				
	$K_2O + Na_2O$	Na_2O	Al_2O_3	Fe_2O_3	$CaO + MgO$
Top grade	12.00 or higher	< 4	> 17.00	0.15 or less	< 2
Grade I	11.00 or higher	< 4	> 17.00	0.20 or less	< 2
Grade II	11.00 or higher	–	> 17.00	0.50 or less	< 2

Table 8.8 Quality requirements for feldspar of potash fertilizer industry (%)

Category	K_2O	Na_2O	$Fe_2O_3 + TiO_2$	Al_2O_3	SiO_2
Indicators	11.00 or higher	< 1–3	1 or less	Around 15	70 or less

8.2.4 Uses and Quality Standards of Quartz

As early as the Stone Age, humans used quartz to make simple production tools such as stone axes and arrows to get food and fight enemies. With the progress of the times, quartz was used more and more widely [24, 25]. For example, after melting into glass, it can be used to make optical instruments, glasses, glass tubes and other products; quartz clocks, electronic equipment in the piezoelectric quartz sheet as the standard frequency; In addition, quartz can also be used as grinding materials, bearings of precision instruments, glass ceramics and other industrial raw materials. In short, quartz is an important industrial mineral raw material, which is widely used in casting, construction, glass, ceramics, metallurgy, refractory materials, chemical industry and so on. It plays an important role in the national economy [26].

8.2.4.1 Glass Industry Raw Material

Quartz raw material is used in the glass manufacturing industry to improve the transparency, chemical stability, mechanical strength and other aspects of the performance of glass and its products [27]. The content of silica in the raw material is also the main factor that determines its application field and product performance. Ordinary silicate glass generally requires 65–75% silica; Modern float glass generally requires a silica content of more than 96%; High-grade glass has more stringent requirements for quartz feedstock, namely that the silica content should be above 99%. The purity requirements for quartz raw materials used in the glass industry in different countries are also different due to the different origin of the raw materials, different technologies and other factors, the specific requirements are given in Table 8.9 [28].

8.2.4.2 Mechanical Casting Industrial Raw Materials

Because of the high melting point of quartz sand, it also plays an important role in the smelting process by serving as a vital constituent for warming, desulfurization, deoxidation, and other essential functions. It is widely used as casting mold, casting mold core, molding sand and so on in the mechanical casting industry. The amount of quartz sand in the mechanical casting industry is also relatively large, accounting for about 33% of its total production in the world, usually every production of a ton of castings will consume a ton of clay raw materials and sand [29]. With the continuous improvement of the degree of precision of mechanical castings, the mechanical casting industry also has higher and more strict product requirements for quartz raw

Table 8.9 Quality requirements of quartz raw materials in glass industry in different countries (%) [28]

Country	Types of glass	Quality requirements				
		SiO_2	Fe_2O_3	Al_2O_3	Cr_2O_3	TiO_2
China	Specialty glass	> 99.5	> 0.05	< 0.5	< 0.001	0.05
	Plate glass	> 98.5	< 0.1	–	–	–
The United States	Optical glass	99.8	0.02	< 0.1	–	–
	Plate glass	> 98.5	< 0.35	< 0.5	–	< 0.2
Britan	Premium optics	99.5	0.008	–	0.0001	0.03
	Plate glass	> 98.5	< 0.03	< 0.5	< 0.001	< 0.05
Japan	First class optics	99.8	0.02	0.1	–	–
	Plate glass	> 97	< 0.06	> 0.1	–	–

Table 8.10 Quality requirements of quartz sand by foundry industry (%)

Grade		Quality requirements	
		SiO_2	Fe_2O_3
Casting build industry	Grade I	> 97.0	< 0.75
	Grade II	> 97.5	< 1.0
	Grade III	> 94.0	< 1.5
	Grade IV	> 90.0	–

materials. The general requirements of quartz sand quality in the casting industry are shown in Table 8.10 [27].

8.2.4.3 Ceramic Industry Raw Materials

As a kind of barren raw material, quartz can enhance the plasticity and bonding of ceramic body in the process of burning, so that the contraction of the body disappeared. In addition, SiO_2 and Al_2O_3 in quartz form the structural skeleton of ceramic at high temperature, which greatly improves the mechanical strength [27, 30] of its products. The quality requirements of quartz sand for general ceramic products are shown in Table 8.11.

Table 8.11 Quality requirements of quartz sand for ceramic products (%)

Types of ceramics	Quality requirements		
	SiO_2	Fe_2O_3	Al_2O_3
Building, sanitary ceramics	> 98.5	< 0.02	< 0.5
Radio ceramics	> 98.5	< 0.05	< 2.0

Table 8.12 Quality requirements of different refractories for quartz raw materials (%)

Grade		Quality requirements		
		SiO_2	Fe_2O_3	Al_2O_3
Fire resistance materials	Premium	> 98	< 0.5	< 0.5
	Grade I	> 97.5	< 1.0	< 1.0
	Grade II	> 96	< 1.5	< 1.5

8.2.4.4 Refractory Industrial Raw Materials

Quartz is one [22] of the main raw materials used in the refractory industry to produce silicon lightweight silicon brick, refractory brick, fused quartz products and other products. In order to ensure the high quality of silica refractory, mechanical properties and load softening properties, it is necessary to strictly control the content of silica in quartz raw materials and the proportion of different grain sizes in the process of producing refractory materials. Generally speaking, the raw material of silica refractory requires the content of silica in quartz to be more than 93%. If the crystal state of quartz in the raw material is poor, the particle size of the product should be limited to 2 mm. In addition, the degree of crystallization of quartz particles in the raw material will also have a very important impact on the refractory performance (especially the refractories) of the refractory. Generally speaking, the better the crystallinity of quartz particles, the better the performance of the prepared refractory. The quality requirements of different grades of refractories for quartz raw materials are shown in Table 8.12.

8.2.4.5 Raw Materials for Other Industrial Products

At present, quartz has a wide range of applications [31, 32] in both traditional industrial fields and high-tech industries. The application of quartz in other industrial fields and the corresponding product quality requirements are shown in Table 8.13.

Table 8.13 Application of quartz in other industrial fields and its corresponding raw material quality requirements

Names of industries	Product	Quality requirements
Semiconductor industry	Packaging materials, etc.	SiO_2 > 99.9%, Fe_2O_3 < 5 ppm, total impurities < 300 ppm
Chemical industry	Pauline, etc.	Liquid: SiO_2 > 98%, Fe_2O_3 < 0.2%, particle size: 0.2–2 mm Solid: SiO_2 > 99%, Fe_2O_3 < 0.03%
Industrial fillers	Plastic, etc.	SiO_2 > 99%, Fe_2O_3 < 0.05%, improve the compressive and tensile strength of plastics

8.3 Flotation Separation Method of Feldspar and Quartz

In the crystal structure of feldspar, a portion of the aluminum atoms replace the silicon atoms to form a tetrahedral feldspar lattice co-angled by AlO_4^{5-} and SiO_4^{4-}, which are connected in an infinite three-dimensional structure, and thus feldspar is defined as a structural silicate or "frame silicate". Feldspar potassium, sodium and other metal ions to compensate for the aluminum replacement of silicon caused by the negative electricity price of the crystal, and the ionic bond formed by these metal ions and oxygen is weak, easy to dissociate in water, so that the feldspar surface appears the phenomenon [33] of negative charge. Quartz is a mineral with a continuous tetrahedral structure composed of oxygen atoms and silicon atoms, each oxygen atom shared by two tetrahedrons, forming an infinite extension of the frame structure [34]. Its structure is highly similar to that of feldspar, and hence their physical properties and chemical composition are also very similar, and it cannot be separated effectively using conventional gravity separation, magnetic separation and other methods. At present, flotation is the most effective way to separate feldspar and quartz. As early as the 1940s, many experts carried out research on flotation separation of feldspar and quartz, initially using hydrofluoric acid method [24]. Although the hydrofluoric acid method is currently the most mature flotation method of feldspar and quartz, and can better separate quartz and feldspar, the use of hydrofluoric acid has a lot of disadvantages, such as: it corrodes the equipment severely and is also more harmful to the human body and the environment [35]. With continuous research, the current flotation separation technology of feldspar and quartz is mainly summarized into three methods: hydrofluoric acid method, fluorine-free acid method and fluorine-free acid method. Due to the toxicity of hydrofluoric acid, fluorine-free acid method is the most widely used separation method at present. However, separation under strong acidic conditions not only has serious corrosion problems for equipment, but also the treatment of acid-containing wastewater is a major problem in industrial production. Therefore, the fluorine-free acid-free method is the focus of current research.

8.3.1 Hydrofluoric Acid Method

The hydrofluoric acid method is the most well-established method for separating feldspar from quartz. This method is to add hydrofluoric acid as the activator of feldspar in feldspar and quartz pulp, using cationic collector (amine), in the pulp pH of 2–3 under the condition of priority to select feldspar, so as to realize the flotation separation of quartz and feldspar [36].

Although hydrofluoric acid is harmful to the human body and the environment, it is of great significance to the flotation separation of feldspar from quartz. After hydrofluoric acid treatment, the zeta potential of feldspar and quartz is very different in the strong acid pulp, the zeta potential of feldspar surface becomes very negative, and the zeta potential of quartz tends to zero. The dissociation equilibrium of feldspar and quartz surface will be broken with the decrease of pH value of the pulp, that is, moving in the direction of decreasing electronegativity [37]. When pH is in the range of 2–3, the zeta potential of quartz tends to zero, and the surface Al^{3+} becomes the active center due to the corrosion of the silico-oxygen bond of feldspar by hydrofluoric acid. At the same time, the $[SiF_6]^{2-}$ complex ions formed by fluorine and silicon in the solution will form complexes with Al^{3+}, K^+ and Na^+ on the surface of feldspar, which will be attached to the surface of feldspar, making the surface potential more negative. When a cationic collector is added to the system, it will be adsorbed on the surface of feldspar through electrostatic action, so that the feldspar surfacing first.

8.3.2 *Fluorine-Free Acid Method*

Due to the toxicity of hydrofluoric acid, since the 1970s of the last century, the flotation separation method of feldspar-quartz has tended to dispense with the use of fluorine [38, 39]. Fluorine-free acid method refers to the method of flotation separation of feldspar by flotation with single amine collector or mixed collector without adding hydrofluoric acid under the condition of using strong acid to adjust the pH value of flotation system. The separation principle is mainly based on the different zeta potential of feldspar and quartz, with strong acid to adjust pH to near the zero electric point of quartz, that is, at this time the quartz surface is not charged, feldspar surface under negative charge, the use of cationic collector or mixed collector so as to make it float and separate. In many studies on flotation processes, it is found that the flotation effect of different types of mixed collectors is better than that of a single component. In recent years, mixed cationic, anionic and non-ionic collectors are used for selective flotation separation of target minerals [40].

Vidyadhar et al. found in their study on the separation of feldspar from quartz by using cationic-anionic collector: the separation of feldspar from quartz can be realized only when the pH of pulp is about 2, that is, the difference in surface charges of quartz and feldspar is the basis [41] of mixed collector flotation. The research shows that when pulp pH is 2, the amine cationic collector will not be adsorbed on quartz with a surface potential of about 0, but only partially adsorbed on feldspar with a surface potential of negative through electrostatic interaction. When the ratio of sulfonates to amine collectors is 1:1, they are adsorbed on the surface of feldspar in the form of a complex, and this complex has no effect on quartz, that is, the presence of sulfonates increases the adsorption of amine collectors, in addition to its co-adsorption, but also reduces the electrostatic repulsion of adjacent surface amines, thus increasing the tail hydrophobicity. Gao et al., using fluorine-free acid flotation process for feldspar ore purification research, in pulp pH = 2–3 conditions, the ratio of octadecylamine and sodium dodecylbenzene sulfonate 1:4, after closed-circuit test to get the feldspar products with K_2O grade of 10.24% and K_2O recovery of 51.79% [42]. Compared with the test index of fluorine-free acid method, it was found that the K_2O grade of the concentrate with fluorine-free acid method was slightly lower, but the recovery rate was not different. Although there is no fluorine-containing substance added to this kind of method, the environmental pollution caused by the discharge of acid waste liquid and the corrosion of equipment during process operation have been the important reasons limiting its popularization.

8.3.3 Fluorine-Free Acid-Free Method

Fluorine-free acid-free method means that fluorine-containing substances are not added, and single or mixed collector is added under neutral or alkaline pulp system to separate feldspar and quartz by flotation [43].

The mechanism by which feldspar is separated from quartz under neutral conditions is generally believed to be as follows. Feldspar and quartz surfaces are negatively charged under neutral conditions. After the addition of the cationic collector, although the cationic collector is adsorbed on the surface of quartz and feldspar, the anionic collector also forms a characteristic adsorption due to the presence of Al^{3+} on the surface of feldspar [44]. When the two kind of collectors are added in a specific ratio, they will form a thin layer of hydrophobic layer on the surface of feldspar. Quartz surface adsorption is by electrostatic adsorption of a single type of collector, the adsorption of this collector will gradually decrease with the progress of mixing pulp adsorption, so that feldspar and quartz are separated. Qiu Yangru et al. conducted a study on fluoroacid-free flotation separation of feldspar and quartz for a feldspar mine in Henan province. Through the flotation condition test, the optimal combination and ratio of cationic and anionic collector, the amount of collector and the amount of inhibitor were determined, and the high-quality feldspar concentrate with K_2O grade of 14.68% and Al_2O_3 grade of 15.88% was obtained [43].

Under the condition of pulp pH 11–12, with anionic surfactant (alkyl sodium sulfonate, etc.) as collector and alkali earth metal ion (Ca^{2+}) as activator, quartz can be first floated in the system, and non-ionic surfactant can increase the flotation difference between the two minerals [45]. The basic principle is: sodium alkyl sulfonate is an anionic collector, can not be directly adsorbed on the surface of negative charge feldspar and quartz surface, it forms a complex with the external alkali earth metal ions, common adsorption on the quartz surface, and make feldspar and quartz separation. This method can make the production process inevitably ions have a positive effect on the pharmaceutical system, and has a good application prospect.

8.4 Flotation Reagents for Feldspar and Quartz and Their Flotation Mechanism

8.4.1 Hydrofluoric Acid and Cationic Collector

In the traditional flotation process, it is necessary to use hydrofluoric acid as the activator of feldspar and cationic surfactant as the collector in acidic pH value range to separate feldspar from quartz by flotation. Many researchers have studied the role of hydrofluoric acid in feldspar flotation, and put forward several mechanisms:

(1) Hydrofluoric acid acts on aluminum points on the surface to form aluminum—fluorine complex on the surface, and the amine cationic collector is adsorbed on these particles.
(2) Hydrofluoric acid reacts with silicon to form fluorosilicate ions (hexafluorosilicate ions), which are absorbed to the surface aluminum to produce negative particles, and the amine cation is then adsorbate on.
(3) Hydrofluoric acid in feldspar surface action, low solubility of sodium or potassium fluosilicate deposition to the surface, making the amine cation more easily adsorbed up.
(4) Quartz is inhibited by HF, which is believed to be due to the silicon surface by the base and fluoride ion exchange, resulting in the negative particle number of silicon surface to reduce and prevent the adsorption of amine.

The use of hydrofluoric acid, in addition to its inherent high price and chemical corrosion to the flotation tank, also has its pollution of the environment and harmful to human health.

8.4.2 Mixed Cationic-Anionic Collector

Cationic collector (such as amines) and anionic collectors (such as petroleum sulfonates) in a certain molar concentration ratio as a mixed collector, using strong acids (generally sulfuric acid) to adjust the pulp pH to 2–3, or under neutral conditions by adding sodium haccetaphosphate to inhibit quartz flotation of feldspar. Commonly used anion collector is: ether amine, dodecylamine, octadecylamine, etc. Commonly used anionic collectors are: sodium dodecyl sulfate, sodium tetradecyl dihydroxyvinylsulfate, sodium cetyldihydroxyvinylsulfate, sodium octadecyl dihydroxyvinylsulfate, etc.

Many researchers have studied the selective mechanism of flotation in feldspar and quartz systems through solution chemistry, electrokinetic potential and flotation tests when using mixed cationic-anionic collector.

(1) With a mixed cationic-anionic collector, a complex is formed through charge neutralization, and precipitation occurs at the same molar ratio. According to the

conductivity and surface tension, the maximum interaction between monoamine and SDS is found when the molar ratio is 1:1 and the maximum molar ratio between diamine and SDS is 1:2. The surface activity of these complexes was demonstrated by the drop in surface tension effect to be higher than that of the single reagent forming the complex.

(2) At negative charged pH values of feldspar and quartz, the adsorption properties of the complexes on both minerals are the same. The cationic collector is adsorbed on the negatively charged surface, so the anionic collector is co-adsorbed due to the complexation between the negative and the cation.

(3) The molar ratio of the mixture plays an important role in the orientation of the molecules on the mineral surface. When the anionic/cationic molar ratio is less than 1, the adsorbed complex increases the hydrophobicity of the mineral particles, thus improving its floatability. At this molar ratio, the charge of the ore particle is dominated by amine cations and is positively charged. When the anionic/cationic molar ratio is equal or higher than 1, there is an excess of anionic collector, and the adsorption of the complex makes the ore particles become hydrophilic, thus inhibiting flotation. At this time, the surface of the ore particle is highly negative, and the charge is controlled by the excess anion adsorbed.

(4) The success of flotation selectivity mainly depends on the flotation pH value, that is, at this pH value, feldspar and quartz surface charges are different, so that amines can be adsorbed on feldspar but not on quartz. The resulting partial hydrophobicity of feldspar is enhanced by co-adsorption of complexes formed by anionic collectors and cationic amines.

8.4.3 Alkali Metal Ions and Anionic Collectors

When metal ions are used as activators, fatty acid collectors are used to separate quartz and feldspar from tailings under alkaline conditions, and the types of activated ions, ion concentration, anion concentration and pulp pH are all important factors affecting the flotation efficiency of quartz.

Figure 8.2 shows the flotation of quartz and microclinic feldspar as a function of pH value [45] when 1×10^{-3} mol/L sodium dodecyl sulfonate (SDSO) is used as collector in pulp with calcium (Ca^{2+}), barium (Ba^{2+}), strontium (Sr^{2+}) and magnesium (Mg^{2+}) as activators. In an alkaline environment, no alkaline earth cations have a significant activation effect on the flotation of microcline. But at high pH, with metal ions acting as activators, quartz can float up significantly.

In alkaline systems, ($MOH^+RSO_3^-$) is an important component of the activation system. It is suggested that neutral complex $M(OH)RSO_3$ is formed by ionic hydrolysis in alkaline solution and acts as a regulator of semi-micellar formation on mineral surface. The hydrolysis of metal ions in aqueous solution is very important for mineral flotation, they associate with free sulfonate ions and adsorb on quartz surface. The adsorption of these molecules is achieved by hydrolysis and is formed

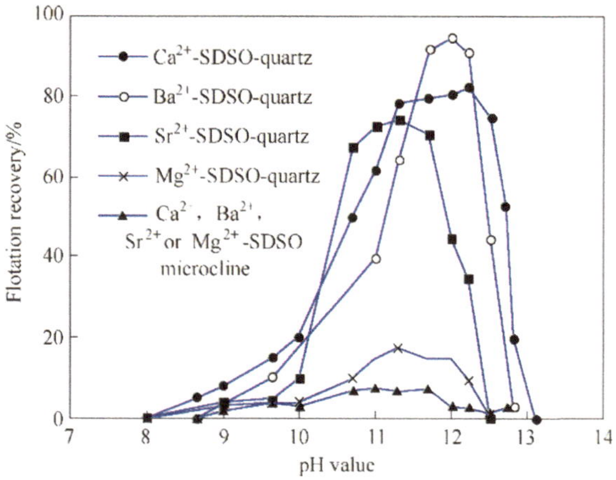

Fig. 8.2 Flotation recovery of quartz and microplagioclase by different ions as a function of pH, reprinted from Ref. [45], copyright 1993, with permission from Minerals Engineering

by combining hydrogen ions in the complexes with adsorbent hydroxide ions, and the effectiveness of activation in these systems is related to the ability of these systems to produce neutral complexes that can co-adsorb with the sulfonyl anions on the quartz surface and act as semi-micellar accelerators. The formation of these species on the quartz surface will vary due to differences in potential, permittivity of water, solvation energy, and interfacial viscosity. In calcium ion solutions, solution chemistry results indicate that the flotation of quartz will be affected when calcium sulfite precipitates are not formed. When the ionic activity on the surface of the mineral is higher than the ionic activity in solution, precipitation of the neutral complex $Ca(OH)RSO_3$ occurs on the quartz surface, and surface precipitation is very important for actual flotation. Conversely, magnesium dodecyl sulfonate has a higher solubility and magnesium hydroxide has a lower solubility; In a 10^{-3} mol/L magnesium solution, when the pH is about 10.1, Mg(OH)s begins to form, which is reduced with the $MgOH^+$ available for sulfonate formation $[Mg(OH)RSO_{33}]$ (aq) or [Mg(OH)RSO] (s). The absence of a neutral complex, known as $Mg(OH)RSO_3$, results in the absence of semi-micelles or surface precipitates. The reduced critical micellar concentration (CMC) of the surfactant in the presence of inorganic salts reflects the degree to which these ions (as counter-ions) bind to the micelle. The degree of binding counter ions decreases with increasing hydration radius. The ability of alkali earth metals to reduce CMC follows the reverse order of hydration ion size ($Ba^{2+} > Sr^{2+} > Ca > Mg^{2+}$).

Neutral non-ionic surfactants as auxiliary collectors in the flotation separation of quartz from feldspar under alkaline conditions, can effectively increase the quartz floatability, such as dodecyl alcohol, terpineol and so on. Figure 8.3 shows the flotation recovery of quartz and plagioclase after adding different molar ratios of dodecyl alcohol in Mg-SDS system. These neutral organic molecules can promote the formation of semi-micelles on the quartz surface, because they are more likely to be adsorbed on the quartz surface with sulfonate ions. There is a synergistic effect between 1-dodecyl alcohol and sodium dodecyl sulfonate. In the case of long chain alkyl sulfonates, these neutral molecules can act as dispersants, to prevent the colloid

Fig. 8.3 Effects of different molar ratios of 8–3 dodecyl alcohol with alkyl sulfonates of different carbon chain lengths (C10–C16) on the recovery rates of quartz and microclinite, reprinted from Ref. [45] with permission from Elsevier

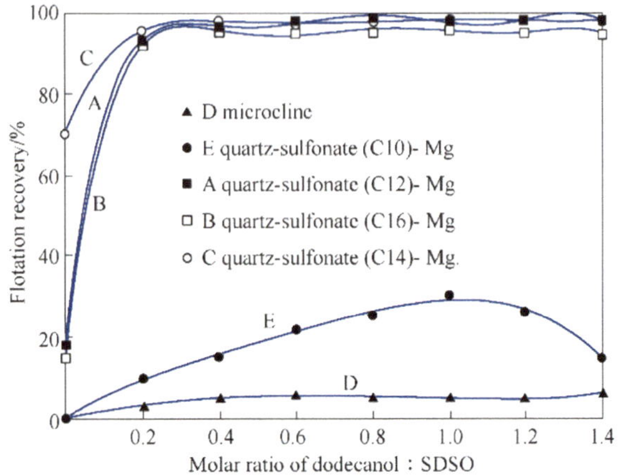

precipitation of cationic sulfonate complexes. Dodecyl alcohol has a dispersive effect on the magnesium sulfonic acid salt, thus suspending the magnesium sulfonic acid precipitated particles. The possible mechanism is the formation of mixed micelles under low total concentrations of surfactants, thus stabilizing the suspension. Other researchers have confirmed this phenomenon in cetyl sulfate-nonionic surfactant mixtures used for the flotation of scheelite and calcite by microthermal methods.

Sodium oleate is also commonly used as an anionic collector in the separation of feldspar and quartz under alkaline conditions. Under the condition of sodium oleate dosage of 1×10^{-4} mol/L, without adding metal ions, the zero electric point of feldspar and quartz is low, in the alkaline pulp, the surface of the two is negatively charged. Sodium oleate is anion collector, adding a single sodium oleate is difficult to overcome the electrostatic effect of adsorption on the surface of mineral particles. Therefore, the recovery rate of feldspar and quartz is almost zero in the measured pH range. When the pH value of pulp is less than 10, the addition of calcium ion has little effect on the flotation behavior of feldspar and quartz by sodium oleate, and the recovery of both is very small. When the pH value of pulp continues to increase to 10.50–12.00, the recovery rate of quartz increases rapidly (the recovery rate is about 95%), and then becomes stable. The recovery rate of feldspar is very small when pH value of pulp is less than 11.0. When pH value of pulp is 11.00–12.00, the recovery rate of feldspar increases rapidly (the recovery rate is about 70%), and then tends to be stable. When the pH value of pulp is 11.00, the recovery rate of feldspar is very low, at this time, the recovery rate of feldspar and quartz difference is about 40%. Therefore, to sum up, the optimal pH value for flotation separation of feldspar and quartz is 11.00 under the fluoro-acid-free, Ca^{2+} and sodium oleate pharmaceutical systems. Calcium ion dosage gradually increased to 2.5×10^{-3} mol/L, quartz recovery rapidly increased to about 80%, continue to increase the amount of calcium ion, quartz recovery little change. The recovery of feldspar has been growing slowly with the increase of calcium ion dosage, when calcium ion dosage of 2.5×10^{-3} mol/L, feldspar recovery is about 5%, at this time, feldspar and quartz flotation separation effect is the best. In summary, under the Ca^{2+} and sodium oleate reagent system, the best dosage of calcium ion is 2.5×10^{-3} mol/L. With the increase of sodium oleate dosage to 1×10^{-4} mol/L when adding calcium ion, the recovery rate of quartz rapidly increases to about 80%. Increasing the dosage of sodium oleate, quartz flotation process began to produce vacuole, and the larger the dosage of reagent, the larger the vacuole amount, and the recovery of quartz with sodium oleate continued to increase and began to decline rapidly. When the dosage of sodium oleate is less than 2.5×10^{-4} mol/L, feldspar basically does not float. Increasing the amount of sodium oleate, the flotation process of feldspar also began to produce vacuole, the same as the quartz flotation phenomenon, the greater the amount of sodium oleate, the larger the amount of vacuole, feldspar began to float. When the dosage of sodium oleate is 1×10^{-4} mol/L, the recovery rate of feldspar is very low, and the recovery rate of quartz is about 80%, and the difference between the two recovery is the largest. In summary, under the fluoro-acid-free, Ca^{2+} and sodium oleate pharmaceutical system, the best dosage of sodium oleate flotation separation of feldspar and quartz is 1×10^{-4} mol/L (Fig. 8.4).

Feldspar and quartz are prepared into binary artificial mixed ore by mass ratio of 1:1. The pH test of flotation separation of feldspar and quartz in mixed ore is carried out. The SiO_2 content of quartz in mixed ore is more than 99%, and the content of

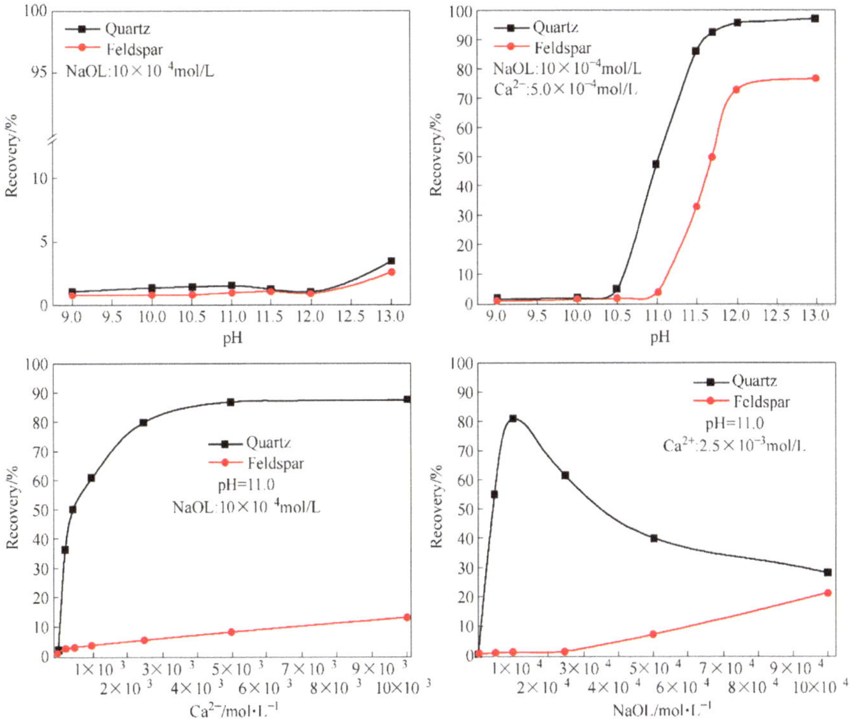

Fig. 8.4 Flotation behavior of quartz and feldspar under alkaline conditions

other elements is very small, which can be ignored. Therefore, the unique elements K and Na of the feldspar in the mixed ore may be used as a basis for judging the effect of flotation on the separation of the feldspar from the quartz. The recovery rate of feldspar in its products is expressed by $K_2O + Na_2O$ in foam products or bottom products. In order to carry out a further study on the flotation separation of feldspar and quartz, calcium ion is selected according to the single mineral test. Under the conditions of sodium oleate in the best dosage, the influence of different pulp pH value on the flotation separation of feldspar and quartz in mixed ore is explored. The flotation separation results of feldspar and quartz in mixed ore at different pH values are shown in Table 8.14.

Figure 8.5 shows the pH variation trend of $K_2O + Na_2O$ grade in mixed flotation concentrate and tailings and Selectivity Index SI (Selectivity Index) to intuitively characterize the selectivity of calcium ion and sodium oleate for mineral flotation separation in mixed ores. The calculation method can be found in Eq. (8.1). The greater the SI index, the better the selectivity index. That is, the greater the flotation difference between quartz and feldspar; The closer it is to 1, the smaller the difference in flotation behavior of minerals.

$$SI = \sqrt{\frac{\varepsilon_{Q-K}}{\varepsilon_{F-K}} \cdot \frac{\varepsilon_{F-X}}{\varepsilon_{Q-X}}} \tag{8.1}$$

where, respectively represent the recovery rate of quartz in mixed flotation foam products and tank bottom products; $\varepsilon_{Q-K}\varepsilon_{Q-X}\varepsilon_{F-K}$, respectively represents the recovery rate of feldspar in mixed flotation foam products and tank bottom products.

The yield of mixed flotation foam products increases with the increase of pH value, which is the same as the trend of quartz single mineral flotation test. The recovery rate of quartz in mixed flotation foam products increases rapidly in the range of pH value 10.00–11.00, and then its recovery rate decreases slightly but basically remains unchanged. Different from feldspar single mineral flotation test, the recovery rate of feldspar in mixed flotation foam products starts to rise rapidly when pH value is greater than 11.00, and the selectivity of flotation separation becomes worse and worse. When the pH value of pulp increases to 12.00, the recovery rate of feldspar and quartz in flotation foam products is close to that of flotation. At this time, there is almost no selectivity in flotation. When the pH value of pulp is 10.00, the yield of mixed flotation foam products is very small, and the content of $K_2O + Na_2O$ in foam products and tank bottom products is not much different, that is, when the pH value is 10.00, the pharmaceutical system of sodium oleate and magnesium ion has no effect on the flotation behavior of two kinds of minerals. When the pH value of pulp is greater than 10.00, the grade of $K_2O + Na_2O$ in mixed flotation foam products increases with the increase of pH, and the grade reaches 14.15% when pH = 11.00. As can be seen from Fig. 8.5, at pH values of 10.50, 10.75 and 11.00, the selectivity index of calcium ion and sodium oleate to the flotation separation of feldspar and quartz in mixed ore is large, but at pH values of 10.50 and 10.75, the yield of mixed flotation foam products is too low. In summary, under the flotation conditions of pulp pH value of 11.00, sodium oleate dosage of 1×10^{-4} mol/L and

Table 8.14 Flotation separation results of feldspar and quartz in mixed ore at different pH values (%)

Product name		Yield	KO_2 grade	Na_2O grade	$K_2O + Na_2O$/feldspar recovery rate	Quartz recovery
pH = 10.00	Foam products	3.27	3.93	2.07	2.37	4.17
	Bottom product	96.73	5.18	3.18	97.63	95.83
	Raw ore	100.00	5.14	3.14	100.00	100.00
pH = 10.50	Foam products	26.94	0.61	0.37	3.18	50.70
	Bottom product	73.06	6.82	4.16	96.82	49.30
	Raw ore	100.00	5.15	3.14	100.00	100.00
pH = 10.75	Foam products	36.59	0.66	0.36	4.50	68.68
	Bottom product	63.41	7.76	4.73	95.50	31.32
	Raw ore	100	5.16	3.13	100.00	100.00
pH = 11.00	Foam products	48.38	1.33	0.71	11.91	84.85
	Bottom product	51.62	8.73	5.42	88.09	15.15
	Raw ore	100.00	5.15	3.14	100.00	100.00
pH = 11.25	Foam products	57.82	3.18	1.63	33.51	82.13
	Bottom product	42.18	7.85	5.23	66.49	17.87
	Raw ore	100.00	5.15	3.15	100.00	100.00
pH = 11.50	Foam products	66.20	4.12	2.12	49.83	82.57
	Bottom product	33.80	7.14	5.17	50.17	17.43
	Raw ore	100.00	5.14	3.15	100.00	100.00
pH = 12.00	Foam products	74.36	5.32	2.71	72.03	76.69
	Bottom product	25.64	4.66	4.39	27.97	23.31
	Raw ore	100.00	5.15	3.14	100.00	100.00

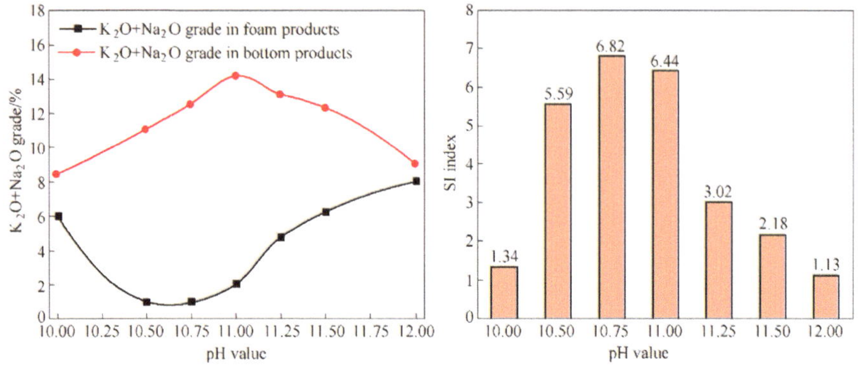

Fig. 8.5 Flotation separation behavior of feldspar and quartz mixed ore

calcium chloride dosage of 1×10^{-4} mol/L, feldspar and quartz in mixed ore can achieve a good flotation separation, and the yield can be 48.38%. Quartz products with about 95.5% SiO_2 grade and feldspar products with 51.62% yield and 14.15% $K_2O + Na_2O$ grade are obtained.

The slight sagging of feldspar recovery observed in all these systems is attributed to the nature of the hydration layer on the surface of feldspar in highly alkaline media. More than a decade ago, many studies have begun to investigate the dissociation mechanism of feldspar. According to the hydration layer hypothesis, in acidic solutions, the surface of feldspar is reduced in aluminum and alkali ions (such as K), but rich in silicon; While in highly alkaline conditions, the silica content on the surface is reduced and the alkali ion content is increased. Compared with quartz, the adsorption density of calcium ions on the surface of feldspar is lower. In highly alkaline media, mobile cations such as aluminum and potassium may exchange with Na or NaOH, and NaOH competes with $CaOH^+$. Although quartz is a three-dimensional skeleton, silicates do not form hydration layers or any exchangeable components.

8.5 Flotation Techniques and Practices of Feldspar and Quartz for Tungsten Tailings

8.5.1 New Flotation Technology and Practice of Feldspar and Quartz from Xingluokeng Tungsten Tailings

Xingluokeng tungsten deposit is a tungsten deposit in China. Orebodies occur in the early Yanshanian biotite granite rock strain, and a few orebodies are located in the metamorphic outside contact zone of granite rock strain, feldspar-bearing quartz sandstone and siltstone. The main body of the deposit is the fine vein type ore, a small amount is the quartz large vein type ore, and the superficial part is the weathering zone ore type. The ore structure is predominantly eukaryotic and semi-eukaryotic, and the main structures are massive, vein-like strip, geode structure, etc. Our group carried out a test on fluorine-free and acid-free separation of feldspar and quartz from the tailings of gravity separation. High quality feldspar products and quartz products were recovered from the tailings, which greatly reduced the discharge of tailings, improved the utilization rate of resources, and created a new profit growth point for enterprises.

8.5.1.1 Research on Process Mineralogy of Tungsten Tailings

Table 8.15 shows the multi-element chemical analysis results of gravity separation tailings of Xingluokeng tungsten deposit. From the chemical analysis results, it can be seen that the tailings of gravity separation have high recovery value of non-metallic minerals, more K and Na elements, and almost no metal elements that can be recycled. The composition of the ore is mainly composed of non-metallic minerals, and the analysis results are shown in Table 8.16. The main minerals in the sample are quartz and potassium feldspar, followed by soda feldspar, muscovite and plagioclase, with a small amount of biotite, dolomite and montmorillonite. It can be also found some trace amounts of minerals, such as calcite, iron dolomite, siderite, kaolinite, chlorite, fluorite, rutile, monazite, hematite, apatite, tourmaline, pyrite and arsenopyrite, etc. Three minerals, quartz, potassium feldspar and soda feldspar, account for 80.99% of the total minerals.

The particle size composition of minerals is an important basis for the development of grinding technology. Therefore, systematic screening analysis was carried out on the tailings of gravity separation, and the distribution of K_2O, Na_2O, Al_2O_3 and Fe_2O_3 in different particle sizes of gravity separation tailings were investigated respectively. The screening results are shown in Table 8.17. The mineral yield of – 0.038 mm particle size is 6.28%, Fe_2O_3 grade is up to 4.60%, Fe_2O_3 distribution rate is 28.11%, that is, the element iron is mainly distributed in the argillated aluminosilicate. Comparing the grade and distribution rate of K_2O and Na_2O, it can be seen that the variation trend of potassium and sodium is the same, and the distribution of potassium and sodium in each grain size is more uniform, but does not enrich in a certain grain size.

Table 8.15 Results of multi-element analysis of gravity separation tailings (%)

Elements	K_2O	Na_2O	Al_2O_3	SiO_2	Fe_2O_3	CaO	MgO
Content	5.22	1.45	11.24	75.41	1.54	1.73	0.54
Elements	P_2O_5	MnO	SO_3	TiO_2	Rb_2O	WO_3	PbO
Content	0.06	0.06	0.12	0.25	0.03	0.03	0.01

Table 8.16 Statistical table of mineral composition and content in gravity separation tailings (%)

Minerals	Mineral	Content
Quartz	Quartz	52.69
Soda feldspar	Soda feldspar	7.16
Potassium feldspar	Potassium feldspar	21.14
Microcline feldspar	Microcline feldspar	3.54
Muscovite	Muscovite	8.19
Biotite	Biotite	1.81
Calcite	Calcite	0.26
Dolomite	Dolomite	1.15
Ankerite	Ankerite	0.06
Siderite	Siderite	0.19
Montmorillonite	Montmorillonite	1.78
Kaolinite	Kaolinite	0.58
Chlorite	Chlorite	0.59
Fluorite	Fluorite	0.28
Apatite	Apatite	0.09
Hematite	Hematite	0.04
Rutile	Rutile	0.12
Monazite-(Ce)	Monazite-(Ce)	0.10
Wolframite	Wolframite	0.03
Schorl	Schorl	0.04
Pyrite	Pyrite	0.03
Arsenopyrite	Arsenopyrite	0.01
Others	Others	0.12
Total	Total	100.00

8.5.1.2 Tailings Pretreatment

Based on the process mineralogy analysis of the Xingluokeng tungsten tailings sample, a pre-processing process of milling-magnetic separation-desliming has been developed to ensure the degree of dissociation of quartz and feldspar, product quality and flotation efficiency, as shown in Fig. 8.6. After pre-processing the tungsten tailings prior to flotation, the product indicators are shown in Table 8.18.

8.5.1.3 Flotation Process Design

Under alkaline conditions, highly efficient anion association collector is used to selectively collect mica from tailings to achieve efficient enrichment of mica, and clean and efficient flotation separation of quartz and feldspar without fluorine and acid is

Table 8.17 Distribution of different elements in gravity separation tailings of different particle grades

Grain size/mm	Yield/%	K_2O/%		Na_2O/%		Al_2O_3/%		Fe_2O_3/%	
		Grade	Distribution rate	Grade	Distribution rate	Grade	Distribution rate	Grade	Distribution rate
> 0.250	39.63	5.05	40.45	1.33	40.27	10.53	41.41	1.45	37.80
0.250–0.150	29.64	4.81	28.81	1.29	29.21	9.99	25.64	1.18	19.05
0.150–0.074	18.4	4.73	17.59	1.26	17.71	10.33	16.28	1.41	14.45
0.074–0.038	6.05	5.13	6.27	1.34	6.19	11.18	6.45	1.91	7.76
0.038	6.28	5.42	6.88	1.38	6.62	19.10	9.71	4.60	28.11
Total	100.00	4.95	100.00	1.31	100.00	10.91	100.00	1.59	100.00

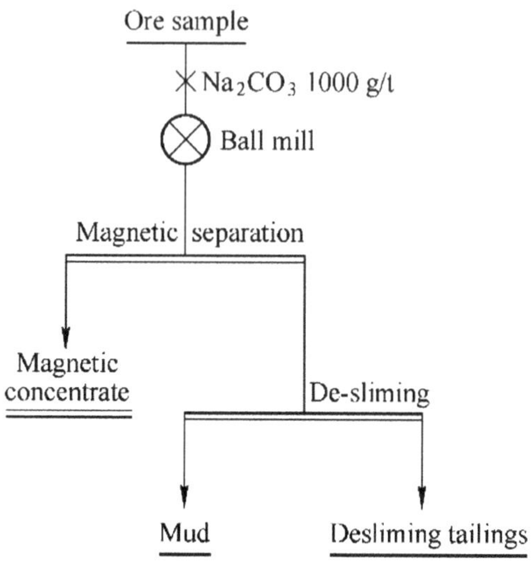

Fig. 8.6 Flow chart of tungsten tailings pretreatment in Xingluokeng

Table 8.18 Indicators of each product after ore pretreatment (%)

Product	Yield	$K_2O + Na_2O$		Fe_2O_3	
		Grade	Distribution rate	Grade	Distribution rate
Magnetic concentrate	9.81	5.39	8.21	12.93	69.78
Mud	19.27	6.33	18.93	2.31	25.66
Desliming tailings	70.92	6.62	72.86	0.45	19.00
Raw ore	100.00	6.44	100.00	1.78	100.00

carried out. After open-circuit experiments, such as dosage of reagent, concentration of pulp, ratio of reagent and utilization of backwater, the optimal process flow is designed (as shown in Fig. 8.7). Finally, the yield of feldspar product is 25.43%, the grade of KO_2 and Na_2O is 10.51% and 3.11%, the recovery rate is 48.47% and 48.50%, the grade of Al_2O_3 is 17.02%, the grade of Fe_2O_3 is 0.14%, the grade of SiO_2 is 68.05%, and the whiteness is 71.50%. The yield of quartz products is 27.74%, the grade of SiO_2 is 98.51%, the grade of Fe_2O_3 is 0.09%, the grade of K_2O and Na_2O is 0.37%, 0.17%, the grade of Al_2O_3 is 0.61%, and the whiteness is 93%. The yield of by-product feldspar is 8.29%, the grade of K_2O and Na_2O is 7.02% and 2.17%, and the recovery rate is 10.55% and 11.02%, respectively. Table 8.19 shows the product indicators of closed-circuit flotation separation of feldspar and quartz from tungsten tailings.

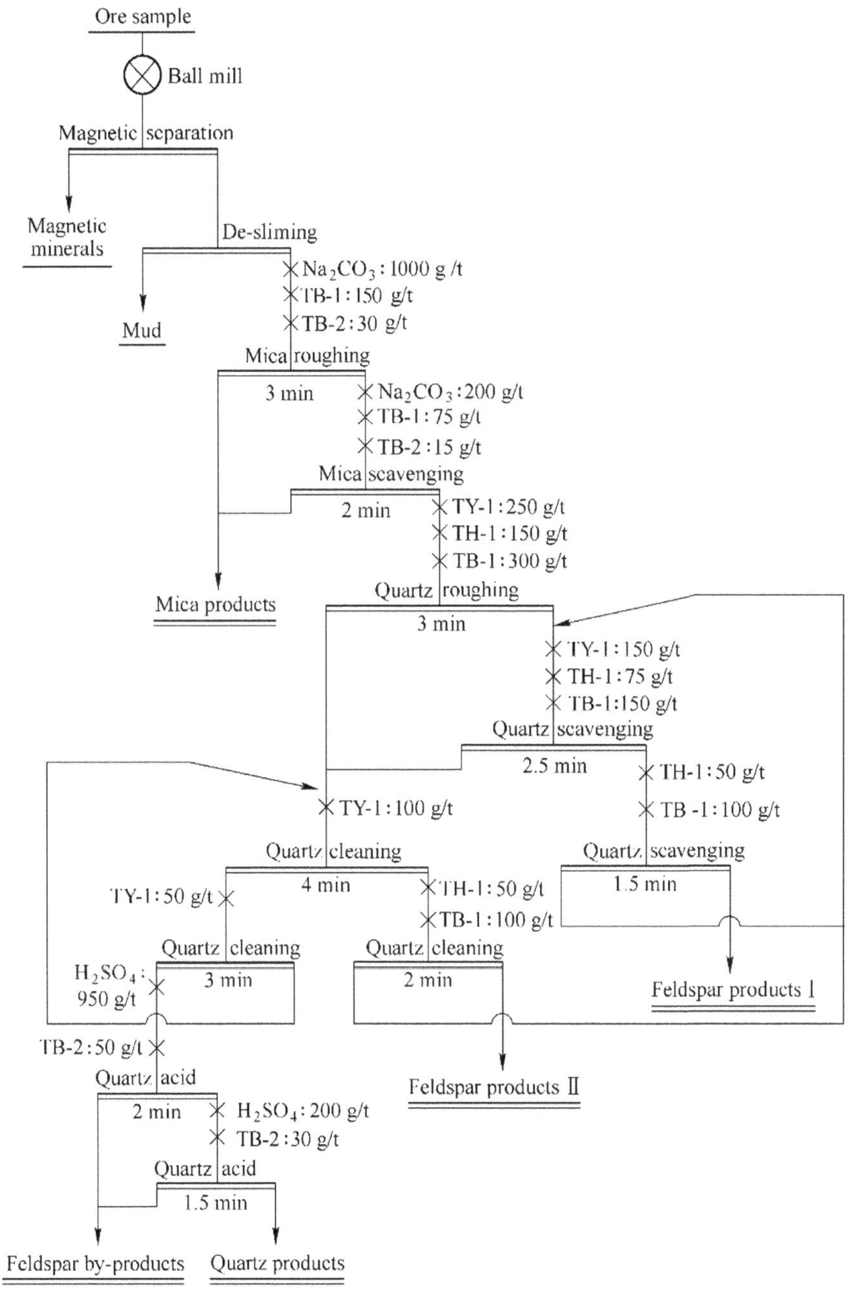

Fig. 8.7 Xingluokeng tungsten tailings comprehensive recycling flow chart of feldspar and quartz

Table 8.19 Feldspar and quartz flotation separation of closed-circuit test product index (%)

Product	Yield	K_2O		Na_2O		Al_2O_3	Fe_2O_3	SiO_2
		Grade	Recovery rate	Grade	Recovery rate			
Magnetic minerals	8.97	4.27	6.95	1.15	6.32	9.13	14.23	62.67
Mud	19.38	5.32	18.70	1.46	17.34	14.68	2.35	72.45
Mica products	10.19	7.29	13.47	2.23	13.93	16.40	4.93	59.62
Feldspar products I	18.58	10.57	35.61	3.12	35.53	17.03	0.15	68.09
Feldspar products II	6.85	10.35	12.86	3.09	12.97	16.99	0.12	67.93
Feldspar by-products	8.29	7.02	10.55	2.17	11.02	11.42	0.43	77.96
Quartz	27.74	0.37	1.86	0.17	2.89	0.61	0.09	98.51
Raw ore	100.00	5.51	100.00	1.63	100.00	10.78	2.33	76.83

8.5.2 *Recovery Process of Quartz Feldspar from a Tungsten-Tin Tailings in Guangxi*

One of the tungsten-tin tailings from Guangxi, province, China was too low in valuable metallic elements to be of any value for recovery, but it contained 35% quartz, 26% potassium feldspar, 18% soda feldspar, and 15% muscovite, and was rich in nonmetallic ores. The pretreatment process of magnetic separation and mechanical slime removal was adopted, sulfuric acid was used as the pH regulator of pulp, dodecylamine was used as collector, mica was separated by flotation process of one roughing, three scavenging and two cleaning. The flotation of mica tailings and then sulfuric acid as pulp pH value regulator, octadecylamine (DDA) and sodium dodecyl sulfonate (SDS) as a mixed anionic-cationic collector, using one roughing, two cleaning, two scavenging process flotation separation of feldspar and quartz, to achieve fluorine-free separation (Fig. 8.8). The closed-circuit test of the whole process obtained a feldspar concentrate with a yield of 44.32%, $K_2O + Na_2O$ grade of 12.19%, K_2O recovery of 70.15% and Na_2O recovery of 73.24%, as well as a quartz concentrate with a yield of 36.04% and SiO_2 grade of 98.14% (Table 8.20). Mica, feldspar and quartz have reached the standards for the use of building materials.

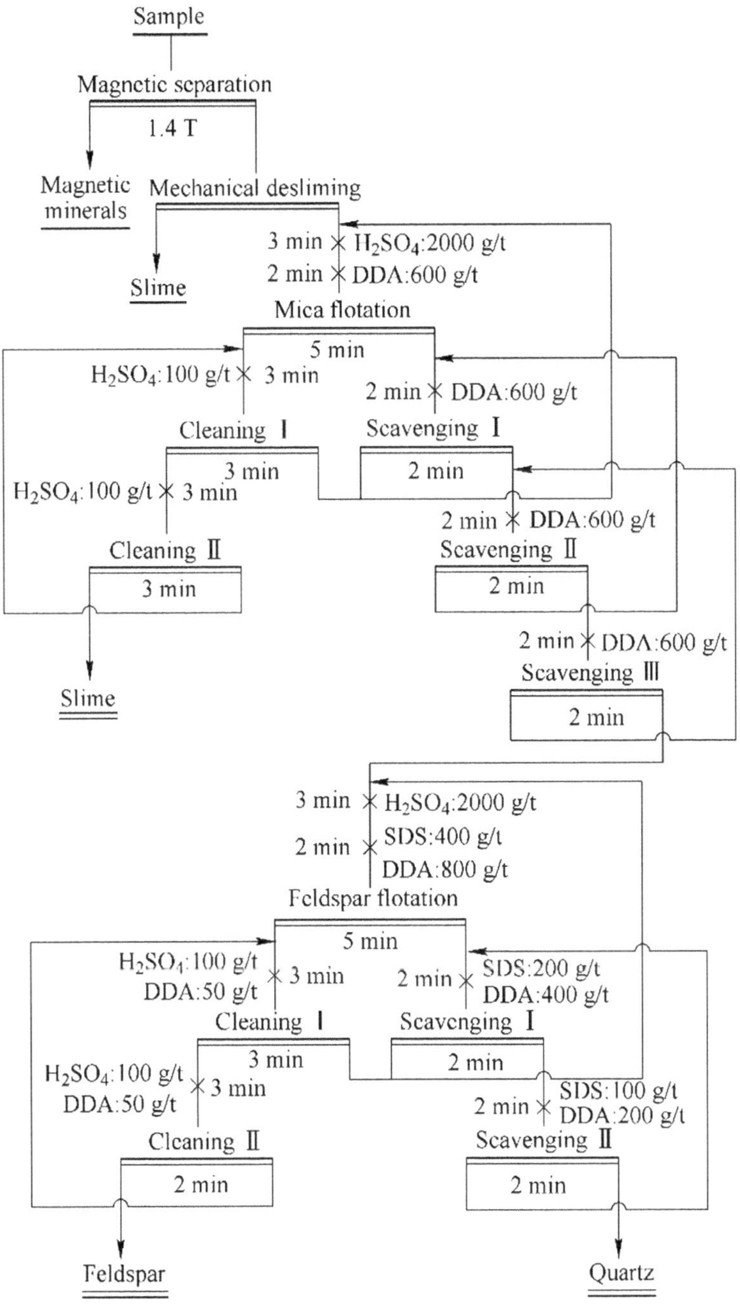

Fig. 8.8 Flow chart of comprehensive recovery of feldspar and quartz from a tungsten tailings in Guangxi

Table 8.20 Product Index of a tungsten tailings closed-circuit test in Guangxi (%)

Product name	Yield	Grade				Recovery			
		K_2O	Na_2O	SiO_2	Al_2O_3	K_2O	Na_2O	SiO_2	Al_2O_3
Magnetic	5.52	3.37	4.14	55.70	6.81	4.24	7.19	4.01	5.23
Slime	8.89	5.35	5.21	65.16	11.40	10.85	14.58	7.55	14.11
Mica	5.23	10.58	1.65	47.27	30.52	12.62	2.72	3.22	22.23
Feldspar	44.32	6.94	5.25	67.66	8.58	70.15	73.24	39.10	52.95
Quartz	36.04	0.26	0.20	98.14	1.09	2.14	2.27	46.12	5.47
Feed mine	100.00	4.38	3.18	76.70	7.18	100.00	100.00	100.00	100.00

8.6 Summary

There are rich feldspar and quartz resources in the dolomite and quartzite tungsten ores, and the comprehensive recovery and utilization of them is an important link in the construction of green mine. A large number of scientific research and engineering practice has proved the feasibility of fluorine-free acid-free quartz-feldspar separation and recovery technology and process. Under neutral or weak acid conditions, mixed anionic-cationic collector is used for priority flotation of feldspar, and under alkaline conditions, metal ion activation is used for priority flotation of quartz. Qualified and high-quality feldspar and quartz products can be produced, so as to maximize the utilization of tailings resources. This technology provides technical support for the future development of resource-saving and environment-friendly mining industry.

Acknowledgements This work was supported financially by the National Natural Science Foundation of China (52274285), the National Key R&D Program of China (Nos. 2022YFC2904401), the Outstanding Youth Fund of Hunan Natural Science Foundation (2021JJ20064), National key research center and development Program of the 14th Five-Year Plan (No. 2022YFC2905105 and No. 2022YFC2905104), National Natural Science Foundation of China (NSFC) (No. 52122406and No. 52374291), Hunan High-tech Industry Technology Innovation Leading plan (No. 2022GK4056), Hunan Innovative Province construction special Project (No. 2020RC3001).

References

1. Wu S, Chen J, Huang L (2017) Separation of mica, feldspar and quartz from a tungsten-tin tailing in Dayu, Jiangxi. Mod Min 000(011):135–137
2. Wu F, Liu Z (2016) Recovery of feldspar and quartz from tungsten-tin tailings in Guangxi. Min Res Dev 7:18–21
3. Gong E, Zhong W, Zhou X (2011) Study on process mineralogy of Shirenzhang dolomite ore in Guangdong Province. China Min 20(03):71–73+85
4. Hong Q, Li B, Liang D et al (2019) Study on process mineralogy of a Quartz vein type tungsten-tin ore and its beneficiability analysis. China Tungsten Ind 34(2):29–33
5. Wu S (2005) Study on comprehensive utilization of tungsten tailing resources in Pangushan, Jiangxi Province. China University of Geosciences, Wuhan

6. Wen R (2014) Research on exploitation and utilization of a tungsten tailings in Jiangxi and its application practice. Proceedings of the 7th nonferrous metal industry development forum in Central and Western China, pp 678–681
7. Peng G, Zhang J, Zhu X et al (2013) Process mineralogy and comprehensive utilization of mineral Resources in Limu Jinzhuyuan deposit. Nonferrous Met Eng 3(5):36–39
8. Li X (1991) Production, marketing and application of feldspar and nepheline syenite in the world. Build Mater Ind Inf 09:6
9. Cai W, Zhao Y (1989) Feldspar—the most common gemstone mineral among rock-forming minerals. Jilin Geol 2:84–88
10. Gu X, Gu L (1994) On the nature, application and comprehensive utilization of feldspar. Guangdong Build Mater 04:36–39
11. Zhang R, Wang H, Xu H (2013) Ceramic technology. Chemical Industry Press
12. Peng S, Chen Z (2008) Current situation and development trend of siliceous raw material industry in China. Foreign Build Mater Sci Technol 02:40–46
13. Zhou Z (2010) Inorganic nonmetallic material technology. China Light Industry Press
14. Yu S, Lin B (2001) Advances in machining technology of advanced ceramic materials at home and abroad. Diam Abras Tools Eng 04:36–39
15. Lu X (2005) Ceramic technology. Hunan University Press
16. Ma T (2011) Ceramic technology. China Light Industry Press
17. Zong P (2013) Application technology and industry prospect of potassium feldspar for ceramic industry. China Non Metallic Miner Ind Guide 000(006):1–3
18. Lines M, Lu Z (2005) Production and consumption of feldspar in Asia. Non Metallic Min 03:7–9
19. Zhang B, Zhang Y (2012) Research on mineral processing of feldspar. China Non Metallic Miner Ind Guide 03:36–39
20. He Z (2006) Study on super smooth ceramic glaze. Hunan University
21. Zhu W (2016) Characteristics and Industrial application of Indian potassium feldspar. Foshan Ceram 2:32–36
22. Qu J, Zhao F, Fu S (2010) Research status and application prospect of insoluble potassium. Mod Chem Ind 30(6)
23. Huang W (2012) Experimental study on flotation separation of feldspar and quartz. Wuhan University of Technology (in Chinese)
24. Ding YZ (2010) Study on preparation of high purity quartz from low grade quartz ore. Northeastern University (in Chinese)
25. Xu H (1992) Industrial utilization and processing technology of Quartz sand. Conserv Utilization Miner Resour 000(001):21–26
26. Zheng S (2009) Processing and application of non-metallic ore. Chemical Industry Press
27. Jing H (2004) Effect of high temperature roasting on process performance of quartz sand for casting. Xi'an University of Technology
28. Zhang J (2006) Development of glass industry and its demand for quartz raw materials. China Glass Industry Annual Conference and Technical Seminar
29. Hu T (2013) Experimental study on beneficiation of Kaolin associated quartz sand ore in Beihai. Wuhan University of Technology
30. Hua H (2017) Preparation and properties of green mullite whisker skeleton porous ceramics
31. Wang L, Li C, Wang Y et al (2011) Current status and development suggestions of high purity quartz processing technology in China. J Mineral Petrol 04:112–116
32. Tian J (1999) Study on purification technology of high purity quartz sand. China Min 03:59–62
33. Skorina T, Allanore A (2015) Aqueous alteration of potassium-bearing aluminosilicate minerals: from mechanism to processing. Green Chem 17:2123–2136
34. Zhang Y, Hu Y, Sun N et al (2018) Systematic review of feldspar beneficiation and its comprehensive application. Miner Eng 128:141–152
35. Hu B, Han X, Xiao Z, Lu Y, Chen M (2005) Distribution, development and utilization, problems and counter measures of potassium feldspar mineral resources of our country. Chem Miner Geol 01:25–32

36. Jia M, Sun C-Y (2001) Analysis of zero electric point, floatability and bond valence of some silicate minerals. Nonferrous Met Min Process 000(006):1–9
37. Chen L (2014) Study on flotation separation of quartz and feldspar. Hunan University of Technology
38. Liu Y, Gong H, Zhang K (1993) Adsorption of sodium oleate and dodecylamine hydrochloride on the surface of feldspar and quartz. Min Metall Eng 02:29–32+35
39. Yu F (2005) Research status of fluorine-free flotation separation technology of quartz feldspar. Prot Utilization Min Resour 3:52–54
40. Tian M, Li H, Zhang H et al (2012) Study on fluorine-free separation process of potassium feldspar quartz and its industrial experiment. Chem Miner Process 11:13–17
41. Vidyadhar A (2007) Role of mixed collector systems in selective separation of soda feldspar from Greek Stefania feldspar ore. Int Min Process Technol
42. Gao W, Lu C, Xiao J et al (2016) Experimental study on recovery of potassium feldspar from molybdenum tailings by flotation. China Molybdenum Ind 040(003):4–8
43. Qiu Y-R, Zhang L-Y, Song Y-H et al (2014) Study on fluoro-acid-free flotation separation of feldspar and quartz. Conserv Utilization Miner Resour 3:47–51
44. Nie Y, Liu S, Wang S et al (2015) Research status and progress of fluorine-free flotation separation of quartz feldspar. Chem Miner Process 044(007):51–54
45. El-Salmawy MS, Nakahiro Y, Wakamatsu T (1993) The role of alkaline earth cations in flotation separation of quartz from feldspar. Miner Eng 6(12):1231–1243

Chapter 9
Treatment and Recycling of Tungsten Ore Flotation Wastewater

Xiangsong Meng, Haisheng Han, and Wei Sun

Abstract Flotation serves as the primary beneficiation method for tungsten ore, generating a significant volume of wastewater in the process. Due to the presence of various pollutants in this wastewater, it often requires appropriate treatment to meet discharge standards or facilitate recycling. This chapter begins by analyzing the wastewater generation in tungsten ore beneficiation, offering a systematic overview of the key pollutants influencing the treatment of tungsten ore flotation wastewater. These pollutants include sodium silicate, suspended solids, and chemical oxygen demand (COD). Subsequently, corresponding removal technologies are introduced. For sodium silicate and suspended solids, due to the dispersion effect of sodium silicate, it is highlighted that removing sodium silicate first is beneficial for the subsequent elimination of suspended solids. For COD removal, the principle of "mass treatment and classification recycling" is suggested to conserve flotation agents and alleviate the pressure of wastewater treatment. Finally, the simple, effective, and cost-efficient treatment process of typical tungsten ore flotation wastewater is introduced. For wastewater primarily containing fatty acid collectors, treatment with industrial waste acid rich in calcium chloride, followed by chemical precipitation, is recommended. Meanwhile, for wastewater dominated by metal–organic complex reagent, the raw pulp coagulation technology has been proposed.

Keywords Tungsten ore flotation wastewater · Chemical oxygen demand · Silicate removal · Suspend solids

X. Meng (✉) · H. Han · W. Sun
Central South University, Changsha, China

Engineering Research Center of Ministry of Education for Carbon Emission Reduction in Metal Resource Exploitation and Utilization, Central South University, Changsha, China

State Key Laboratory of Mineral Processing, Changsha, China

H. Han
Technology Research Center of Hunan Province for Comprehensive Utilization of Associated Fluorite and Fluorine Chemical Engineering, Chenzhou, China

W. Sun et al., *Flotation Chemistry of Tungsten Minerals and Its Application*,
https://doi.org/10.1007/978-981-96-5077-4_9

9.1 Introduction

During the flotation process, tungsten ore will produce a large amount of wastewater and the wastewater contains a variety of pollutants, which usually cannot be directly discharged or reused, and the wastewater must be treated. After wastewater treatment, it is divided into standard discharge or recycling, in which the standard discharge requires all the pollutants in the wastewater to meet the discharge standard requirements, often requiring advanced treatment, which is difficult and costly; Recycling, on the other hand, only requires moderate treatment of wastewater, removal of specific pollutants in the wastewater, and return it to the appropriate process of flotation without reducing the flotation index, but in practice there are often many factors that limit the recycling rate of wastewater. Tungsten mining enterprises usually implement the first level standard of *Comprehensive Sewage Discharge Standard GB 8978–1996*, and the maximum allowable discharge concentration of some pollutants in wastewater is shown [1] in Table 9.1. This chapter analyzes and discusses the production and pollution factors of tungsten flotation wastewater, the removal technology of typical pollutants in wastewater, the key factors of wastewater recycling and the treatment process of typical tungsten mine wastewater, aiming to provide references for the treatment and recycling of tungsten mine flotation wastewater for the better development and utilization of tungsten resources.

9.2 Production and Pollution Factors of Tungsten Ore Flotation Wastewater

The beneficiation of mixed ores of scheelite and wolframite is mainly based on flotation method. Flotation 1 t of raw ore consumption 4–7 t of water [2]. Tungsten flotation wastewater is mainly derived from the flotation process of each middling ore and concentrate density overflow water, concentrate drying dehydration, ground flushing water, and tailings water, where tailings water occupies the main body. Because a variety of flotation reagents are added in the flotation process, such as adjuster, collector, foaming agent, etc., the composition of tailings water is particularly complex, and the wastewater treatment and recycling is more difficult [3]. At present, the main process of tungsten flotation includes fatty acid method and metal–organic complex method, with fatty acid as the main collector, sodium silicate as inhibitor in the tungsten flotation process, and the tailings pulp is difficult to settle naturally. The main pollutants in wastewater are residual fatty acids, water glasses, and solid suspensions; Metal–organic complex method tungsten flotation process uses metal–organic complex as the main collector ping agent, no water glass is added in tungsten roughing, and a small amount of salted water glass is added in tungsten cleaning, and the tailing pulp is easier to settle naturally. The main pollutants in wastewater are residual metal–organic complexes.

Table 9.1 The maximum allowable emission concentration of pollutants in wastewater according to the National Standard (some indicators, unit mg/L)

Pollutants	Total cadmium	Total chromium	Total arsenic	Total lead	pH	Suspended solids	COD	Ammonia nitrogen	Fluoride	Total zinc
Concentration	0.1	1.5	0.5	1.0	6–9	70	100	15	10	2.0

In addition, tungsten minerals are usually associated with other useful minerals, such as molybdenum, bismuth tungsten fluorite ore type [4], molybdenum tungsten ore type [5], copper tungsten ore type [6], copper molybdenum tungsten ore type [7], so the flotation process of tungsten polymetallic ore is usually the first selection of sulfide ore, then tungsten ore, and finally selected fluorite and other associated resources. Each section of tungsten polymetallic ore flotation process will produce wastewater, and the chemical composition of the wastewater residues are different, so that in the final tailings wastewater, there are all the reagents of the whole flotation process remaining, which increases the difficulty of tailings wastewater treatment.

The flotation process of tungsten polymetallic ore is shown in Fig. 9.1. The main pollutants affecting recycling or standard discharge in wastewater are sodium silicate (sodium silicate), suspended solids and residual organic flotation reagents. Because the types and contents of organic flotation reagents in wastewater are difficult to be detected quantitatively, chemical oxygen demand (COD) is usually used to characterize the concentration of organic agents in wastewater. Because there are all reagents of the whole flotation process remaining in the final tailings' wastewater, if it is not processed directly back to the source grinding, it will usually deteriorate the polymetallic flotation index. Research teams at home and abroad put forward the technical idea of "mass treatment and classification recycling" for polymetallic ore flotation wastewater. In other words, for the wastewater generated from different flotation processes, moderate treatment is carried out for specific pollutants and then returned to the appropriate flotation process for recycling, providing a direction for the low-cost treatment and recycling of tungsten polymetallic ore flotation wastewater.

9.3 Typical Pollutant Removal Techniques in Wastewater

9.3.1 Water Glass Removal Technology

Water glass generally refers to sodium silicate, which is often used as a tungsten flotation vein mineral inhibitor, especially when the "Petrov" method is used to heat flotation, a large amount of water glass needs to be added, however, too much water glass can also affect the flotation of tungsten minerals, because too much water glass will compete with the collector for adsorption sites on the surface of tungsten minerals, and reduce the adsorption of the collector [8]. When Kang Jianhua et al. studied Luanchuan scheelite tailing wastewater, they found that the wastewater contained more silicate ions (content up to 1200 mg/L), which was directly used for scheelite roughing would worsen the flotation index [9]. In practice, calcium ions are often used to combine with silicate to form calcium silicate precipitation to remove sodium silicate. However, when lime is used to treat wastewater containing high concentration sodium silicate tungsten flotation, the pH of tailings pulp is high due to the large amount of lime. Under the condition of high pH, free calcium ions will hydrolyze to form calcium hydroxide, which limits its combination with silicate ions.

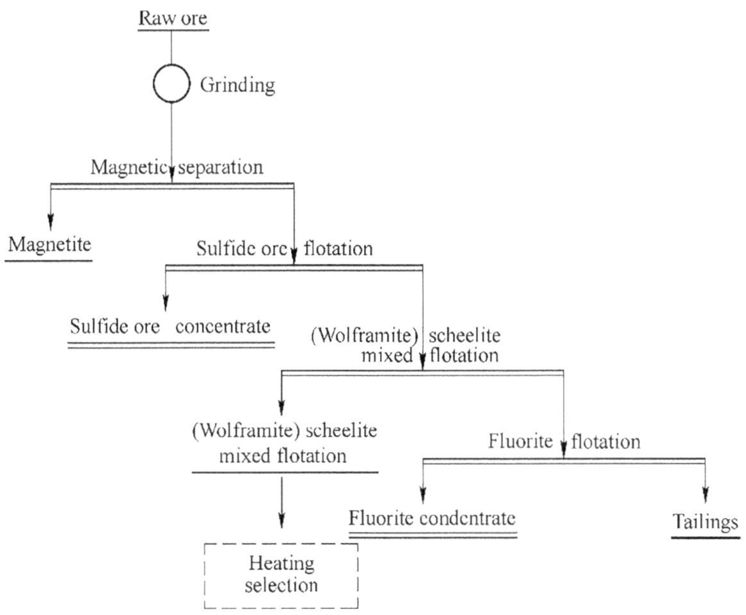

Fig. 9.1 Flotation process flow of tungsten polymetallic ore

As a result, lime alone cannot effectively remove the silicate ion. The composition distribution of silicate and calcium ions in aqueous solution is shown in Fig. 9.2. It can be seen that pH affects the existence form of silicate and calcium ions in aqueous solution.

Deng Shuangli et al. used conventional wastewater treatment reagents (ferric chloride, hydrated lime, polyferric sulfate, polyacrylamide) and different acids (sulfuric acid, hydrochloric acid, nitric acid) to treat scheelite heating and cleaning wastewater, and found that sulfuric acid had the best removal effect on silicate ions [10].

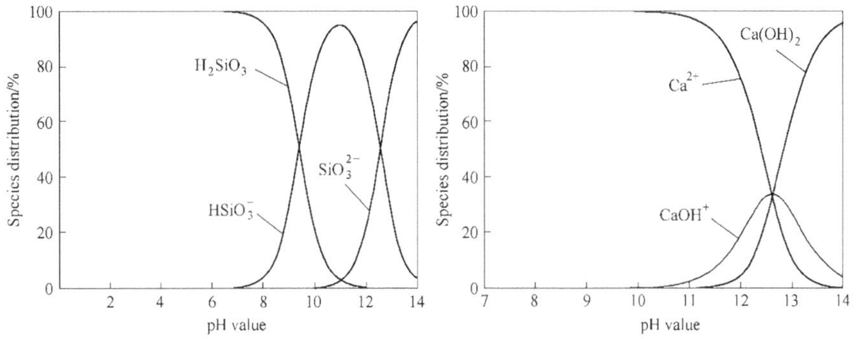

Fig. 9.2 Composition distribution of silicate and calcium ions in aqueous solution, adapted from Ref. [9], Copyright 2017, with permission from Elsevier

Chen Qian used the "combined acid–base process" to effectively remove water glass from scheelite flotation wastewater. The principle is that adding sulfuric acid to the wastewater can gradually hydrolyze water glass into silicic acid, and then adding lime milk to the wastewater, the silicic acid is converted into calcium silicate precipitation, which can be removed from the wastewater [11]. Wang Shuiyun used electric flocculation technology to treat scheelite mineral processing wastewater, the wastewater pH is adjusted to 9.5, after primary electric flocculation treatment, the removal rate of silicate can reach 89.2%, and after secondary electric flocculation treatment, the removal rate of silicate can reach 98.1% [12].

9.3.2 Removal Technology of Suspended Solids

After the flotation of tungsten minerals, fine particles in the pulp are difficult to settle naturally under the action of water glass, resulting in a relatively high concentration of suspended solids in the wastewater. After the flotation is completed, the water glass residue in the wastewater adsorbed on the surface of fine particles, so that the particles are negatively charged, and the particles repel each other to form a stable dispersion system, resulting in the natural settlement of suspended solids. On the one hand, the wastewater containing high concentration of suspended solids is directly used for flotation, which will lead to the slime cover on the surface of useful minerals and affect the floating of useful minerals. At the same time, slime will also consume a lot of flotation reagents, deteriorate the flotation index and increase the cost of reagents; on the other hand, the content of suspended solids in wastewater is high and it is difficult to meet the discharge standards.

The commonly used technology to remove suspended solids in tungsten flotation wastewater is to add lime milk to remove silicate ions in wastewater (applicable to the pH of raw water, silicate ion concentration is not too high), so that the wastewater is de-stabilized, and then the solid suspended solid is further removed by coagulation and precipitation technology. Chen Ming et al. took a tungsten tailings dam overflow water (suspended solids 3070 mg/L) to study the removal method of suspended solids, in the wastewater first add lime milk to adjust the pH of the wastewater to 11.5, after standing settlement take supernatant, add acid to adjust the pH to 8.5 plus polyacrylamide 2 mg/L settlement can reduce the wastewater suspended solids to 128 mg/L [13]. Chen Houxing et al. took the overflow water of a tungsten tailings dam in southern Jiangxi to study the removal process of suspended solids. By adding lime milk to the wastewater to adjust the pH to about 10 and then adding 15 mg/L of ferric chloride and stirring to settle, the suspended solids in the wastewater could be reduced from 415.5 to 5.4 mg/L [14].

9.3.3 Removal Technology of Organics

The effluent COD of tungsten polymetallic tailings dam is a strictly monitored index in the discharge standard, and it is also the most difficult pollution factor to purify. A typical characteristic of tungsten polymetallic ore flotation wastewater is its large volume. A processing plant with a daily processing capacity of 3500 t can produce up to 20,000 t/d of tailing wastewater, which usually needs to be treated in depth by oxidation technology if the COD is to be treated to standard. However, in addition to requiring good purification effect, mine wastewater treatment technology also requires simple process and low cost, so some high-cost advanced oxidation technology is difficult to realize industrial application. Based on the principle of "mass treatment and classification recycling", the wastewater is returned to proper flotation operation after moderate treatment to make full use of the residual reagents in the wastewater, which not only saves resources but also reduces the pressure of wastewater discharge.

In recent years, Chinese mineral processing workers have made a lot of exploration on the treatment and recycling of tungsten polymetallic ore flotation wastewater, and some progress has been made. Jiang Zhichao et al. studied the overflow water of a tungsten-bismuth polymetallic tailing dam in Hunan using the process of "oxidant ME22 + PAM coagulation + acid conditioning" and concluded that the oxidant ME22 could hydrolyze hypochlorite in the wastewater and the COD of the wastewater could be reduced from 118 to 40.6 mg/L When the oxidant reagent dosage is 760 mg/L and the oxidation time is 45 min, the COD of wastewater can be reduced from 118 to 40.6 mg/L [15]. Shen Yi et al. studied a tungsten-molybdenum-copper polymetallic flotation wastewater and concluded that the copper-molybdenum mixed flotation and copper-molybdenum separation wastewater could be directly returned to the original process for reuse after precipitation and clarification, while the tungsten system wastewater could be returned to the tungsten system or copper-molybdenum system for reuse after "lime coagulation and precipitation—chlorine dioxide deep oxidation—pH adjustment" [16]. Jia Pengfei et al. studied the flotation wastewater of a scheelite polymetallic ore and concluded that the untreated tailings water can be directly used for tungsten flotation at room temperature, which can reduce the amount of agents; The water treated by the "acid–base combination—pressure soluble air flotation" process can be used back for grinding and sulfide ore beneficiation process [17]. Feng Zhangbiao studied the purification process of mineral processing wastewater of Shizhuyuan polymetallic mine. The wastewater treated by "lime precipitation + calcium removal agent" can be used back in the process of tungsten or fluorite separation, which can save part of the reagents, but cannot be used back in sulfide ore separation, otherwise it will deteriorate the flotation index of sulfide ore, so that a lot of tungsten and fluorite loss in sulfide ore concentrate; After further treatment by electrooxidation, it can be used back for sulfide ore beneficiation [18]. Meng Xiangsong et al. used Fenton oxidation method, polymeric sulfate coagulation method and activated carbon adsorption method to treat the effluent water from the molybdenum, bismuth, tungsten and fluorite polymetallic tailings dam in Shizhuyuan, and found

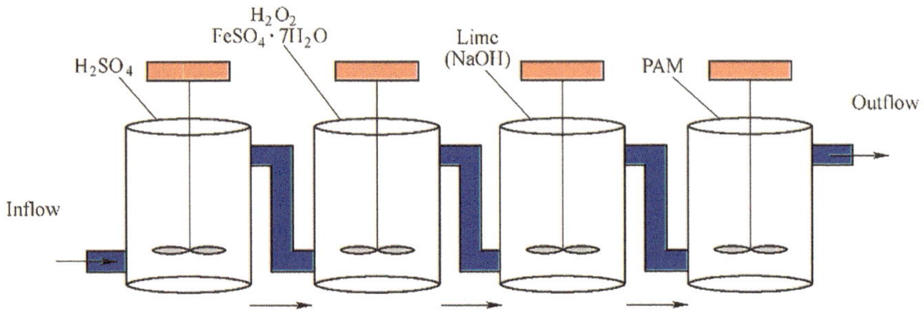

Fig. 9.3 Fenton oxidation process flow chart

that the contribution of polymeric sulfate coagulation method to the COD removal of dissolved organics was limited, and could not reduce the COD of tailings wastewater to the standard. While Fenton oxidation method and granular activated carbon adsorption method can effectively remove the COD of tailings wastewater and bring it down to the standard, but the cost of activated carbon adsorption method is [19] too high. The Fenton oxidation process flow chart is shown in Fig. 9.3.

Fenton oxidation involves several chemical reactions, among which the widely recognized core reaction is shown in Eq. (9.1). Under acidic conditions, Fe^{2+} catalyzes H_2O_2 to produce OH with strong oxidation, OH can degrade most of the stubborn organic pollutants in wastewater quickly and without selectivity to produce carbon dioxide and water. The mechanism of Fenton oxidation reaction is shown in Fig. 9.4 [20]. Based on this principle, Fenton oxidation is widely used in the treatment of various organic wastewater. In Fenton oxidation process, the degradation efficiency of organic pollutants depends on the pH value of wastewater, the concentration of Fenton reagent, the initial concentration of organic pollutants and other operating parameters, among which the pH value of wastewater is a very important parameter. When the pH value is too low or too high, ferrous catalyst will be deactivated and affect the degradation efficiency of organic pollutants. Many studies have shown that Fenton oxidation treatment has the best effect when the reaction pH is 2–4, especially when the reaction pH is 3. When the reaction is complete, it is necessary to add alkali to adjust the pH of the wastewater to 6–9, so as to coagulate and precipitate the iron salt in the wastewater. Therefore, the Fenton oxidation process needs to adjust acid first and then alkali, which will increase the cost of wastewater treatment. Traditional Fenton reagents consist of homogeneous solutions of Fe^{2+} and H_2O_2, but both are chemically unstable and easily inactive, resulting in waste of reagents. However, their chemical properties are unstable and easy to lose activity, resulting in reagent waste. In addition, high concentrations of H_2O_2 are explosive and toxic, which is harmful to human health. T Therefore, the traditional Fenton oxidation process has three obvious defects: narrow reaction pH range, high cost and risk of transportation and storage of Fenton reagent (H_2O_2 and Fe^{2+} homogeneous solution), and a large amount of iron-containing sludge will be produced after the

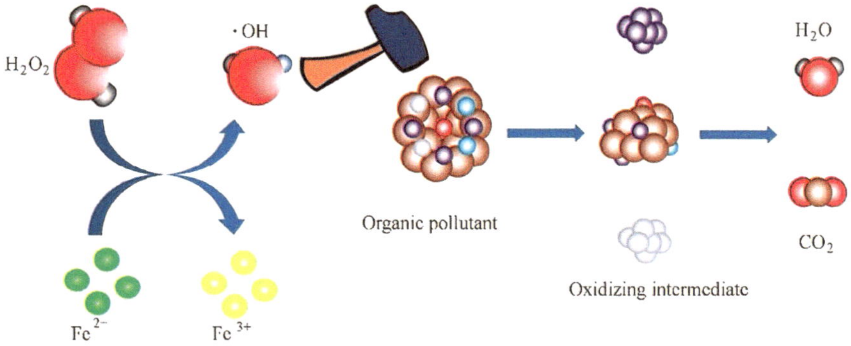

Fig. 9.4 Mechanism diagram of Fenton oxidation reaction, adapted from Ref. [20], Copyright 2019, with permission from Elsevier

reaction. In order to overcome the shortcomings of the traditional Fenton oxidation process, the traditional process has been continuously optimized and improved, forming heterogeneous Fenton process, light Fenton process and electrochemical Fenton process.

$$Fe^{2+} + H_2O_2 + H^+ \rightarrow Fe^{3+} + H_2O + {}^{\cdot}OH \quad (9.1)$$

Wu Meirong et al. studied the tailings wastewater from Shizhuyuan Bismuth desulphurization operation by using electric flocculation technology. The results showed that the removal rates of COD, sulfate and calcium ions in the treated wastewater could reach 98%, 93% and 94%, respectively, and that the wastewater returned to grinding after treatment had no negative effect on the flotation index of molybdate bismuth [21]. Electro flocculation is a kind of wastewater treatment technology which integrates the advantages of coagulation, flotation and electrochemical technology. It can remove many pollutants in wastewater, such as COD, fluoride, suspended solids and heavy metals. The electro flocculation device usually consists of an electrolytic cell whose anode and cathode plates are immersed in the wastewater and connected to an external DC power supply through wires. A schematic diagram of the basic treatment unit of electric flocculation is shown in Fig. 9.5 [22]. Iron plate and aluminum plate are the most widely used electro flocculating electrode plates due to their easy availability, non-toxic and reliable performance.

The complex physicochemical action in the electro flocculation process can be summarized as the following successive stages: (1) "Sacrificial anode" oxidation to cationic coagulant; (2) The reduction of water molecules to hydrogen bubbles and hydroxide ions at the cathode; (3) The cationic and cationic ions generated during electrolysis process migrate to anode and cathode respectively under the action of electric field, and the migration movement leads to the instability of pollutants in wastewater; (4) Cationic coagulant and hydroxide ion combine to produce metal hydroxide with good adsorption performance, metal hydroxide gathers together to form a larger structure and diffuses in solution; (5) The pollutants in the wastewater

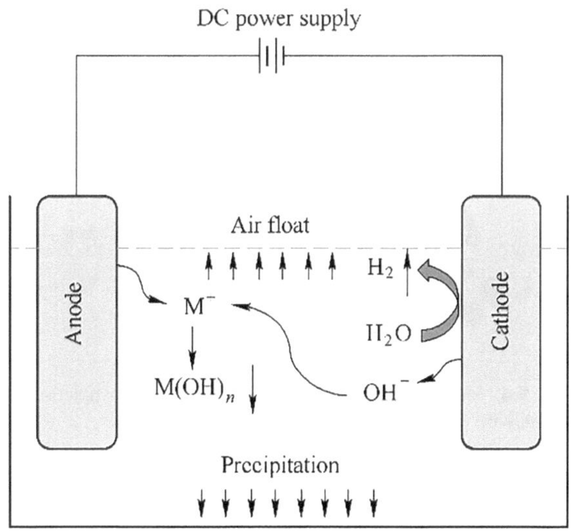

Fig. 9.5 Schematic diagram of the basic electro flocculation processing unit, adapted from Ref. [22], Copyright 2017, with permission from Elsevier

are adsorbed to the metal hydroxide structure to form larger aggregates; (6) The pollutants in the wastewater may be converted into less toxic forms through REDOX reactions; (7) Aggregates float on the surface of the wastewater through bubble air or precipitate under the action of gravity [23]. In the process of electro flocculation, the anode reaction is shown as Eqs. (9.2)–(9.6), and the cathode reaction is shown as Eq. (9.7). There are two main factors affecting the effect of electro flocculation treatment. One is the reaction device factors, such as electrode material, electrode spacing, electrode arrangement, electrode shape, power supply type, and reactor form. The second is the operating factors, such as electrolytic current, electrolytic time, wastewater pH, temperature, stirring, and pollutant types and concentrations.

$$\mathrm{Fe\ (s)} \rightarrow \mathrm{Fe^{2+}\ (aq)} + 2\mathrm{e^-} \tag{9.2}$$

$$4\mathrm{Fe^{2+}\ (aq)} + 10\mathrm{H_2O} + \mathrm{O_2\ (aq)} \rightarrow 4\mathrm{Fe(OH)_3\ (s)} + 8\mathrm{H^+} \tag{9.3}$$

$$\mathrm{Fe^{2+}\ (aq)} + 2\mathrm{OH^-} \rightarrow \mathrm{Fe(OH)_2\ (s)} \tag{9.4}$$

$$\mathrm{Al\ (s)} \rightarrow \mathrm{Al^{3+}\ (aq)} + 3\mathrm{e^-} \tag{9.5}$$

$$\mathrm{Al^{3+}\ (aq)} + n\mathrm{H_2O} \rightarrow \mathrm{Al(OH)}_n^{3-\mathrm{n}} + n\mathrm{H^+} \tag{9.6}$$

$$2\mathrm{H_2O} + 2\mathrm{e^-} \rightarrow \mathrm{H_2} + 2\mathrm{OH^-} \tag{9.7}$$

9.4 Factors Affecting Wastewater Recycling

The recycling of tungsten polymetallic ore flotation wastewater can be divided into external cycle and internal cycle. The external cycle refers to the large cycle, that is, the recycling of tailings dam effluent water. And the internal cycle refers to the small cycle, that is, the recycling of the overflow water of the flotation middling ore and concentrate thickener in the concentrator. Tungsten polymetallic mine tailings wastewater accounts for the main part of flotation wastewater, in the absence of internal circulation tailings wastewater with other flotation process wastewater along with the tailings pulp to tailings dam settlement. Tailings dams are usually far away from the processing plant, if the tailings dam outflow pumps are sent to the processing plant for recycling, on the one hand, long distance pipelines need to be laid, consuming more electricity and higher operating costs; on the other hand, the tungsten polymetallic ore flotation process is complex, and different types and effects of flotation chemicals are added to each section of flotation, resulting in a very complex chemical composition of the final tailings dam outflow water, which usually cannot be returned directly to the source mill. The main reason is that the outflow water of tailings dam contains all the residual reagents of the whole flotation process of the concentrator. If it is directly returned to the source for grinding, the residual collector in the wastewater will cause the loss of minerals in the latter section to the former section, and also affect the recovery of minerals in the former section.

Compared with the external circulation of tailings dam outflow water, the internal circulation of the ore, concentrate thickener overflow water and concentrate drying dehydration in the concentrator is simpler. These wastewaters can be directly returned to the original process after settlement and clarification, and because the residual agents in these wastewaters come from the original process, the reagent dosage of the original process can be reduced accordingly when recycling. Lin Shangyong et al. studied the mineral processing wastewater from molybdenum and bismuth flotation section of Shizhuyuan returning to the original process for utilization. The results showed that recycling wastewater not only had no bad effect on the grade and recovery rate of molybdenum and bismuth concentrate, but also could reduce the amount of flotation reagent needed by more than 10%, especially the amount of sodium sulfide could reduce by 18%. The recycling of wastewater can also reduce the addition of fresh water and reduce the pressure [24] of high concentration COD wastewater treatment.

9.5 Treatment Process of Typical Tungsten Ore Flotation Wastewater

9.5.1 Treatment Process of Tungsten Ore Flotation Wastewater by Fatty Acid Method

The typical scheelite mountains in our country are Xiang Lushan ore, Luanchuan low grade scheelite, Yaogangxian scheelite etc. At present, the main flotation process used in scheelite mine is fatty acid method, and the main pollutants in the flotation wastewater are sodium silicate, solid suspended substance and fatty acid reagent. Because sodium silicate and fatty acid agents can combine with calcium ions to form precipitation, so in practice fatty acid flotation wastewater (tailings pulp) is usually treated by low-cost lime precipitation process. When lime is added to the tailings pulp and transported to the settlement of the tailings dam, the outflow water from the tailings dam can usually be clarified, and the concentration of sodium silicate and fatty acid reagent is greatly reduced. At this time, the wastewater can be returned to the concentrator for recycling after the pH is adjusted by adding acid.

When applying fatty acid process, the reagent dosage of sodium silicate in scheelite roughing operation is about 5–10 kg/t, and the amount of sodium silicate in heating and cleaning operation is more, and the amount of sodium silicate in the relative operation to the mine is up to 100 kg/t. When lime is used to settle tailings pulp, the settlement effect of roughing tailings pulp is generally better, but the settlement effect of heating and cleaning tailings pulp is poor. The reason is that the pH value of the refined tailings pulp by heating is high. After adding lime, the pH value of the refined tailings pulp is further increased, which can reach more than 12.5. As can be seen from Fig. 9.2, when the pH value is above 12.5, calcium ions hydrolyze into calcium hydroxide, which restricts the combination of calcium ions and silicate ions to produce calcium silicate, thus making the effect of lime removing silicate from the heated and purified tailings worse and the pulp settling effect worse. In practice, the heated and refined tailings pulp is finally transported to the settlement of tailings dam after being combined with roughing and refined tailings pulp. When the effect of lime settlement is poor, the settlement effect of the final tailings pulp will be affected. Especially in the north of China, in winter, the tailings dam wastewater freezes, the retention time of the wastewater in the tailings dam becomes shorter, the removal effect of sodium silicate becomes worse, and the content of suspended solids in the wastewater increases. At this time, the recycling of the tailings dam effluent water will worsen the scheelite roughing selection index.

In view of this problem, Central South University put forward the technical scheme [25] of using industrial waste acid to treat scheelite tailings pulp by heating and refining. The industrial waste acid is the waste acid produced in the acid leaching process of scheelite wet smelting. The heated tailings contain a large amount of calcite and fluorite, the grade of calcite is about 45%, fluorite grade is about 25%. See Table 9.2 for the chemical composition of the wastewater from the heated tailings and industrial waste acid. It can be seen from the table that the pH value of the wastewater is high, which is 11.64. At the same time, the silicon content is very high, indicating that the residual water glass is very high concentration. While the industrial waste acid has a low pH value of 3.14, which contains high concentration of chloride and calcium ions.

The industrial waste acid was added to the scheelite heated and refined tailings pulp, and the influence of the amount of waste acid on the removal rate of silicate ion and the concentration of suspended solids in the supernatant after the settlement of heated and refined tailings pulp was explored, as shown in Fig. 9.6. When the amount of waste acid is 50 g/L, the removal rate of silicate ion reaches 90%, and the concentration of suspended solids is reduced to 200 mg/L. Figure 9.7 shows the effect of waste acid on the treatment of heated and selected tailings pulp: When the amount of waste acid exceeds 50 g/L, the supernatant liquid after pulp settlement is clear and transparent, and the content of suspended solids is low. This indicates that industrial waste acid can effectively settle scheelite heated and refined tailings pulp, and effectively remove silicate ions and suspended solids in wastewater after pulp settlement.

The essence of industrial waste acid is that it can effectively remove sodium silicate from scheelite heated tailing wastewater. On the one hand, waste acid contains a large number of calcium ions, which can react with silicate ions in the pulp to produce calcium silicate hydrate precipitate (Eq. 9.8); On the other hand, after the waste

Table 9.2 Results of ICP-OES analysis of tailings wastewater and industrial waste acid by heating (mg/L)

Index	pH	Na	Mg	Al	Si	P	S	Cl	K	Ca	Mn	Fe	Mo	W
Waste water	11.64	2156	16	2.91	2040	5.88	330	176	9.91	105	3.62	18	32	12
Waste acid	3.14	144	177	107	142	4076	220	34,792	67	27,296	386	765	169	438

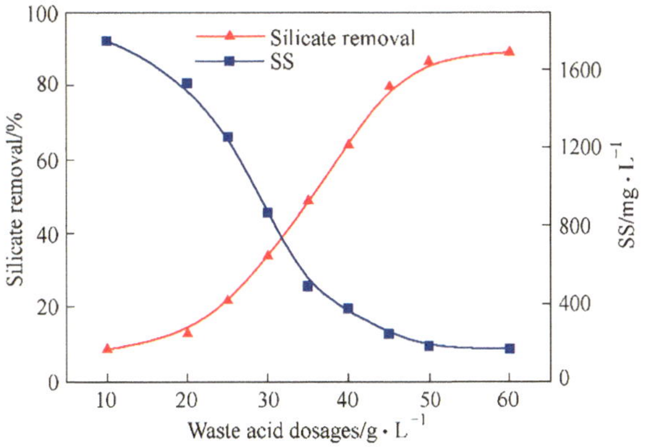

Fig. 9.6 Effect of the amount of waste acid on the removal rate of silicate ions, adapted from Ref. [9], Copyright 2017, with permission from Elsevier

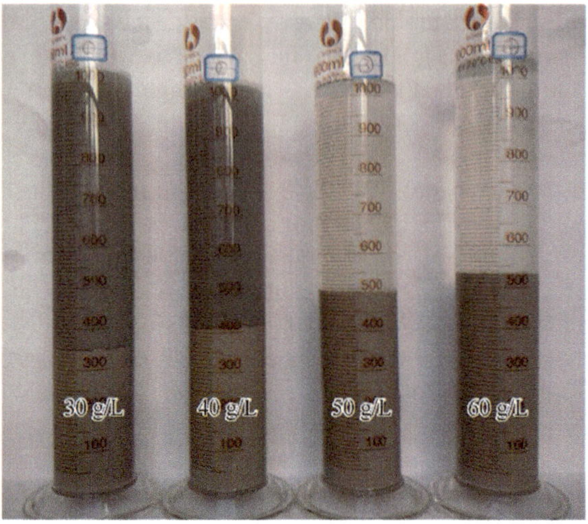

Fig. 9.7 Treatment effect of tungsten flotation wastewater with different amounts of waste acid, adapted from Ref. [9], Copyright 2017, with permission from Elsevier

acid is added, the silicate ions in the tailings pulp can be polymerized to produce polymeric silicate, and the equation is shown as Eqs. (9.9) and (9.10).

$$Ca^{2+} + H_3SiO_4^- \rightarrow CaO \cdot mSiO_2\ (s) \tag{9.8}$$

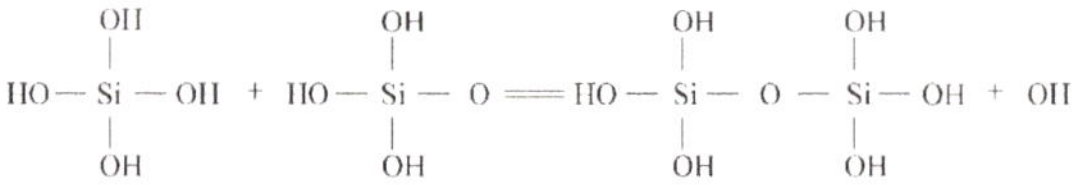

(9.9)

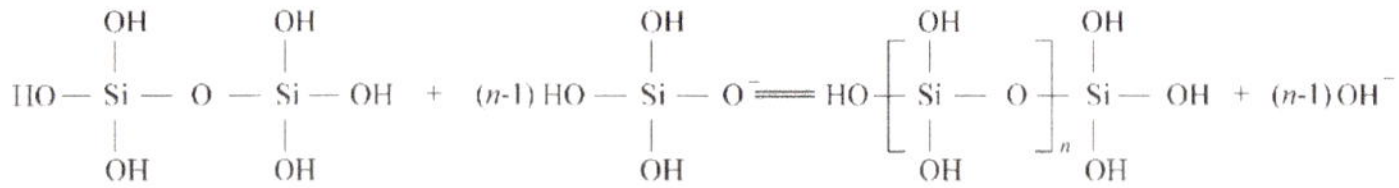

(9.10)

This technology is applied in the scheelite separation plant of Luoyang Molybdenum Industry Group. The amount of waste acid is about 50–60 t/d, and the removal rate of silicate ion and the content of suspended solids in the effluent water of tailings dam are greatly reduced, as shown in Fig. 9.8. As the wastewater has a certain cycle in the tailings dam, with the continuous addition of waste acid, the removal rate of silicate ion in the outflow water of the tailings dam gradually increases to 60–70%, and the concentration of suspended solids gradually decreases to 150–200 mg/L. During the industrial test, the tungsten recovery rate of scheelite roughing workshop (2015) increased by 5–10% [25] compared with the results of the same period in previous years (2013 and 2014) (as shown in Fig. 9.9). The implementation of this technical scheme effectively guarantees the flotation index of scheelite in winter, and provides a reference for the treatment and recycling of scheelite mountain wastewater at home and abroad.

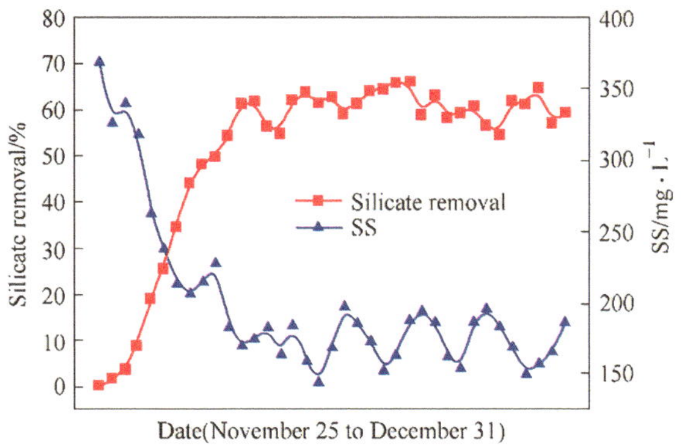

Fig. 9.8 Removal rate of silicate ion and content of suspended solids in backwater during industrial test, adapted from Ref. [9], Copyright 2017, with permission from Elsevier

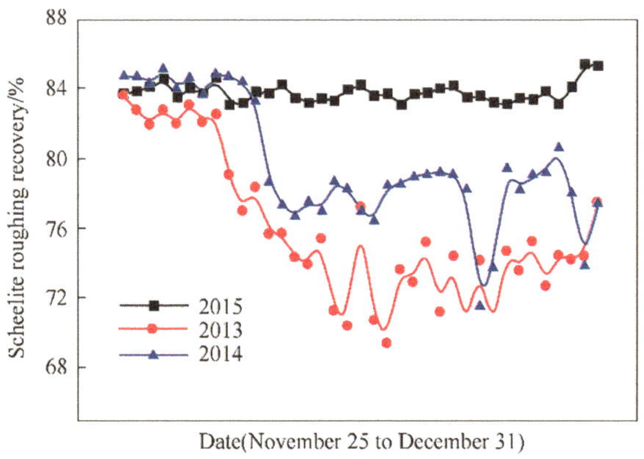

Fig. 9.9 Comparison of tungsten recovery rate (2015) and previous years (2013, 2014) in roughing workshop during industrial test, adapted from Ref. [9], Copyright 2017, with permission from Elsevier

9.5.2 Treatment of Tungsten Ore Flotation Wastewater by Metal Complex Method

Typical mixed ores of scheelite and wolframite in China includes Shizhuyuan tungsten ore and Xingluokeng tungsten ore, etc. At present, the main flotation process used in mixed ores of scheelite and wolframite is metal–organic complex method. In this process, no water glass is added in tungsten roughing process, and the pulp of tungsten roughing tailings is easier to settle naturally, so the main residues of metal–organic complex chemicals in the wastewater of tungsten roughing tailings are generated. However, the flotation process of polymetallic ore of mixed ores of scheelite and wolframite is usually to select sulfide ore first, then select tungsten ore, and finally select associated resources such as fluorite. In the flotation process of sulfide ore and associated resources such as fluorite, sodium silicate is usually added as an inhibitor of gangue minerals such as silicate. Therefore, the main pollutants in the tailing's wastewater of the final flotation of the polymetallic ore are still sodium silicate, suspended solids, and residual organic flotation reagents. In practice, the polymetallic separation plant of the mixed ores of scheelite and wolframite mainly uses the lime method to settle the final tailings pulp. After the tailings pulp settles in the tailings dam, the effluent can generally be clarified, and the water glass and suspended solids in the wastewater can be effectively removed. However, the COD in the wastewater is still high, so the wastewater is not suitable for recycling directly wolframite to the tungsten polymetallic separation plant, nor does it meet the discharge requirements. Therefore, the effluent water from the tailings dam needs to be further treated by the subsequent wastewater treatment station. The wastewater treatment station often adopts the coagulation-precipitation process of iron salt/ aluminum salt. However, the effect of this process on removing COD from wastewater is limited, and the COD can not be stably removed to the standard.

To solve this problem, Central South University put forward a low-cost COD removal technology scheme of flotation wastewater of mixed ores of scheelite and wolframite polymetallic ore based on original pulp coagulation. The technical scheme uses polyferric sulfate instead of lime settling tailings pulp, and uses gangue minerals in the pulp and polyferric sulfate to remove residual metal–organic complex chemicals in the pulp [26]. Figure 9.10 shows the test results of COD and pH of the supernatant after settling the tailings pulp of Shizhuyuan tungsten polymetallic ore under different reagent dosage conditions. As can be seen from Fig. 9.10, with the increase of lime reagent dosage, from 0 to 0.5 g/L, the COD of the supernatant of pulp sedimentation increased from 195 to 226 mg/L of natural sedimentation, and then increased the lime reagent dosage, the COD basically did not change. However, with the increase of the reagent dosage of polyferric sulfate, from 0 to 1 g/L, the COD of the supernatant of pulp settlement decreases from the natural settlement of 195 to 155 mg/L, and then increases the reagent dosage of polyferric sulfate, and the COD tends to rise to a certain extent. Therefore, the COD of the supernatant of the traditional lime settling tailings pulp is higher than that of the natural settlement, while the COD of the supernatant of the polyferric sulfate settling tailings pulp is lower than that of the natural settlement. Therefore, the polyferric sulfate has a significant advantage.

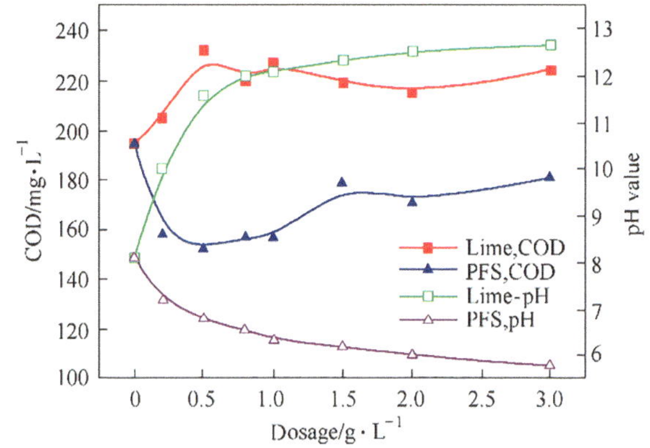

Fig. 9.10 Influence of the amount of lime or polyferric sulfate on the COD and pH of the supernatant of tailings pulp settling. adapted from Ref. [26], Copyright 2019, with permission from Elsevier

Among the remaining organic flotation reagents in tailings pulp of Shizhuyuan tungsten polymetallic ore flotation, it is difficult to be removed by lime sedimentation process for the collector of mixed ores of scheelite and wolframite, lead benzohydroxamic acid (BHA-Pb) complex agents. Figure 9.11 shows the COD results of the "supernatant" after 15 h precipitation of BHA-Pb reagent solution under different end pH conditions. As shown in the figure, COD of Bha-Pb solution with BHA concentration of 80 mg/L and 100 mg/L respectively showed a law of first decreasing and then increasing with the increase of pH, and reached the lowest when pH was 9.0. When BHA concentration is 80 mg/L, COD can be reduced from 177 to 32 mg/L. When the concentration of BHA is 100 mg/L, the COD energy decreases from 234 to 30 mg/L, which is corresponding to the experimental phenomenon that precipitation turbidity first appears in the BHA-Pb solution in the alkali regulation process, and then the pH continues to increase after reaching the critical point, and the turbidity gradually becomes clear. Therefore, the addition of a large amount of lime is not conducive to the purification of organic agents, but aggravates the rise of COD.

Figure 9.12 shows the COD results of "supernatant" after 15 h precipitation of coagulation BHA-Pb solution with calcium ions, iron ions and aluminum ions under different reaction pH conditions. In BHA-Pb solution precipitated by calcium ions, the variation law of COD in supernatant with pH is similar to that in BHA-Pb solution, indicating that calcium ions have no precipitation effect on BHA-Pb under test conditions. When the pH of aluminum ion coagulate BHA-Pb solution is 7.5, the COD of supernatant is higher than that of BHA-Pb solution, indicating that aluminum ion has antagonistic effect on the removal of BHA-Pb. Compared with calcium and aluminum ions, the COD of the supernatant of the BHA-Pb solution of iron ion coagulation is the lowest. When the pH is 7.0, the COD of the BHA-Pb solution can be reduced from 223 to 70 mg/L, and then with the increase of pH, the COD slowly decreases. When the pH increases to 9.0. The result of COD is equivalent to that

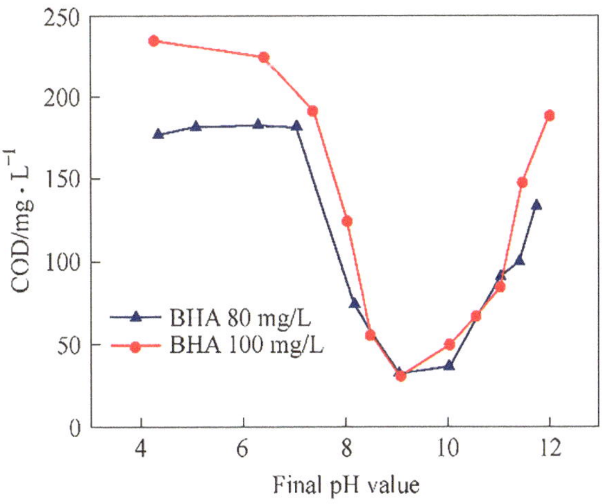

Fig. 9.11 Effect of terminal pH on COD in BHA-Pb complex reagent solution system (test condition: molar ratio of benzohydroxamic acid to lead ion 1:1), adapted from Ref. [27], Copyright 2023, with permission from Elsevier

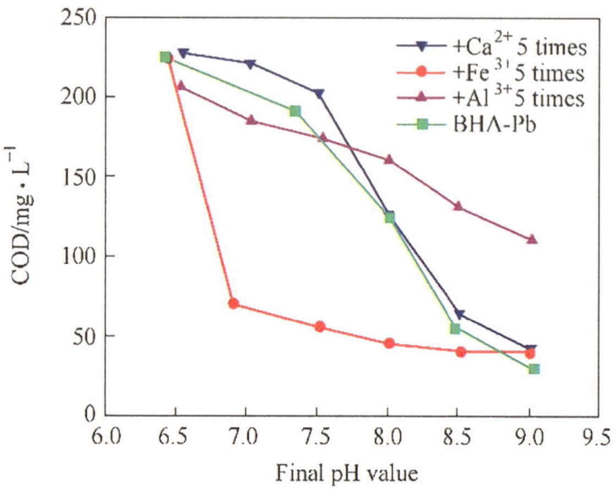

Fig. 9.12 Effect of reaction pH on COD of BHA-Pb solution precipitated by calcium, iron and aluminum ions (test condition: BHA 100 mg/L, theoretical reagent dosage of calcium chloride, ferric chloride and aluminum sulfate are 5 times, precipitation 15 h), adapted from Ref. [27], Copyright 2023, with permission from Elsevier

of BHA-Pb solution. Therefore, iron ion can effectively reduce COD in BHA-Pb solution, while calcium ion and aluminum ion cannot.

Figure 9.13 shows the photos of BHA-Pb solution with different precipitation methods after precipitation for 15 h. Three beakers from left to right in the photo represent BHA-Pb solution containing fluorite, BHA-Pb solution containing fluorite in theory with 4 times the reagent dosage of ferric chloride, BHA-Pb solution containing fluorite in theory with 4 times the reagent dosage of ferric chloride in theory (the theoretical reagent dosage is the reagent dosage of corresponding compounds calculated by chemical valence, For example, one iron ion combined with three benzohydroxamic acid ions to produce iron benzohydroxamic acid, at this time, the amount of iron ion is 1 times the theoretical amount). As can be seen from Fig. 9.13, in the test of BHA-Pb solution containing fluorite, the supernatant of the BHA-Pb solution containing fluorite was colorless and clarified after the addition of fluorite for precipitation, and the supernatant of the BHA-Pb solution containing fluorite and the test phenomenon of the BHA-Pb solution containing fluorite were basically the same. However, the supernatant of BHA-Pb solution precipitated by iron ion coagulation alone has a darker color and higher turbidity than the previous two. Therefore, it can be concluded that after adding gangue minerals in BHA-Pb solution, adding coagulant is conducive to the precipitation and clarification of BHA-Pb solution, that is, in tailings pulp, the presence of gangue minerals is conducive to strengthening the effect of iron ion coagulation settlement.

In order to reveal the essential reason that iron ions can effectively remove COD from BHA-Pb solution, while calcium ions and aluminum ions cannot purify BHA-Pb, Gibbs free energy variation of BHA-Pb, BHA-Ca, BHA-Fe and BHA-Al formation reactions was further calculated, and the results are shown in Fig. 9.14. The Gaussian 09 software is used and the density functional PBE0 is adopted. def2tzvp group is used for all elements in the calculation. As can be seen from Fig. 9.14, the Gibbs free energy variations of BHA-Pb, BHA-Ca, BHA-Fe and BHA-Al formation reactions are − 372.52 kJ/mol, − 340.96 kJ/mol, − 841.26 kJ/mol and − 826.36 kJ/

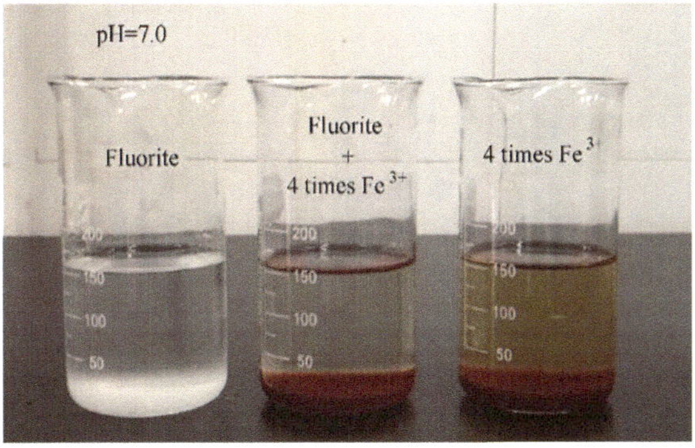

Fig. 9.13 Different precipitation methods to precipitate BHA-Pb solution (test conditions: BHA 100 mg/L, theoretical reagent dosage of ferric chloride 4 times, reaction pH = 7.0, fluorite 3 g/L, precipitation 15 h), adapted from Ref. [27], Copyright 2023, with permission from Elsevier

mol, respectively, all of which are negative values. Indicating that all the formation reactions can occur spontaneously. By comparing the ΔG values of the four elements, it can be concluded that BHA-Fe is the smallest, followed by BHA-Al, BHA-Pb and BHA-Ca, which indicates that the reaction trend of BHA-Fe formation is the largest, followed by BHA-Al, BHA-Pb and BHA-Ca. It can be concluded that when iron, aluminum or calcium ions are added into BHA-Pb solution, BHA-Pb can be converted to BHA-Fe first, followed by BHA-Al, but not to BHA-Ca.

The change rule of COD in BHA-Pb solution with pH was studied in Fig. 9.11. In order to explain the rule of COD removal in BHA-Pb solution by calcium, iron or aluminum ions (Fig. 9.12), the conversion sequence of BHA-Pb with BHA-Ca, BHA-Fe and BHA-Al complexes was combined. It is necessary to further study the change rule of COD with pH in BHA-Ca, BHA-Fe and BHA-Al solutions. Referring to the test conditions in Fig. 9.12, the relationship between COD and pH in BHA-Ca, BHA-Fe and BHA-Al solutions is further studied, and the results are shown in Fig. 9.15. As can be seen from Fig. 9.15, in the range of pH 5–8, the COD of BHA-Fe solution is lower than that of BHA-Pb solution, while the COD of BHA-Al solution is equal to that of BHA-Pb solution, while the COD of BHA-Ca solution is higher than that of BHA-Pb solution and does not change with the change of pH. For example, when pH is 7, the COD of BHA-Pb solution is 210 mg/L, that of BHA-Ca solution is 230 mg/L, that of BHA-Fe solution is 80 mg/L and that of BHA-Al solution is 180 mg/L. Because the relationship between COD and pH of hydroxamic acid complex solution is determined by its own properties, combined with the conversion rule between hydroxamic acid complex. Therefore, iron ion can effectively reduce the COD of BHA-Pb solution.

In conclusion, although the traditional lime method can make the tailings pulp of BHA-Pb complex flotation settle quickly, it will increase the pH value of tailings water, and also increase the COD of tailings water. When polyferric sulfate is used to replace lime settlement of tailings pulp, the pH of tailings water does not change much and the COD of tailings water decreases greatly. Figure 9.16 shows the three-dimensional fluorescence spectra of tailings water after the settlement of tailings pulp by three different sedimentation methods (natural settlement, lime settlement,

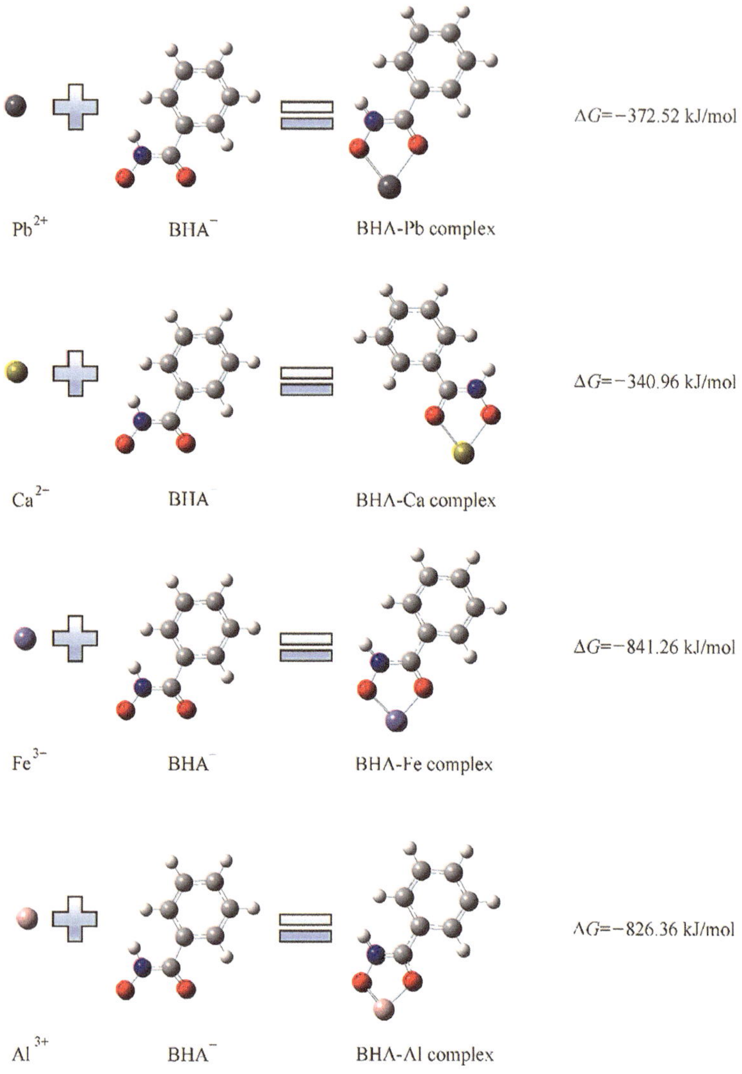

Fig. 9.14 Gibbs free energy variation of BHA-Pb, BHA-Ca, BHA-Fe, and BHA-Al formation reactions, adapted from Ref. [27], Copyright 2023, with permission from Elsevier

and polyferric sulfate settlement) (which can indicate the change of organics concentration in wastewater). It can be seen that the concentration of organics in tailings water after lime settlement is higher than that of natural settlement, which corresponds to high fluorescence intensity. However, the concentration of organics in the tailings water after the settlement of polyferric sulfate is lower than that of natural sedimentation, and the corresponding fluorescence intensity is lower. This indicates that the COD of tailings water does not fall but rises after lime is added because

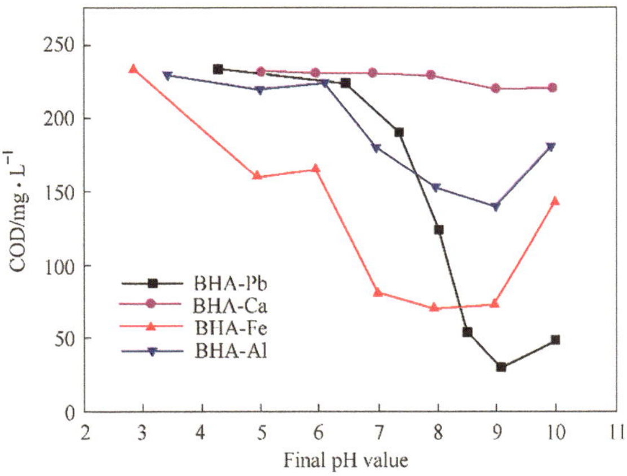

Fig. 9.15 Relationship between COD and pH of BHA-Pb, BHA-Ca, BHA-Fe, and BHA-Al solutions (test conditions: BHA 100 mg/L, molar ratio of BHA to Pb 1:1, theoretical reagent dosages of calcium chloride, ferric chloride and aluminum sulfate are all 5 times, 15 h precipitation), adapted from Ref. [27], Copyright 2023, with permission from Elsevier

part of the residual precipitated BHA-Pb complex reagent in tailings pulp dissolves into tailings water after pH increases, increasing the concentration of organics in wastewater. On the one hand, the pH value of the tailings pulp from the settlement of polyferric sulfate does not change much, on the other hand, iron ion can effectively precipitate part of the dissolved BHA-Pb complex reagent, and then reduce the COD of the tailings water.

The low cost COD removal technology of flotation wastewater from polymetallic ore of mixed ores of scheelite and wolframite based on original pulp coagulation has been completed in Hunan Shizhuyuan polymetallic separation plant [26]. The industrial test process is shown in Fig. 9.17. Before the industrial test, the monthly average pH and COD of the effluent water from the tailings dam are 11.8 and 110 mg/L respectively, which does not meet the requirements of the first class standard of *Comprehensive Wastewater Discharge Standard* 8978–1996 (pH 6–9, COD < 100 mg/L). Further treatment by the subsequent wastewater treatment station is also required. After the industrial test, the monthly average pH and COD of the effluent

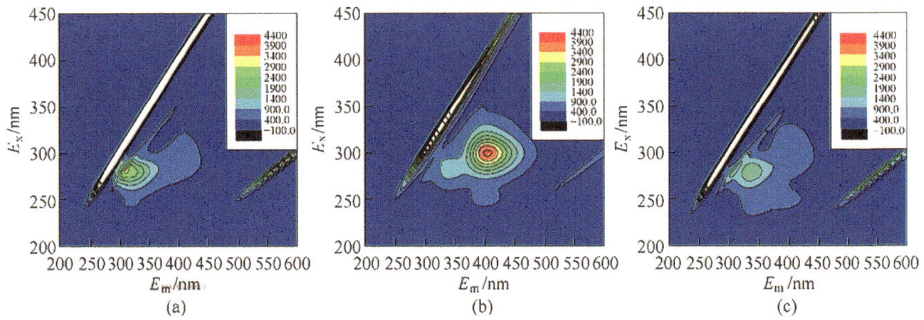

Fig. 9.16 Three-dimensional fluorescence spectra of natural deposition **a**, lime deposition **b** and polyferric sulfate deposition **c** of tailings pulp supernatant, adapted from Ref. [26], Copyright 2019, with permission from Elsevier

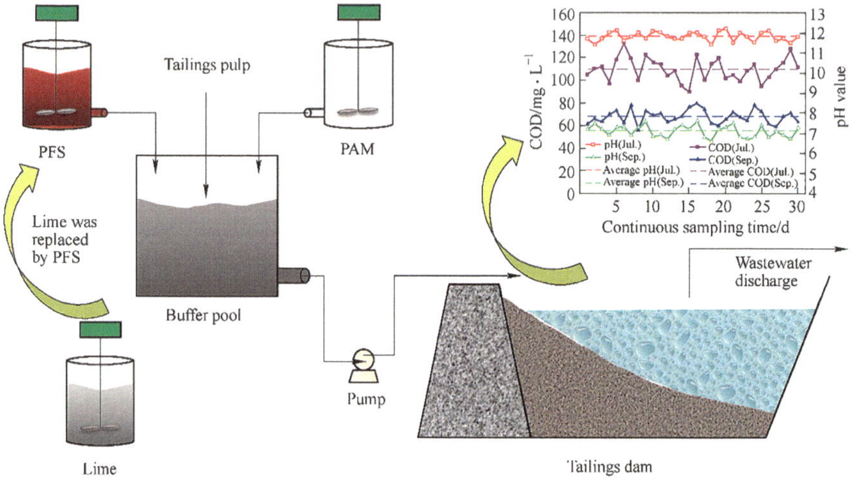

Fig. 9.17 Industrial test flow chart of raw pulp coagulation process, adapted from Ref. [26], Copyright 2019, with permission from Elsevier

water from the tailings dam are 7.1 and 68 mg/L respectively, and the indicators are continuous and stable, so the effluent water from the tailings dam can be discharged up to the standard without follow-up treatment. At present, the technical scheme has been applied in Shizhuyuan mixed ores of scheelite and wolframite mixed mine and Huangshaping mixed ores of scheelite and wolframite. The original pulp coagulation process greatly simplifies the wastewater treatment process and reduces the cost of wastewater treatment.

9.6 Summary

In the flotation of tungsten ore, the main pollutants affecting the standard discharge or recycling of wastewater are water glass, suspended solids, and COD. The removal of sodium silicate and suspended solids is closely related, it is because of the dispersion of sodium silicate, resulting in fine suspended solids is difficult to settle, so it is generally necessary to remove sodium silicate before removing suspended solids. The removal of water glass is usually removed by combining with calcium ions to produce calcium silicate precipitation, which is industrially treated by adding lime or other waste streams containing calcium ions; suspended solids can be more easily removed by coagulation and precipitation with iron or aluminum salts. The degradation of organics in tungsten ore flotation wastewater is the most difficult step. If the wastewater can be returned to the original process or similar process after moderate treatment through the "mass treatment and classification recycling" technology, the flotation reagent can be saved on the one hand and the treatment pressure of wastewater can be reduced. For the tungsten flotation process with fatty acid as the main collector, industrial waste acid rich in calcium chloride can be used to treat and heat the fine tailings, and chemical precipitation can be used to remove pollutants. For the tungsten flotation process with metal–organic complex reagent as the main collector, the raw pulp coagulation technology can be used based on the conversion principle between the complex, the process is simple and the cost is low.

Acknowledgements We acknowledge the financial assistance from the China National Postdoctoral Program for Innovative Talents (No. BX2021380), the China Postdoctoral Science Foundation (No. 2021M703635), Natural Science Foundation of Hunan Province (2023JJ40740), Open Foundation of State Key Laboratory of Mineral Processing (BGRIMM-KJSKL-2023-26), Open Foundation of State Environmental Protection Key Laboratory of Mineral Metallurgical Resources Utilization and Pollution Control (HB202207), Key Laboratory of Hunan Province for Clean and Efficient Utilization of Strategic Calcium-containing Mineral Resources (No. 2018TP1002), National key research center and development Program of the 14th Five-Year Plan (No. 2022YFC2905105 and No. 2022YFC2905104), National Natural Science Foundation of China (NSFC) (No. 52122406and No. 52374291), Hunan High-tech Industry Technology Innovation Leading plan (No. 2022GK4056), Hunan Innovative Province construction special Project (No. 2020RC3001).

References

1. Department of Science and Technology Standards, State Environmental Protection Agency (1996) Comprehensive wastewater discharge standard. State Environmental Protection Bureau, Beijing
2. Guangyuan XIE (2010) Beneficiation. China University of Mining and Technology Press, Xuzhou
3. Hu Y, Feng Q (2006) Mineral resources processing technology and equipment. Science Press, Beijing (in Chinese)
4. Meng X, Khoso SA, Lyu F et al (2019) Study on the influence and mechanism of sodium chlorate on COD reduction of minerals processing wastewater. Miner Eng 134:1–6
5. Hu G, Zhang J (2009) Research on treatment of Scheelite flotation tailings and utilization of backwater. 2009 China Mineral Processing Technology Summit Forum and Equipment Exhibition, Harbin
6. Weilian L, Xinbiao L (2009) Experimental study on recycling of tungsten mine wastewater. Mod Min 25(12):45–47 (in Chinese)
7. Yi S, Xinping P, Kun O et al (2018) Experimental study on treatment and reuse of flotation wastewater from a tungsten-molybdenum-copper polymetallic mine. Ind Water Treat 38(01):62–64
8. Kang J, Fan R, Hu Y et al (2018) Silicate removal from recycled wastewater for the improvement of scheelite flotation performance. J Clean Prod 195:280–288
9. Kang J, Chen C, Sun W et al (2017) A significant improvement of scheelite recovery using recycled flotation wastewater treated by hydrometallurgical waste acid. J Clean Prod 151:419–426
10. Shuangli D, Yanfeng W, Jianhua K (2017) Research and application of scheelite heating wastewater treatment technology. China Tungsten Ind 32(01):75–78 (in Chinese)
11. Chen Q (2020) Analysis on recycling of scheelite wastewater. World Nonferrous Metals 01:58–59 (in Chinese with English abstract)
12. Wang S (2019) Hunan Nonferrous Metals 35(02):57–60 (in Chinese with English abstract)

13. Ming C, Yichun Z, Wanfu H (2007) Study on the treatment of wastewater from a tungsten mine tailings pond by lime destabilization-flocculant sedimentation method. China Tungsten Ind 05:43–46
14. Houxing C, Lanping L (2014) Experimental study on the treatment of suspended solids from a tungsten mine wastewater in southern Jiangxi Province. China Tungsten Ind 29(05):48–51
15. Zhichao J, Yang G, Fu X et al (2019) Separation test of 5000 t/d tungsten-bismuth mineral processing wastewater. Min Metall Eng 39(03):77–80
16. Shen Y (2017) Study on treatment and reuse of flotation wastewater from a Tungsten-Molybdenum-copper polymetallic ore. Hunan University of Science and Technology
17. Jia P, Tian C, Xu F et al (2018) Research and application of scheelite mineral processing wastewater purification technology. Nonferrous Metals (Miner Process Sect 02:60–61+73 (in Chinese)
18. Feng Z (2017) Study on New process and Mechanism of treatment and reuse of wastewater from Shizhuyuan Tungsten polymetallic mine. Jiangxi University of Science and Technology
19. Meng X, Wu J, Kang J et al (2018) Comparison of the reduction of chemical oxygen demand in wastewater from mineral processing using the coagulation–flocculation, adsorption and Fenton processes. Miner Eng 128:275–283
20. Zhang M-H, Dong H, Zhao L et al (2019) A review on Fenton process for organic wastewater treatment based on optimization perspective. Sci Total Environ 670:110–121
21. Wu M, Hu Y, Liu R et al (2019) Electrocoagulation method for treatment and reuse of sulphide mineral processing wastewater: characterization and kinetics. Sci Total Environ 696
22. Moussa DT, El-Naas MH, Nasser M et al (2017) A comprehensive review of electrocoagulation for water treatment: potentials and challenges. J Environ Manage 186:24–41
23. Al-Qodah Z, Al-Shannag M (2017) Heavy metal ions removal from wastewater using electrocoagulation processes: a comprehensive review. Sep Sci Technol 52(17):2649–2676
24. Lin S, Liu R, Wu M et al (2019) Minimizing flotation wastewater through internal reuse of process water in flotation circuit. J Clean Prod 118898
25. Kang J, Sun W, Hu Y et al (2017) The utilization of waste by-products for removing silicate from mineral processing wastewater via chemical precipitation. Water Res 125:318–324
26. Meng X, Khoso SA, Kang J et al (2019) A novel scheme for flotation tailings pulp settlement and chemical oxygen demand reduction with polyferric sulfate. J Clean Prod 241:118371
27. Meng X, Jiang M, Lin S et al (2023) Removal of residual benzohydroxamic acid-lead complex from mineral processing wastewater by metal ion combined with gangue minerals. J Clean Prod 396:136578

GPSR Compliance

The European Union's (EU) General Product Safety Regulation (GPSR) is a set of rules that requires consumer products to be safe and our obligations to ensure this.

If you have any concerns about our products, you can contact us on ProductSafety@springernature.com

In case Publisher is established outside the EU, the EU authorized representative is:

Springer Nature Customer Service Center GmbH
Europaplatz 3
69115 Heidelberg, Germany

Batch number: 10704305

Printed by Printforce, the Netherlands